Sustainable Weed Management

Sustainable Weed Management

Editors

Alessia Restuccia
Aurelio Scavo

MDPI • Basel • Beijing • Wuhan • Barcelona • Belgrade • Manchester • Tokyo • Cluj • Tianjin

Editors
Alessia Restuccia
University of Catania
Catania, Italy

Aurelio Scavo
University of Catania
Catania, Italy

Editorial Office
MDPI
St. Alban-Anlage 66
4052 Basel, Switzerland

This is a reprint of articles from the Special Issue published online in the open access journal *Plants* (ISSN 2223-7747) (available at: https://www.mdpi.com/journal/plants/special_issues/Sustainable_Weed_Management).

For citation purposes, cite each article independently as indicated on the article page online and as indicated below:

LastName, A.A.; LastName, B.B.; LastName, C.C. Article Title. *Journal Name* **Year**, *Volume Number*, Page Range.

ISBN 978-3-0365-8290-0 (Hbk)
ISBN 978-3-0365-8291-7 (PDF)

Cover image courtesy of Aurelio Scavo

Contents

About the Editors

Alessia Restuccia

Alessia Restuccia is a biologist specializing in Environmental and Applied Botany and a Ph.D. researcher in Agricultural, Food, and Environmental Science at the Dpt of Agriculture, Food and Environment (Di3A), University of Catania (UC), Italy. Since 2004, she has received different research contracts at the Science and Technology Park of Sicily, and at the Dpt of Botany (UC). Since 2005, she has taught Plant Biology, Plant morphology and Physiology, Environmental Botany, and Landscape Ecology at UC. She has supervised numerous degree theses, been involved in national and international research projects, and participated in different national and international scientific congresses. She is author and/or co-author of scientific papers in national and international peer-reviewed journals and book chapters mainly relating to soil seedbank evaluation, the germination ecology of spontaneous species, the allelopathic effects on weeds for biological control, and the bio-agronomic behavior of cover crops in different agroecosystems.

Aurelio Scavo

Aurelio Scavo is Ph.D. researcher in Agricultural, Food, and Environmental Science in the Department of Agriculture, Food and Environment (Di3A) at the University of Catania (UC), Italy. At UC, he has lectured for master students and supervised B.Sc./M.Sc. theses. He is involved or was involved in national and international research projects, and has received several scientific awards for his research activity. Currently, his main research lines are 1) the set-up of sustainable weed control practices; 2) the role of allelopathy in weed management, both in laboratory and field conditions; 3) the role of cover crops in agroecosystems; 4) the effects of mycorrhization on field crops; 5) agronomical and technical practices to improve crop productivity in the Mediterranean area. He regularly participates in various national/international scientific conferences and acts as reviewer for many ISI/Scopus peer-reviewed journals. He is the author/co-author of more than 30 papers in international scientific peer-reviewed journals, 1 book chapter, and more than 40 contributions to national or international congresses.

plants

MDPI

ditorial

'ustainable Weed Management

lessia Restuccia and Aurelio Scavo *

Department of Agriculture, Food and Environment (Di3A), University of Catania, 95123 Catania, Italy; a.restuccia@unict.it
* Correspondence: aurelio.scavo@unict.it

1. Introduction

Weeds are the most important biological constraint determining yield losses for field crops. For this reason, after World War II, synthetic herbicides have been largely adopted in developed countries in order to enhance yields and reduce the costs of cultivation. Unfortunately, their irrational use has caused environmental pollution, the development of herbicide-resistant weeds and shifts in weed communities, thus making cropping systems herbicide-dependent. Hence, following the 'Zero Hunger' goal of the United Nations Sustainable Development Goals and the strategies of the European Commission 'Green Deal', a weed management system based on cultural, mechanical, physical, biological and ecological methods to prevent or reduce the use of synthetic herbicides has become of outstanding importance in agricultural systems. Different techniques (cover cropping, the use of high-competitive cultivars, the choice of plant arrangement and seeding time, tillage systems, allelochemicals, etc.) and methodological approaches (e.g., soil seedbank analysis, weed adaptation along environmental gradients, and the analysis of weed abundance and diversity) have shown effectiveness in managing weeds from an eco-friendly perspective. The current Special Issue, entitled 'Sustainable Weed Management', was born within this context. It is a compilation of eighteen papers, including a review article related to the recent advancements in sustainable weed control methods and to biotic and abiotic factors affecting weed adaptation. The main topics covered by the Special Issue are:

- The effects of weed control practices on weed density and diversity;
- Cultural methods;
- Cover cropping and mulching;
- The use of allelopathic plant extracts and allelochemicals;
- Innovative chemical weeding methods.

Citation: Restuccia, A.; Scavo, A. Sustainable Weed Management. Plants 2023, 12, 1673. https://doi.org/10.3390/plants12081673

Received: 7 March 2023
Revised: 6 April 2023
Accepted: 14 April 2023
Published: 17 April 2023

2. Description of the Special Issue Main Findings

2.1. Weed Adaptation and Assemblages

Prior to analyzing the latest advancements in the wide area of weed control practices, nowadays, weed scientists are faced with the indirect effects of climate change on weed adaptation. Climate warming is inducing a high phenotypic plasticity in several weed species that may facilitate their invasive ability along environmental gradients. For this reason, Gentili et al. [1] used the seeds of the annual plant invader common ragweed (*Ambrosia artemisiifolia* L.) to determine variation in phenology and bio-morphological traits when grown along a 1000 m altitudinal gradient in Northern Italy, and under different temperature conditions in the growth chamber. They found that common ragweed may shift toward higher elevations and, at the same time, may improve the in situ (pre)adaptation of populations currently abundant at low elevations in the invasive European range. Another central topic in weed science is the determination of the processes that shape weed assemblages in farmlands. Studying the effects of crop competition on weeds, nitrogen input, weed control and landscape on both weed diversity and abundance in the margins and centers of 115 oilseed rape (*Brassica napus* L.) fields in Western France, Berquer et al. [2]

found that landscape is the main driver of weed assemblages in field margins. In particular, they indicated crop height (i.e., competition) as the main driver for weed assemblages in field cores, and the number of meadows in the landscape (i.e., spatial dispersal) for weed assemblages in field margins.

2.2. Preventive Methods for Weed Management

In integrated weed management systems, indirect or preventive methods have a key role in reducing the impact and improving the effectiveness of direct control methods. Essentially, prevention is based on the management of the soil seedbank and an improvement in crop competitiveness against weeds. Preventive methods include crop rotation, cover cropping, mulching, the choice of row spacing and seeding rate, etc. Their combination is often associated with a higher weed-suppressive ability than a single method. For instance, the study by Naeem et al. [3] evaluated the impact of different weed management options (i.e., false seedbeds, allelopathic water extracts, chemical control, weed-free and weedy check) on weed flora in various barley-based cropping systems. From this study it emerged that including mungbean (*Vigna radiata* (L.) R. Wilczek) or mainly sorghum (*Sorghum* spp.) in rotation with barley (*Hordeum vulgare* L.) and applying allelopathic water extracts could suppress weeds, similarly to herbicides. Hence, the combination of crop rotation and allelopathic water extracts was demonstrated as a valid alternative to herbicides in barley crop. Barroso and Genna [4] studied the effect of row spacing (18 or 36 cm) and seeding rate (73 or 140 kg ha^{-1}) on Russian thistle (*Salsola tragus* L.) in spring barley and spring wheat crops in the Pacific Northwest. They concluded that increasing seeding rates or planting spring crops in narrow rows may be effective for yield increase in low-rainfall years of the zone under study, while no effect may be observed in years with higher rainfall than the normal trend. Concerning the role of highly competitive cultivars, Scavo et al. [5] conducted research over 10 farms in central–eastern Sicily on the weed-suppressive ability of old durum wheat landraces vs. modern cultivars in order to study the indirect effect of old landraces in sustainably reducing weed pressure without the adoption of chemical weed control. They reported that old durum wheat landraces were associated with a 47% reduction in the soil seedbank size and to a 64% decrease in the aboveground weed biomass compared to modern cultivars. Moreover, the weed species compositions of modern and old cultivars were quite separated for both soil seedbank and real flora, with the latter showing few specific associations with major weeds. The authors attributed the higher weed-suppressive ability of old durum wheat landraces to a combined competition–allelopathy effect.

2.3. Cover Cropping

Among the well-recognized ecosystem services provided by cover crops (i.e., non-harvested crops grown in addition to the primary cash crop with the aim of improving soil fertility and enhancing yields), the limitation of weeds is receiving more and more attention from the scientific community and stakeholders. Recently, Restuccia et al. [6] investigated the 5-year effect of subterranean clover (*Trifolium subterraneum* L.) and spontaneous flora, both with and without burying dead mulch into the soil, on weed abundance and diversity in a Mediterranean apricot orchard. They found that weed biomass was significantly reduced by subterranean clover, especially with burying dead mulch into the soil, with the cover crop biomass that was negatively correlated to weed biomass. Furthermore, compared to conventional apricot management, subterranean clover decreased the size of the soil seed bank by 57%. In a similar study, Las Casas et al. [7] studied the role of conservation agriculture and living mulches in a young Mediterranean olive orchard. The authors reported that the use of sage (*Salvia officinalis* L.) and lemongrass (*Cymbopogon citratus* (DC) Stapf) as living mulches combined to minimize soil disturbance, reduce the need for weed management, and promote the complexity of the Arthropod fauna in terms of both the number of species and the taxonomic complexity. Another technique related to cover cropping, i.e., mulching, was studied in this Special Issue by Ryan et al. [8] in winter wheat cultivated in central New York (USA). Evaluating a gradient of mulch biomass primarily composed of perennial species such as orchardgrass (*Dactylis glomerata* L.), timothy (*Phleum pratense* L.) and red clover (*T. pratense*), they found that wheat seedling density showed an asymptotic relationship with mulch biomass (no effect at low rates and a gradual decrease from moderate-to-high rates of mulch) and that the highest level of mulch (9000 kg ha^{-1}) selectively suppressed weed biomass without reducing wheat grain yield.

2.4. New Advances in Chemical Weed Control

Herbicides still represent the most popular tool for weed control, mainly in developing countries. However, the study conducted by Pattanayak et al. [9] in an Indian sub-tropical environment highlighted that the chemical control with the herbicides bensulfuron, pretilachlor and bispyribac sodium negatively affected the soil microbial and enzymatic activity, whereas improved microbial populations and enzyme activities were noted in unpuddled transplanted rice (*Oryza sativa* L.) under organic weed management. Another negative effect related to the irrational application of herbicides is the spread of invasive or resistant weed species. Vázquez-García et al [10] studied the resistance to acetyl-coenzyme A carboxylase (ACCase)-inhibiting herbicides in three resistant biotypes of *Phalaris*: *P. brachystachys*, *P. minor* and *P. paradoxa*. From their study, it emerged that cross-resistance in *Phalaris* species is conferred by specific point mutations, with *P. brachystachys* resistance that is due to target site and non-target-site resistance mechanisms, while only an altered target site was found in *P. minor* and *P. paradoxa*.

The present Special Issue pointed out different advances in the field of synthetic herbicides for the control of invasive and resistant weed biotypes. The use of tank-mix herbicides is one of these. For instance, Abu-Nassar and Matzrafi [11] indicated that tank mixes of oxadiazon and oxyfluorfen with different concentrations of surfactant significantly suppressed *Solanum rostratum* Dunal, an important invasive weed in Israel since the 1950s, when applied at a later growth stage (8–9 cm height). Additionally, Campos et al. [12] suggested a methyl-capped polyethylene glycol ester of pelargonic acid (PA-MPEG) in synergism with a non-phytotoxic alkylated seed oil-based adjuvant (i.e., Hasten™) to improve the herbicidal efficacy of this novel fatty acid ester by disintegrating the bio-membranes and, thus, negatively affecting plant transpiration. O'Brien et al. [13] tested the effectiveness of a novel stem implantation system for controlling the woody weed Chinese elm (*Celtis sinensis* Pers.) in a conserved habitat. They found that the encapsulated glyphosate (245 mg/capsule), aminopyralid and metsulfuron-methyl (58.1 and 37.5 mg/capsule) and picloram (10 mg/capsule) achieved a similar herbicidal activity to the benchmark treatment (diesel + triclopyr + picloram + liquid hydrocarbon), because these encapsulated herbicides are immediately sealed into the vascular system of the target species, thus reducing the amount of active agent required and preventing environmental exposure.

2.5. Use of Allelopathy for Weed Management

Allelopathic species can be manipulated for the sustainable management of weeds in different ways such as the introduction of an allelopathic crop into crop rotation schemes [3], the use of an allelopathic cover crop [6], or the identification, isolation and extraction of plant allelochemicals for the possible production of bioherbicides. In this Special Issue, the bioherbicidal potential of the essential oils from Mediterranean Lamiaceae members was reviewed by De Mastro et al. [14]. In addition, Motmainna et al. [15] investigated the allelopathic potential of *Parthenium hysterophorus* L. methanolic extracts at different concentrations under laboratory and glasshouse conditions. They indicated eight amino acids, seven phenolic compounds, three terpenoids and other secondary organic compounds as *P. hysterophorus* allelochemicals in methanolic extract. The *P. hysterophorus* extract was also capable of inhibiting the germination and growth of *Cyperus iria* L. to a similar extent to the synthetic herbicides glyphosate and glufosinate-ammonium. Another study evaluated the thermal allelopathic effect of two coniferous plants (*Pinus densiflora* Siebold & Zucc. and *Pinus koraiensis* Siebold & Zucc.) on oilseed rape (*B. rapus*) germination and seedling growth in order to assess whether high temperatures, generated during composting, d crease allelopathic ability [16]. It was found that the allelopathic capacity of two *Pinu* species showed root-specific inhibition, but the decrease in volatile contents after the the mal process was lesser in *P. koraiensis* than in *P. densiflora*. The authors, therefore, suggeste the application of the two conifer needles as allelopathic compost to control the initial wee growth in horticultural crops thanks to their thermal stability and root-specific inhibitio

Seed meals obtained from allelopathic crops are another allelopathic tool and eco-friendl alternative to synthetic herbicides. Pytlarz and Gala-Czekaj [17] assessed the allelopathi activity of seed meals from *Fagopyrum esculentum* Moench, *Sinapis alba* L., *Phacelia tanacetifoli* Benth., *Lupinus luteus* L., *Raphanus sativus* var. *oleiformis* and *Ornithopus sativus* Brot., at 1 an 3% doses, on herbicide-susceptible and -resistant (to propoxycarbazone-sodium) rye brom (*Bromus secalinus* L.) biotypes in winter wheat. They reported crop- and dose-dependen results. In particular, (1) wheat emergence and initial growth were not affected by th seed meals from *F. esculentum*, *P. tanacetifolia*, and *R. sativus* at 1% concentration in the soi (2) the phytotoxicity of these seed meals was at the same level as the herbicide or highe (3) an increase in seed meal concentration is not recommended due to the reduction i wheat emergence.

Plants' allelopathic potential is known to be influenced by genotype, partly du to the different concentration of allelochemicals. Following the return to local durun wheat landraces demanded by the market, Scavo et al. [18] conducted research on th allelopathic effects of the extracts from three durum wheat landraces ('Timilia', 'Russell and 'Perciasacchi') and a modern variety ('Mongibello'), obtained from three different plan parts (ears, stems and roots), on the weeds *Portulaca oleracea* L. and *Stellaria. media* (L. Vill. It was found that old landraces (mainly 'Timilia' and 'Russello') showed a highe allelopathic activity and that ear extracts were the most active. These results confirmed i the laboratory the findings obtained by Scavo et al. [5] in open-field conditions.

3. Conclusions

This Special Issue involves a wide range of knowledge, methods and practices recentl achieved for sustainable weed management. Altogether, the papers published here demon strate that effective weed control can be performed not only with an indiscriminate use o herbicides, but also with proper chemical weed control and with other eco-friendly meth ods including allelopathy, cover cropping, tillage, etc. It also emerged that the combinatio of different methods often results in an improved weed-suppressive ability.

As Guest Editors, we acknowledge all the authors for their submissions to our Special Issue. We believe that this excellent research is a significant breakthrough for current science and will be made available to farmers and stakeholders.

Author Contributions: A.R. and A.S. contributed equally to this article. All authors have read and agreed to the published version of the manuscript.

Funding: This research received no external funding.

Institutional Review Board Statement: Not applicable.

Informed Consent Statement: Not applicable.

Data Availability Statement: Not applicable.

Conflicts of Interest: The authors declare no conflict of interest.

References

1. Gentili, R.; Ambrosini, R.; Augustinus, B.A.; Caronni, S.; Cardarelli, E.; Montagnani, C.; Müller-Schärer, H.; Schaffner, U.; Citterio, S. High Phenotypic Plasticity in a Prominent Plant Invader along Altitudinal and Temperature Gradients. *Plants* **2021**, *10*, 2144. [CrossRef] [PubMed]
2. Berquer, A.; Martin, O.; Gaba, S. Landscape Is the Main Driver of Weed Assemblages in Field Margins but Is Outperformed by Crop Competition in Field Cores. *Plants* **2021**, *10*, 2131. [CrossRef] [PubMed]

3. Naeem, M.; Farooq, S.; Hussain, M. The Impact of Different Weed Management Systems on Weed Flora and Dry Biomass Production of Barley Grown under Various Barley-Based Cropping Systems. *Plants* **2022**, *11*, 718. [CrossRef] [PubMed]
4. Barroso, J.; Genna, N.G. Effect of Row Spacing and Seeding Rate on Russian Thistle (*Salsola tragus*) in Spring Barley and Spring Wheat. *Plants* **2021**, *10*, 126. [CrossRef] [PubMed]
5. Scavo, A.; Restuccia, A.; Bannò, M.; Mauromicale, G. Differentiated Weed-Suppressive Ability of Modern and Old Durum Wheat Cultivars after Long-Term Cultivation under Semi-Arid Climate. *Plants* **2022**, *11*, 3368. [CrossRef] [PubMed]
6. Restuccia, A.; Scavo, A.; Lombardo, S.; Pandino, G.; Fontanazza, S.; Anastasi, U.; Abbate, C.; Mauromicale, G. Long-Term Effect of Cover Crops on Species Abundance and Diversity of Weed Flora. *Plants* **2020**, *9*, 1506. [CrossRef] [PubMed]
7. Las Casas, G.; Ciaccia, C.; Iovino, V.; Ferlito, F.; Torrisi, B.; Lodolini, E.M.; Giuffrida, A.; Catania, R.; Nicolosi, E.; Bella, S. Effects of Different Inter-Row Soil Management and Intra-Row Living Mulch on Spontaneous Flora, Beneficial Insects, and Growth of Young Olive Trees in Southern Italy. *Plants* **2022**, *11*, 545. [CrossRef] [PubMed]
8. Ryan, M.R.; Wayman, S.; Pelzer, C.J.; Peterson, C.A.; Menalled, U.D.; Rose, T.J. Winter Wheat (*Triticum aestivum* L.) Tolerance to Mulch. *Plants* **2021**, *10*, 2047. [CrossRef] [PubMed]
9. Pattanayak, S.; Jena, S.; Das, P.; Maitra, S.; Shankar, T.; Praharaj, S.; Mishra, P.; Mohanty, S.; Pradhan, M.; Swain, D.K.; et al. Weed Management and Crop Establishment Methods in Rice (*Oryza sativa* L.) Influence the Soil Microbial and Enzymatic Activity in Sub-Tropical Environment. *Plants* **2022**, *11*, 1071. [CrossRef] [PubMed]
10. Vázquez-García, J.G.; Torra, J.; Palma-Bautista, C.; Alcántara-de la Cruz, R.; Prado, R.D. Point Mutations and Cytochrome P450 Can Contribute to Resistance to ACCase-Inhibiting Herbicides in Three *Phalaris* Species. *Plants* **2021**, *10*, 1703. [CrossRef] [PubMed]
11. Abu-Nassar, J.; Matzrafi, M. Effect of Herbicides on the Management of the Invasive Weed *Solanum rostratum* Dunal (Solanaceae). *Plants* **2021**, *10*, 284. [CrossRef] [PubMed]
12. Campos, J.; Bodelon, L.; Verdeguer, M.; Baur, P. Mechanistic Aspects and Effects of Selected Tank-Mix Partners on Herbicidal Activity of a Novel Fatty Acid Ester. *Plants* **2022**, *11*, 279. [CrossRef] [PubMed]
13. O'Brien, C.J.; Mellor, V.; Galea, V.J. Controlling Woody Weed Chinese Elm (*Celtis sinensis* Pers.) with Stem-Implanted Herbicide Capsules. *Plants* **2022**, *11*, 444. [CrossRef] [PubMed]
14. De Mastro, G.; El Mahdi, J.; Ruta, C. Bioherbicidal Potential of the Essential Oils from Mediterranean Lamiaceae for Weed Control in Organic Farming. *Plants* **2021**, *10*, 818. [CrossRef] [PubMed]
15. Motmainna, M.; Juraimi, A.S.; Uddin, M.K.; Asib, N.B.; Islam, A.K.M.M.; Ahmad-Hamdani, M.S.; Hasan, M. Phytochemical Constituents and Allelopathic Potential of *Parthenium hysterophorus* L. in Comparison to Commercial Herbicides to Control Weeds. *Plants* **2021**, *10*, 1445. [CrossRef] [PubMed]
16. Gil, C.S.; Hong, D.; Duan, S.; Eom, S.H. Volatile and Non-Volatile Allelopathic Characteristics in Thermally Processed Needles of Two Conifers. *Plants* **2022**, *11*, 1003. [CrossRef] [PubMed]
17. Pytlarz, E.; Gala-Czekaj, D. Possibilities of Using Seed Meals in Control of Herbicide-Susceptible and -Resistant Biotypes of Rye Brome (*Bromus secalinus* L.) in Winter Wheat. *Plants* **2022**, *11*, 331. [CrossRef] [PubMed]
18. Scavo, A.; Pandino, G.; Restuccia, A.; Caruso, P.; Lombardo, S.; Mauromicale, G. Allelopathy in Durum Wheat Landraces as Affected by Genotype and Plant Part. *Plants* **2022**, *11*, 1021. [CrossRef] [PubMed]

plants
MDPI

rticle

‿andscape Is the Main Driver of Weed Assemblages in Field Margins but Is Outperformed by Crop Competition in Field Cores

drien Berquer [1], Olivier Martin [2] and Sabrina Gaba [1,3,4,*]

1 Centre d'Etudes Biologiques de Chizé, UMR7372, CNRS & La Rochelle Université, 79360 Villiers-en-Bois, France; adrien.berquer@cebc.cnrs.fr
2 UR 0546 Biostatistiques et Processus Spatiaux, INRAE, CEDEX 9, 84914 Avignon, France; olivier.martin@inrae.fr
3 LTSER, «Zone Atelier Plaine & Val de Sèvre», 79360 Villiers-en-Bois, France
4 Centre d'Etudes Biologiques de Chizé, USC 1339 Agripop, INRAE, 79360 Villiers-en-Bois, France
* Correspondence: sabrina.gaba@inrae.fr

Abstract: Weeds are considered a major pest for crops, and as such have been intensively managed by farmers. However, weeds, by providing resources, also support farmland biodiversity. The challenge for sustainable weed management is therefore to maintain weed diversity without compromising crop production. Meeting this challenge requires determining the processes that shape weed assemblages, and how agricultural practices and landscape arrangement affect them. In this study, we assess the effects of crop competition on weeds, nitrogen input, weed control and landscape on both weed diversity and abundance in the margins and centres of 115 oilseed rape fields in Western France. We show that weed assemblages in field cores were mainly shaped by crop height, a proxy of crop competition. By contrast, weed assemblages in field margins increased with the number of meadows in the landscape, revealing the role of spatial dispersal. Using structural equation modelling, we further show that in the field core, weed assemblages were also indirectly shaped by landscape through spatial dispersal from the field margin. Overall, our study gives empirical support for crop competition as a way to reduce the intensity of chemical weeding, and for meadows as a way to enhance biodiversity in the landscape.

Keywords: agroecology; competition; dispersal; landscape; oilseed rape; sustainable weed management

Citation: Berquer, A.; Martin, O.; Gaba, S. Landscape Is the Main Driver of Weed Assemblages in Field Margins but Is Outperformed by Crop Competition in Field Cores. *Plants* **2021**, *10*, 2131. https://doi.org/10.3390/plants10102131

Academic Editors: Alessia Restuccia and Aurelio Scavo

Received: 15 August 2021
Accepted: 4 October 2021
Published: 7 October 2021

Publisher's Note: MDPI stays neutral with regard to jurisdictional claims in published maps and institutional affiliations.

1. Introduction

Taking into account the challenges of sustainable food for a growing human population, the preservation of biodiversity and natural resources and the mitigation of climate change requires a profound transition in our agricultural and food system [1,2]. Weed management in arable crops is typical of this issue. Weeds are recognized as a major pest in agriculture, resulting in yield loss of up to 30% [3]. For decades, they have been intensively managed to reduce their competition for resources with crop plants. This has resulted in the decline of at least 20% of weed species over the past 30 years [4], and an overall decline in rare flagship species [5]. However, by providing food and shelter for birds, insects and small mammals [6,7], weeds are also an important component in the maintenance of farmland biodiversity and agroecosystem functioning [8,9]. To meet agricultural production demand while conserving weed diversity and enhancing its related ecological functions, promotion of diverse weed assemblages has been suggested, assuming that increasing species richness would ensure for weed functions without selecting for few dominant species [10,11]. Designing management strategies that ensure for diverse weed assemblages therefore requires strengthening our understanding of the processes that shape weed species richness and abundance.

Weed species assemblages can be understood in terms of a complex scheme including interactions between ecological processes (e.g., competition, spatial dispersal) operating over various scales and management through disturbance regimes (e.g., weeding operations) and resource levels (e.g., light, nitrogen) [12–16]. While there is substantial evidence showing that crop type and farming practices influence weed species richness [17–19], weed abundance [20,21], or crop–weed competition [22], only recently have studies explored the interactive effects of competition and farming practices on weed assemblages [23 Crop competition has however been acknowledged as a way to regulate weed species [24 The effect of landscape on weed abundance is also less documented compared to weed species richness, and when studied was shown to have either no effect [25] or an indirect effect [26] through an interaction with farmer management intensity. Indeed, evidence on the interplay of local and landscape effects on weeds have recently been revealed [13,25,27 For instance, Henckel et al. [28] demonstrated that the presence of organic farming in the surrounding landscape of conventional fields could balance the negative effect of conventional management through species dispersal. The diversity of crop types [12] and the amount of seminatural habitats [29,30] in the landscape also benefit in-field weed species richness. However, the effect of landscape varies with field position (i.e., field core versus field margin [12,13,31]) revealing the complex interplay between spatial dispersion and local processes. These differences can indeed be attributed to the variation in farming practices (crop density, fertilization and weed control) as well as to their distances to source habitats. However, whether landscape effects interact with competition with crop plants disturbances induced by weed control or both remains to be established.

In this study, we evaluated the interactive effects of crop–weed competition, farming practices and landscape on both weed diversity and abundance in the margins and centres of 115 oilseed rape (*Brassica napus* L.) fields in South-West France. We used a new approach to evaluate the effects of landscape variables without specifying *a priori* distances of spatial extents of their effects [32]. To our knowledge, this is the first study to address the combined effects of competition, farming practices and landscape on both weed species richness and abundance in the margins and centres of arable fields, considering that the spatial extent of the landscape variables can vary with the landscape variables, the weed metrics and the field compartment. As a first step, we assessed the effects of competition farming practices and landscape on weed species richness and abundance in the two field compartments. Then, we investigated whether local dispersal from field margin to field core could compensate for a loss of weed diversity through an indirect effect of spatial dispersal from the landscape, as highlighted by Bourgeois et al. [13]. We expected the contribution of competition to be higher in field cores due to a higher crop density. We also expected the contribution of competition to increase with the amount of nitrogen, because oilseed rape plants are nitrophilous plants [33], and decrease with higher weed management due to the selection of specialist species [34]. We further expected a higher response of weed abundance to competition compared to weed species richness, especially in field cores. Finally, we expect landscape effect to act predominantly indirectly across the field margin on in-field weed assemblages.

2. Results

A total of 158 weed species was identified across the 115 oilseed rape fields sampled from 2014 and 2018. We identified 131 weed species in the field cores and 143 in the field margins. *Mercurialis annua* L. was the most abundant and common weed species occurring in 92 fields. A total of 90 species (57.0% of all species) occurred in fewer than 10% of the sampled fields. Mean species richness per field was 28.85 $\pm$ 8.73 (min = 10, max = 53) species, and mean abundance was 267.60 $\pm$ 150.97 (min = 48, max = 956). Weed species richness was on average higher in the field margin, with 19.97 $\pm$ 8.11 (min = 3, max = 42) species, than in field core with an average of 11.09 $\pm$ 4.89 (min = 3, max = 40) species when accounting for the same sampling effort using a 5000 times bootstrap of five quadrats in the field core. In the same way, the abundance was higher in the field margin (88.84 $\pm$ 45.47) than in field core (48.72 ± 33.18; on average in five quadrats using a bootstrap). Crop height was significantly lower in the field margin (54.4 cm ± 47.1 cm) than in the field core (145.9 cm ± 23.7 cm; Wilcoxon paired-test, V = 17, *p*-value < 0.001). In 3.5% of the fields, there was no crop plant in the margin.

2.1. Competition and Weed Management Highly Affect Weed Species Richness in Field Core

For weed species richness, the selection procedure retained the variables related to competition (crop height and nitrogen) and chemical disturbances as well as the interaction between nitrogen and herbicides (Table 1A). These effects explained 15% of the variance of weed species richness in field cores when using Treatment Frequency Index (TFI) as a proxy of herbicide use (Figure 1; 11% with amount of herbicide active substances (QA) Figure S1). Weed species richness significantly decreased with crop height (Figure 2A) but not with herbicides (Figure 2B) nor with nitrogen. Contrary to our expectation, we did not find any significant interaction between crop height and the amount of nitrogen or the quantity of herbicide use. Rather, we found a significant positive effect of the 'nitrogen × herbicides' interaction on weed species richness, suggesting a higher efficiency of weed chemical control in nitrogen-rich fields.

Table 1. Statistics of the models for weed (A) species richness and (B) abundance in field cores. Herbicide intensity is expressed with the Treatment Frequency Index (TFI). Abundance was log-transformed, and management practice variables were centred and reduced. The estimated buffer radii are indicated for each landscape variable. Landscape variables have two degrees of freedom because both their spatial extent (i.e., estimation of buffer radius) and their effect were estimated. Significant effects are indicated in bold. R-squared values were 27.4% and 18.0% using TFI.

		Estimated Buffer Radius (m)	Estimate	df	t-Value	*p*-Value
(A) Weed species richness	Intercept		16.314	1	11.204	**<0.001**
	Crop height		−1.407	1	−2.400	**0.018**
	Nitrogen		1.058	1	1.090	0.278
	Herbicides		−1.787	1	−1.788	0.077
	Hedge density	17	48.506	2	1.418	0.159
	Number of meadows	500	−4.439	2	−0.366	0.715
	Amount of organic farming	777	−6.513	2	−0.979	0.330
	Amount of oilseed rape	600	6.913	2	0.714	0.477
	Nitrogen × Herbicides		2.017	1	3.316	**0.001**
	Nitrogen × Number of meadows		−47.839	1	−3.061	**0.003**
	Nitrogen × Amount of organic farming		0.420	1	0.099	0.921
	Herbicides × Number of meadows		−5.072	1	−0.281	0.779
	Herbicides × Amount of organic farming		16.105	1	2.051	**0.043**
(B) Weed Abundance						
	Intercept		2.103	1	42.803	**<0.001**
	Crop Height		−0.088	1	−3.101	**0.002**
	Herbicides		−0.075	1	−2.223	**0.028**
	Hedge density	68	6.262	2	1.345	0.181
	Number of meadows	26	−0.167	2	−0.429	0.669
	Amount of organic farming	140	0.159	2	0.814	0.418
	Amount of oilseed rape	26	0.199	2	1.075	0.285
	Herbicides × Number of meadows		0.938	1	1.674	0.097
	Herbicides × Amount of organic farming		0.353	1	1.394	0.166

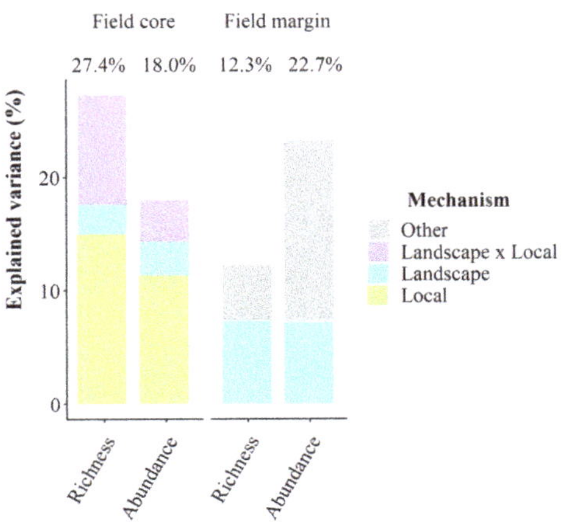

Figure 1. Percentage of explained variance by local factors (crop height, the amount of nitrogen, the intensity of herbicide use and number of mechanical operations), landscape (amount of organic farming, meadows, oilseed rape and hedge density) and weather conditions (rainfall and temperature) on weed species richness and abundance in field cores and field margins. The intensity of herbicide use is expressed using the herbicide TFI. The buffer radii at which the amount of each landscape variable was estimated are shown in Table 1 for field cores and Table 2 for field margins. R-squared computed from the type III ANOVAs of respective models are indicated above each corresponding bar plot.

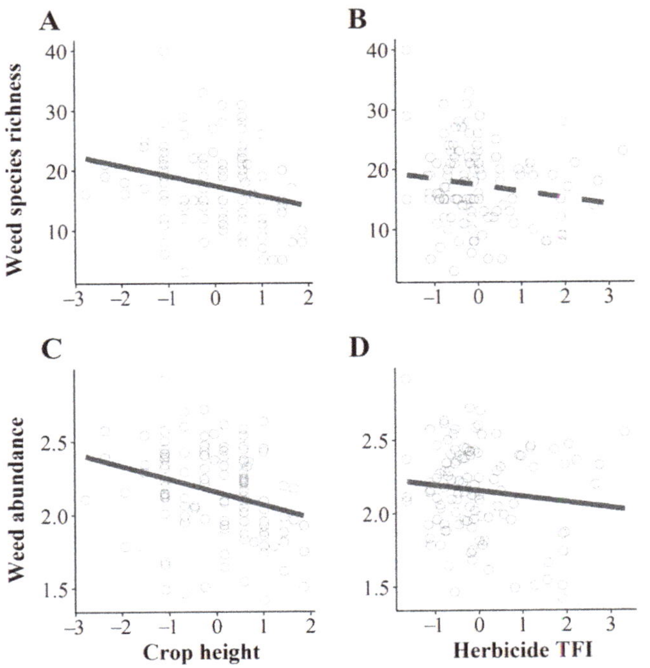

Figure 2. Relationship between weed species richness (**A**,**B**) and abundance (**C**,**D**) in field cores with crop height (**A**,**C**) and the intensity of herbicide applications (**B**,**D**). Abundance was log-transformed and explanatory variables were scaled. The intensity of herbicide use is expressed using the Treatment Frequency Index. Dashed line indicates a nonsignificant relationship.

Adding landscape variables improved the model, which explained 27.4% of the variance (Figure 1; 22.4% with QA Figure S1). However, the contribution of landscape variables alone to weed species richness was lower compared to the contribution of the local variables. The estimated spatial extent of the effects of the landscape variables was always lower than 1000 m, ranging from a very small scale for hedge density (17 m from the border of the field), to medium scales for meadows (500 m), oilseed rape (600 m) and organic farming (780 m). Weed species richness was generally unaffected by landscape variables. The interplay between local and landscape variables was revealed by the significant interaction between the number of meadows and amount of nitrogen (Table 1 and Table S1), suggesting a lower positive effect of nitrogen on weed species richness in fields surrounded by a high number of meadows. We also found a significant positive interaction between the amount of organic farmed fields and herbicides when using TFI as a proxy of herbicide intensity (Table 1; i.e., the relationship is almost significant with QA, $p = 0.054$; Table S1). This suggests that herbicide use significantly decreased weed species richness in oilseed rape fields in landscapes rich in organic farming.

2.2. *Competition and Weed Management Strongly Affect Weed Abundance in Field Cores*

The pattern for weed abundance in field cores was mostly consistent with the pattern of weed species richness: environmental variables and mechanical weed control were discarded as was nitrogen (Table 1B). Weed abundance in the centre of oilseed rape fields significantly decreased with crop height (Figure 2C) and herbicide use (Figure 2D), with a higher effect attributed to crop height. Adding the landscape variables improved the model (Figure 1), although we found no significant effect of landscape variables on weed abundance (Table 1B).

2.3. *Landscape Is a Major Driver of Weed Assemblages in Field Margins*

Diversity and abundance patterns showed a contrasted situation in field margins, revealing that weed species assemblages in field margins were mainly affected by environmental conditions and landscape (Figure 1). The selection procedure removed crop height and farming practices for both weed species richness and abundance, while several environmental variables were kept. Weed species richness and abundance significantly decreased with rainfall (Table 2A), while only weed abundance increased with temperature (Table 2B). Landscape effect was mainly due to the number of meadows, which had a significant positive effect on both weed species richness (Figure 3A) and weed abundance (Figure 3B) at small scale, i.e., for 140 m from the border of the field for weed species richness and 265 m for weed abundance.

Table 2. Statistics of the models for weed (A) species richness and (B) abundance in field margins. Abundance was log-transformed, and environmental variables were centred and reduced. The estimated buffer radii are indicated for each landscape variable. Landscape variables have two degrees of freedom because both their spatial extent and their effect were estimated. Significant effects are indicated in bold. R-squared of respective models computed with type III ANOVAs were 12.3% and 22.7%.

(A)		Estimated Buffer Radius (m)	Estimate	df	t-Value	*p*-Value
	Intercept		20.262	1	10.818	**<0.001**
	Rainfall		−1.805	1	−2.467	**0.015**
	Hedge density	37	−86.517	2	−1.083	0.281
	Number of meadows	138	18.731	2	2.184	**0.031**
	Amount of organic farming	20	8.255	2	1.642	0.103
	Amount of oilseed rape	980	−11.572	2	−0.711	0.478
(B)						
	Intercept		1.892	1	60.673	**<0.001**
	Rainfall		−0.087	1	−4.060	**<0.001**
	Temperature		0.051	1	2.310	**0.023**
	Hedge density	5	−0.990	2	−1.892	0.061
	Number of meadows	266	0.664	2	1.991	**0.049**
	Amount of organic farming	22	0.168	2	1.244	0.216
	Amount of oilseed rape	8	−0.099	2	−0.755	0.452

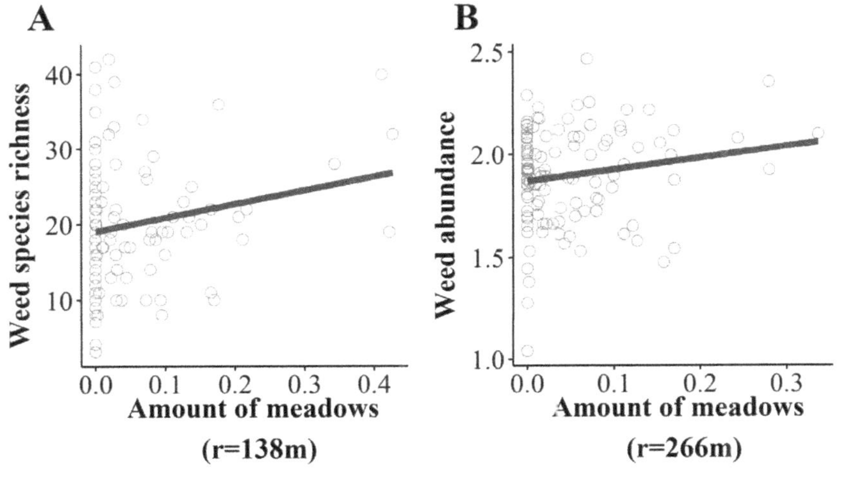

Figure 3. Relationship between weed species richness (**A**) and abundance (**B**) in field margins with the number of meadows in the surrounding landscape at respective buffer radii of 138 and 266 m. Abundance was log-transformed.

2.4. Multiscale Processes Shape Weed Assemblages in Field Cores

Because landscape affects weed assemblages in field margins and previous studies revealed local dispersal from field margins to field cores, we performed an SEM to assess the joint effect of local and landscape processes when considering weed assemblages in the two field compartments. The best model shown by BIC-based selection for both weed species richness and abundance was the SEM considering an indirect effect of landscape on weed assemblages in the field core through to the field margin. Competing models with either a direct link between landscape variables and weeds in field cores, or no link between the margin and centre of the fields were never retained (Tables S2 and S3). Accounting for local dispersal from the field margin strongly increased the part of variance explained in the field core, with R-squared increasing from 27% to 30% for weed species richness and from 18% to 30% for weed abundance using TFI (22.4% to 33% and 16% to 32% when using QA). The strength of local dispersal was similar for weed species richness and abundance (Figure 4). These analyses suggest that weed assemblages in the centre of oilseed rape fields were shaped by local factors (mainly crop competition and chemical weeding) and local dispersal from field margins, its relative importance being related to the number of meadows in the surrounding landscape. Interestingly, when accounting for spatial dispersal across the field margin, herbicide applications had a significant negative effect on both species richness and abundance. This effect was found when using linear models (without incorporating spatial dispersal across the field margin) for analysing weed abundance in the field core, but this was not the case for weed species richness. These results are, however, in line with the significant positive interaction between herbicide use and amount of organically farmed fields and suggest that herbicides decrease weed species richness in oilseed rape fields located in more diversified landscapes (i.e., a higher number of meadows and amount of organic farming).

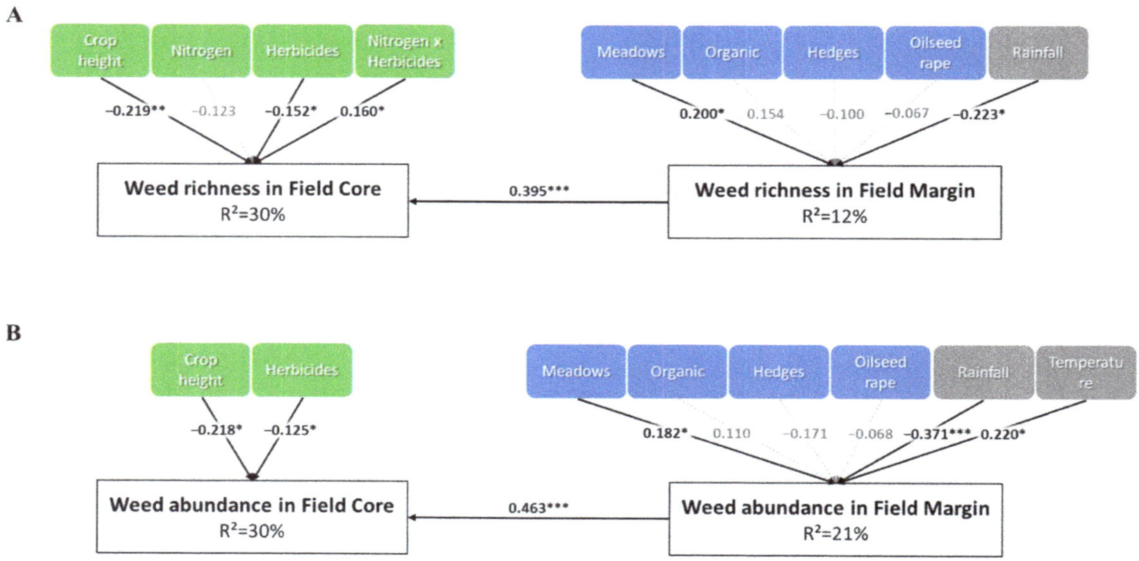

Figure 4. Structural Equation Models for weed species richness (**A**) and abundance (**B**) where the link between field margin and field core was specified. Arrows represent the directionality of the effect, and the coefficients indicate the standardized estimates. Dashed lines and grey estimates represent nonsignificant effects. *: p-value < 0.05; **: p-value < 0.01; ***: p-value < 0.001. FC: field core, FM: field margin. The intensity of herbicide use is expressed using the Treatment Frequency Index. The buffer radius at which the amount of each landscape variable was estimated is shown in Table 2 for field margin.

3. Discussion

Weed species assemblages are the result of the complex interplay between weed–crop competition, farming practices and landscape. In this study, we aimed at determining their relative contribution on both weed species richness and abundance in the margins and centres of 115 oilseed rape fields. As expected, our results highlighted that the mechanisms shaping weed assemblages differed between field cores and margins. Using crop height as a proxy for weed–crop competition, we found that competition strongly affected weed species richness and abundance in field cores while low or no effects could be detected for farming practices and landscape. Conversely, crop competition had almost no effect on weed assemblages in field margins, where we found a strong effect of landscape, suggesting a predominant role of spatial dispersal. Although landscape had no direct effect on weed species richness and abundance in field cores, the use of structural equation modelling revealed that landscape arrangement may affect weed assemblages in field cores indirectly through field margins.

As expected, the main driver of weed assemblages in field cores was the presence of the crop itself, since its height was positively related to a decrease of weed species richness and abundance. Competition with the crop had a higher effect on weed abundance than on weed species richness. Taller and denser crop plants in field cores are more prone to take up resources as light and nutrients, leaving lower amounts of resources available for weeds [35]. These results confirm the competitive ability of oilseed rape against weeds [36–38], but to our knowledge, our study is the first to demonstrate it in oilseed rape farmers' fields and with a natural flora. We expected a higher importance of competition in high nitrogen conditions because oilseed rape plants are nitrophilous plants and because reduced nitrogen amount might delay canopy closure [39]. No effect of nitrogen alone or in interaction with crop height was found, however. While mechanical weeding did not affect weed assemblages in field cores, we found a significant decrease of weed abundance with herbicide use, although the effect was lower compared with the effect of the competition with crop plants. We also found a significant negative effect of herbicide applications on weed species richness, but only when interacting with landscape or accounting for spatial dispersal across the field margin. Herbicide application directly affects weeds and is generally related to a decrease in weed species richness, mostly due to removal of rare species [40,41]. Fried et al. [34] found that weed species from the same family as oilseed rape (*Brassicaceae*) had higher densities in treated plots and suggested a phylogenetic convergence of weeds [42]. Such specialization of weed assemblages may explain the low effect of herbicides on weed species richness in the centres of oilseed rape fields located in landscapes with low numbers of organic farmed fields and meadows. Conversely, oilseed rape fields in more diversified landscapes may shelter more rare or unspecialized species because of spatial dispersal, explaining the significant effect of herbicides on weed species richness in these fields.

Accounting for landscape in our analysis resulted in an increase of the goodness-of-fit of our models on both weed species richness and abundance in field cores. However, the contribution of landscape variables was low and weed assemblages were generally unaffected by landscape variables. We did however find that a higher numbers of meadows weakened the positive effect of nitrogen on weed species richness. Our results therefore contrast with previous studies conducted in winter cereals, which revealed a positive effect of a higher number of organic farming [28] or seminatural habitats [29] on weed species richness. They are however in accordance with a previous study conducted in oilseed rape fields on the same study site [13], although more generally the literature on the effects of landscape on weed assemblages in oilseed rape fields is still lacking. Although no direct effect of landscape was found, by using structural equations models (SEMs), we revealed a strong indirect effect of landscape on weed assemblages in field cores and field margins. Accounting for the indirect effect greatly increased the goodness-of-fit of the models, especially for weed abundance. We acknowledge that this pattern may arise due to high correlation between weed species richness (or abundance) in field cores and field margins. However, the SEMs showed higher goodness-of-fits with the directionality from field margin towards field core than the contrary. In addition, Bourgeois et al. [13], showed that the similarity of weed assemblages in field core decreased with the distance from the field margin. Therefore, it is likely that landscape affects weed assemblages in the centre of oilseed rape fields across the margins.

Spatial dispersal was the main mechanism shaping weed species richness and abundance in field margins. Among landscape variables, the number of meadows had the strongest effect on weed species richness and abundance. The margins of oilseed rape fields surrounded with a higher number of meadows showed greater weed species richness and weed abundance. The spatial extent of the effects was, however, different: the number of meadows increased weed species richness in field margins at a lower scale (138 m) than weed abundance (266 m). Meadows are seminatural elements of agricultural landscapes that contribute to the maintenance of biodiversity in the agroecosystem by providing food and nesting habitats [43]. Meadows, acting as source habitats, can thus increase weed diversity in field margins. Our results highlighted that spatial dispersal might be the predominant process affecting weeds in field margins, since the variables related to crop competition and farming practices were discarded by the model selection procedure. Such a result is in accordance with our expectations. Indeed, field margins are generally managed at a lower intensity compared to field cores, and crop plants are smaller, present at lower density or even absent. Interestingly, climatic variables had significant effects on weed assemblages in field margins, contrary to field cores. This suggests that in absence of strong filtering factors such as competition or disturbances, climate has a higher filtering effect on weed assemblages. Indeed, a recent study investigating trait–climate relationships in plant assemblages revealed that these relationships were much weaker in croplands compared to grasslands, suggesting a reduced sensitivity of plant assemblages to bioclimatic variations in intensively managed habitats [44].

Our findings suggest that weed assemblages in field cores and margins are shaped by different mechanisms acting at different spatial scales. However, a large part of the variance remained unexplained (around 70% when accounting for dispersal from field margin to field core for both weed richness and abundance). This suggests that other factors may shape weed assemblage such as temporal dispersal [15]. Arable weed species are mainly therophyte species [45], which can persist for long periods as dormant seeds in the seed bank. Such a strategy may allow weed species, and especially those with long, persistent dormant seeds, to avoid unsuitable environmental conditions through delayed emergence (i.e., temporal storage effect, [46]). Further studies should therefore consider the respective roles of temporal dispersal, together with competition, environmental filtering (effect of farming practices) and spatial dispersal.

In conclusion, our study emphasizes the critical importance of crop competition in shaping weed assemblages in field cores, and spatial dispersal in shaping weed assemblages in field margins of oilseed rape fields. Herbicides had a lower effect than crop competition on weed abundance and were shown to reduce weed species richness in oilseed rape fields located in landscapes with higher numbers of extensively managed fields (i.e., organic farmed fields or meadows). Our findings give empirical support for crop competition as a way to reduce the intensity of chemical weeding, and for meadows as a way to enhance biodiversity in agricultural landscapes

4. Materials and Methods

4.1. Study Area

The study was conducted on the Long-Term Social-Ecological Research (LTSER) site Zone Atelier 'Plaine & Val de Sèvre' [47], a 435 km^2 agricultural landscape located in the Deux-Sèvres district, central western France. Climatic conditions are a mild, temperature, Atlantic oceanic climate (mean annual temperature 12.5 °C and precipitation 867.2 mm). Land use is dominated by cereal production, mostly winter cereals (41.3%), maize (9.6%), sunflower (8.8%) and rapeseed (7.6%). Meadow cover represents around 13%.

4.2. Weed Sampling

We surveyed weeds in 115 oilseed rape fields, managed by local farmers, from 2014 to 2018 (23 in 2014, 25 in 2015, 25 in 2016, 20 in 2017 and 22 in 2018). Field size averaged 6.5 ha and ranged from 0.8 ha to 23.1 ha (Electronic supplementary material Table S4). The annual survey spanned from the end of March to late April. Weed flora was monitored in 25 × 1 m^2 plots per field, each plot being subdivided into four 0.5 × 0.5 m subplots. A total of 20 plots were placed in field core (at least 10 m from the field edge) along two 100 m-long parallel transects (10 plots per transect). The two transects were separated by 50 m and were orthogonal to crop rows. Five plots were placed in the field margin and spaced 10 m apart [48]. We recorded the occurrence of weed species in each subplot and inventoried 157 plant taxa overall (species list in Electronic supplementary material, Table S5).

We computed weed species richness (i.e., the number of species) and abundance separately in the two field compartments. Weed abundance was the sum of individual presence in the 20 or 80 subplots within the field margin and field core, respectively. We did not account for the difference in sampling effort in the study because we conducted the statistical analysis in the two field compartments.

4.3. Local, Landscape and Environmental Variables

We used crop height (cm) as a proxy for crop competition because height is related to the plant's ability to intercept light [49]. During the weed survey, we measured the average canopy height of crop plants in each compartment. In four fields, crop heights in the field margin were missing. We estimated crop height in these fields by averaging the crop height values in fields in which crop height in the field core was 10 cm smaller or greater than the crop height value measured in the field core of the field in which we had a missing observation.

Local management practices related to the level of resources (nitrogen fertilizer) and disturbances (chemical and mechanical weed control) were recorded through farmers' interviews. The amount of nitrogen input (kg·ha^{-1}) was calculated from the fertilizer composition and the quantity applied. The intensity of herbicide applications was assessed using two quantitative indicators [50]: (i) the amount of active substances, which is the sum of the amount of active substances applied, and (ii) the Treatment Frequency Intensity (TFI), which is a measure of the intensity of herbicide application related to the recommended application. The intensity of mechanical weed control was estimated using the average depth of the soil operations. Nitrogen inputs, herbicide applications and mechanical weeding were considered from harvest of the previous crop to the weed sampling date. Data are summarized in the electronic supplementary material (Table S4).

Landscape information was obtained from the land-use database of the LTSER Zone Atelier Plaine & Val de Sèvre [48]. We considered four landscape variables previously shown to affect weed species assemblages, i.e., organic farming [28], seminatural habitats (including meadows and fallows [29,30]), hedgerows measured as a linear [51], but converted of surface of one metre width, and oilseed rape fields [12]. Proportions of landscape variables were computed from the field edge within buffer areas around each field. The scale of buffers for each landscape variable was estimated using the Siland approach [32] which is based on an optimization procedure of the likelihood, without any a priori information on the buffer extent value. For each weed metric, we estimated the buffer radius for the four landscape variables in the two compartments (see below).

4.4. Statistical Analysis

The statistical analysis consisted of three main steps. In a first step, we investigated the relative contributions of competition with the crop, resource levels (i.e., amount of nitrogen) and disturbances induced by weed control on weed species richness and abundance in both field cores and field margins. We included crop height and the amount of nitrogen input as proxies of crop competition for resources, and the intensity of herbicide applications and the intensity of soil mechanical operations as proxies for disturbances induced by weed control. To account for interactive effects, we added two-way interactions. Here, we also considered confounding factors acting on weed species richness and abundance, namely field area (in ha), date of sampling (in Julian day as quadratic polynomial), as well as temperature and rainfall, which vary among years. We included temperature (sum of growing degree days, °C) and rainfall (mean precipitation, mm) during the growing period of weeds, rather than throughout the year because these two variables are directly related to plant growth [52,53]. All these variables were included in linear models (LMs) for weed species richness and abundance in the two field compartments (i.e., four LM models were built). We used a variable selection procedure comparing models based on minimizing the Bayesian Information Criterion (BIC [54]) using the dredge function of MuMIn package [55] in R software version 4.0.3 [56]. All explanatory variables involved in at least one of the models with a BIC difference lower than two, from the model with the lowest BIC, were kept for the second step.

In a second step, we examined how the landscape context affects the importance of competition and disturbances on both weed species richness and abundance in field cores and field margins. We built LMs (one for each weed metric in each field compartment) that included the variables retained in the model selection procedure performed in the first step and landscape variables, i.e., hedgerow density and the amount of organic farming, seminatural habitats and oilseed rape fields. We also included the interactions between each landscape variable and the retained variables. The effect and spatial extent of each landscape variable were simultaneously estimated using the Bsiland function of the R package Siland [32]. We used a type III analysis of variance ('car' R package, third version [57]).

Finally, the third and last step of the analysis consisted of testing for the effect of local spatial dispersal from the field margin to the field core. We built a Structural Equation Modelling (SEM) where we considered the field margin flora, as an endogenous variable as well as the variables retained at the second step. Three competing models were tested. The first one incorporated the local and landscape variables included in the linear models built for each metric in step two, without any link relating weed assemblages in the field core and the field margin. In the second model, we tested for an indirect effect of landscape on the weed assemblage in the field core across the field margin. The third model extended the second one by including a direct effect of landscape variables on the weed assemblage in the field core. We considered the strength and directionality of the effect only for the SEM minimizing the BIC, this criterion being relevant to compare SEMs [58]. We assured that the SEMs respected four conditions of well structuration and goodness-of-fit: a p-value of the Fischer's C test > 0.05, a Comparative Fit Index (CFI) > 0.9, a Root Mean Square Error of Approximation (RMSEA) < 0.08 and a Standardized Root Mean Square Residuals (SRMR) < 0.08. We performed SEMs using the package 'piecewiseSEM' on R software [59].

All models were run using either one of the two quantitative indicators used to estimate the intensity of herbicide applications, i.e., the amounts of active substances (QA) and the Treatment Frequency Index (TFI). Because the goodness-of-fit of the models using TFI was higher for weed species richness (not for weed abundance) compared to those of the model using QA, only results with TFI are presented here (results with QA are shown in Table S1 and Figures S1–S3). Using TFI and QA generally did not change the general patterns (except for a significant interaction in the weed species richness, see Results section).

When analysing weed abundance in field cores, we found an outlier which affected the outcome of the model. We therefore removed this field from all the analysis conducted with weed abundance (data not shown).

Weed abundance was log10 transformed and explanatory variables were scaled (i.e., transformed using a z-score) using the "scale" function on R software before analysis. We also checked for each model the Variance Inflation Factor (VIF) to control for collinearity between the explanatory variables [60] using the "vif" function in the Car R package [57]. All VIF scores were below 5, showing the absence of problematic collinearity between variables. R-squared values were calculated from the best model determined.

Supplementary Materials: The following are available online at https://www.mdpi.com/article/10.3390/plants10102131/s1: Figure S1: Percentage of variance explained by local, landscape and environmental variables when using the quantity of active substances as a proxy for herbicide intensity; Figure S2: Model predictions for weed abundance in field cores using the quantity of active substances as a proxy for herbicide intensity; Figure S3: Structural Equation Models for weed species richness (A) and abundance (B) using the quantity of active substances; Table S1: Statistics of the models of weed (A) species richness and (B) abundance in field core using the quantity of active substances; Table S2: Statistics of the competing structural equation models with the type of link specified between field margin and field core floras using the quantity of active substances; Table S3: Statistics of the competing structural equation models with the type of link specified between field margin and field core floras using the treatment frequency intensity; Table S4: Descriptive statistics for local and environmental variables, Table S5: List of the taxa identified in the 115 oilseed rape fields of the study.

Author Contributions: Conceptualization, S.G.; Methodology, O.M. and S.G.; Formal Analysis, A.B. and O.M.; Writing—Original Draft Preparation, A.B. and S.G.; Writing—Review and Editing, A.B., S.G. and O.M. All authors have read and agreed to the published version of the manuscript.

Funding: This project was supported by the French Ministry of Ecology project (2017–2020 "Pollinisateurs"), the ANR IMAgHO (ANR-18-CE32-002) and the INRAE MP SumCrop project (2020–2021 "INITIATE"). This project also received funding from the European Union's Horizon 2020 research and innovation programme under grant agreement SHOWCASE No 862480. AB is supported by CNRS and the Nouvelle Aquitaine Region project (2018–2022 "Multistress"). SG and OM are funded by INRAE.

Institutional Review Board Statement: Not Applicable.

Informed Consent Statement: Not Applicable.

Data Availability Statement: Data will be provided upon request to the corresponding author.

Acknowledgments: We would like to express our thanks to Stanislas de Closets, Thierry Fanjas-Mercere, Brice Livoir, Jonathan Millaud and Solenne Rousselet for conducting the weed surveys and to Ambroise Lahu for checking the agricultural data and computing the herbicide and fertilizer indices and the number of soil operations. We also sincerely thank the farmers of the LTSER Zone Atelier Plaine & Val de Sèvre for their involvement in our research programs.

Conflicts of Interest: The authors declare no conflict of interest. The funders had no role in the design of the study; in the collection, analyses, or interpretation of data; in the writing of the manuscript, or in the decision to publish the results.

References

1. Altieri, M.A.; Nicholls, C.I. Scaling up Agroecological Approaches for Food Sovereignty in Latin America. *Development* **2008**, *51*, 472–480. [CrossRef]
2. Gliessman, S.R. 3 Sustainable agriculture: An agroecological perspective. In *Advances in Plant Pathology*; Andrews, J.H., Tommerup, I.C., Eds.; Academic Press: Cambridge, MA, USA, 1995; Volume 11, pp. 45–57. ISBN 0736-4539.
3. Oerke, E.-C. Crop Losses to Pests. *J. Agric. Sci.* **2006**, *144*, 31–43. [CrossRef]
4. Richner, N.; Holderegger, R.; Linder, H.P.; Walter, T. Reviewing change in the arable flora of Europe: A meta-analysis. *Weed Res.* **2015**, *55*, 1–13. [CrossRef]
5. Albrecht, H.; Cambecèdes, J.; Lang, M.; Wagner, M. Management options for the conservation of rare arable plants in Europe. *Bot. Lett.* **2016**, *163*, 389–415. [CrossRef]
6. Bretagnolle, V.; Gaba, S. Weeds for Bees? A Review. *Agron. Sustain. Dev.* **2015**, *35*, 891–909. [CrossRef]
7. Marshall, E.J.P.; Brown, V.K.; Boatman, N.D.; Lutman, P.J.W.; Squire, G.R.; Ward, L.K. The role of weeds in supporting biological diversity within crop fields. *Weed Res.* **2003**, *43*, 77–89. [CrossRef]
8. Gaba, S.; Cheviron, N.; Perrot, T.; Piutti, S.; Gautier, J.-L.; Bretagnolle, V. Weeds enhance multifunctionality in arable lands in south-west of France. *Front. Sustain. Food Syst.* **2020**, *4*, 71. [CrossRef]
9. Sánchez-Bayo, F.; Wyckhuys, K.A.G. Worldwide decline of the entomofauna: A review of its drivers. *Biol. Conserv.* **2019**, *232*, 8–27. [CrossRef]
10. Storkey, J.; Neve, P. What good is weed diversity? *Weed Res.* **2018**, *58*, 239–243. [CrossRef]
11. MacLaren, C.; Storkey, J.; Menegat, A.; Metcalfe, H.; Dehnen-Schmutz, K. An ecological future for weed science to sustain crop production and the environment. A review. *Agron. Sustain. Dev.* **2020**, *40*, 24. [CrossRef]
12. Alignier, A.; Solé-Senan, X.O.; Robleño, I.; Baraibar, B.; Fahrig, L.; Giralt, D.; Gross, N.; Martin, J.; Recasens, J.; Sirami, C.; et al. Configurational crop heterogeneity increases within-field plant diversity. *J. Appl. Ecol.* **2020**, *57*, 654–663. [CrossRef]
13. Bourgeois, B.; Gaba, S.; Plumejeaud, C.; Bretagnolle, V. Weed diversity is driven by complex interplay between multi-scale dispersal and local filtering. *Proc. R. Soc. B Biol. Sci.* **2020**, *287*, 20201118. [CrossRef]
14. Gaba, S.; Fried, G.; Kazakou, E.; Chauvel, B.; Navas, M.-L. Agroecological weed control using a functional approach: A review of cropping systems diversity. *Agron. Sustain. Dev.* **2014**, *34*, 103–119. [CrossRef]
15. Mahaut, L.; Fried, G.; Gaba, S. Patch dynamics and temporal dispersal partly shape annual plant communities in ephemeral habitat patches. *Oikos* **2018**, *127*, 147–159. [CrossRef]
16. Metcalfe, H.; Milne, A.E.; Coleman, K.; Murdoch, A.J.; Storkey, J. Modelling the effect of spatially variable soil properties on the distribution of weeds. *Ecol. Model.* **2019**, *396*, 1–11. [CrossRef]
17. Fried, G.; Norton, L.R.; Reboud, X. Environmental and management factors determining weed species composition and diversity in France. *Agric. Ecosyst. Environ.* **2008**, *128*, 68–76. [CrossRef]
18. Gunton, R.M.; Petit, S.; Gaba, S. Functional traits relating arable weed communities to crop characteristics: Traits relating weed communities to crops. *J. Veg. Sci.* **2011**, *22*, 541–550. [CrossRef]
19. Gandía, M.L.; Casanova, C.; Sánchez, F.J.; Tenorio, J.L.; Santín-Montanyá, M.I. Arable weed patterns according to temperature and latitude gradient in central and southern Spain. *Atmosphere* **2020**, *11*, 853. [CrossRef]
20. Borgy, B.; Gaba, S.; Petit, S.; Reboud, X. Non-random distribution of weed species abundance in arable fields: Distribution of abundances in weed communities. *Weed Res.* **2012**, *52*, 383–389. [CrossRef]
21. Melander, B.; Munier-Jolain, N.; Charles, R.; Wirth, J.; Schwarz, J.; van der Weide, R.; Bonin, L.; Jensen, P.K.; Kudsk, P. European perspectives on the adoption of nonchemical weed management in reduced-tillage systems for arable crops. *Weed Technol.* **2013**, *27*, 231–240. [CrossRef]
22. Kaur, S.; Kaur, R.; Chauhan, B.S. Understanding crop-weed-fertilizer-water interactions and their implications for weed management in agricultural systems. *Crop Prot.* **2018**, *103*, 65–72. [CrossRef]
23. Gaba, S.; Caneill, J.; Nicolardot, B.; Perronne, R.; Bretagnolle, V. Crop competition in winter wheat has a higher potential than farming practices to regulate weeds. *Ecosphere* **2018**, *9*, e02413. [CrossRef]

24. Sardana, V.; Mahajan, G.; Jabran, K.; Chauhan, B.S. Role of competition in managing weeds: An introduction to the special issue. *Crop Prot.* **2017**, *95*, 1–7. [CrossRef]
25. Petit, S.; Gaba, S.; Grison, A.-L.; Meiss, H.; Simmoneau, B.; Munier-Jolain, N.; Bretagnolle, V. Landscape scale management affects weed richness but not weed abundance in winter wheat fields. *Agric. Ecosyst. Environ.* **2016**, *223*, 41–47. [CrossRef]
26. Bohan, D.A.; Haughton, A.J. Effects of local landscape richness on in-field weed metrics across the Great Britain scale. *Agric. Ecosyst. Environ.* **2012**, *158*, 208–215. [CrossRef]
27. Roschewitz, I.; Gabriel, D.; Tscharntke, T.; Thies, C. The effects of landscape complexity on arable weed species diversity in organic and conventional farming: Landscape complexity and weed species diversity. *J. Appl. Ecol.* **2005**, *42*, 873–882. [CrossRef]
28. Henckel, L.; Börger, L.; Meiss, H.; Gaba, S.; Bretagnolle, V. Organic fields sustain weed metacommunity dynamics in farmland landscapes. *Proc. R. Soc. B Biol. Sci.* **2015**, *282*, 20150002. [CrossRef]
29. Kovács-Hostyánszki, A.; Batáry, P.; Báldi, A.; Harnos, A. Interaction of local and landscape features in the conservation of hungarian arable weed diversity: Weed richness in Hungarian cereal fields. *Appl. Veg. Sci.* **2011**, *14*, 40–48. [CrossRef]
30. Medeiros, H.R.; Thibes Hoshino, A.; Ribeiro, M.C.; de Oliveira Menezes Junior, A. Landscape complexity affects cover and species richness of weeds in brazilian agricultural environments. *Basic Appl. Ecol.* **2016**, *17*, 731–740. [CrossRef]
31. Pallavicini, Y.; Bastida, F.; Hernández-Plaza, E.; Petit, S.; Izquierdo, J.; Gonzalez-Andujar, J.L. Local factors rather than the landscape context explain species richness and functional trait diversity and responses of plant assemblages of mediterranean cereal field margins. *Plants* **2020**, *9*, 778. [CrossRef]
32. Carpentier, F.; Martin, O. Siland a R Package for Estimating the Spatial Influence of Landscape. *Sci. Rep.* **2021**, *11*, 7488. [CrossRef]
33. Bouchet, A.-S.; Laperche, A.; Bissuel-Belaygue, C.; Snowdon, R.; Nesi, N.; Stahl, A. Nitrogen use efficiency in rapeseed. A review. *Agron. Sustain. Dev.* **2016**, *36*, 38. [CrossRef]
34. Fried, G.; Chauvel, B.; Reboud, X. Weed flora shifts and specialisation in winter oilseed rape in France. *Weed Res.* **2015**, *55*, 514–524. [CrossRef]
35. Smith, R.G.; Mortensen, D.A.; Ryan, M.R. A new hypothesis for the functional role of diversity in mediating resource pools and weed–crop competition in agroecosystems. *Weed Res.* **2010**, *50*, 37–48. [CrossRef]
36. Lemerle, D.; Verbeek, B.; Coombes, N. Losses in grain yield of winter crops from lolium rigidum competition depend on crop species, cultivar and season. *Weed Res.* **1995**, *35*, 503–509. [CrossRef]
37. Holman, J.D.; Bussan, A.J.; Maxwell, B.D.; Miller, P.R.; Mickelson, J.A. Spring wheat, canola, and sunflower response to persian darnel (*Lolium Persicum*) interference. *Weed Technol.* **2004**, *18*, 509–520. [CrossRef]
38. Hashem, A.; Borger, C.P.D.; Riethmuller, G. *Weed Suppression by Crop Competition in Three Crop Species in Western Australia*; New Zealand Plant Protection Society: Christchurch, New Zealand, 2010; pp. 26–30.
39. Rathke, G.; Behrens, T.; Diepenbrock, W. Integrated nitrogen management strategies to improve seed yield, oil content and nitrogen efficiency of winter oilseed rape (*Brassica napus* L.): A review. *Agric. Ecosyst. Environ.* **2006**, *117*, 80–108. [CrossRef]
40. Fried, G.; Cordeau, S.; Metay, A.; Kazakou, E. Relative importance of environmental factors and farming practices in shaping weed communities structure and composition in French vineyards. *Agric. Ecosyst. Environ.* **2019**, *275*, 1–13. [CrossRef]
41. Gaba, S.; Gabriel, E.; Chadœuf, J.; Bonneu, F.; Bretagnolle, V. Herbicides do not ensure for higher wheat yield, but eliminate rare plant species. *Sci. Rep.* **2016**, *6*, 30112. [CrossRef]
42. Cavender-Bares, J.; Kozak, K.H.; Fine, P.V.A.; Kembel, S.W. The merging of community ecology and phylogenetic biology. *Ecol. Lett.* **2009**, *12*, 693–715. [CrossRef]
43. Sullivan, E.R.; Powell, I.; Ashton, P.A. Long-term hay meadow management maintains the target community despite local-scale species turnover. *Folia Geobot.* **2018**, *53*, 159–173. [CrossRef]
44. Bourgeois, B.; Munoz, F.; Gaba, S.; Denelle, P.; Fried, G.; Storkey, J.; Violle, C. Functional biogeography of weeds reveals how anthropogenic management blurs trait–climate relationships. *J. Veg. Sci.* **2021**, *32*, e12999. [CrossRef]
45. Bourgeois, B.; Munoz, F.; Fried, G.; Mahaut, L.; Armengot, L.; Denelle, P.; Storkey, J.; Gaba, S.; Violle, C. What makes a weed a weed? A large-scale evaluation of arable weeds through a functional lens. *Am. J. Bot.* **2019**, *106*, 90–100. [CrossRef] [PubMed]
46. Chesson, P. Mechanisms of maintenance of species diversity. *Annu. Rev. Ecol. Syst.* **2000**, *31*, 343–366. [CrossRef]
47. Bretagnolle, V.; Berthet, E.; Gross, N.; Gauffre, B.; Plumejeaud, C.; Houte, S.; Badenhausser, I.; Monceau, K.; Allier, F.; Monestiez, P.; et al. Towards sustainable and multifunctional agriculture in farmland landscapes: Lessons from the integrative approach of a French LTSER platform. *Sci. Total Environ.* **2018**, *627*, 822–834. [CrossRef]
48. Bretagnolle, V.; Berthet, E.; Gross, N.; Gauffre, B.; Plumejeaud, C.; Houte, S.; Badenhausser, I.; Monceau, K.; Allier, F.; Monestiez, P.; et al. Description of long-term monitoring of farmland biodiversity in a LTSER. *Data Brief* **2018**, *19*, 1310–1313. [CrossRef]
49. Violle, C.; Garnier, E.; Lecoeur, J.; Roumet, C.; Podeur, C.; Blanchard, A.; Navas, M.-L. Competition, traits and resource depletion in plant communities. *Oecologia* **2009**, *160*, 747–755. [CrossRef]
50. Möhring, N.; Gaba, S.; Finger, R. Quantity based indicators fail to identify extreme pesticide risks. *Sci. Total Environ.* **2019**, *646*, 503–523. [CrossRef]
51. Boutin, C.; Baril, A.; Martin, P. Plant diversity in crop fields and woody hedgerows of organic and conventional farms in contrasting landscapes. *Agric. Ecosyst. Environ.* **2008**, *123*, 185–193. [CrossRef]
52. Erwin, J. Temperature and light effects on seed germination. *Minn. Flower Grow. Bull.* **1991**, *40*, 16–23.
53. Vidotto, F.; De Palo, F.; Ferrero, A. Effect of short-duration high temperatures on weed seed germination: High temperatures affecting weed seeds. *Ann. Appl. Biol.* **2013**, 454–465. [CrossRef]

54. Burnham, K.P.; Anderson, D.R. Multimodel inference: Understanding AIC and BIC in model selection. *Sociol. Methods Res.* **2004**, *33*, 261–304. [CrossRef]
55. Barton, K. MuMIn: Multi-Model Inference. R Package Version 1.43.17. 2020. Available online: https://CRAN.R-project.org/package=MuMIn (accessed on 15 August 2021).
56. R Core Team, R. *A Language and Environment for Statistical Computing*; R Foundation for Statistical Computing: Vienna, Austria, 2020.
57. Fox, J.; Weisberg, S. *An R Companion to Applied Regression*; Third Version; Sage: Thousand Oaks, CA, USA, 2019; Available online: https://socialsciences.mcmaster.ca/jfox/Books/Companion/ (accessed on 15 August 2021).
58. Hertzog, L.R. How robust are structural equation models to model misspecification? A simulation study. *arXiv* **2019**, arXiv:1803.06186v3 [stat.AP].
59. Lefcheck, J.S. PiecewiseSEM: Piecewise structural equation modeling in R for ecology, evolution, and systematics. *Methods Ecol. Evol.* **2016**, *7*, 573–579. [CrossRef]
60. Zuur, A.F.; Ieno, E.N.; Walker, N.J.; Saveliev, A.A.; Smith, G.M. *Mixed Effects Models and Extensions in Ecology with R*; Springer Science & Business Media: New York, NY, USA, 2009; ISBN 978-0-387-87457-9.

plants

MDPI

Article

High Phenotypic Plasticity in a Prominent Plant Invader along Altitudinal and Temperature Gradients

Rodolfo Gentili [1,*], Roberto Ambrosini [2], Benno A. Augustinus [3,4,5], Sarah Caronni [1], Elisa Cardarelli [1], Chiara Montagnani [1], Heinz Müller-Schärer [4], Urs Schaffner [5] and Sandra Citterio [1]

1 Department of Earth and Environmental Sciences, University of Milano-Bicocca, Piazza della Scienza 1, 20126 Milano, Italy; sarah.caronni@unimib.it (S.C.); elisa.cardarelli@unimib.it (E.C.); chiara.montagnani@unimib.it (C.M.); sandra.citterio@unimib.it (S.C.)
2 Department of Environmental Science and Policy, University of Milano, Via Celoria 2, 20133 Milano, Italy; roberto.ambrosini@unimi.it
3 Swiss Federal Institute for Forest, Snow and Landscape Research WSL, Zuercherstrasse 111, CH-8903 Birmensdorf, Switzerland; benno.augustinus@wsl.ch
4 Department of Biology, University of Fribourg, Chemin du Musée 10, CH-1700 Fribourg, Switzerland; heinz.mueller@unifr.ch
5 CABI, Rue des Grillons 1, CH-2800 Delémont, Switzerland; u.schaffner@cabi.org
* Correspondence: rodolfo.gentili@unimib.it; Tel.: +39-02-64482700

Abstract: Studies on plant growth and trait variation along environmental gradients can provide important information for identifying drivers of plant invasions and for deriving management strategies. We used seeds of the annual plant invader *Ambrosia artemisiifolia* L. (common ragweed) collected from an agricultural site in Northern Italy (226 m. a.s.l; Mean Annual Air Temperature: 12.9 °C; precipitations: 930 mm) to determine variation in growth trajectories and plant traits when grown along a 1000-m altitudinal gradient in Northern Italy, and under different temperature conditions in the growth chamber (from 14/18 °C to 26/30 °C, night/day), using a non-liner modeling approach. Under field conditions, traits related to plant height (maximum height, stem height, number of internodes) followed a three-parameter logistic curve. In contrast, leaf traits (lateral spread, number of leaves, leaf length and width) followed non-monotonic double-Richards curves that captured the decline patterns evident in the data. Plants grew faster, reaching a higher maximum plant height, and produced more biomass when grown at intermediate elevations. Under laboratory conditions, plants exhibited the same general growth trajectory of field conditions. However, leaf width did not show the recession after the maximum value shown by plants grown in the field, although the growth trajectories of some individuals, particularly those grown at 18 °C, showed a decline at late times. In addition, the plants grown at lower temperatures exhibited the highest value of biomass and preserved reproductive performances (e.g., amount of male inflorescence, pollen weight). From our findings, common ragweed exhibits a high phenotypic plasticity of vegetative and reproductive traits in response to different altitudes and temperature conditions. Under climate warming, this plasticity may facilitate the shift of the species towards higher elevation, but also the in situ resistance and (pre)adaptation of populations currently abundant at low elevations in the invasive European range. Such results may be also relevant for projecting the species management such as the impact by possible biocontrol agents.

Keywords: growth curve; plant traits; elevation gradient; climate change; invasive plant species; *Ophraella communa*; invasive species management

Citation: Gentili, R.; Ambrosini, R.; Augustinus, B.A.; Caronni, S.; Cardarelli, E.; Montagnani, C.; Müller-Schärer, H.; Schaffner, U.; Citterio, S. High Phenotypic Plasticity in a Prominent Plant Invader along Altitudinal and Temperature Gradients. *Plants* **2021**, *10*, 2144. https://doi.org/10.3390/plants10102144

Academic Editors: Alessia Restuccia and Aurelio Scavo

Received: 14 September 2021
Accepted: 27 September 2021
Published: 9 October 2021

1. Introduction

Natural climatic variations associated with altitude are widely used to infer possible plant trait adaptations to temporal climate change and their phenotypic plasticity [1]. In response to increasing altitude and decreasing temperature, plants may modulate their functional traits including morphology, reproduction, and physiology [2–5]. With increasing altitude, the need of plants to survive and maintain reproductive success may result in either local adaptation or plastic responses [6,7]. To face sudden environmental changes, generally plastic response is very important over the short period whereas genetic responses is observed over longer-term periods [8]. Species showing larger plasticity may be better prepared to persist in new or stressing environments, helping the expansion of their geographical or altitudinal range across different environmental conditions [9,10], finally favoring local adaptation. Usually, along an altitudinal gradient, plants on higher sites invest a larger amount of resources in vegetative growth, with a possible reduction of reproductive output [7,11]. Plants often cope with resource deficit due to short vegetative seasons allocating biomass to resource-capturing organs (i.e., leaves and roots) [12]. Such ability has been demonstrated in plants in response to different levels or deficits of light, nutrient, water, and CO_2 [7]. Along an altitudinal gradient, the study of these plant traits, which are highly sensitive to climate changes, can be also used to phenotypically 'track' the observed climatic variations [1,12].

In biological invasion studies, a high phenotypic plasticity of alien plants is widely acknowledged to contribute to invasion success even in harsh environments, often outcompeting native species [13,14]. Trait plasticity can explain the reason why some invasive species show better ability to establish in a wide range of environments, thanks to their aptitude to increase and/or maintain fitness in both favorable and stressful situations [15,16]. For instance, fast growth rate and modulation of reproductive periods allow alien plant species to establish over wide altitudinal and temperature ranges [14]. Recent studies in this direction have shown that an increasing number of alien species occur at higher altitudes in temperate regions, favored by warmer temperatures [17]. Particularly, acclimation to stressful conditions in adverse climatic circumstances is considered a key factor of the success of alien species colonization [18]. While a higher tolerance of warmer temperatures is supposed to be the key for a successful alien invader compared to native species, a growing number of studies also report better abilities of aliens to cope with low temperatures [19,20]. Alien species tend to acclimate to new areas also under harsh environmental conditions by modulating their growth traits such as plant height, number of vegetative shoots, and number of flowers [21]. For such reasons, some invasive species may be more adapted to climate change due to traits that facilitate rapid range shifts (e.g., resource allocations, short time to maturity) and their wide climatic tolerances [22].

Incorporating impacts of climate change into invasive species management has been identified as a priority for land managers [23]. Such a goal can be achieved by implementing prevention actions and strategic planning, adjusting control actions, and by information exchanges between researchers and managers [23,24]. Some specific plant traits modulated by temperature changes or new environmental conditions (for species colonizing new territories or higher altitudes) via phenotypic plasticity can be responsible for the invasion success of alien species. In a changing climate, trait plasticity could confer a strong competitive advantage to alien invaders compared to native species, therefore augmenting their impact on ecosystems [25]. For instance, plasticity to new environmental conditions is expected to influence management strategies using biocontrol agents that are generally dependent on plant phenology [26]. Indeed, life cycle and timing of releases of host-specific insects may greatly vary in relation to climatic condition, becoming asynchronous with respect to reproductive events, possibly reducing the success of biocontrol [27].

Despite investigations on how invasive alien plants adapt to altitudinal gradients can be important to the understanding of processes involved in their establishment and spreading, relatively little is known about their growth patterns, especially for herbaceous species. Most studies have focused on factors affecting primary production (e.g., respiration, photosynthesis, carbon fluxes), with scarce attention dedicated to resource allocation and turnover [28]. Recently, Kühn et al. [29] investigated the variability of plant functional traits along elevation gradients. They found that within-population variability of leaf traits decreased with altitude. March-Salas and Pertierra [14] monitored the phenological development of two invasive alien species (*Poa annua* and *Cerastium fontanum*) at different altitudes in a sub-Antarctic region; the species showed great acclimation (growth) and reproductive ability also under limiting conditions. Alexander et al. [30] investigated growth trends and reproductive traits in native and invaded ranges of eight invasive Asteraceae forbs along an altitudinal gradient; plants exhibited smaller size and fewer inflorescences towards higher altitudes.

Among invasive alien species, common ragweed (*Ambrosia artemisiifolia* L.) is a successful invader of great concern in Europe and around the world [31]. Since the 19th century, this species native to North America has been inadvertently introduced in Europe (and then in other continents) where it has established and has become a serious threat to both agriculture, economy, and human health due to the production of large amounts of highly allergenic pollen [31,32]. In Europe, some 13.5 million people suffer from ragweed-induced allergies, with an annual economic cost of approximately 7.4 billion Euros [33]. It is a fast-growing annual weed that, thanks to its wide ecological amplitude and high within-population genetic diversity, colonizes a large variety of open-disturbed habitats including crop and abandoned fields, roadsides, and ruderal areas [34–36]. In addition to flat temperate areas, the species has been also reported to colonize a wide range of climates along latitudinal and altitudinal gradients, from the sea level to mountains [37,38].

In this study, we aimed to investigate how variation in altitude and temperature affects phenotypic expression of growth-related and reproductive traits of this invasive alien plant. We grew common ragweed plants both in the field along an altitudinal gradient and in the laboratory under controlled conditions and used prediction models to estimate the species performance in relation to altitude and temperature. In particular, by fitting parametric growth curves with nonlinear mixed models (NLMMs), we tested for trait size and reproductive performance reduction of the individuals along to a ~1000 m altitudinal (at several sites) and a decreasing temperature gradient.

2. Results

2.1. Growth Trajectories and Biomass of Plant Grown in Field Conditions

Under field conditions, maximum height, stem height, and the number of internodes grew monotonically along the whole time span considered and their growth trajectories were best described by three-parameter logistic curves. In contrast, lateral spread, number of leaves, and leaf length and width followed non-monotonic double-Richards curves that captured the recession patterns evident in the data (Table 1; Figure 1). With the only exception of the number of leaves, all measured features showed significant differences among sites in the upper asymptote (K parameter).

For the number of leaves, differences appeared in the decrease after the maximum number, which was less marked in the lowest-altitude site A than in the other sites. Generally, plants at the highest altitude sites D and E showed lower values than those at the lower sites (Table 1; Figure 1). Leaf length and width were exceptions in this pattern, as plants at the lowest altitude site A showed lower values than those at sites D and E, while sites at intermediate altitude, particularly site B, showed higher values. (Table 1; Figure 1). Growth trajectories of stem height also showed differences in the scale parameter (reciprocal of the growth rates), generally denoting a trend toward slower growth rates at sites at increasing altitudes (Table 1; Figure 1). Notably, the lateral spread showed a very complex pattern of variation among sites, with all the parameters of the double-Richards curve differing significantly among sites (Table 1). However, plants at the highest-altitude site E showed a generally slower growth and reached lower maximum sizes than those at sites A, B, and C. Plants at site D initially grew similarly to plants at sites A, B, and C, but then they reached lower maximum and final values, similar to those of plants in site E (Figure 1).

Dry and wet biomass of plants showed significant differences among sites at different times. The growth patterns also differed among sites, as indicated by the significant site by time interaction (Table 2).

Table 1. Final NLMMs of the growth trajectories of plants grown in field conditions. Coefficients of each parameter of the growth curves are reported in Table S1. The growth curve fitted and the number of observations and plants included in each analysis are reported as well as the residual degrees of freedom and the values of the temporal autocorrelation coefficient (φ) and of the Akaike information criterion corrected for a small sample size (AICc).

Parameter	F	df	p
Maximum height (three-parameter logistic curve, 867 observations and 89 plants)			
K	14.539	4	<0.001
i	1.7	4	0.148
s	1.914	4	0.162
Residual df = 764; φ = 0.578; AICc = 3702.3			
Stem height (three-parameter logistic curve, 502 observations and 65 plants)			
K	6.229	4	<0.001
i	2.281	4	0.060
s	4.192	4	0.002
Residual df = 423; φ = 0.609; AICc = 2242.9			
Number of internodes (four-parameter logistic curve, 849 observations and 89 plants)			
L	0.951	4	0.434
K	6.303	4	<0.001
i	0.483	4	0.748
s	0.432	4	0.785
Residual df = 741; φ = 0.558; AICc = 2438.2			
Lateral spread (double-Richards curve #31, 740 observations and 86 plants)			
K	11.949	4	<0.001
r	3.600	4	0.006
i	11.191	4	<0.001
K'	2.699	4	0.030
Residual df = 635; φ = 0.761; AICc = 2791.1			
$m = 1.233$, $r' = 2.073$, $i' = 10$, $m' = 0.963$			
Number of leaves (double-Richards curve #31, 453 observations and 59 plants)			
K	1.379	4	0.241
r	1.864	4	0.116
i	1.035	4	0.389
K'	4.844	4	0.001
Residual df = 375; φ = 0.737; AICc = 3050.7			
$m = -0.123$, $r' = 1.438$, $i' = 12$, $m' = 0.944$			
Leaf length (double-Richards curve #34, 773 observations and 89 plants)			
K	6.145	4	<0.001
r	1.677	4	0.086
i	0.104	4	0.606
r'	2.271	4	0.229
i'	1.283	4	0.275
Residual df = 660; φ = 0.637; AICc = 1351.5			
$i = -0.924$, $K' = -4.645$, $m' = 0.880$			
Leaf width (double-Richards curve #31, 773 observations and 89 plants)			
K	12.171	4	<0.001
r	2.047	4	0.086
i	0.68	4	0.606
K'	1.14	4	0.229
Residual df = 665; φ = 0.792; AICc = 855.3			
$m = 0.938$, $r' = 2.580$, $i' = 10$, $m' = 0.819$			

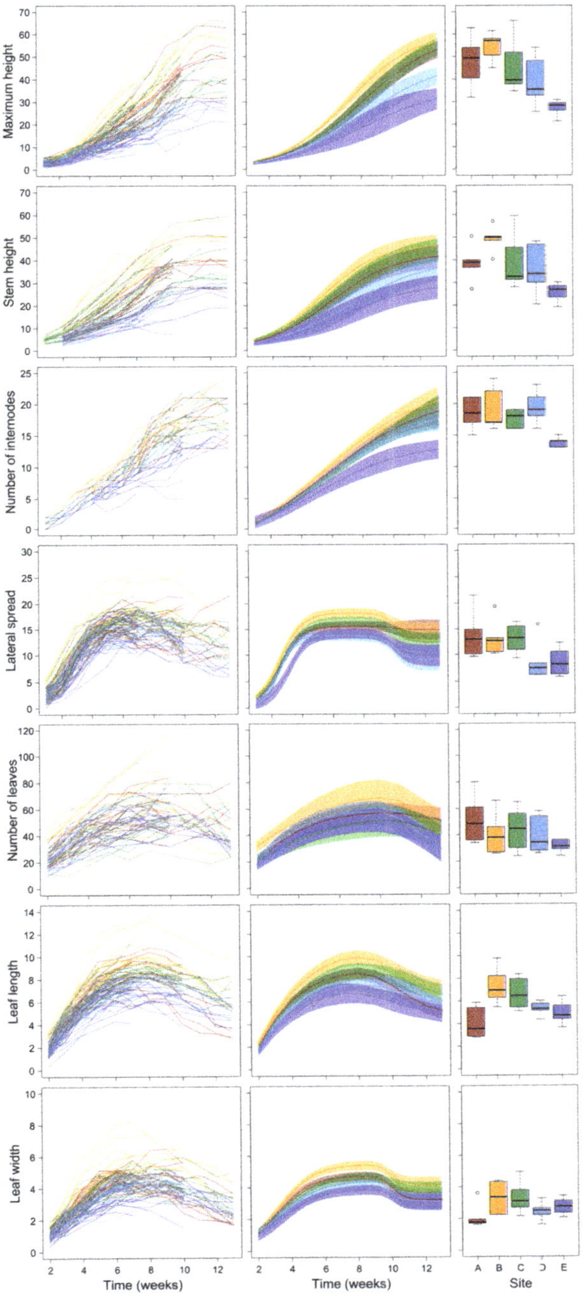

Figure 1. Left panels show growth trajectories of individual plants grown in the field (after the removal of deviant data points). Each line represents an individual plant. Central panel shows the interpolated growth curves and their 95% confidence intervals. Right panels show boxplots of the values of each plant measured at the last date (week 13). In all panels, colors represent sites (brown = site A, yellow = site B, green = site C, turquoise = site D, blue = site E).

Table 2. GLS models of the wet and dry biomass according to time, site, and their interaction.

Parameter	F	df	*p*
	Wet biomass ($n = 120$)		
Time	38.318	3	<0.001
Site	2.464	4	0.05
Time × Site	3.087	12	0.001
	Dry biomass ($n = 120$)		
Time	51.068	3	<0.001
Site	2.755	4	0.032
Time × Site	4.143	12	<0.001

Both dry and wet biomass did not differ among sites at week 4, but plants at site E showed lower values than site A already at week 7, while the other sites showed intermediate values (Figure 2). The same general pattern remained until week 13. For dry biomass, difference among sites were not significant at week 10 due to large spread of values, although the median values showed the same pattern as above (Figure 2).

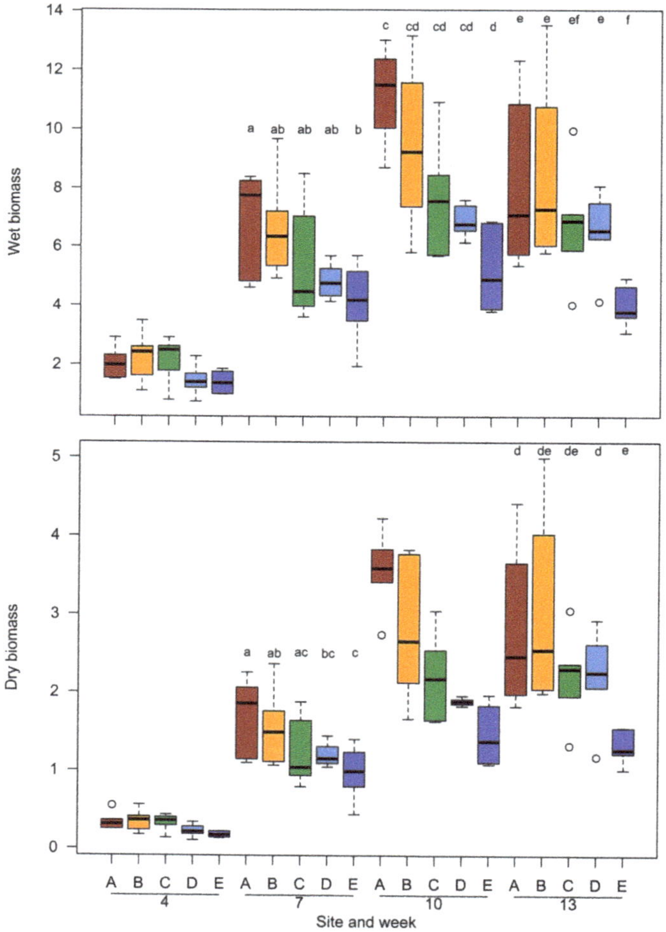

Figure 2. Boxplot of wet (upper panel) and dry (lower panel) biomass of plants grown at different sites (4, 7, 10, and 13 weeks). Different letters above bars denote significant differences between sites at that time. Circles represent outliers.

2.2. Growth Trajectories of Plants Grown in Laboratory Conditions

Maximum height and lateral spread of plants grown in the laboratory at different temperatures followed the same general growth pattern as plants grown in the field, as indicated by the fact that the same curves interpolated the data best (Table 3). Leaf length also showed a non-monotonic growth pattern, with a recession at later growth stages like plants grown in the field, but according to a slightly different parameterization of the double-Richards curves (compare Table 3 with Table 1). In contrast, in laboratory conditions, leaf width did not show the recession after the maximum value shown by plants grown in the field, although the growth trajectories of some individuals, particularly those grown at 18 °C, showed a decline at late times (see the left panel of Figure 3).

Table 3. Final NLMMs of the growth trajectories of plants grown in laboratory conditions. Coefficients of each parameter of the growth curves are reported in Table S2. The growth curve applied and the number of observations and plants included in each analysis are reported as well as the residual degrees of freedom and the values of the autocorrelation coefficient (φ) and of the Akaike information criterion corrected for small sample size (AICc).

Parameter	F	df	p
Maximum height (three-parameter logistic curve, 982 observations and 76 plants)			
K	83.642	2	<0.001
i	78.978	2	<0.001
s	25.544	2	<0.001
Residual df = 898; φ = 0.904; AICc = 3974.1			
Number of internodes (three-parameter logistic curve, 1096 observations and 85 plants)			
K	64.147	2	<0.001
i	5.007	2	0.007
s	3.484	2	0.003
Residual df = 741; φ = 0.558; AICc = 2438.2			
Lateral spread (double-Richards curve #31, 1182 observations and 96 plants)			
K	16.213	2	<0.001
r	26.439	2	<0.001
i	17.121	2	<0.001
K'	13.381	2	<0.001
Residual df = 1075; φ = 0.726; AICc = 3398.5 m = 1.228, r' = 0.542, Ri = 7.739, m' = 1.000			
Leaf length (double-Richards curve #31, 453 observations and 59 plants)			
K	11.727	2	<0.001
r	49.162	2	<0.001
i	16	2	<0.001
K'	7.138	2	0.001
Residual df = 997; φ = 0.704; AICc = 1376.1 m = 0.572, r' = 1.372, i' = 10, m' = 0.998			
Leaf width (three-parameter logistic curve, 1119 observations and 96 plants)			
K	46.151	2	<0.001
i	48.659	2	<0.001
s	27.652	2	<0.001
Residual df = 1015; φ = 0.615; AICc = 421.4			

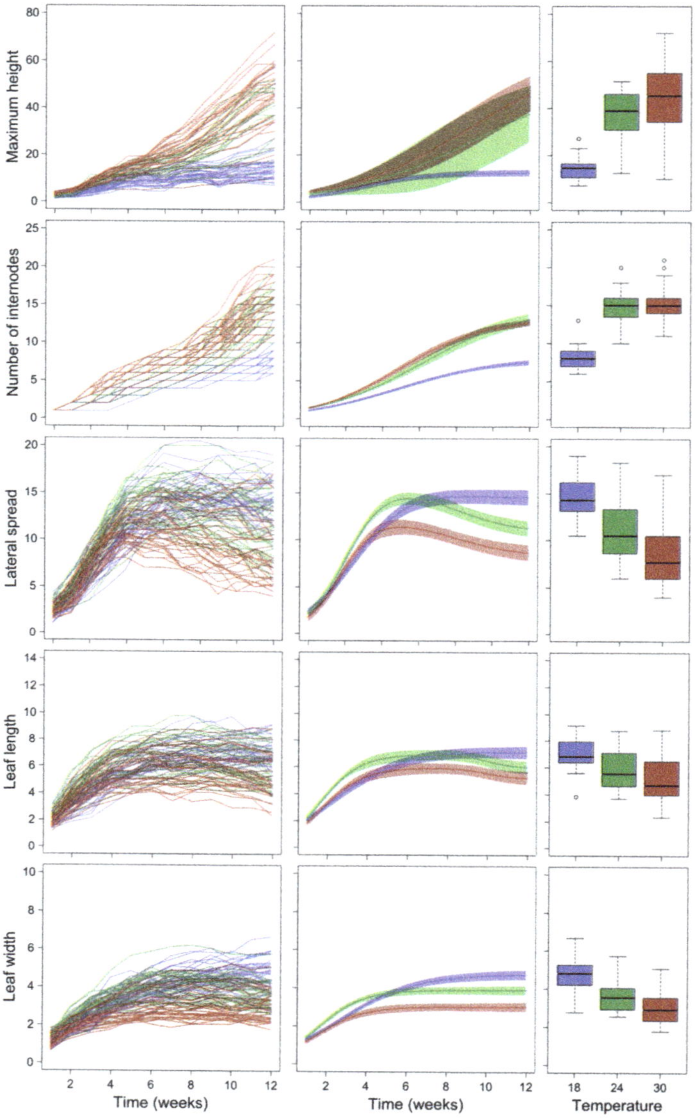

Figure 3. Left panels show growth trajectories of individual plants grown in the lab (after the removal of deviant data points). Each line represents an individual plant. Central panels show the interpolated growth curves and their 95% confidence intervals. Right panels show boxplots of the measured values of each plant at the last measure (week 12). Circles in the right panels represent outliers. In all panels, colors represent temperatures (blue = 18 °C, green = site 24 °C, red = 30 °C).

Generally, parameters describing growth trajectories of plants grown at 18 °C differed significantly from those of plants grown at 30 °C, while those of plants grown at 24 °C showed intermediate values. The only exceptions were parameters *i* and *s* of the growth trajectories of the number of internodes that did not differ significantly between 18 °C and 30 °C (Table 3). However, the general shape of the growth trajectory of plants grown at 24 °C was very similar to that of plants grown at 30 °C, while that of plants grown at 18 °C markedly differed (Figure 3). Generally, plants grown at 18 °C were shorter and with fewer internodes than those grown at 30 °C but had larger lateral spread and larger and longer leaves (Table 3, Figure 3). Plants grown at 24 °C showed a maximum height and a number of internodes similar to those of plants grown at 30 °C, while for lateral spread and leaf length and width they showed intermediate patterns between those of plants grown at 18 °C and 30 °C (Table 3, Figure 3).

2.3. Biomass and Reproductive Parameters of Plants Grown in Laboratory Conditions

Dry biomass of plants grown in the laboratory was higher for plants grown at 18 °C than for plants grown at 30 °C, while plants grown at 24 °C showed intermediate values (Table 4, Figure 4).

Plants grown at 18 °C also emitted both male and female flowers later than those grown at 24 °C and 30 °C (Table 4, Figure 4). On the day of plant collection, both male and female flower emissions did not covary significantly with dry biomass (Table 4). Pollen weight was on average lower for plants grown at 30 °C than for those grown at lower temperature. In addition, pollen weight did not change with plant biomass for plants grown at 30 °C, while it increased significantly for those grown at 18 °C and 24° C. In contrast, spike dry weight increased significantly with plant dry biomass, but did not differ among growth temperatures (Table 4, Figure 4).

Table 4. Linear models of biomass and reproductive parameters of plants grown in laboratory conditions except for the number of male flowers. Coefficients of each parameter of the growth curves are reported in Table S3. Number of plants included in each analysis is reported as well as the residual degrees of freedom and the value of the Akaike information criterion corrected for small sample size (AICc).

Parameter	F	df	p
Dry biomass (76 plants)			
Temperature	4.841	2	0.011
Residual df = 73; AICc = 148.4			
Day of emission of female flowers (71 plants)			
Temperature	105.800	2	<0.001
Centred dry biomass	0.421	2	0.519
Temp. x c. dry biomass	0.007	2	0.992
Residual df = 65; AICc = 277.8			
Day of emission of male flowers (72 plants)			
Temperature	110.462	2	<0.001
Centred dry biomass	2.851	2	0.096
Temp. x c. dry biomass	1.027	2	0.364
Residual df = 66; AICc = −225.7			
Spike dry weight (72 plants)			
Temperature	1.872	2	0.162
Centered dry biomass	294.895	2	<0.001
Temp. x c. dry biomass	1.347	2	0.267
Residual df = 65; AICc = −141.6			
Pollen weight (70 plants)			
Temperature	34.639	2	<0.001
Centered dry biomass	24.125	2	<0.001
Temp. x c. dry biomass	7.147	2	<0.001
Residual df = 64; AICc = −345.0			

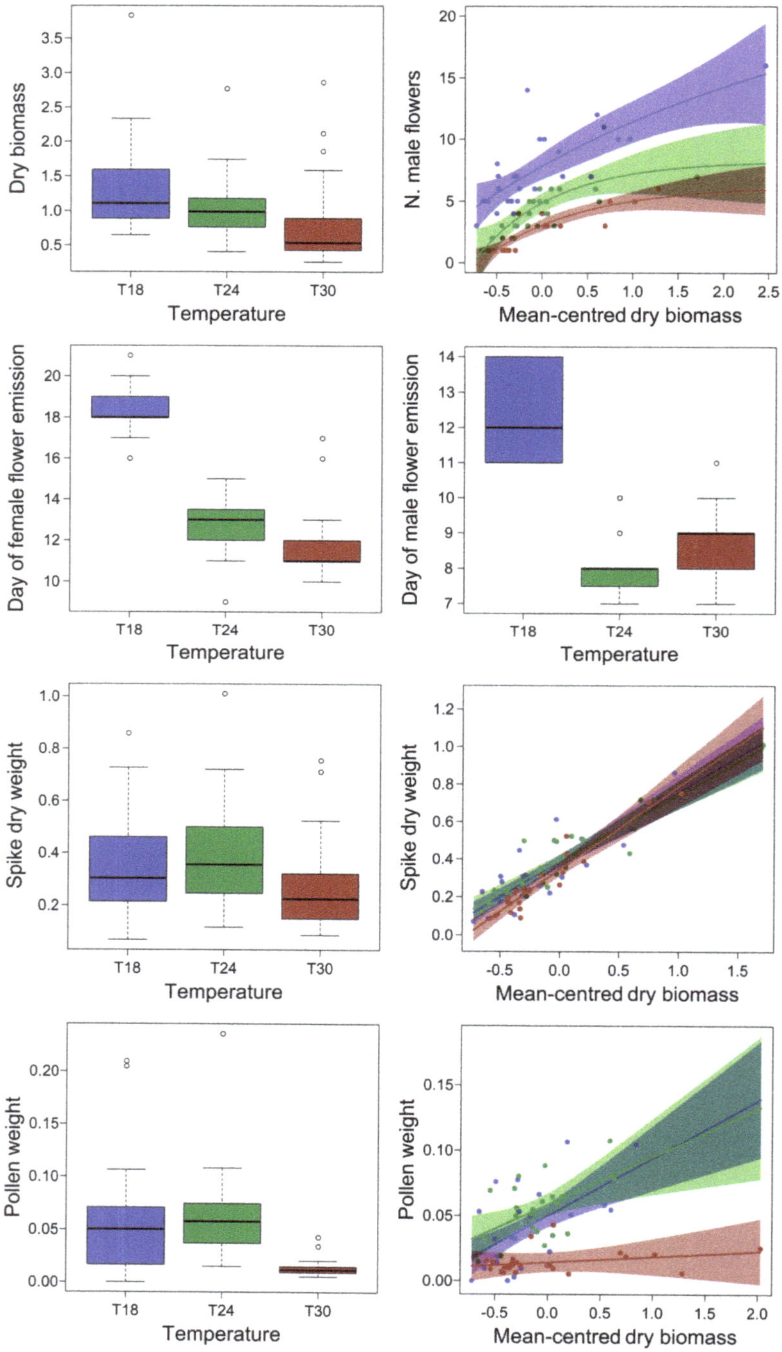

Figure 4. Boxplot and regression curves of dry biomass and reproductive parameters of plants grown in laboratory conditions. The shaded area represents the 95% confidence interval of the curve. Colors represent different temperatures (blue = 18 °C; green = 24 °C; red = 30 °C).

The number of male flowers changed non-linearly with plant biomass, following an asymptotic regression model (Table 5). Only parameter L' of this model differed significantly among temperatures. Since biomass values were centered within group before the analysis to reduce the collinearity among predictors (see Methods), the significant differences in L' indicate that a plant of average size among those grown at 30 °C produced fewer male flowers that a plant of average size among those grown at 24 °C, and that this latter also produced significantly fewer male flowers than a plant of average size among those grown at 18 °C (Table 5, Figure 4).

Table 5. Final NLM of the number of male flowers produced by plants grown in laboratory under different temperature conditions. The growth curve applied and the number of plants included in the analysis are reported as well as the residual degrees of freedom and the value of the Akaike information criterion corrected for small sample size (AICc). Coefficients of each parameter of the growth curves are reported in Table S4.

Parameter	F	df	p
	Number of male flowers (Asymptotic regression 73 plants)		
K	0.75	2	0.477
L'	37.75	2	<0.001
r	1.211	2	0.305
	Residual df = 64; AICc = 237.3.8		

3. Discussion

Our study revealed considerable phenotypic plasticity in terms of growth and reproductive performances of the invasive *Ambrosia artemisiifolia* L. (common ragweed) along altitude and temperature gradients, in field and controlled conditions. Overall, common ragweed reduced its size (plant height, stem, internodes, etc.) along the studied altitudinal gradient but with different strengths or patterns, especially when considering leaf traits. As a general rule, trait variability of common ragweed tended to reduce at higher altitude, likely due to environmental filters, as already observed for other alien species, and dissimilarly to what happens to native species that generally exhibit an increase of trait variability [28].

The ability of the common ragweed to modulate its phenotypic traits according to environmental gradients has been highlighted in previous studies performed along latitudinal (and temperature) gradients [39,40]. In addition, the modulation of traits in common ragweed within a single generation in response to increased temperatures in its invaded range has been recently explained as rapid evolution [41]. Previously, in the native range of the species, different "ecotypes" were observed to preadapt to local conditions, reducing plant height and increasing width in response to day length and temperature reductions [39].

Similarly, we found that plant height (stem, maximum height, and number of internodes) was the main contributor to the growth of common ragweed towards middle and low altitudes and the highest temperatures. Conversely, leaf traits increased in importance and size toward the highest altitude and the lowest temperatures.

With regard to the lab experiment, results showed similar growth trajectories to those observed in the field for plant height and leaf traits, but with contrasting trends. In fact, the species seems to mostly invest in vegetative vigor (i.e., biomass) and flower abundance at the lowest temperature tested (18 °C).

3.1. Field Experiment

In the field experiment, common ragweed exhibited different growth trajectories for traits related to height and leaf. Maximum height, stem height, and number of internodes grew according to a logistic curve over time, indicating that the plants elongate monotonically toward a plateau, with a trajectory likewise described for other species [42]. The plateau of the curve started in correspondence with the emission of flower buds (then removed in our study). Despite that we were not able to collect data on flowering time due to local authority restrictions, a previous study highlighted that the flowering phenology and growth pattern of all traits are associated with maximum plant height in herbaceous plant species [43].

The maximum value of plant height was not reached, as expected, at the lowest altitude of the gradient (site A: 130 m and 23.4 °C of mean temperature), but at the following growth station (i.e., site B: 250 m and 22.9 °C). With plant height being a key determinant of a species ability to compete for light [44], in the studied altitudinal gradient, this trait reached the maximum value at the site B, located at the base of the Prealpine slopes. Conversely, the minimum plant height was observed at the growth station with the highest altitude (site E: 1242 m and 16.6 °C of mean temperature) that also exhibited the lowest growth velocity.

On the other hand, leaf length, leaf width, and lateral spread (i.e., canopy) grew hump-shaped, according to a double-Richard curve, with a decrease of the trait value, likely in correspondence of the bud emission (then removed due to authority restrictions). In fact, previous authors observed that the maximum leaf size decreases early in the season, presumably prior to the reproductive onset [45,46]. Analogously, during the experiment, common ragweed tended to lose the lower leaves, indicating the tendency to allocate resource for reproduction tissues. In accordance with such observations, the timing of leaf senescence has been found to be in relation with the flowering phenology of numerous species along altitudinal gradients [47].

Leaf traits had better performances towards higher elevations (i.e., bigger size in relation to plant height) with lower mean temperatures, indicating that along the altitudinal gradient the species tends to invest more resources in photosynthesis and light capturing than in competition [48]. Indeed, larger individuals are expected to produce larger seeds, conferring higher competitive ability to seedlings [49,50]. This tendency is also mirrored in the leaf number. At the lowest and middle altitudes, common ragweed tended to maintain a higher number of smaller leaves.

All traits exhibited a tendency to reduce their variability along the altitudinal gradient. Similarly, the environmental constraints along altitudinal gradients (temperature, growth season, competitive ability, etc.) have been described to gradually limit the functional suitability of non-native species [29].

3.2. Laboratory Experiment

The laboratory experiment showed similar trends as the field one with respect to the growth trajectories (logistic and double-Richards curves) of common ragweed traits investigated at different temperature conditions. On the other hand, the reduction of the leaf traits (i.e., hump-shaped trajectory) along the observation period were less evident, probably due to the quite different light conditions of the growth chamber in comparison with those of the growth stations in the field. Additionally, in this case, a clear countertendency of the traits relating to height and those of leaves (lower temperatures) was observed along the temperature gradient: At the lowest temperatures (18 °C) the plants exhibited larger leaves and lower height, with the opposite trend at the highest temperatures (30 °C). In common ragweed, increasing temperature has been observed to increase transpiration rates [51]. An increased transpiration rate (and water loss) towards the highest temperatures can explain both the reduction of the leaf size and the lower biomass.

Concerning reproductive traits, our results indicated that at the lowest temperature common ragweed preserved its fitness (conservation of the pollen weight, highest number of male flowers, and same spike dry weight than the other temperatures), in turn supported by the highest biomass value. In this species, dry weight has already been observed to support reproductive performances at different growth conditions according to pH (i.e., inflorescence size and number of inflorescences; see [52]). These findings are surprising and in countertendency to those performed in the native range of the species, where common ragweed reduced its biomass and reproductive performance toward higher latitudes and lower temperatures [39]. In addition, taller individuals of common ragweed grown at the highest temperature (30 °C) tended to reach their maximum height and flowering (both male and female flowers) earlier than shorter ones; this trend is divergent with those observed by Sun and Frelich [43] on several herbaceous (native) grass species.

3.3. Implication for the Invasion Syndrome of Common Ragweed

The ability to reallocate biomass and plasticity in several phenotypic traits (e.g., plant height, number of leaves, etc.) is known to contribute to the invasion of alien species in new environments [53]. Our results are significant to understand how an invasive alien plant like the common ragweed may migrate and adapt to mountain ecosystems. Consequently, common ragweed would potentially compete with migrating native flora, since the distribution of mountain biota is moving upward in response to increasing temperature at the continental scales [54]. In any case, plants during migration are subject to new biotic and abiotic conditions that could favor selection in the migrating population of ragweed [41].

The difference in the trait growth trajectories and the shifts in biomass allocation found in the common ragweed at different altitudes and temperatures can certainly reflect its adaptation ability to new environmental conditions and, therefore, the potential to invade and compete for resources toward higher altitudes.

In the European invaded range of the species, the adaptability of its natural populations and their invasion potential are supported by high levels of intra-population genetic variation [34]. Whereas phenotypic plasticity may increase the ecological niche breadth of the species, post-introduction or post-colonization rapid evolution is known to produce genetically based phenotypic variations and adaptations, which can increase plant invasiveness. Therefore, phenotypic plasticity and rapid adaptation may be considered key components of the release–naturalization–invasion continuum [55]. In recent studies, common ragweed has been observed to grow in new areas at high altitude (over 1200 m) as a casual species; however, it seems that the species is not still able to establish at the current climatic conditions [38].

Although the plants seemed to have decreasing fitness (biomass) towards higher elevation, the laboratory experiment showed an opposite trend. In fact, the species conserved good fitness and reproductive ability in term of biomass and male and female flower production at low temperature. There are several explanations for these contrasting trends found in the field and laboratory experiments. First, findings of lab experiments need to be carefully assessed when they are translated to understand phenomena linked to natural populations [56]. Influences of environmental variables (abiotic and biotic) on the different responses between plant grown in the laboratory and in the field should be considered [57]. The increasing light intensity (especially UV-B radiation) at higher altitude may have limited plant size in terms of tissue formation and need for protecting the photosynthetic system with shorter individuals close to the soil [58]. Finally, the exposition of field plants to a variable regime of relative humidity and temperature peaks with drought periods may have played a role in shaping such differences. Indeed, during the last two decades, drought episodes have become stronger in terms of frequency and length [59] in the Po valley, with a subsequent increased evapotranspiration and lower relative humidity for the field plants. For this reason, in our study, the growth station at the lowest altitude could not be considered as optimal climatic conditions for developing biomass. Indeed, the better performances (in term of biomass and plant height) were found at the intermediate growth station (stations B and C) in the Prealps. This pattern may have repercussions in the future invasion trends of the species in hill/mountain areas currently presenting lower levels of disturbance (i.e., new colonization spaces) in comparison to the flat areas of the Po valley and, therefore, lower occurrence.

In perspective, our results can be used to inform future management decisions regarding common ragweed [60]. First, it is very likely that the expected increasing temperatures due to climate change will push plastic populations of the species to invade higher altitudes.

So, to implement measures of early detection and eradication in the mountain areas falling within the invaded range of the species is recommended. On the other hand, considering possible biocontrol actions for the future, the onset of flowering at different temperature regimes will clearly set the limits of action of biocontrol agents of the species [26]. Indeed, if plants held at 24 °C and 30 °C opened their flowers at the same time, it would mean that for biocontrol agents like the ragweed leaf beetle *Ophraella communa*, there would be fewer day degrees available at 24 °C to build up high enough density population to suppress pollen release by common ragweed.

4. Material and Methods

4.1. Plant Material and Preliminary Germination

All the experiments were performed using common ragweed seeds collected from an agriculture area in Lombardy (45.597811 N; 8.869912 E). Seeds were cold-stratified at 4 °C for 3 months to overcome seed dormancy and then planted in a tray containing autoclaved natural soil for germination. Seedlings were transplanted in pots of 2.5 L and 0.75 L for the field and lab experiments, respectively (Figure 5). The pots contained the same standard soil made of potting soil (VigorPlant©, pH 6) and sand in the proportions of 60–40%.

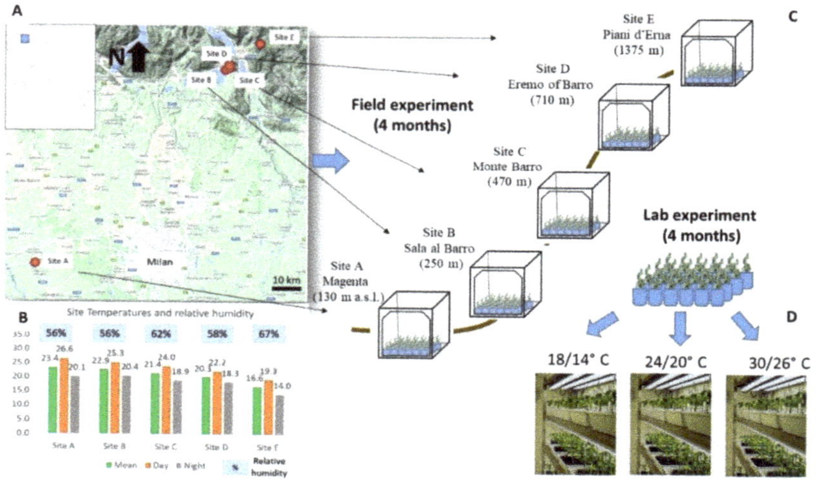

Figure 5. Experimental design of the study. (**A**) Study area of the field experiment and (**B**) mean temperatures and relative humidity of the selected sites (**C**) where a growth station of common ragweed (*A. artemisiifolia*) was set up. (**D**) In the laboratory experiment, common ragweed individuals were grown in three growth chambers set up at different temperature ranges (day/night): 18/14 °C, 24/20 °C, and 30/26 °C.

4.2. Field Experiment

In order to study the trends of functional traits of common ragweed along an altitudinal gradient, five sites at increasing altitudes were selected in an area between the Po Valley and the Prealps in Lombardy, where the species has been established for several decades. As common ragweed is included in the regional blacklist of species to be controlled and whose spread is restricted [61], the following sites were selected in agreement with local authorities (Lombardy region, phytosanitary service): Site A at Magenta (130 m a.s.l.; lat. 45.459, lon. 8.874); site B at Sala al Barro (250 m; lat. 45.822, lon. 9.361); site C at the Monte Barro slope (470 m; lat. 45.825, lon. 9.372); site D at the Eremo of Barro (710 m; lat. 45.831, lon. 9.371); and site E at the Piani d'Erna (1242 m; lat. 45.870, lon. 9.449). Overall, the altitudinal gradient considered was about 1000 m.

At each site, a field cage with the species individuals was set up. Within the cages of 1 m^3, 24 seedlings were grown in pots of 2.5 L for about 4 months, from June to September 2015. Each cage was placed in a flat area and covered with a soft insect net to avoid herbivory, especially by *Ophraella communa* (ragweed leaf beetle), which has been accidentally introduced in the region since 2013 and preferentially feeds on common ragweed [62]. The plants were watered weekly within plant pot saucers.

During plant growth, the following data about vegetative and reproductive traits were collected weekly: maximum plant height (cm), stem height (cm), number of internodes (*n*), lateral spread (cm; vertical projection of the area covered by the plant), number of leaves (*n*), leaf length (cm), and leaf width (cm). Measurements of the biomass of individuals over time were also conducted. To do this, six plants were removed from each site every 3 weeks and their dry and fresh weights of the aboveground portion were measured.

Due to the regional phytosanitary restrictions implemented to prevent allergic syndromes to human populations, flower buds were repeatedly removed from the plants during the flowering season in order to avoid the dispersion of the highly allergenic pollen of this species [32].

4.3. Lab Experiment

Three growth chambers with identical photoperiods, light intensity (15:9 h light:dark 150 μmol m^{-2} s^{-1}) and humidity (65%) but different temperatures (LT: 18/14 °C light-dark, IT: 24–20 °C and HT: 30–26 °C) were set up. For each temperature, 51 seedlings within pots were grown for about 4 months (summer 2015), until seed production. During plant growth, the following data about vegetative and reproductive traits were collected weekly or at the end of the experiment: maximum plant height (cm), stem height (cm), number of internodes (*n*), lateral spread (cm), number of leaves (*n*), leaf length (cm) and leaf width, dry biomass at the end of the experiment (g), number of male flowers (*n*), day of male and female flower emission (*n* weeks of first flower open), spike dry weight (g), and pollen weight (g).

At maturity, pollen was collected from the plants by enveloping each spike with a transparent plastic collector, according to Ghiani et al. [32].

4.4. Data Analysis

Growth trajectories of morphological traits of common ragweed grown at different sites were modelled by fitting parametric growth curves with nonlinear mixed models (NLMMs) using the nlme procedure in the nlme package [63] of R.

Fitting parametric curves is computationally efficient and allows the estimation of parameters of biological significance [64]. In addition, NLMMs are very flexible statistical tools as they allow modelling any parameter of the growth curves as a function of different predictors. This flexibility extends also to the random part of the model because it is possible to enter different random structures for each parameter of the growth curve. However, fitting NLMMs is challenging. To reduce the complexity of these models, we ran preliminary analyses to assess (1) which growth curve best fitted the growth trajectory of the morphological trait under scrutiny and (2) which parameters of the growth curves showed large variability among individual plants, in order to properly parameterize the random part of the NLMMs.

Different morphological traits may show growth trajectories that can be described by different curves. Among the growth curves widely used in modelling plant growth trajectories [65–67], we selected five, including the double-Richards curves, which allow modelling growth trajectories for those morphological traits that change non-monotonically and decline after having reached a peak [64]. Equations for the five growth curves used in the analyses are reported in Table 6. To assess the growth curve that best fit the growth trajectory of each morphological trait, we fitted non-linear models (NLMs) to all data and compared the values of the small-sample Akaike's information criterion (AICc) of each model. We note that these preliminary models did not account for repeated measures collected at each individual plant; however, we considered this approximation reasonable given the explorative nature of these analyses, which were only aimed at assessing the general shape of the growth trajectory of each morphological trait. The double-Richards curves were fitted with the *FlexParamCurve* v. 1.5-3 package [64].

Table 6. Equations and parameters of the five growth curves used in the analyses. The morphological traits whose growth trajectory was best modelled by each curve are also reported. All curves are function of time (t) except for the asymptotic curve, which was used to model the number or male flowers of plants grown under laboratory conditions and was a function of plant dry biomass (x). See Figure S1 for an illustration of these parameters.

Growth Curve	Equation	Morphological Trait
(1) Linear	$y = y_0 + rt$	Reproductive parameters except for the number of male flowers
(2) Asymptotic	$y = K + (L' - K)e^{-rx}$	Number of male flowers
(3) Three-parameter logistic	$y = \frac{K}{1+e^{\frac{i-t}{s}}}$	Plant height, stem height
(4) Four-parameter logistic	$y = L + \frac{K-L}{1+e^{\frac{i-t}{s}}}$	Number of leaves
(5) Double-Richards	$y = \frac{K}{\left(1+me^{r(i-t)}\right)^{\frac{1}{m}}} + \frac{K'}{\left(1+m'e^{r'(i'-t)}\right)^{\frac{1}{m'}}}$	Lateral spread, leaf length, leaf width
Parameter	Description	
y_0	Intercept, corresponding to mean value if t is centred	
r and r'	Growth rates	
s	Scale parameter replacing the growth rate r ($s = 1/r$) in the parameterization of Equations (2) and (3) of *SSlogis* and *SSfpl* (used to fit them)	
K	Upper asymptote	
L *and* L'	Lower asymptote or initial value	
m and m'	Shape parameters of the generalized logistic curves, values > 1, imply that the inflection points are realized sooner than i or i' and the growth rates at i or i' are lower than r or r'; values < 1 imply the opposite	
i and i'	Inflection points, i.e., time at which the fastest growth/recession is attained	
K'	Difference between asymptotes of the curve before and after recession	

To assess the structure of the random part of the NLMM, we followed the procedure described in Morganti et al. [68]. First, we interpolated the selected growth curves to data of each plant separately. Then, we plotted the range of parameters from curves fitted to individual plants and noted those that, at a visual inspection, showed large heterogeneity (see [68,69] for a similar approach). It should be noted that repeated measures of the same plant often showed temporal autocorrelation, and trait variance also usually increased with their size. In the final NLMMs, we, therefore, assumed (1) a random variation of those parameters showing large heterogeneity, (2) a first-order residual temporal autocorrelation, and (3) a change of the variance with time according to an exponential function, as suggested in Oswald et al. [64]. In the fixed part of the model, we allowed for variation of all model parameters among sites and, when we detected significant variation, we performed post-hoc tests for differences between each pair of sites with the Tukey method.

Growth curves of plants grown under laboratory conditions were fitted with the same procedure, but temperature (of the growth chambers) was included as a three-level predictor in the fixed part of the models, allowing for variation in all model parameters among temperatures.

Data collected in the field showed sometimes unreasonably large variations between consecutive measures that inflated non-linear model variance and caused convergence difficulties. We thus applied an in-home procedure (fully described in the supporting information) to remove deviant measures, which were replaced by missing values. Application of this procedure greatly improved non-linear model fit (details not shown).

Analyses of wet and dry biomass of plants were conducted using generalized least square (GLS) models that included as predictors: site (five-level factor), time (four-level factor), and the interaction among them for field-grown plants, while the predictors for laboratory-grown plants were temperature (three-level factor), time (four-level factor), and the interaction among them. The model also accounted for inhomogeneity of variances among groups, which was detected during preliminary model checks. Since each individual was measured only once in this part of the study, we entered no random term in the model.

From those laboratory-grown individuals that were kept until the end of the experiment, we also collected data on traits related to the dry biomass and the reproductive investment of the plant. These data were collected only at the endpoint of the experiment. Dry biomass was analyzed in an ANOVA model with temperature of the growth chamber (three-level factor) as predictor. Data on the reproductive investment of the plants were modelled using linear models (LMs) assuming a Gaussian error distribution with the temperature of the growth chamber (three-level factor), the dry biomass of the plant (covariate), and their interaction as predictors. Since dry biomass differed significantly among plants grown at different temperatures, we preliminarily centered dry biomass values while removing from the value of each plant the mean value of all the plants grown at that temperature. This procedure strongly reduced the collinearity among predictors (details not shown).

In the analysis of dry biomass, we removed potentially influential data (i.e., data that may strongly condition the results of the model) based on their Cook's distance being >4/N, where N is the number of data [70]. In some cases, we also found small deviations from model assumptions. We, therefore, further checked significance of LMs by a randomization procedure performed with the *permuco* package [71], which, however, always confirmed the results of parametric tests (details not shown). The only exception was the number of male flowers that were modelled using an asymptotic growth curve in a NLM. Additionally, for these models, post hoc tests were performed with the Tukey method using the *emmeans* package [72].

All the analyses were performed in R 3.6.2 [73] with packages *FlexParamCurve* [64], *nlme* [67], *emmeans* [72], and *permuco* [71].

5. Conclusions

Common ragweed (*Ambrosia artemisiifolia* L.) was found to exhibit a high phenotypic plasticity in response to different altitude and temperature conditions. This ability may support future shifts of the species to higher altitude (and likely latitudes) under climate warming. In the face of climate change and the expected northward shift of the species [36], our study suggests a great ability of common ragweed to survive in a wide range of temperatures and the in situ resistance and adaptation of the natural populations currently abundant at low altitudes and latitudes in the European invaded range. Therefore, the expansion of the species toward higher altitude should be intensively monitored in the short and middle term. Our results are also directly relevant for projecting impact by *Ophraella communa* (ragweed leaf beetle) and probably other biocontrol agents.

In the near future, climate change may modify the altitudinal ranges of several invasive alien plants, increasing their impacts; therefore, control and management plans for plant invaders should take this aspect into account.

Supplementary Materials: The following are available online at https://www.mdpi.com/article/10.3390/plants10102144/s1, Table S1: Final NLMMs of the growth trajectories of plants grown in field conditions, Table S2: Final NLMMs of the growth trajectories of plants grown in laboratory conditions, Table S3: Linear models of biomass and reproductive parameters of plants grown in laboratory conditions except for the number of male flowers, Table S4: Final NLM of the number of male flowers produced by plants grown in laboratory under different temperature conditions.

Author Contributions: Conceptualization, R.G. and S.C. (Sandra Citterio); methodology, R.G. and R.A.; formal analysis, R.A.; investigation, R.G., B.A.A., S.C. (Sarah Caronni) and C.M.; resources, S.C. (Sandra Citterio) and R.G.; data curation, R.G. and E.C.; writing—original draft preparation, R.G. and R.A.; writing—review and editing, R.G., R.A., B.A.A., S.C. (Sarah Caronni), C.M., E.C., H.M.-S., U.S. and S.C. (Sandra Citterio); supervision, S.C. (Sandra Citterio); funding acquisition, Sa.Ci., U.S. and H.M.-S. All authors have read and agreed to the published version of the manuscript.

Funding: This research was funded by: (1) Fondazione Banca del Monte di Lombardia, "Invasione biologica delle specie allergeniche del genere *Ambrosia* L. in Lombardia: distribuzione dettagliata, pericolosità e metodologie finalizzata contrastarne la diffusione", grant number 2016-NAZ-0144"; (2) EU COST Action FA1203 "Sustainable management of *Ambrosia artemisiifolia* in Europe (SMARTER)"; (3) LIFE financial instrument of the European Commission to Lombardy Region "LIFE14 IPE IT 018GESTIRE2020-Nature Integrated Management in 2020"; (4) Fondazione Cariplo, project "Tataricum".

Acknowledgments: We would like to thank Silvia Ciappetta, Federico Colombo, Alessandra Ghiani, and Maria Guarino for their technical support during field and laboratory work.

Conflicts of Interest: The authors declare no conflict of interest.

References

1. Pfennigwerth, A.A.; Bailey, J.K.; Schweitzer, J.A. Trait variation along elevation gradients in a dominant woody shrub is population-specific and driven by plasticity. *AoB Plants* **2017**, *9*, plx027. [CrossRef] [PubMed]
2. Körner, C. The use of 'altitude' in ecological research. *Trends Ecol. Evol.* **2007**, *22*, 569–574. [CrossRef]
3. Pellissier, L.; Fournier, B.; Guisan, A.; Vittoz, P. Plant traits covary with altitude in grasslands and forests in the European Alps. *Plant. Ecol.* **2010**, *211*, 351–365. [CrossRef]
4. Šťastná, P.; Klimešová, J.; Doležal, J. Altitudinal changes in the growth and allometry of *Rumex alpinus*. *Alp. Bot.* **2012**, *122*, 35–44. [CrossRef]
5. Midolo, G.; De Frenne, P.; Hölzel, N.; Wellstein, C. Global patterns of intraspecific leaf trait responses to elevation. *Glob. Change Biol.* **2019**, *25*, 2485–2498. [CrossRef]
6. Jösson, K.I.; Tuomi, J. Costs of reproduction in a historical perspective. *Trends Ecol. Evol.* **1994**, *9*, 304–307. [CrossRef]
7. Von Arx, G.; Edwards, P.J.; Diez, H. Evidence for life history changes in high-altitude populations of three perennial forbs. *Ecology* **2006**, *87*, 665–674. [CrossRef]
8. Gienapp, P.; Teplitsky, C.; Alho, J.S.; Mills, J.A.; Merilä, J. Climate change and evolution: Disentangling environmental and genetic responses. *Mol. Ecol.* **2008**, *17*, 167–178. [CrossRef] [PubMed]
9. Bakhtiari, M.; Formenti, L.; Caggìa, V.; Glauser, G.; Rasmann, S. Variable effects on growth and defense traits for plant ecotypic differentiation and phenotypic plasticity along elevation gradients. *Ecol. Evol.* **2019**, *9*, 3740–3755. [CrossRef]
10. Spencer, W.E.; Teeri, J.; Wetzel, R.G. Acclimation of photosynthetic phenotype to environmental heterogeneity. *Ecology* **1994**, *75*, 301–314. [CrossRef]
11. Roach, D.A.; Gampe, J. Age-specific demography in *Plantago*: Uncovering age-dependent mortality in a natural population. *Am. Nat.* **2004**, *164*, 60–69. [CrossRef] [PubMed]
12. Xiang, S.; Wu, N.; Sun, S. Within-twig biomass allocation in subtropical evergreen broad-leaved species along an altitudinal gradient: Allometric scaling analysis. *Trees* **2009**, *23*, 637–647. [CrossRef]
13. Alexander, J.M.; Kueffer, C.; Daehler, C.C.; Edwards, P.J.; Pauchard, A.; Seipel, T.; Arevalo, J.; Cavieres, L.; Dietz, H.; Jakobs, G.; et al. Assembly of nonnative floras along elevational gradients explained by directional ecological filtering. *Proc. Natl. Acad. Sci. USA* **2011**, *108*, 656–661. [CrossRef]
14. March-Salas, M.; Pertierra, L.R. Warmer and less variable temperatures favour an accelerated plant phenology of two invasive weeds across sub-Antarctic Macquarie Island. *Austral. Ecol.* **2020**, *45*, 572–585. [CrossRef]
15. Davidson, A.M.; Jennions, M.; Nicotra, A.B. Do invasive species show higher phenotypic plasticity than native species and if so, is it adaptive? A meta-analysis. *Ecol. Lett.* **2011**, *14*, 419–431. [CrossRef]
16. Knop, E.; Reusser, N. Jack-of-all-trades: Phenotypic plasticity facilitates the invasion of an alien slug species. *Proc. R. Soc. B* **2012**, *279*, 4668–4676. [CrossRef] [PubMed]
17. Pyšek, P.; Jarošík, V.; Pergl, J.; Wild, J. Colonization of high altitudes by alien plants over the last two centuries. *Proc. Natl. Acad. Sci. USA* **2011**, *108*, 439–440. [CrossRef] [PubMed]
18. Kelley, A.L. The role thermal physiology plays in species invasion. *Conserv. Physiol.* **2014**, *2*, cou045. [CrossRef]
19. Hou, Q.-Q.; Chen, B.-M.; Peng, S.-L.; Chen, L.-Y. Effects of extreme temperature on seedling establishment of nonnative invasive plants. *Biol. Invasions* **2014**, *16*, 2049–2061. [CrossRef]

20. Fenollosa, E.; Munné-Bosch, S. Increased chilling tolerance of the invasive species *Carpobrotus edulis* may explain its expansion across new territories. *Conserv. Physiol.* **2019**, *7*, coz075. [CrossRef]
21. Galera, H.; Chwedorzewska, K.J.; Wo´dkiewicz, M. Response of *Poa annua* to extreme conditions: Comparison of morphological traits between populations from cold and temperate climate conditions. *Polar Biol.* **2015**, *38*, 1657–1666. [CrossRef]
22. Hellmann, J.J.; Byers, J.E.; Bierwagen, B.G.; Dukes, J.S. Five potential consequences of climate change for invasive species. *Conserv. Biol.* **2008**, *22*, 534–543. [CrossRef]
23. Beaury, E.M.; Fusco, E.J.; Jackson, M.R.; Laginhas, B.B.; Morelli, T.L.; Allen, J.; Pasquarella, V.J.; Bradley, B.A. Incorporating climate change into invasive species management: Insights from managers. *Biol. Invasions* **2020**, *22*, 233–252. [CrossRef]
24. Morelli, T.L.; Brown-Lima, C.J.; Allen, J.M.; Beaury, E.M.; Fusco, E.J.; Barker-Plotkin, A.; Laginhas, B.B.; Quirion, B.R.; Griffin, B.; McLaughlin, B.; et al. Translational invasion ecology: Bridging research and practice to address one of the greatest threats to biodiversity. *Biol. Invasions* **2021**. [CrossRef]
25. Morais, M.C.; Cabral, J.A.; Gonçalves, B. Seasonal variation in the leaf physiology of co-occurring invasive (*Hakea sericea*) and native (*Pinus pinaster*) woody species in a Mediterranean-type ecosystem. *For. Ecol. Manag.* **2021**, *480*, 118662. [CrossRef]
26. Augustinus, B.; Sun, Y.; Beuchat, C.; Schaffner, U.; Müller-Schärer, H. Predicting impact of a biocontrol agent: Integrating distribution modeling with climate-dependent vital rates. *Ecol. Appl.* **2019**, *30*, e02003. [CrossRef] [PubMed]
27. Wood, A.R.; Fourie, A.; Krug, R.M.; Gordon, A.J. A guide to the biological control of alien invasive hakea species. In *Plant Protection Research Institute Handbook No. 22*; Agricultural Research Council: Pretoria, South Africa, 2021; pp. 1–68.
28. Falster, D.S.; Duursma, R.A.; FitzJohn, R.G. How functional traits influence plant growth and shade tolerance across the life cycle. *Proc. Natl. Acad. Sci. USA* **2018**, *115*, E6789–E6798. [CrossRef] [PubMed]
29. Kühn, P.; Ratier Backes, A.; Römermann, C.; Bruelheide, H.; Haider, S. Contrasting patterns of intraspecific trait variability in native and non-native plant species along an elevational gradient on Tenerife, Canary Islands. *Ann. Bot.* **2021**, *127*, 565–576. [CrossRef]
30. Alexander, J.M.; Edwards, P.J.; Poll, M.; Parks, C.G.; Dietz, H. Establishment of parallel altitudinal clines in traits of native and introduced forbs. *Ecology* **2009**, *90*, 612–622. [CrossRef]
31. Montagnani, C.; Gentili, R.; Smith, M.; Guarino, M.; Citterio, S. The Worldwide Spread, Success, and Impact of Ragweed (*Ambrosia* spp.). *Crit. Rev. Plant Sci.* **2017**, *36*, 139–178. [CrossRef]
32. Ghiani, A.; Ciappetta, S.; Gentili, R.; Asero, R.; Citterio, S. Is ragweed pollen allergenicity governed by environmental conditions during plant growth and flowering? *Sci. Rep.* **2016**, *6*, 30438. [CrossRef]
33. Schaffner, U.; Steinbach, S.; Sun, Y.; Skjøth, C.; de Weger, L.A.; Lommen, S.T.; Augustinus, B.; Müller-Schärer, H. Biological weed control to relieve millions of allergy sufferers in Europe. *Nat. Comm.* **2020**, *11*, 1745. [CrossRef]
34. Ciappetta, S.; Ghiani, A.; Gilardelli, F.; Bonini, M.; Citterio, S.; Gentili, R. Invasion of *Ambrosia artemisiifolia* in Italy: Assessment via analysis of genetic variability and herbarium data. *Flora* **2016**, *223*, 106–111. [CrossRef]
35. Fumanal, B.; Girod, C.; Fried, G.; Bretagnolle, F.; Chauvel, B. Can the large ecological amplitude of *Ambrosia artemisiifolia* explain its invasive success in France? *Weed Res.* **2008**, *48*, 349–359. [CrossRef]
36. Gentili, R.; Montagnani, C.; Gilardelli, F.; Guarino, M.F.; Citterio, S. Let native species take their course: *Ambrosia artemisiifolia* replacement during natural or "artificial" succession. *Acta Oecol.* **2017**, *82*, 32–40. [CrossRef]
37. Cunze, S.; Leiblein, M.C.; Tackenberg, O. Range expansion of *Ambrosia artemisiifolia* in Europe is promoted by climate change. *ISRN Ecol.* **2013**, *2013*, 610126. [CrossRef]
38. Gentili, R.; Gilardelli, F.; Bona, E.; Prosser, F.; Selvaggi, A.; Alessandrini, A.; Martini, F.; Nimis, P.L.; Wilhalm, T.; Adorni, M.; et al. Distribution map of *Ambrosia artemisiifolia* L. (Asteraceae) in Italy. *Plant Biosyst.* **2017**, *151*, 381–386. [CrossRef]
39. Dickerson, C.T.; Sweet, R.D. Common ragweed ecotypes. *Weed Sci.* **1971**, *19*, 64–66. [CrossRef]
40. Leiblein-Wild, M.C.; Tackenberg, O. Phenotypic variation of 38 European *Ambrosia artemisiifolia* populations measured in a common garden experiment. *Biol. Invasions* **2014**, *16*, 2003–2015. [CrossRef]
41. Sun, Y.; Bossdorf, O.; Grados, R.D.; Liao, Z.; Müller-Schärer, H. Rapid genomic and phenotypic change in response to climate warming in a widespread plant invader. *Glob. Change Biol.* **2020**, *26*, 6511–6522. [CrossRef]
42. Kawano, T.; Wallbridge, N.; Plummer, C. Logistic Models for Simulating the Growth of Plants by Defining the Maximum Plant Size as the Limit of Information Flow. *Plant Signal. Behav.* **2020**, *15*, 1709718. [CrossRef] [PubMed]
43. Sun, S.; Frelich, L.E. Flowering phenology and height growth pattern are associated with maximum plant height, relative growth rate and stem tissue mass density in herbaceous grassland species. *J. Ecol.* **2011**, *99*, 991–1000. [CrossRef]
44. Moles, A.T.; Warton, D.I.; Warman, L.; Swenson, N.G.; Laffan, S.W.; Zanne, A.E.; Pitman, A.; Hemmings, F.A.; Leishman, M.R. Global patterns in plant height. *J. Ecol.* **2009**, *97*, 923–932. [CrossRef]
45. Nabeshima, E.; Hiura, T. Size-dependency in hydraulic and photosynthetic properties of three *Acer* species having different maximum sizes. *Ecol. Res.* **2008**, *23*, 281–288. [CrossRef]
46. Thomas, S.C. Photosynthetic capacity peaks at intermediate size in temperate deciduous trees. *Tree Physiol.* **2010**, *30*, 555–573. [CrossRef]
47. Bucher, S.F.; Römermann, C. The timing of leaf senescence relates to flowering phenology and functional traits in 17 herbaceous species along elevational gradients. *J. Ecol.* **2021**, *109*, 1537–1548. [CrossRef]
48. Falster, D.S.; Westoby, M. Plant height and evolutionary games. *Trends Ecol. Evol.* **2003**, *18*, 337–343. [CrossRef]

49. Tracey, A.J.; Aarssen, L.W. Competition and body size in plants: The between-species trade-off for maximum potential versus minimum reproductive threshold size. *J. Plant Ecol.* **2011**, *4*, 115–122. [CrossRef]
50. Turnbull, L.A.; Rees, M.; Crawley, M.J. Seed mass and the competition/colonization trade-off: A sowing experiment. *J. Ecol.* **1999**, *87*, 899–912. [CrossRef]
51. Bazzaz, F.A. Ecophysiology of *Ambrosia artemisiifolia*: A successional dominant. *Ecology* **1974**, *55*, 112–119. [CrossRef]
52. Gentili, R.; Ambrosini, R.; Montagnani, C.; Caronni, S.; Citterio, S. Effect of soil pH on the growth, reproductive investment and pollen allergenicity of *Ambrosia artemisiifolia* L. *Front. Plant Sci.* **2018**, *9*, 1335. [CrossRef] [PubMed]
53. Liao, Z.-Y.; Scheepens, J.F.; Li, W.-T.; Wang, R.-F.; Zheng, Y.-L.; Feng, Y.-L. Biomass reallocation and increased plasticity might contribute to successful invasion of *Chromolaena odorata*. *Flora* **2019**, *256*, 79–84. [CrossRef]
54. Alexander, J.M.; Lembrechts, J.J.; Cavieres, L.A.; Daehler, C.C.; Haider, S.; Kueffer, C.; Liu, G.; McDougall, K.; Milbau, A.; Pauchard, A.; et al. Plant invasions into mountains and alpine ecosystems: Current status and future challenges. *Alp. Bot.* **2016**, *126*, 89–103. [CrossRef]
55. Clements, D.R.; Jones, V.L. Rapid Evolution of Invasive Weeds Under Climate Change: Present Evidence and Future Research Needs. *Front. Agron.* **2021**, *3*, 664034. [CrossRef]
56. Poorter, H.; Fiorani, F.; Pieruschka, R.; Wojciechowski, T.; van der Putten, W.H.; Kleyer, M.; Schurr, U.; Postma, J. Pampered inside, pestered outside? Differences and similarities between plants growing in controlled conditions and in the field. *New Phytol.* **2016**, *212*, 838–855. [CrossRef] [PubMed]
57. Xu, X.; Yan, L.; Xia, J. A threefold difference in plant growth response to nitrogen addition between the laboratory and field experiments. *Ecosphere* **2019**, *10*, e02572. [CrossRef]
58. Pescador, D.S.; de Bello, F.; Valladares, F.; Escudero, A. Plant Trait Variation along an Altitudinal Gradient in Mediterranean High Mountain Grasslands: Controlling the Species Turnover Effect. *PLoS ONE* **2015**, *10*, e0118876. [CrossRef] [PubMed]
59. Baronetti, A.; González-Hidalgo, J.C.; Vicente-Serrano, S.M.; Acquaotta, F.; Fratianni, S. A weekly spatio-temporal distribution of drought events over the Po Plain (North Italy) in the last five decades. *Int. J. Climatol.* **2020**, *40*, 4463–4476. [CrossRef]
60. Gentili, R.; Schaffner, U.; Martinoli, A.; Citterio, S. Invasive alien species and biodiversity: Impacts and management. *Biodiversity* **2021**, *22*, 1–3. [CrossRef]
61. Banfi, E.; Galasso, G. *La Flora Esotica Lombarda*; Museo di Storia Naturale di Milano—Regione Lombardia, Sistemi Verdi e Paesaggio: Milano, Italy, 2010; pp. 1–139.
62. Müller-Schärer, H.; Lommen, S.T.E.; Rossinelli, M.; Bonini, M.; Boriani, M.; Bosio, G.; Schaffner, U. *Ophraella communa*, the ragweed leaf beetle, has successfully landed in Europe: Fortunate coincidence or threat? *Weed Res.* **2014**, *54*, 109–119. [CrossRef]
63. Pinheiro, J.; Bates, D.; DebRoy, S.; Sarkar, D.; R Core Team. *nlme: Linear and Nonlinear Mixed Effects Models*; R Package Version 3.1-153; R Foundation for Statistical Computing: Vienna, Austria, 2021. Available online: https://CRAN.R-project.org/package=nlme (accessed on 26 September 2021).
64. Oswald, S.A.; Nisbet, I.C.T.; Chiaradia, A.; Arnold, J.M. FlexParamCurve: R package for flexible fitting of nonlinear parametric curves. *Methods Ecol. Evol.* **2012**, *3*, 1073–1077. [CrossRef]
65. Yin, X.; Goudriaan, J.; Lantinga, E.A.; Vos, J.; Spiertz, H.J. A flexible sigmoid function of determinate growth. *Ann. Bot.* **2003**, *91*, 361–371. [CrossRef] [PubMed]
66. Paine, C.E.T.; Marthew, T.R.; Vogt, D.R.; Purves, D.; Rees, M.; Hector, A.; Turnbull, L.A. How to fit nonlinear plant growth models and calculate growth rates: An update for ecologists. *Methods Ecol. Evol.* **2012**, *3*, 245–256. [CrossRef]
67. Vasseur, F.; Fouqueau, L.; de Vienne, D.; Nidelet, T.; Violle, C.; Weigel, D. Nonlinear phenotypic variation uncovers the emergence of heterosis in *Arabidopsis thaliana*. *PLoS Biol.* **2019**, *17*, e3000214. [CrossRef] [PubMed]
68. Morganti, M.; Rubolini, D.; Åkesson, S.; Bermejo, A.; De La Puente, J.; Lardelli, R.; Liechti, F.; Boano, G.; Tomassetto, E.; Ferri, M.; et al. Effect of light-level geolocators on apparent survival of two highly aerial swift species. *J. Avian Biol.* **2017**, *49*, jav-01521. [CrossRef]
69. Sicurella, B.; Caprioli, M.; Romano, A.; Romano, M.; Rubolini, D.; Saino, N.; Ambrosini, R. Hayfields enhance colony size of the barn swallow *Hirundo rustica* in northern Italy. *Bird Conserv. Int.* **2014**, *24*, 17–31. [CrossRef]
70. Bollen, K.A.; Jackman, R.W. Regression Diagnostics: An Expository Treatment of Outliers and Influential Cases. In *Modern Methods of Data Analysis*; Fox, J., Long, J.S., Eds.; Sage Publications: Newbury Park, CA, USA, 1990; pp. 1–446.
71. Fossard, J.; Renaud, O. Permutation Tests for Regression, ANOVA and Comparison of Signals: The Permuco Package. R Package Version 1.1.0. 2019. Available online: https://cran.r-project.org/web/packages/permuco/index.html (accessed on 3 July 2021).
72. Lenth, R.V. emmeans: Estimated Marginal Means, aka Least-Squares Means. 2020. Available online: https://CRAN.R-project.org/package=emmeans (accessed on 3 July 2021).
73. R Core Team. *R: A Language and Environment for Statistical Computing*; R Foundation for Statistical Computing: Vienna, Austria, 2021. Available online: https://www.R-project.org/ (accessed on 26 September 2021).

plants

MDPI

Article

Long-Term Effect of Cover Crops on Species Abundance and Diversity of Weed Flora

Alessia Restuccia, Aurelio Scavo *, Sara Lombardo, Gaetano Pandino, Stefania Fontanazza, Umberto Anastasi, Cristina Abbate and Giovanni Mauromicale

Department of Agriculture, Food and Environment (Di3A), University of Catania, 95123 Catania, Italy; a.restuccia@unict.it (A.R.); saralomb@unict.it (S.L.); g.pandino@unict.it (G.P.); stefania.fontanazza@yahoo.it (S.F.); umberto.anastasi@unict.it (U.A.); cristina.abbate@unict.it (C.A.); g.mauromicale@unict.it (G.M.)

* Correspondence: aurelio.scavo@unict.it; Tel.: +39-0954783415

Received: 13 October 2020; Accepted: 4 November 2020; Published: 6 November 2020

Abstract: Cover crops are gaining in popularity as an eco-friendly tool for weed control in organic and low-input agricultural systems. A 5-year study was carried out in a Mediterranean environment (Sicily, south Italy) to (1) quantify cover crop biomass production and (2) evaluate the effects on weed soil seed bank, aboveground biomass, species richness, species composition and associations between communities. Cover crop treatments included subterranean clover (*Trifolium subterraneum* L.) and spontaneous flora, both with and without burying dead mulch into the soil, compared to a conventional management treatment. Weed biomass was significantly reduced by subterranean clover, contrariwise to spontaneous flora, with season-dependent results. Cover crop biomass, which ranged from 44 to more than 290 g DW m^{-2}, was negatively correlated to weed biomass. Moreover, subterranean clover decreased the size of the soil seed bank and species richness. Based on relative frequency, a low similarity was found between the conventional management and cover crop treatments. In addition, no significant differences in species composition across treatments were observed, whereas principal component analysis highlighted some associations. The results suggest that subterranean clover cover cropping is a good option for weed management in Mediterranean agroecosystems.

Keywords: cover crop; weed management; seed bank; weed associations; species richness; multivariate analysis; sustainability

1. Introduction

Specialized orchards of the arid or semiarid regions of the Mediterranean basin are often characterized by low levels of soil organic matter and severe weed infestations, which need a frequent use of chemical inputs for their management [1]. In these agroecosystems, weeds represent the most serious constraint to agricultural production, causing serious yield losses due to their highly competitive capacity and allelopathic activity [2,3]. For many decades, they have been controlled almost exclusively through an irrational use of herbicides that, in addition to the negative effects on the environment, humans and animals [4,5], caused a significant reduction of biodiversity [6]. Low biodiversity in agroecosystems is associated not only to the development of a selective weed flora more difficult to manage, but also to a greater vulnerability to new invasive species [7]. Both weed abundance and diversity are closely influenced by agricultural practices, mainly soil tillage systems, crop rotation and fertilization [8], with a central role played also by environmental conditions [9,10]. The effects (positive or negative) of agronomic techniques on weed diversity are unclear and contradictory, depending on the specific conditions of field experiments, while conservation tillage systems are commonly reported to increase weed abundance [11]. Nowadays, given the increasing interest in limiting the dependence on herbicides, weed control in croplands is addressing to find ecologically-based practices (e.g., crop rotation, stale seedbed, cover cropping, mechanical and physical methods, etc.) under an integrated approach in a medium–long-term strategy [3]. The basic principle is that weeds are an integral part of the agroecosystem and, thus, they should be managed to reduce their harmful effects and increase benefits [12]. Integrated weed management systems are not absolute, but may vary in relation to the context-specific requirements and from year to year.

One of the most common eco-friendly practices, commonly adopted in organic and low-input agricultural systems, is cover cropping, which in the present study is going to include the techniques of mulching, intercropping and green manuring. Indeed, cover crops can be used as living mulches when intercropped between rows in herbaceous crops or on the whole field surface in tree crops, as well as dead mulches either on the soil surface or buried into the soil [3]. In both cases, they prevent weed germination and emergence physically by increasing the competition with weeds and chemically through allelopathic mechanisms [13]. In addition to the phytotoxic activity, cover crops are referred to increase soil fertility by reducing erosion and nutrient leaching, while improving the organic matter content, soil structure and microbial activities [14]. Among the high number of cover crops used in agroecosystems, the *Trifolium* genus and subterranean clover (*T. subterraneum*) in particular, play a key role in Mediterranean orchards thanks to N-fixation ability, rapid growth, rusticity, allelopathic activity and resistance to low radiation levels [15,16]. Subterranean clover originated in the Mediterranean basin, from where it spread throughout western Europe, northern Africa and other world regions with Mediterranean-type climates including Americas, New Zealand and mainly in southern Australia, where it is actually the major pasture legume [17]. It is a free-seeding annual legume, diploid (2n = 16) and predominantly self-pollinated, with remarkable geocarpism. Despite the dispute about the intraspecific taxonomy of *T. subterraneum*, recent genetic studies have confirmed the original classification provided by Katznelson and Morley [18], according to whom the species includes three subspecies with different ecological behavior: subsp. *subterraneum*, subsp. *yanninicum* and subsp. *brachycalycinum*.

In a recent study, Scavo et al. [1] demonstrated that *T. subterraneum* cover cropping significantly reduced the size of the weed seed bank, enhanced the amount of soil nitrogen bacteria and increased the levels of ammoniacal and nitric soil nitrogen. Developed as a continuation of the above-mentioned study, in this research we hypothesized that the observed changes in the potential weed flora (soil seed bank) could reflect on the real one in terms of abundance, richness and diversity, all key aspects for the development of an optimal integrated weed management strategy. Therefore, the objective of this work was to determine the long-term effect of *T. subterraneum* and spontaneous flora cover crops, with respect to a conventional management, on the aboveground weed populations and species composition in an apricot orchard.

2. Results

The real weed flora analysis showed that 38 weed species or genera were present in total throughout the study (Table 1), although most of them were not high frequent enough to be analyzed for principal component analysis (PCA). Seventeen botanical families were observed, the most representative of which was Asteraceae (32%), followed by Brassicaceae (10%) and Poaceae (8%). Concerning the life cycle, 55% were annuals, 29% perennials and 16% biennials. Moreover, 55% of weeds were therophytes and 39% hemicryptophytes, with only two geophytes: *Cirsum arvense* (L.) Scop. and *Convolvulus arvensis* L. (Table 1). Weed communities were dominated by dicotyledonous species (92%) and indifferent (26%) or spring–summer-germinating weeds.

Table 1. Botanical family, life cycle, ecophysiological (EG) and biological groups (BG), frequency (F) and relative frequency (RF) of weed population among 5 cropping systems and 5 seasons.

Weed Species	Botanical Family	Life Cycle	EG	BG	F (%) [1]	RF (%) [1]
Avena sp.	Poaceae	Annual	Sp	T	5.0	2.34
Adonis annua L. subsp. *cupaniana* (Guss.) Steinberg	Ranunculaceae	Annual	Sp	T	8.0	2.13
Anagallis arvensis L.	Primulaceae	Annual	Au-Wi	T	18.0	5.84
Beta vulgaris L.	Amaranthaceae	Perennial	Su	H	11.0	3.43
Brassica rapa L. subsp. *campestris* (L.) A.R. Clapham	Brassicaceae	Perennial	Sp-Su	H	1.2	0.17
Capsella bursa-pastoris (L.) Medik.	Brassicaceae	Biennial	Ind	H	1.0	0.18
Chenopodium opulifolium Schrad. ex W.D.J. Koch & Ziz	Chenopodiaceae	Annual	Su	T	1.0	0.44
Chenopodium sp.	Chenopodiaceae	Annual	Su	T	5.0	1.09
Cichorium intybus L.	Asteraceae	Perennial	Ind	H	14.0	3.23
Cirsium arvense (L.) Scop.	Asteraceae	Perennial	Su	G	2.0	1.0
Convolvulus arvensis L.	Convolvulaceae	Perennial	Ind	G	5.0	2.44
Conyza canadensis L.	Asteraceae	Annual	Sp-Su	T	3.0	1.20
Daucus carota L.	Apiaceae	Biennial	Sp-Su-Au	H	2.2	0.54
Diplotaxis erucoides (L.) DC.	Brassicaceae	Annual	Ind	T	8.0	2.54
Dittrichia viscosa (L.) Greuter subsp. *viscosa*	Asteraceae	Perennial	Au	H	1.0	0.22
Ecballium elaterium (L.) A. Rich.	Cucurbitaceae	Annual	Su	T	22.0	7.86
Erigeron sumatrensis Retz.	Asteraceae	Annual	Su	T	9.0	2.58
Foeniculum vulgare Mill.	Apiaceae	Perennial	Su	H	1.0	0.50
Fumaria officinalis L.	Fumariacee	Annual	Sp-Su-Au	T	10.2	2.52
Galactites elegans (All.) Soldano	Asteraceae	Biennial	Sp-Su	H	1.0	0.24
Galium aparine L.	Rubiaceae	Annual	Sp-Su-Au	T	15.8	3.68
Glebionis coronaria (L.) Spach	Asteraceae	Annual	Sp-Su	T	2.0	0.80
Helminthotheca echioides (L.) Holub	Asteraceae	Annual	Su-Au	T	48.6	13.16
Hypochaeris radicata L.	Asteraceae	Perennial	Sp	H	6.6	1.61
Lamium amplexicaule L.	Lamiaceae	Annual	Ind	T	2.0	0.36
Malva sylvestris L.	Malvaceae	Perennial	Ind	H	4.2	0.99
Medicago polymorpha L.	Fabaceae	Annual	Sp	T	2.2	0.35
Papaver rhoeas L.	Papaveraceae	Annual	Wi	T	9.2	2.39
Reichardia picroides (L.) Roth	Asteraceae	Perennial	Ind	H	9.8	1.77
Setaria verticillata (L.) P. Beauv.	Poaceae	Annual	Su	T	31.0	7.41
Setaria italica subsp. *viridis* (L.) Thell.	Poaceae	Annual	Su-Au	T	27.0	8.28
Silene sp.	Caryophyllaceae	Perennial	Sp-Su	H	3.8	0.64
Sinapis arvensis L.	Brassicaceae	Annual	Sp	T	51.0	14.79
Sonchus asper (L.) Hill	Asteraceae	Biennial	Ind	H	61.0	16.87
Sonchus oleraceus L.	Asteraceae	Biennial	Ind	H	11.0	2.76
Stellaria media (L.) Vill.	Caryophyllaceae	Biennial	Ind	H	1.0	0.40
Trigonella foenum-graecum L.	Fabaceae	Annual	Sp	T	11.0	3.59
Vicia faba L.	Fabaceae	Annual	Sp	T	5.0	1.25

Note: T: therophytes; H: hemicryptophytes; G: geophytes; Su, Au, Wi, Sp: summer, autumn, winter, spring species; In: indifferent species; [1] averaged over all treatments.

2.1. *Effect of Cover Cropping on Weed Diversity*

ANOVA demonstrated that weed species richness varied in relation to both cover cropping and season, while their interaction was not significant (Table 2). The relationship between species richness and cover cropping was consistent at $p \leq 0.05$, with only *Trifolium subterraneum* cover cropping leaving dead mulch on the soil surface (TCC-S) showing a significant reduction with respect to conventional apricot management (CM), contrary to *T. subterraneum* cover cropping burying dead mulch in the soil (TCC-B) that showed the highest value (10.2). Season had the greatest influence on the number of species ($p \leq 0.01$). Overall, except for season III, weed species richness increased among years (+154% from season I to season V).

Table 2. Mean weed species richness found in 5 cropping systems and 5 seasons.

Treatment	No.	Season	No.
TCC-S	6.8 ^{b}	I	4.4 ^{b}
TCC-B	10.2 ^{a}	II	9.0 ^{a}
SCC-S	8.0 ^{a}	III	7.4 ^{ab}
SCC-B	8.0 ^{a}	IV	9.6 ^{a}
CM	8.6 ^{a}	V	11.2 ^{a}
F-test	*	*F*-test	**
SED [1]	1.98	SED [1]	1.37

Values within a column followed by the different letters are significant at $p \leq 0.05$ (Tukey's HSD test). SED: standard error of difference. ** and * indicate significance at $p \leq 0.01$ and $p \leq 0.05$, respectively. [1] 20 d.f. TCC-S: *Trifolium subterraneum* cover cropping leaving dead mulch on the soil surface; TCC-B: *T. subterraneum* cover cropping burying dead mulch in the soil; SCC-S: spontaneous flora cover cropping leaving dead mulch on the soil surface; SCC-B: spontaneous flora cover cropping burying dead mulch in the soil; CM: conventional apricot management; I: 2015/2016; II: 2016/2017; III: 2017/2018; IV: 2018/2019; V: 2019/2020.

Jaccard and Sørensen's indices were used to compare the similarity in terms of species composition between weed communities. Both showed very similar tendencies, with Sørensen's coefficient always presenting higher values than Jaccard's (Table 3). In total, following the trend of species richness, similarity increased across years (excluding season III), with values of +141% (J) and +98% (S) from the first to the last season. Regardless of season, SCC-S × SCC-B and TCC-S × TCC-B showed very high similarity (48.9% and 42.3% for J, 64.7% and 57.8% for S, respectively), while low values were determined between control and treatments (33.5% and 48% for TCC-S × CM, 30% and 44.3% for TCC-B × CM, respectively). The highest similarity was found between TCC-B and SCC-B in season V (92.9% and 96.3% for J and S, respectively), and the lowest one between TCC-B and SCC-S in season I (11.1% and 20% for J and S, respectively).

Table 3. Jaccard's (J, %) and Sørensen's (S, %) similarity coefficients of β-diversity for a 5-cover cropping × 5 seasons system in an apricot orchard.

Treatments	I		II		III		IV		V	
	J	S	J	S	J	S	J	S	J	S
TCC-S × TCC-B	22.2	36.4	66.7	80.0	30.0	46.2	42.9	60.0	50.0	66.7
TCC-S × SCC-S	12.5	22.2	50.0	66.7	30.0	46.2	55.6	71.4	42.9	60.0
TCC-S × SCC-B	14.3	25.0	23.1	37.5	25.0	40.0	36.4	53.3	53.8	70.0
TCC-S × CM	12.5	22.2	50.0	66.7	16.7	28.6	42.9	60.0	45.5	62.5
TCC-B × SCC-S	11.1	20.0	66.7	80.0	16.7	28.6	33.3	50.0	50.0	66.7
TCC-B × SCC-B	28.6	44.4	38.5	55.6	23.1	37.5	23.5	38.1	92.9	96.3
TCC-B × CM	11.1	20.0	42.9	60.0	15.4	26.7	36.8	53.8	43.8	60.9
SCC-S × SCC-B	40.0	57.1	33.3	50.0	60.0	75.0	66.7	80.0	44.4	61.5
SCC-S × CM	14.3	25.0	50.0	66.7	66.7	80.0	42.9	60.0	37.5	54.5
SCC-B × CM	40.0	57.1	33.3	50.0	70.0	82.4	50.0	66.7	37.5	54.5

TCC-S: *Trifolium subterraneum* cover cropping leaving dead mulch on the soil surface; TCC-B: *T. subterraneum* cover cropping burying dead mulch in the soil; SCC-S: spontaneous flora cover cropping leaving dead mulch on the soil surface; SCC-B: spontaneous flora cover cropping burying dead mulch in the soil; CM: conventional apricot management; I: 2015/2016; II: 2016/2017; III: 2017/2018; IV: 2018/2019; V: 2019/2020.

2.2. Effect of Cover Cropping on Weed Abundance

The biomass production of subterranean clover did not statistically differ between TCC-S and TCC-B, although the incorporation of dead mulch into the soil resulted in a higher cover biomass for each season, except for the last one (Table 4). However, the effect of season was highly significant (*F*-Fisher = 63.6, $p \leq 0.001$), with a general trend of season II > III > IV > V > I, suggesting a good establishment of the cover crop. ANOVA indicated no significance of the two-way interaction. On the contrary, weed biomass was significantly affected at $p \leq 0.001$ by the interaction "cover cropping × season", with 51.7% of the total variance explained by the latter factor. Overall, the biomass increased by 75.1% from season I to season IV and then decreased in the last season. Averaged over seasons, TCC-S and TCC-B decreased the weed biomass by 40.9% and 32.3%, respectively, as compared to CM (Table 4). The mean decrease highlighted by *T. subterraneum* cover cropping was marked in seasons IV (−5.5%), II (−70.6%) and V (−63%) (Figure 1). These seasons, with the exception of the third, were those with the major production of cover crop biomass. Indeed, a significant and negative relationship was found between these two parameters ($r = -0.953$, $p = 0.0122$), demonstrating that the lower the subterranean clover biomass is, the higher the weed biomass is. On the contrary, there was no correlation between species richness and weed biomass ($r = -0.043$, $p = 0.944$).

Table 4. Effect of cover cropping (CC) and season (S) on aboveground dry biomass of *Trifolium subterraneum*, weeds and their sum (total) with analysis of variance (ANOVA, *F*-values).

Treatments		Aboveground Biomass (g DW m^{-2})		
		Trifolium Subterraneum	Weeds	Total
CC	TCC-S	155.5 (43.1) ^{a}	82.9 (15.2) ^{b}	238.4 (40.7) ^{a}
	TCC-B	171.7 (32.7) ^{a}	88.3 (9.5) ^{b}	260.0 (36.7) ^{a}
	SCC-S	0.0	120.4 (28.0) ^{a}	120.4 (28.0) ^{b}
	SCC-B	0.0	119.2 (31.8) ^{a}	119.2 (31.8) ^{b}
	CM	0.0	116.8 (17.2) ^{a}	116.8 (17.2) ^{b}
S	I	44.4 (4.1) ^{d}	84.7 (6.4) ^{b}	102.5 (6.5) ^{c}
	II	283.1 (99.9) ^{a}	95.1 (24.8) ^{b}	208.3 (35.6) ^{a}
	III	239.2 (24.5) ^{a}	100.3 (24.1) ^{b}	195.8 (31.8) ^{a}
	IV	159.4 (32.1) ^{b}	148.3 (28.3) ^{a}	212.0 (38.2) ^{a}
	V	92.0 (29.1) ^{c}	99.4 (18.2) ^{b}	136.2 (19.7) ^{b}
ANOVA				
	CC	0.4 NS	9.2 ***	64.1 ***
	S	63.6 ***	14.2 ***	31.6 ***
	CC × S	1.3 NS	4.0 ***	4.2 ***

Values are means with standard deviation (in brackets). Values within a column followed by different letters are significant at $p \leq 0.05$ (Tukey's HSD test). TCC-S: *Trifolium subterraneum* cover cropping leaving dead mulch on the soil surface; TCC-B: *T. subterraneum* cover cropping burying dead mulch in the soil; SCC-S: spontaneous flora cover cropping leaving dead mulch on the soil surface; SCC-B: spontaneous flora cover cropping burying dead mulch in the soil; CM: conventional apricot management; I: 2015/2016; II: 2016/2017; III: 2017/2018; IV: 2018/2019; V: 2019/2020. *** and NS indicate significance at $p \leq 0.001$ and not significance, respectively.

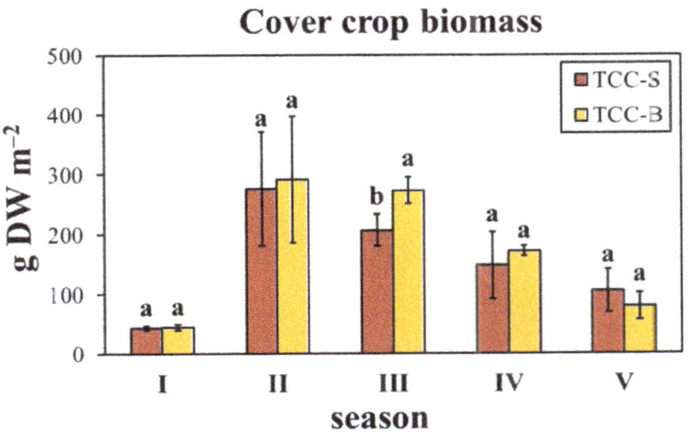

(a)

Figure 1. *Cont.*

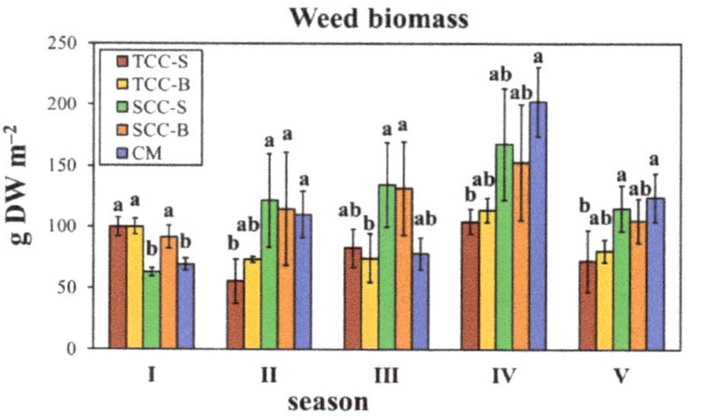

(**b**)

Figure 1. (**a**) Cover crop and (**b**) weed aboveground dry biomass production over five seasons in an apricot orchard. Bars are standard deviation ($n = 4$). Within each season, different letters indicate statistical significance at $p \leq 0.05$ (Tukey's HSD test). TCC-S: *Trifolium subterraneum* cover cropping leaving dead mulch on the soil surface; TCC-B: *T. subterraneum* cover cropping burying dead mulch in the soil; SCC-S: spontaneous flora cover cropping leaving dead mulch on the soil surface; SCC-B: spontaneous flora cover cropping burying dead mulch in the soil; CM: conventional apricot management. I: 2015/2016; II: 2016/2017; III: 2017/2018; IV: 2018/2019; V: 2019/2020.

Results on weed biomass were consistent with the potential flora. Indeed, all the cover cropping systems significantly lowered the number of weed seeds in the soil with respect to the conventional management (Figure 2). After 5-years, TCC-S and TCC-B had the lowest seed bank size, showing a reduction of 40.5% and 57%, respectively, compared to CM, in concordance with the aboveground weed biomass. However, the size of the soil seed bank was not correlated to the mean cover crop biomass ($r = -0.827$, $p = 0.084$), nor to the mean weed biomass ($r = 0.767$, $p = 0.131$). Anyway, despite the lack of significance, the seed bank decreased with increasing subterranean clover biomass; at the same time, average weed biomass was lower in TCC-S and TCC-B plots, where seed bank densities where the lowest.

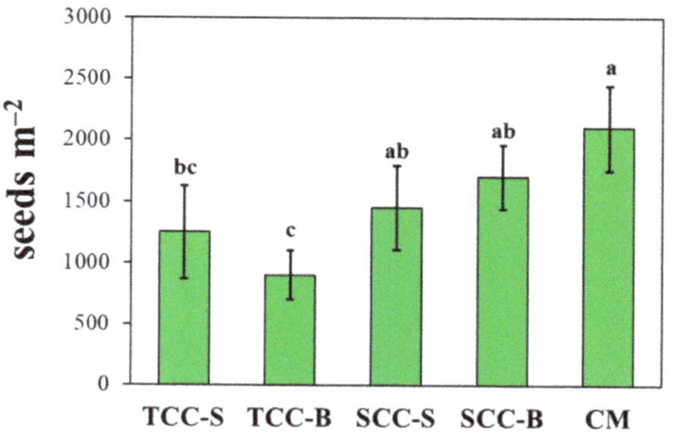

Figure 2. Cumulative effect on the soil weed seed bank after 5 years of different cover cropping systems. Bars are standard deviation ($n = 4$). Different letters indicate statistical significance at $p \leq 0.05$ (Tukey's HSD test). TCC-S: *Trifolium subterraneum* cover cropping leaving dead mulch on the soil surface; TCC-B: *T. subterraneum* cover cropping burying dead mulch in the soil; SCC-S: spontaneous flora cover cropping leaving dead mulch on the soil surface; SCC-B: spontaneous flora cover cropping burying dead mulch in the soil; CM: conventional apricot management.

2.3. Aboveground Weed Species Composition

Among the 38 taxa recorded throughout the 5-year period, only 11 weeds, predominantly annual seed-propagated species, showed a F ≥ 11% and a RF ≥ 3%: *Anagallis arvensis* L., *Beta vulgaris* L., *Cichorium intybus* L., *Ecballium elaterium* (L.) A. Rich., *Galium aparine* L., *Helminthotheca echioides* (L.) Holub, *Setaria verticillata* (L.) P. Beauv., *S. italica*, *Sinapis arvensis* L., *Sonchus asper* (L.) Hill and *Trigonella foenum-graecum* L. (Table 1). *Sonchus asper* was the most prominent weed species in all the seasons with mean values ranging from 8% to 31%, followed by *S. arvensis* with values in the range 15–22%. Moreover, averaged over seasons, this species showed the highest RF in TCC-S (12%), SCC-S (13%) and CM (9%) plots. Weed communities of TCC-B and SCC-B, instead, were dominated by *H. echioides* (F = 100, RF = 10%) and *E. elaterium* (F = 100, RF = 13%), respectively. However, the two-way ANOVA performed on RF data outlined that neither cover cropping nor season, except for some species, affected major weeds at $p \leq 0.05$; for this reason, they have not been shown.

Among the 11 major weed species selected for PCA, the scree plot for standardized variables (correlation matrix) highlighted that only the first four PCs contributed to variance, while PC5-PC11 were insignificant (Figure 3). The cumulative variance explained by the first two eigenvalues together was 75.2%, which is an acceptable percentage for weed communities, thus suggesting a consideration of PC1 and PC2. The weeds *S. viridis*, *S. italica*, *A. arvensis*, *C. intybus* and *G. aparine* showed jointly the majority of variance (49%) in PC1; *E. echioides*, *T. foenum-graecum*, *S. arvensis* and *E. elaterium* added an additional 26% in PC2; *B. vulgaris* explained a further 16% of variance for PC3, while the eigenvector associated with PC4 corresponded to an eigenvalue <1, in which *S. asper* had the highest weight (Table 5). Table 5 showed also that PC1 was positively correlated to *G. aparine* and *A. arvensis* (right side of the biplot) and negatively by the two *Setaria* species and *C. intybus* (left side). A positive association was found between PC2 and *T. foenum-graecum*, *S. arvensis* and *E. elaterium* (top of biplot), and a negative one with *E. echioides* (bottom). Interesting associations were observed by PCA of weed species and cover cropping (Figure 4). *Setaria viridis*, *S. italica* and *C. intybus* were associated with SCC-S, whereas SCC-B was associated with *G. aparine* and TCC-S with *A. arvensis*. The other weeds were discriminated mainly along PC2, with a correlation observed between *S. arvensis* and CM, while TCC-B was not associated with any species, thus confirming a lower infestation in terms of weed biomass, soil seed bank and species composition. No relevant differences were observed between cover cropping systems in terms of botanical family, biological or ecophysiological groups, indicating no clear patterns of the weed flora.

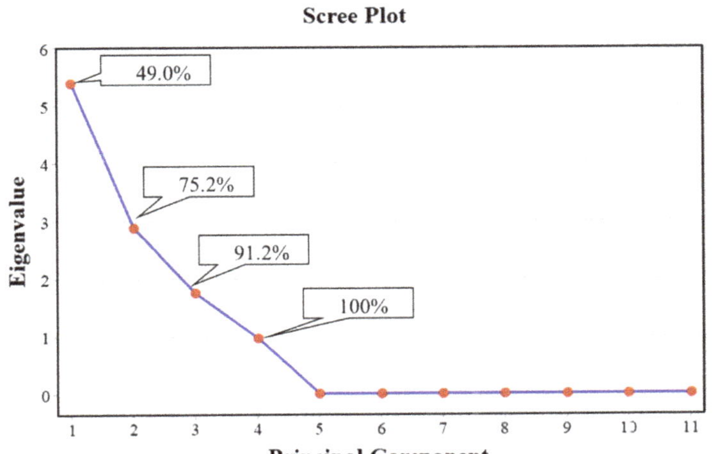

Figure 3. Scree plot of eigenvalues and cumulative variance of principal components from the correlation matrix.

Table 5. Eigenvectors defining the linear combination of variables (11 major weeds) and principal components from the correlation matrix (PC5-PC11 were insignificant). The variables with the largest influence for each principal component are in bold.

Variable	PC1	PC2	PC3	PC4
ANAAR	**0.383**	−0.211	−0.167	0.182
BEVULG	−0.035	−0.015	**−0.752**	−0.056
CICIN	**−0.363**	−0.015	0.079	0.534
ECBAL	0.317	**0.382**	−0.056	−0.181
GALAP	**0.352**	0.050	0.319	0.388
HELEC	0.124	**−0.549**	−0.042	0.211
SETVER	**−0.430**	0.030	0.019	−0.027
SETIT	**−0.404**	−0.070	0.206	0.182
SINAR	−0.290	**0.426**	−0.094	−0.088
SONCAS	0.043	0.329	−0.410	**0.628**
TRIGO	0.215	**0.458**	0.273	0.121

ANAAR (*Anagallis arvensis*); BEVULG (*Beta vulgaris*); CICIN (*Cichorium intybus*); ECBAL (*Ecballium elaterium*); GALAP (*Galium aparine*); HELEC (*Helminthotheca echioides*); SETVER (*Setaria viridis*); SETIT (*Setaria italica*); SINAR (*Sinapis arvensis*); SONCAS (*Sonchus asper*); TRIGO (*Trigonella foenum-graecum*).

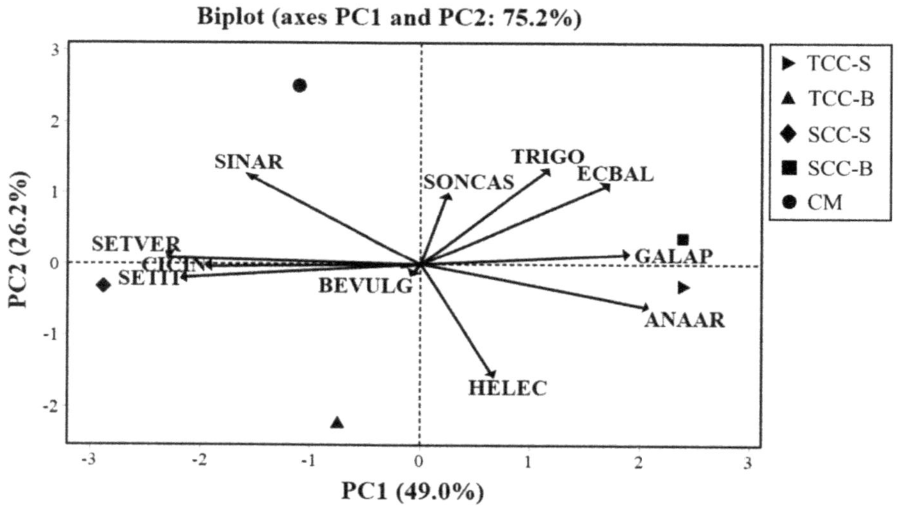

Figure 4. Principal components analysis ordination biplot from the correlation matrix with the 11 most frequent weed species, averaged over seasons, in different cover cropping treatments. Weeds: ANAAR (*Anagallis arvensis*); BEVULG (*Beta vulgaris*); CICIN (*Cichorium intybus*); ECBAL (*Ecballium elaterium*); GALAP (*Galium aparine*); HELEC (*Helminthotheca echioides*); SETVER (*Setaria viridis*); SETIT (*Setaria italica*); SINAR (*Sinapis arvensis*); SONCAS (*Sonchus asper*); TRIGO (*Trigonella foenum-graecum*). Treatments: TCC-S: *Trifolium subterraneum* cover cropping leaving dead mulch on the soil surface; TCC-B: *T. subterraneum* cover cropping burying dead mulch in the soil; SCC-S: spontaneous flora cover cropping leaving dead mulch on the soil surface; SCC-B: spontaneous flora cover cropping burying dead mulch in the soil; CM: conventional apricot management.

3. Discussion

The present study aimed to evaluate the influence of 5 years of cover cropping, by subterranean clover and spontaneous flora, both buried and living dead mulches on the soil surface, on diversity and abundance of the real weed flora. In our previous research [1], we found a 70% reduction of the weed soil seed bank, compared to CM, after 3-years of *T. subterraneum* green manuring (TCC-B).

Given that the real weed flora generally reflects the spectrum of the potential one, the effects on the emerged weeds were evaluated for a further two years on a medium–long-term period. We found that subterranean clover, in some seasons, significantly decreased the mean weed biomass up to 86%, contrariwise to spontaneous flora cover crop. The intensity of such a decrease was season-dependent, likely due to a combined effect of climatic conditions and cover crop biomass. In contrast with Moonen and Bàrberi [19], in our study cover crop biomass highly varied between the seasons from 44 to more than 290 g of DW m^{-2}. Weed biomass decrease caused by subterranean clover was higher in seasons when cover crop biomass was higher (seasons II, IV and V), except for season III. Our results are similar to those obtained by the study of Bàrberi and Mazzoncini [20], in which subterranean clover was found to reduce weed biomass from 21% to 67%, with a positive correlation between weed growth suppression and cover crop biomass and with seasonal effects. The results obtained on the real flora were corroborated by the effects on the soil seed bank, in which all the cover cropping systems decreased the number of weed seeds. TCC-S and TCC-B showed the highest weed suppressive ability after a further two years, although with a lower degree than the third year [1].

Weed suppressive ability of subterranean clover may be attributed to competitive or allelopathic effects, or even to a combination of them. *Trifolium subterraneum*, in fact, competes well with weeds thanks to its rapid growth, developed canopy, length of biological cycle and development of root system [21]. Generally, weed suppression increases with increasing cover crop biomass and cycle length, as found in the present study. Furthermore, subterranean clover is recognized as allelopathic species and allelochemicals responsible for such phytotoxic effects have been indicated as phenols and isoflavonoids [22]. These secondary metabolites can be directly exuded into the soil or released by decomposition of plant residues. Once present into the rhizosphere, allelochemicals interact with the complex of physical, chemical and biological soil characteristics, which altogether fix their availability [23]. Unfortunately, competitive and allelopathic effects are very difficult to distinguish in field experiments.

The emerged flora reflected the composition of the seed bank, given that weed communities were dominated by Asteraceae members, therophytes and annual spring–summer weeds. As previously observed on the seed bank [1], weed species richness was significantly affected by TCC-S, while, interestingly, TCC-B increased it, suggesting no clear influence of cover cropping. On the contrary, the effect of season was more noticeable, with a much higher number of weed species detected in season V, showing an increase in weed biodiversity. Conflicting reports have been provided by authors concerning the effects of cover crops on species richness. Ngouajio et al. [24], for example, observed no significant relationships, with results depending on cover crop type and season, while a reduction of weed density was found by Moonen and Bàrberi [19] using rye (*Secale cereale* L.) cover crop.

Since the contradictory results, many authors agree in not considering species richness as the only parameter to evaluate the herbicidal activity of cover crops. In this regard, the composition of weed communities plays a key role in shifting the phytotoxic effects. It should be pointed out, in fact, that the sensitivity of weed species to cover crop residue is highly variable, mainly depending on weed community structure. On one side, annual weeds with small seed sizes are more susceptible to surface residues than large seeded species, and on the other side, large seeds have a greater metabolic capacity for allelochemical detoxification [25]. In this study, the β-diversity indices of Jaccard and Sørensen were applied the compare the areas in terms of composition of the weed communities [26]. These indices are closely influenced by agronomic practices. Here, the highest similarity was found between spontaneous flora (SCC-S × SCC-B) and between subterranean clover (TCC-S × TCC-B) cover crop, often with values across seasons higher than 50%, at which an elevated similarity can be interpreted. Instead, a general low similarity was found between the conventional management and cover cropping systems. Therefore, it is reasonable to assume that *T. subterraneum* treatments (TCC-S and TCC-B) determined similar weed communities based on presence/absence, as well as spontaneous flora cover crops (SCC-S and SCC-B), both different with respect to CM. ANOVA performed on RF data of single species, however, pointed out any significant effect among treatments under study, demonstrating that weeds were able to establish independently of cover type and season. To overcome the complexity of weed data, species composition was studied by PCA on major weed species. In addition to a reduction in weed seed bank density and aboveground biomass, TCC-B did not show any association with weeds, contrariwise to SCC-S and SCC-B. No evident weed patterns emerged in this study, as observed also in the seed bank [1]. Overall, treatments were quite similar also with reference to frequency, botanical families, life cycle, biological and ecophysiological groups. The lack of consistent associations between cover crop and weeds has been reported in many other studies [19,27], since species composition can be influenced by abiotic and biotic factors. In the 9-year research study carried out by Shrestha et al. [28] on winter wheat and three beans, rye and maize cover crop were also indicated to have differential effects on weed densities, species composition and associations depending on crop type and interaction with agronomic management.

In conclusion, this research suggests that long-term changes in weed flora are linked to the soil seed bank. On one hand, the adoption of 5 years of cover cropping with subterranean clover was found to reduce not only the number of weed seeds in the soil, but also the aboveground weed biomass and the number of species, with significant variations by season. On the other hand, instead, no clear shifts in weed populations were observed. These results are very useful in view of reducing intensive tillage and the frequent application of herbicides, thus allowing multiple benefits for the environment. The benefits in using subterranean clover in Mediterranean agroecosystems are further increased considering its self-reseed capacity, N-fixation ability and high adaptability in such contexts [1]. Future studies may consider the evaluation of subterranean clover cover cropping in combination with other control techniques under an integrated weed management system, as well as a better knowledge of the mechanisms involved in its phytotoxicity.

4. Materials and Methods

4.1. Experimental Site and Set-Up

A field experiment was conducted over five growing seasons (from 2015/2016 to 2019/2020, hereafter named season I, II, III, IV and V) in an apricot (*Prunus armeniaca* L.) orchard sited in central Sicily (37°13′ N, 14°05′ E, 290 m a.s.l., Italy). The zone is subjected to a semiarid-Mediterranean climate, characterized by mean annual precipitations of ~500 mm, hot-rainless summers and mild winters. Based on the Rivas-Martinez bioclimatic classification, the area belongs to the thermo-Mediterranean inferior bioclimatic belt, with upper dry ombrotype. The experimental soil, Regosoil type according to the USDA soil taxonomy classification [29], at the beginning of the experiment presented an average soil texture of 25.7% sand, 30.6% silt and 43.7% clay, an average organic matter content of 1.9%, and an amount of 1.1‰, 13 mg kg^{-1} and 422 mg kg^{-1} of total nitrogen, assimilable P_2O_5 and exchangeable K_2O, respectively, with pH 8.0.

For each growing season, the experiment was set-up in a randomized block design with four replicates (plot size = 10 × 8.7 m) including five treatments: four cover cropping systems compared to a conventional management (CM) as control following the standard commercial practices (−0.15 cm winter disc ploughing and three instances of shallow chopping per year for weed control). Cover cropping treatments were: (a) *T. subterraneum* cover cropping leaving dead mulch on the soil surface (TCC-S); (b) *T. subterraneum* cover cropping burying dead mulch in the soil (TCC-B); (c) spontaneous flora cover cropping leaving dead mulch on the soil surface (SCC-S), and (d) spontaneous flora cover cropping burying dead mulch in the soil (SCC-B). The experiment therefore included 20 plots and a net plot size of 1740 m^2 (348 m^2 per treatment), with a distance of 2 m between treatments.

The apricot orchard, composed by cv. Wonder and two pollinators (cvs. Pinkcot® and Big Red®), was planted on January 2012 by using a 3.5 × 4.5 m arrangement. Subterranean clover cv. Seaton Park, a cheap and common Australian early-mid season genotype showing high adaptability in Mediterranean orchards [15], was hand-seeded on November 2015 at 2–3 cm depth with 2000 germinable seeds m^{-2}. Detailed information about orchard management, fertilization, irrigation, weed and pest control were already reported in Scavo et al. [1]. Moreover, Table 6 summarizes the biological cycle of subterranean clover during the five growing seasons.

Table 6. Emergence, flowering and length of the biological cycle of subterranean clover among the five growing seasons under study.

Season	Emergence	Flowering	Length of the Biological Cycle [1]
I	15 December 2015	22 May 2016	~200 days
II	22 November 2016	12 April 2017	~220 days
III	10 October 2017	29 April 2018	~250 days
IV	13 October 2018	26 April 2019	~240 days
V	5 November 2019	4 April 2020	~230 days

[1] From the beginning of emergence until all the plants within a plot had completely dried up (first decade of July for all the seasons).

4.2. Monitoring, Sampling and Aboveground Biomass Determination

Monitoring was carried out by field scouting to visualize the weed spatial distribution, obtain a representative view of the weed flora and locate the sampling units. For each treatment, the sampling zone was chosen excluding the outer 3 m of each plot and the non-homogeneous areas. Within each zone, four permanent 1.0 m^2 quadrats were randomly placed. The aboveground biomass of both weeds and subterranean clover was obtained by clipping in April for each season at soil surface from four 0.25 m^2 patches per quadrat. In the laboratory, for TCC-S and TCC-B, cover crop biomass was separated from weed species and samples were dried at 55 °C in a forced-air oven up to constant weight for dry biomass determination. For the weed flora analysis, clipped weeds were identified according to Conti et al. [30] and grouped to botanical family and life-form category considering the Raunkiaer system; to obtain the total weed biomass per quadrat, the weed biomass was pooled at the quadrat level. The analysis of the weed soil seed bank was carried out in accordance with Scavo et al. [1]. In summary, soil samples were collected twice (April and September) per season at 10–15 cm depth along the diagonals of the central part of each sampling area and each soil sample was a composite of five soil cores per plot (each of 0.75 dm^3). Then, the inert fraction (stones, pebbles, etc.) was hand-removed and a metal tube (Karcher, K 3500 model, Winnenden, Germany) with a removable cap fitted with steel mesh of 250 μm was used for seed extraction. Finally, the extracted fraction was placed inside Petri dishes for weed counts and identification by using a MS5 Leica stereomicroscope (Leica Microsystems, Wetzlar, Germany).

4.3. Weed Flora Analysis

Following Nkoa et al. [7] and Travlos et al. [31], the weed flora was analyzed by estimating species abundance and diversity. Abundance, describing the quantitative significance of a species in its habitat, was measured considering the total biomass of weeds (B), frequency (F) and relative frequency (RF):

$$F\ (\%) = \left(\frac{\sum Z_i}{n}\right) \times 100 \tag{1}$$

$$RF\ (\%) = \left(\frac{F_i}{\sum F}\right) \times 100 \tag{2}$$

where: $\sum Z_i$ = number of sampling units in which the species i occurred; n = total number of sampling units; F_i = absolute frequency of a species i; $\sum F$ = sum of the absolute frequencies of all species. Despite needing destructive sampling, biomass, expressed as dry weight per unit area, is an accurate and objective index. F and RF are non-destructive indices reflecting the species' spatial distribution across the sampled area and the changes over time.

In addition to species richness, i.e., the total number of species in each plot [19], the β-diversity was measured to estimate the species' composition differences or similarity between communities.

The β-diversity was calculated by using the Jaccard's index of similarity (J) and the Sørensen's coefficient index (S), computed as in Real and Vargas [32] and Nkoa et al. [7], respectively:

$$J\,(\%) = \left(\frac{c}{a + b\,c}\right) \times 100 \quad (3)$$

$$S\,(\%) = \left(\frac{2c}{a + b}\right) \times 100 \quad (4)$$

where: a = total number of species present only in one community; b = total number of species in the second community; c = total number of species common to each community. Both J and S are binary similarity coefficients based on presence/absence data.

4.4. Meteorological Trend

A meteorological station (Mod. Multirecorder 2.40; ETG, Firenze, Italy) located at ~15 m on the experimental site was used to record rainfall and air temperatures every day during the five growing seasons, from November 2015 to April 2020 (Figure 5). Following the typical trend of the zone, summers were always particularly hot and dry, while most of rainfall fell in autumn. In particular, the sum of January 2016 (119 mm), October 2018 (189 mm) and November 2019 (250 mm) accounted for 22.1% of the total rainfall occurred in the whole experimental period. The years 2018 (668 mm) and 2019 (673 mm) experienced higher rainfall levels compared to the 30-year-period trend. The highest maximum temperature was recorded on August 2017 (36.1 °C), while the lowest minima temperatures were noted on January 2017 (3.0 °C) and 2019 (2.8 °C). The temperatures were always within the optimal range for *T. subterraneum* growth, since during the 5-year period, minimum temperatures only fall below 3.0 °C in season IV; the mean maximum temperatures were 25.7 °C in October (emergence) and 21.5 °C at flowering (April).

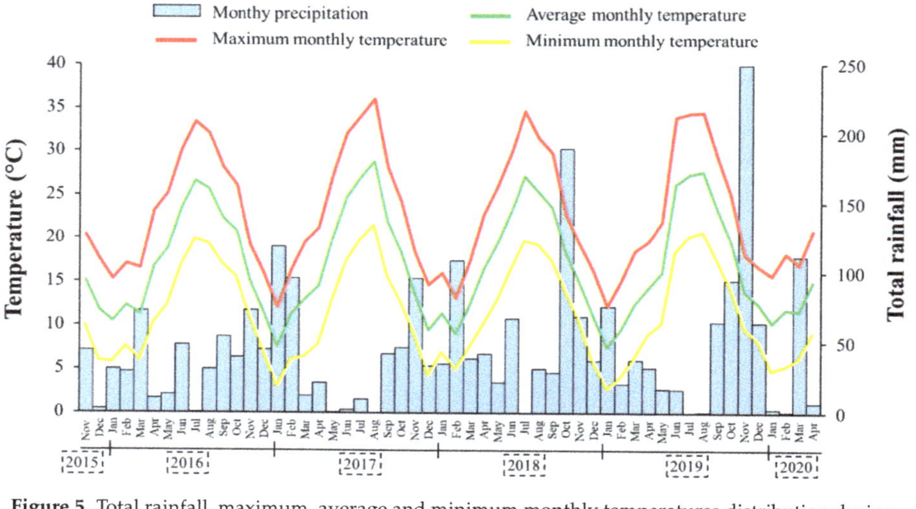

Figure 5. Total rainfall, maximum, average and minimum monthly temperatures distribution during the five cropping seasons.

4.5. Statistical Analysis

Data about aboveground biomass of both subterranean clover and weeds, as well as species richness and soil seed bank, were analyzed through analysis of variance (ANOVA) by using the statistical software CoStat® version 6.003 (CoHort Software, Monterey, CA, USA). Prior to ANOVA, the Bartlett's and the Shapiro–Wilk tests were used to check for homoscedasticity and normality, respectively. Furthermore, to comply with the ANOVA basic assumptions, biomass and seed bank data were $\log_{10}$-transformed (untransformed data are reported), while species richness data did not show any violation and, therefore, they were not transformed. A factorial two-way ANOVA model with "cover cropping × season" as main factors was performed and means were separated with the Tukey's HSD test at $p \leq 0.05$. In some cases, one-way ANOVAs were applied. In accordance with Moonen and Bàrberi [19], correlations between soil seed bank size and *T. subterraneum* biomass, weed biomass, between species richness and weed biomass, and between cover crop and weed biomass, were calculated by using the Pearson Product Moment Correlation Coefficient (r) on mean values for these parameters.

To study the species composition and the interactions between cover cropping treatments and weed flora, a multivariate analysis was performed. Due to the high number of variables (weed species) composing the weed flora, the principal component analysis (PCA) was adopted to reduce the complex multivariate dataset in few orthogonal variables called principal components (PC) [33]. In particular, a PCA on the correlation matrix for 11 major variables (weeds with RF > 3%) was applied, considering the means for each "treatment × season" combination. Before PCA, all variables were standardized through arcsin $\sqrt{x}$ (Bliss transformation), and then the results of the ordination were displayed on "distance" biplots deriving from the PCA by using the first two PCs [34]. The computer package Minitab® version 16 (Minitab Inc., State College, PA, USA) was used to perform the PCA.

Author Contributions: Conceptualization, G.M.; methodology, A.S., A.R., S.L., G.P., S.F., U.A., C.A. and G.M.; investigation and data curation, A.S., A.R., S.L., G.P., S.F., U.A. and C.A.; writing—original draft preparation, A.S. and A.R.; writing—review and editing, A.S., A.R. and G.M.; supervision, G.M. All authors have read and agreed to the published version of the manuscript.

Funding: This research was funded by University of Catania, Project "Piano per la Ricerca 2016–2018—Linea di intervento 2—P5/WP1", grant number 5A722192135.

Conflicts of Interest: The authors declare no conflict of interest.

References

1. Scavo, A.; Restuccia, A.; Lombardo, S.; Fontanazza, S.; Abbate, C.; Pandino, G.; Anastasi, U.; Onofri, A.; Mauromicale, G. Improving soil health, weed management and nitrogen dynamics by *Trifolium subterraneum* cover cropping. *Agron. Sustain. Dev.* **2020**, *40*, 18. [CrossRef]
2. Restuccia, A.; Lombardo, S.; Mauromicale, G. Impact of a cultivation system upon the weed seedbank size and composition in a Mediterranean environment. *Agriculture* **2019**, *9*, 192. [CrossRef]
3. Scavo, A.; Mauromicale, G. Integrated weed management in herbaceous field crops. *Agronomy* **2020**, *10*, 466. [CrossRef]
4. Scavo, A.; Restuccia, A.; Abbate, C.; Mauromicale, G. Seeming field allelopathic activity of *Cynara cardunculus* L. reduces the soil weed seed bank. *Agron. Sustain. Dev.* **2019**, *39*, 41. [CrossRef]
5. Scavo, A.; Pandino, G.; Restuccia, A.; Lombardo, S.; Pesce, G.R.; Mauromicale, G. Allelopathic potential of leaf aqueous extracts from *Cynara cardunculus* L. on the seedling growth of two cosmopolitan weed species. *Ital. J. Agron.* **2019**, *14*, 78–83. [CrossRef]
6. Landis, D.A. Designing agricultural landscapes for biodiversity-based ecosystem services. *Basic Appl. Ecol.* **2017**, *18*, 1–12. [CrossRef]
7. Nkoa, R.; Owen, M.D.K.; Swanton, C.J. Weed abundance, distribution, diversity, and community analyses. *Weed Sci.* **2015**, *63*, 64–90. [CrossRef]
8. Santín-Montanyá, M.I.; Martín-Lammerding, D.; Walter, I.; Zambrana, E.; Tenorio, J.L. Effects of tillage, crop systems and fertilization on weed abundance and diversity in 4-year dry land winter wheat. *Eur. J. Agron.* **2013**, *48*, 43–49. [CrossRef]
9. Fried, G.; Norton, L.R.; Reboud, X. Environmental and management factors determining weed species composition and diversity in France. *Agric. Ecosyst. Environ.* **2008**, *128*, 68–76. [CrossRef]
10. Gresta, F.; Cristaudo, A.; Onofri, A.; Restuccia, A.; Avola, G. Germination response of four pasture species to temperature, light, and post-harvest period. *Plant Biosyst.* **2010**, *144*, 849–856. [CrossRef]

11. Bàrberi, P. Weed management in organic agriculture: Are we addressing the right issues? *Weed Res.* **2002**, *42*, 177–193. [CrossRef]
12. Jordan, N.; Vatovec, C. Agroecological benefits from weeds. In *Weed Biology and Management*; Inderjit, Ed.; Springer: Dordrecht, The Netherlands, 2004; pp. 137–158. [CrossRef]
13. Lemessa, F.; Wakjira, M. Mechanisms of ecological weed management by cover cropping: A review. *J. Biol. Sci.* **2014**, *14*, 452–459. [CrossRef]
14. Gerhards, R.; Schappert, A. Advancing cover cropping in temperate integrated weed management. *Pest Manag. Sci.* **2020**, *76*, 42–46. [CrossRef]
15. Mauromicale, G.; Occhipinti, A.; Mauro, R. Selection of shade-adapted subterranean clover species for cover cropping in orchards. *Agron. Sustain. Dev.* **2010**, *30*, 473–480. [CrossRef]
16. Mauro, R.P.; Occhipinti, A.; Longo, A.M.G.; Mauromicale, G. Effects of shading on chlorophyll content, chlorophyll fluorescence and photosynthesis of subterranean clover. *J. Agron. Crop Sci.* **2011**, *197*, 57–66. [CrossRef]
17. Zohary, M.; Heller, D. *The Genus Trifolium*; Israel Academy of Sciences and Humanities: Jerusalem, Israel, 1984.
18. Katznelson, J.; Morley, F.H.W. A taxonomic revision of sect. *Calycomorphum* of the genus *Trifolium*. I. The geocarpic species. *Isr. J. Bot.* **1965**, *14*, 112–134.
19. Moonen, A.C.; Bàrberi, P. Size and composition of the weed seedbank after 7 years of different cover crop-maize management systems. *Weed Res.* **2004**, *44*, 163–177. [CrossRef]
20. Bàrberi, P.; Mazzoncini, M. Changes in weed community composition as influenced by cover crop and management system in continuous corn. *Weed Sci.* **2001**, *49*, 491–499. [CrossRef]
21. Den Hollander, N.G.; Bastiaans, L.; Kropff, M.J. Clover as a cover crop for weed suppression in an intercropping design: II. Competitive ability of several clover species. *Eur. J. Agron.* **2007**, *26*, 104–112. [CrossRef]
22. Liu, Q.; Xu, R.; Yan, Z.Q.; Jin, H.; Cui, H.Y.; Lu, L.Q.; Zhang, D.H.; Qin, B. Phytotoxic allelochemicals from roots and root exudates of *Trifolium pratense*. *J. Agric. Food Chem.* **2013**, *61*, 6321–6327. [CrossRef]
23. Scavo, A.; Abbate, C.; Mauromicale, G. Plant allelochemicals: Agronomic, nutritional and ecological relevance in the soil system. *Plant Soil* **2019**, *442*, 23–48. [CrossRef]
24. Ngouajio, M.; McGiffen, M.E., Jr.; Hutchinson, C.M. Effect of cover crop and management system on weed populations in lettuce. *Crop Prot.* **2003**, *22*, 57–64. [CrossRef]
25. Mirsky, S.B.; Ryan, M.R.; Teasdale, J.R.; Curran, W.S.; Reberg-Horton, C.S.; Spargo, J.T.; Wells, M.S.; Keene, C.L.; Moyer, J.W. Overcoming weed management challenges in cover crop–based organic rotational no-till soybean production in the eastern United States. *Weed Technol.* **2013**, *27*, 193–203. [CrossRef]
26. Concenço, G.; de Farias, P.M.; Quintero, N.F.A.; Schreiber, F.; Galon, L.; Tomazi, M.; Moisinho, I.S.; Coradini, M.C.; Ceolin, W.C.; Andres, A. Phytosociological surveys in weed science: Old concept, new approach. In *Plant Ecology-Traditional Approaches to Recent Trends*; Yousaf, Z., Ed.; IntechOpen: London, UK, 2017; pp. 121–146. [CrossRef]
27. Swanton, C.J.; Shrestha, A.; Roy, R.C.; Ball-Coelho, B.R.; Knezevic, S.Z. Effect of tillage systems, N, and cover crop on the composition of weed flora. *Weed Sci.* **1999**, *47*, 454–461. [CrossRef]
28. Shrestha, A.; Knezevic, S.Z.; Roy, R.C.; Ball-Coelho, B.R.; Swanton, C.J. Effect of tillage, cover crop and crop rotation on the composition of weed flora in a sandy soil. *Weed Res.* **2002**, *42*, 76–87. [CrossRef]
29. Soil Survey Staff. *Soil Taxonomy: A Basic System of Soil Classification for Making and Interpreting Soil Surveys*, 2nd ed.; U.S. Gov. Print. Office: Washington, DC, USA, 1999.
30. Conti, F.; Abbate, G.; Alessandrini, A.; Blasi, C. *An Annotated Checklist of the Italian Vascular Flora*; Palombi Editore: Rome, Italy, 2005.
31. Travlos, I.S.; Cheimona, N.; Roussis, I.; Bilalis, D.J. Weed-species abundance and diversity indices in relation to tillage systems and fertilization. *Front. Environ. Sci.* **2018**, *6*, 11. [CrossRef]
32. Real, R.; Vargas, J.M. The probabilistic basis of Jaccard's index of similarity. *Syst. Biol.* **1996**, *45*, 380–385. Available online: https://www.jstor.org/stable/2413572 (accessed on 15 August 2020). [CrossRef]

33. Deligios, P.A.; Carboni, G.; Farci, R.; Solinas, S.; Ledda, L. The influence of herbicide underdosage on the composition and diversity of weeds in oilseed rape (*Brassica napus* L. var. *oleifera* DC) Mediterranean fields. *Sustainability* **2019**, *11*, 1653. [CrossRef]
34. Legendre, P.; Legendre, L. *Numerical Ecology*; Elsevier: Amsterdam, The Netherlands, 2012; pp. 429–463.

Publisher's Note: MDPI stays neutral with regard to jurisdictional claims in published maps and institutional affiliations.

plants

MDPI

Article

Effects of Different Inter-Row Soil Management and Intra-Row Living Mulch on Spontaneous Flora, Beneficial Insects, and Growth of Young Olive Trees in Southern Italy

Giuseppina Las Casas [1,†], Corrado Ciaccia [2,†], Valeria Iovino [1], Filippo Ferlito [1,*], Biagio Torrisi [1], Enrico Maria Lodolini [3,*], Alessio Giuffrida [1], Roberto Catania [1], Elisabetta Nicolosi [4] and Salvatore Bella [1]

1 Consiglio per la Ricerca in Agricoltura e l'Analisi dell'Economia Agraria (CREA)—Centro di Ricerca Olivicoltura, Frutticoltura e Agrumicoltura, Corso Savoia 190, 95024 Acireale, Italy; giuseppina.lascasas@crea.gov.it (G.L.C.); iovino.valeri@gmail.com (V.I.); biagiofrancesco.torrisi@crea.gov.it (B.T.); alessio020594@live.it (A.G.); robertocatania1995@libero.it (R.C.); salvatore.bella@crea.gov.it (S.B.)

2 Consiglio per la Ricerca in Agricoltura e l'Analisi dell'Economia Agraria (CREA)—Centro di Ricerca Agricoltura e Ambiente, Via della Navicella 4, 00184 Roma, Italy; corrado.ciaccia@crea.gov.it

3 Consiglio per la Ricerca in Agricoltura e l'Analisi dell'Economia Agraria (CREA)—Centro di Ricerca Olivicoltura, Frutticoltura e Agrumicoltura, Via Fioranello 52, 00134 Roma, Italy

4 Dipartimento di Agricoltura, Alimentazione e Ambiente (Di3A), Università di Catania, Via Valdisavoia 5, 95123 Catania, Italy; elisabetta.nicolosi@unict.it

* Correspondence: filippo.ferlito@crea.gov.it (F.F.); enricomaria.lodolini@crea.gov.it (E.M.L.); Tel.: +39-09571-63111 (F.F.)

† These authors contributed equally to this work.

Abstract: Conservation agriculture (i.e., minimized soil disturbance and permanent soil covering) and living mulches represent two agroecological practices that can improve soil fertility, spontaneous flora, and beneficial insect communities. This research studied the effect of these practices in a young olive orchard in the Mediterranean area. Two Sicilian olive cultivars ('Nocellara del Belice' and 'Nocellara etnea') were used for the field experiment; inter-row minimum and zero tillage and four species of aromatic plants as living mulch along the row were tested. Spontaneous flora and beneficial insect communities, as well as tree growth, were monitored. The inter-row management did not influence the spontaneous flora dynamics. The species adopted for living mulch showed a very different degree of development and soil cover; 69 insect species (pollinators and predators) belonging to five orders (Hymenoptera, Lepidoptera, Diptera, Neuroptera, and Coleoptera) and 17 families were recorded. The growth of the olive trees was not affected by the conservative strategies.: In the inter-row, the growth of the spontaneous flora was limited by the high temperatures during the summer. Among the living mulch species, sage and lemongrass guaranteed an almost full soil cover, reducing the need for weed management along the row, as well as increasing the beneficial insects without influencing the young tree growth.

Keywords: *Olea europaea* L.; Mediterranean basin; agroecological practices; minimum tillage; zero tillage; pollinating and predatory insects; agroforestry; intercropping; consociation

Citation: Las Casas, G.; Ciaccia, C.; Iovino, V.; Ferlito, F.; Torrisi, B.; Lodolini, E.M.; Giuffrida, A.; Catania, R.; Nicolosi, E.; Bella, S. Effects of Different Inter-Row Soil Management and Intra-Row Living Mulch on Spontaneous Flora, Beneficial Insects, and Growth of Young Olive Trees in Southern Italy. *Plants* **2022**, *11*, 545. https://doi.org/10.3390/plants11040545

Academic Editors: Alessia Restuccia, Aurelio Scavo and Adriano Sofo

Received: 16 December 2021
Accepted: 15 February 2022
Published: 18 February 2022

1. Introduction

One of the main goals established by the European Commission during the period 2019–2024 is to lay the foundation for making the European Union the first climate-neutral continent by 2050. To achieve this objective, the Commission presented the European Green Deal policy, the most ambitious package of measures that should enable European citizens and businesses to benefit from sustainable green transition. Concerning the agricultural sector, this objective will be reached by a drastic reduction in farm input (fertilizers, chemical pesticide, hormones), reducing the nutrient losses and preserving and restoring ecosystems and biodiversity [1].

Currently, this policy is mandatory considering the ongoing climate change and its impact on agriculture (increase in average temperature and risk of extreme natural events such as floods and droughts) and the land degradation process (erosion, salinity, soil borne diseases) occurring in large areas of the world and the subsequent loss of biodiversity. Moreover, it is important to consider that the world population will increase in the same period (2050) and will reach about 9.1 billion people [2], consequently increasing the food demand [3]. Therefore, in this scenario, agricultural sectors also need to increase the crop efficiency, since the land availability and productivity will play a central role for the maintenance of several rural contexts [4].

The Mediterranean basin is a representative area in which the abovementioned criticisms are well recognized. Among fruit tree crops, the olive (*Olea europaea* L.), one of the most cultivated species that covers about 9.5 million hectares in Europe [5], is an important crop for its social, economic, and ecological role [6].

Regarding the social aspect, it is able to contrast the depopulation of the countryside, as well as maintain the historical aspect of its cultivation [7], providing healthy and safe food for the population. In addition, olive cultivation connects different generations because most of the farmers cultivating traditional olive orchards are aged or retired people, who are still active in agriculture and share their knowledge with younger people in order to maximize the production only using the potential of the agroecosystem [8].

The economic role is well documented; in fact, the olive production has increased in recent years due to the introduction of new planting models [9], mechanization of some cultural practices, harvest above all [10], precision management technologies [11,12], and the use of high-quality standard propagation material [13]. Moreover, at least 95% of the olive cultivation is located in the Mediterranean basin [14], and it represents about 70% of world's olive production [15], from about 1.9 million olive-growing farms.

In terms of the agroecological value, olive plays a fundamental role in maintaining some fragile areas, preventing soil erosion, as well as loss of water and nutrients, and increasing biodiversity. Moreover, thanks to its historical aspect and adaptation, compared to the other woody crops, olive cultivation does not require high external inputs, thus contributing to reducing environmental pollution [16]. On the other hand, tillage (full or partial) is often realized, while minimum and zero tillage is less adopted. Low-intensity tillage leads to an increase in the number of beneficial insects such as pollinators [17] that sustain wild plant communities providing key ecosystem services (e.g., contributing to control pest and crop disease) [18]. As demonstrated by different studies, various anthropogenic factors, such as the expansion of agriculture and livestock, habitat fragmentation, and irrational use of pesticides and pollution, are causing a global decline in insects [19]. In Europe, 9% of butterflies and 9.2% of wild bees are threatened by conventional agriculture [20]. Conservative agriculture (e.g., minimized soil disturbance, permanent soil covering) has been shown to have an impact on biodiversity and ecosystem service provision [21], reducing the negative effects of conventional tillage and enhancing the number of beneficial insects, as well as improving their role in the agroecosystem. Similarly, diversification strategies in space and time by the inclusion of agroecological infrastructure in agricultural landscape such as hedgerows and cover crops (including living mulch) are considered redesign strategies able to magnify the role of agro-biodiversity in ecosystem service provision [22,23]. Moreover, the management strategies can impact the spontaneous flora community (i.e., the weeds), reducing the selection of competitive flora toward a more service provision-oriented community, by supporting pollinators or beneficial attraction [24]. Then, the introduction of herbaceous species (e.g., intercropping, living mulch) that are not directly aimed at production but provide ecological services, called agroecological service crops (ASC) [25], can positively influence the overall ecosystem functioning by providing pest control and ecological services such as weed control in the row [26], protection of the soil from degradation, an increase in organic carbon content, which improves the soil structure and fertility [27], and a decrease in the concentration of CO_2 in the atmosphere if properly managed [28]. Among the conservative soil management strategies, consociations (annual or perennial intercropping), soil management practices (minimum tillage, zero tillage), and organic fertilization were considered for a comprehensive meta-analysis (187 experiments realized in the Mediterranean basin with several woody crops for a total of 46 papers) [29] that highlighted a general positive effect of the abovementioned strategies in carbon sequestration compared to mono-cropping, conventional tillage, and inorganic fertilization. For olive, since the last century, consociations with herbaceous or woody species have been described [30]. These were due to the extensive olive orchards, as well as the consociations with livestock where possible [31]. For other species such as grapevines, minimum or zero tillage is commonly applied in order to regulate the vegetative and reproductive balance of vines and, in some cases, in order to reduce erosion and land degradation [32–34].

In this context, olive could represent an important source of ecological interest among the numerous Mediterranean species due to its specific characteristics, such as high drought resistance, low chill unit requirement, adaptation to hot and dry climatic conditions, and low pest and disease incidence, all of which are significant characteristics to consider in the establishment of new orchards with an agro-ecological approach [35]. However, it is important to consider that the cultivation of olive trees is very diversified among the Mediterranean countries, and that the social, economic, and agroecological value of the olive orchards is strongly variable according to the different cultivation systems (traditional, intensive, and super-intensive orchards), farming techniques, and genetic resources [36]. In traditional orchards, the social and agroecological characteristics are highly relevant, whereas, in the intensive model, only the olive agroecological importance is essential. In these categories, olive models are in accordance with the main objectives of the agroecological approach, which aims to reinforce the natural strength of the agroecosystem without using external inputs and augment the resilience of the crops, encompassing the social, ecological, and economic dimensions of sustainability [37]. In the super-intensive growing system, the economic factor is of greater importance than the social and agroecological factors.

In our research, we tested the impact of some agroecological practices (i.e., conservative soil management and ASC living mulch introduction) on the wild agro-biodiversity (weed and arthropod communities) and vegetative growth of a newly planted olive orchard. We assumed that different floor management (minimum tillage vs. zero tillage) and intra-row management (different living mulch species vs. no living mulch) would differently influence the dynamics of the monitored agro-biodiversity and the young plant response. In particular, we hypothesized that (i) the zero-tillage floor management would guarantee permanent soil cover without selecting higher competitive flora, (ii) the living mulches would positively influence the presence of beneficial insects, and (iii) different living mulches would have a different impact on both arthropods and weed communities, depending on the introduced species.

2. Results

2.1. Entomological Report

The complete list of the 69 recorded species of beneficial insects, as well as their relation to the spontaneous flora or the consociated ones, in the studied olive orchard is reported in Tables 1 and 2. Specimens of pollinators (61 species) and predators (eight species) were collected in the 2 years of field surveys on the wild and cultivated plants. Regarding pollinators, the 33 species of Apoidea reported belong to five different families, Colletidae (one species), Andrenidae (seven species), Halictidae (four species), Megachilidae (five species), and Apidae (16 species), and 15 genera (Table 1). Most of these species nest by digging into the ground (24 species, 72.72%), while 21.21% (seven species) of the taxa nest in pre-existing cavities in the ground, in the walls, or in dry and hollow vegetables. Two species among the 33 observed (6.06%) belong to the *Nomada* Scopoli genus of brood parasitic bees characterized by the presence of females that lay eggs in the nest of other wild bees. Regarding the behavior, 24 species are solitary (72.72%), five species (15.15%) exhibit a pre-social behavior, two species (6.06%) have a social behavior, and two species (6.06%) are brood parasite species.

Table 1. Hymenoptera Apoidea, Lepidoptera, Diptera, Neuroptera, and Coleoptera collected in the years 2020–2021 in the field inter-rows, and in the year 2021 in the consociated rows. * In this species, the larvae are predators.

Order	Family	Species	Wild Plants in the Inter-Rows	Consociated Plants in the Row
		Pollinators		
Hymenoptera	Colletidae	*Hylaeus cornutus* Curtis, 1831	*Foeniculum vulgare*	*Helichrysum italicum*
	Andrenidae	*Andrena aerinifrons* Dours, 1873	*Sinapis arvensis* *Ranunculus muricatus*	
		Andrena bicolorata (Rossi, 1790)	*Sinapis arvensis*	
		Andrena pilipes Fabricius, 1781	*Senecio vulgaris*	
		Andrena brumanensis Friese, 1899	*Ranunculus muricatus*	*Salvia officinalis*
		Andrena distinguenda Schenck, 1871	*Glebionis coronaria*	
		Andrena labialis (Kirby, 1802)	*Ecballium elaterium*	
		Andrena nigroaenea (Kirby, 1802)	*Sinapis arvensis*	
	Halictidae	*Halictus fulvipes* (Klug, 1817)	*Galactites tomentosa*	*Thymus vulgaris*
		Halictus quadricinctus (Fabricius, 1776)	*Senecio vulgaris*	
		Halictus scabiosae (Rossi, 1790)	*Senecio vulgaris* *Dittrichia viscosa*	*Thymus vulgaris*
		Lasioglossum malachurum (Kirby, 1802)	*Ecballium elaterium*	*Helichrysum italicum*
	Megachilidae	*Heriades rubicola* Pérez, 1890	*Dittrichia viscosa*	*Helichrysum italicum*
		Osmia latreillei (Spinola, 1806)	*Glebionis coronaria*	*Salvia officinalis*
		Osmia signata Erichson, 1835	*Glebionis coronaria*	
		Rhodanthidium siculum (Spinola, 1838)	*Oxalis pes-caprae*	*Salvia officinalis*
		Megachile sicula (Rossi, 1792)	*Galactites tomentosa*	
	Apidae	*Xylocopa violacea* (Linnaeus, 1758)	-	*Salvia officinalis* *Thymus vulgaris*
		Ceratina cyanea Kirby, 1802	*Ecballium elaterium*	*Helichrysum italicum*
		Nomada discrepans Schmiedeknecht, 1882	*Sinapis arvensis*	
		Nomada distinguenda Morawitz, 1874	*Raphanus raphanistrum*	
		Eucera algira Brullé, 1840	*Raphanus raphanistrum*	
		Eucera eucnemidea Dours, 1873	*Galactites tomentosa*	
		Eucera nigrescens Pérez, 1879	*Glebionis coronaria*; *Vicia* sp.	
		Eucera nigrilabris Lepeletier, 1841	*Raphanus raphanistrum*	
		Eucera numida Lepeletier, 1841	*Vicia sativa*	
		Eucera oraniensis Lepeletier, 1841	*Glebionis coronaria* *Galactites tomentosa*	*Salvia officinalis*
		Amegilla garrula (Rossi, 1790)	-	*Salvia officinalis* *Thymus vulgaris*

Table 1. *Cont.*

Order	Family	Species	Wild Plants in the Inter-Rows	Consociated Plants in the Row
		Amegilla quadrifasciata (de Villers, 1789)	-	*Salvia officinalis* *Thymus vulgaris*
		Anthophora dispar Lepeletier, 1841	*Fumaria officinalis*	*Salvia officinalis*
		Anthophora plumipes squalens Dours, 1869	*Fumaria officinalis* *Papaver rhoeas*	
		Bombus pascuorum siciliensis Tkalcu, 1977	*Vicia sativa*	*Salvia officinalis* *Thymus vulgaris*
		Bombus terrestris (Linnaeus, 1758)	*Vicia sativa*	*Salvia officinalisThymus vulgaris*
Lepidoptera	Sphingidae	*Macroglossum stellatarum* (Linnaeus, 1758)	*Convolvulus arvensis*	*Thymus vulgaris*
		Hyles euphorbiae (Linnaeus, 1758)	-	
		Hyles livornica (Esper, 1780)	*Convolvulus arvensis*	*Helichrysum italicum*
	Sesiidae	*Tinthia tineiformis* (Esper, 1789)	*Convolvulus arvensis*	
	Geometridae	*Rhodometra sacraria* (Linnaeus, 1767)	-	*Thymus vulgaris*
		Menophra abruptaria (Thunberg, 1792)	*Dittrichia viscosa*	*Thymus vulgaris*
	Noctuidae	*Heliothis peltigera* (Denis & Schiffermüller, 1775)	*Senecio vulgaris*	*Helichrysum italicum* *Thymus vulgaris*
		Autographa gamma (Linnaeus, 1758)		*Salvia officinalis*
	Hesperiidae	*Carcharodus alceae* (Esper, 1780)	*Lysimachia arvensis*	*Thymus vulgaris*
	Lycaenidae	*Lycaena alciphron* (Rottemburg, 1775)	*Althaea officinalis*	*Thymus vulgaris*
		Lycaena phlaeas (Linnaeus, 1761)	*Polygonum aviculare* *Portulaca oleracea* *Ranunculus muricatus*	*Thymus vulgaris*
	Nymphalidae	*Aglais urticae* (Linnaeus, 1758)	*Althaea officinalis*	*Helichrysum italicum*
		Vanessa atalanta (Linnaeus, 1758)	*Althaea officinalis* *Convolvulus arvensis* *Ecballium elaterium*	*Salvia officinalis* *Thymus vulgaris*
		Vanessa cardui (Linnaeus, 1758)	*Ranunculus muricatus*	*Thymus vulgaris*
		Lasiommata megera (Linnaeus, 1767)	*Polygonum aviculare*	*Salvia officinalis* *Thymus vulgaris*
		Pararge aegeria (Linnaeus, 1758)	-	*Thymus vulgaris*
	Papilionidae	*Iphiclides podalirius* (Linnaeus, 1758)	-	*Helichrysum italicum*
		Papilio machaon Linnaeus, 1758	*Dittrichia viscosa*	
	Pieridae	*Colias croceus* (Geoffroy, 1785)	*Ecballium elaterium*	*Thymus vulgaris*
		Gonepteryx cleopatra (Linnaeus, 1767)	-	*Thymus vulgaris*
			Capsella bursa-pastoris *Raphanus raphanistrum* *Sinapis arvensis*	

Table 1. *Cont.*

Order	Family	Species	Wild Plants in the Inter-Rows	Consociated Plants in the Row
		Pieris mannii (Mayer, 1851)	*Beta vulgaris*	
		Pieris rapae (Linnaeus, 1758)	*Portulaca oleracea* *Sinapis arvensis*	*Thymus vulgaris*
Diptera	Syrphidae	*Episyrphus balteatus* (DeGeer, 1776) *	-	*Thymus vulgaris*
		Eupeodes luniger (Meigen, 1822) *	*Ranunculus muricatus*	
		Eristalinus taeniops (Wiedemann, 1818)	*Beta vulgaris* *Polygonum aviculare*	*Thymus vulgaris*
		Eristalis tenax (Linnaeus, 1758)	-	*Helichrysum italicum* *Thymus vulgaris*
		Syritta pipiens (Linnaeus, 1758)	*Portulaca oleracea*	
Predators				
Neuroptera	Chrysopidae	*Chrysopa viridana* Schneider, 1845	-	*Thymus vulgaris*
		Chrysoperla carnea (Stephens, 1836)	*Ranunculus muricatus* *Beta vulgaris*	*Salvia officinalis* *Helichrysum italicum*
Coleoptera	Coccinellidae	*Chilocorus bipustulatus* (Linnaeus, 1758)	*Ecballium elaterium*	*Thymus vulgaris*
		Coccinella septempunctata Linnaeus, 1758	*Amaranthus retroflexus* *Diplotaxis erucoides*	*Helichrysum italicum* *Salvia officinalis*
		Hippodamia variegata (Goeze, 1777)	*Cerinthe major*	-
		Propylea quatuordecimpunctata (Linnaeus, 1758)	*Althaea officinalis*	-
		Scymnus interruptus (Goeze, 1777)	*Salvia officinalis*	*Senecio vulgaris*
		Scymnus subvillosus (Goeze, 1777)	*Diplotaxis erucoides*	

Table 2. Seasonal presence of Apoidea species in the experimental farm of Palazzelli during the years 2020–2021.

	Hymenoptera	January	February	March	April	May	June	July	August	September	October	November	December	Years 2020	Years 2021
	Colletidae														
1	*Hylaeus cornutus*							√						X	
	Andrenidae														
2	*Andrena aerinifrons*			√											X
3	*Andrena bicolorata*				√									X	
4	*Andrena brumanensis*					√									X
5	*Andrena distinguenda*		√	√										X	X
6	*Andrena labialis*					√		√						X	X
7	*Andrena nigroaenea*		√	√										X	X
8	*Andrena pilipes*				√	√								X	X
	Halictidae														
9	*Halictus fulvipes*					√	√	√		√				X	X
10	*Halictus quadricinctus*						√	√	√					X	X
11	*Halictus scabiosae*					√	√	√	√					X	X
12	*Lasioglossum malachurum*			√	√										X
	Megachilidae														
13	*Heriades rubicola*							√	√	√	√			X	X
14	*Osmia latreillei*				√	√								X	X
15	*Osmia signata*						√							X	
16	*Rhodanthidium siculum*			√	√									X	X
17	*Megachile sicula*		√	√										X	X
	Apidae														
18	*Xylocopa violacea*		√	√	√	√	√			√	√			X	X
19	*Ceratina cyanea*						√	√						X	X
20	*Nomada discrepans*								√					X	
21	*Nomada distinguenda*								√						X
22	*Eucera algira*		√	√											X
23	*Eucera eucnemidea*				√	√	√							X	X
24	*Eucera nigrescens*		√	√	√									X	X
25	*Eucera nigrilabris*				√										X
26	*Eucera numida*				√	√								X	X
27	*Eucera oraniensis*		√	√	√									X	X
28	*Amegilla garrula*					√	√							X	X
29	*Amegilla quadrifasciata*						√	√	√					X	X
30	*Anthophora dispar*	√	√	√	√									X	X
31	*Anthophora plumipes squalens*		√	√	√	√								X	X
32	*Bombus pascuorum siciliensis*						√	√	√	√	√			X	X
33	*Bombus terrestris*	√	√	√	√	√	√			√	√			X	X

The 23 species of Lepidoptera reported belong to nine different families, Sphingidae (three species), Sesiidae (one species), Geometridae (two species), Noctuidae (two species) Hesperiidae (one species), Lycaenidae (two species), Nymphalidae (five species), Papilionidae (two species), and Pieridae (five species), and 17 genera (Table 1).

Five species (and five genera) of Diptera were found belonging to the Syrphidae family. The adults of these species are pollinators of spontaneous plants; however, the larvae have different trophic regimes. For example, larvae of *Episyrphus balteatus* (DeGeer), and *Eupeodes luniger* (Meigen) are predators of aphids, while those of *Eristalinus taeniops* (Wiedemann), *Eristalis tenax* (L.), and *Syritta pipiens* (L.) are scavengers [38].

Furthermore, regarding predator insects, two species of Neuroptera Chrysopidae and six species of Coleoptera Coccinellidae were found; among these, one species feeds mainly on coccids, while the others feed mainly on aphids.

2.2. Spontaneous Flora Distribution and Diversity

The complete list of the spontaneous flora species found in the field, as well as the time (spring or autumn) and the area in which they were recorded (inter-row or intra-row), is reported in Table 3. A total of 40 species of plants are listed. Among these, five species, *Amaranthus retroflexus* L. (AMARE), *Cynodon dactylon* (L.) Pers. (CYNDA), *Cyperus rotundus* L. (CYPRO), *Polygonum aviculare* L. (POLAV), and *Portulaca oleracea* L. (POROL), were found in each period and position. In spring, 28 species of plants were detected: 14 of them both in the intra-row and inter-row, and the remaining 14 exclusively in the intra-row. On the contrary, no exclusive species in the inter-row were observed. In autumn, 26 species were observed, and only eight grew both in inter-row and intra-row. In this period, six species were exclusive in the inter-row while 11 species were found only along the row.

Regarding the weed monitoring achieved in spring in the inter-row, in MT treatments, most of the space (70%) was classified as bare soil, while the predominant spontaneous plants were *Portulaca oleracea* (POROL) (11%) and *Convolvulus arvensis* L. (CONAR) (7%), even though the quantity was lower compared to the ZT treatment. In these areas, the bare soil was in less quantity (18%), and the predominant spontaneous plant was *Papaver rhoeas* L. (PAPRH) (almost 40% of the total space was occupied from this plant), followed by *Beta vulgaris* L. (BEAVX) (almost 25%).

In terms of the distribution of the weed community during spring in the intra-rows, in the MT treatment, the prevalent species found were *Portulaca oleracea* (POROL) (13%) and *Cynodon dactylon* (CYNDA) (10%), while the remaining weeds showed a distribution more or less constant along the intra-rows. Regarding the frequency and distribution of the spontaneous flora community in the intra-rows in ZT treatments, *Papaver rhoeas* (PAPRH) was present in a larger proportion (27%) compared to the others, followed by *Beta vulgaris* (BEAVX) (10%). The presence of other weed species was similar to that observed in the tillage blocks even if, in the control, *Papaver rhoeas* (PAPRH) covered about 60% of the soil.

In the autumn survey, it was observed how the vegetation developed almost exclusively along the rows due to the presence of irrigation, whereas, in the inter-row, a high percentage of bare soil (MT 96%; ZT 77%) was registered. Along the row, there was a significant increase in the space occupied by ASC species, particularly sage and lemongrass, and, for both MT and ZT, the most represented spontaneous species was *Setaria verticillata* (L.) P. Beauv. (SETVE).

Table 3. List of the spontaneous flora species detected in spring and in autumn in both the inter-row and the intra-row of the experimental field 'long-term trial on organic olive (BiOlea)' at Palazzelli.

Spontaneous Flora Species	Family	EPPO Code	Spring			Autumn		
			Inter-Row			Inter-Row		
			Zero Tillage	Minimum Tillage	Intra-Row	Zero Tillage	Minimum Tillage	Intra-Row
Amaranthus retroflexus L.	Amaranthaceae	AMARE	+	+	+	+	+	+
Arum maculatum L.	Araceae	ABGMA	-	-	+	-	-	-
Avena sterilis L.	Poaceae	AVEST	+	-	+	-	-	-
Beta vulgaris L.	Chenopodiaceae	BEAVX	+	+	+	+	+	+
Brassica nigra (L.) W.D.J. Koch	Brassicaceae	BRSNI	-	-	+	+	+	-
Capsella bursa-pastoris (L.) Medik.	Brassicaceae	CAPBP	-	-	+	+	+	-
Convolvolus arvensis L.	Convolvulaceae	CONAR	-	+	+	+	+	-
Cynodon dactylon (L.) Pers.	Poaceae	CYNDA	+	+	+	+	+	+
Cyperus rotundus L.	Cyperaceae	CYPRO	+	+	+	+	+	+
Dactylis glomerata L.	Poaceae	DACGL	+	-	+	-	-	-
Digitaria sanguinalis (L.) Scop.	Poaceae	DIGSA	-	-	+	-	-	+
Dittrichia viscosa (L.) Greuter	Asteraceae	INUVI	-	-	+	-	-	-
Ecballium elaterium (L.) A. Rich.	Cucurbitaceae	ECBEL	-	-	+	-	-	+
Elymus repens (L.) Gould	Poaceae	AGGRE	-	-	-	+	+	-
Erigeron canadensis L.	Asteraceae	ERICA	-	-	-	-	-	+
Euphorbia prostrata Aiton	Euphorbiaceae	EPHPT	-	-	-	-	-	+
Fumaria officinalis L.	Papaveraceae	FUMOF	+	+	+	+	-	-
Lactuca sativa subsp. *serriola* (L.) Galasso, Banfi, Bartolucci & Ardenghi	Asteraceae	LACSE	+	+	+	-	-	-
Lamium amplexicaule L.	Lamiaceae	LAMAM	-	-	-	-	-	+
Lolium perenne L.	Poaceae	LOLPE	+	+	+	-	-	-
Lysimachia arvensis (L.) U. Manns & Anderb.	Primulaceae	LYSAR	-	-	+	-	-	-
Malva sylvestris L.	Malvaceae	MALSY	-	-	+	+	+	-
Myosotis arvensis (L.) Hill	Boraginaceae	MYOAR	-	-	+	-	-	-
Oxalis pes-caprae L.	Oxalidaceae	OXAPC	-	-	-	-	-	+
Papaver rhoeas L.	Papaveraceae	PAPRH	+	+	+	-	-	-
Polygonum aviculare L.	Polygonaceae	POLAV	+	+	+	+	+	+
Portulaca oleracea L.	Portulacaceae	POROL	+	+	+	+	+	+
Ranunculus muricatus L.	Ranunculaceae	RANMU	-	-	+	-	-	-
Raphanus raphanistrum L.	Brassicaceae	RAPRA	-	-	+	+	+	+
Senecio vulgaris L.	Asteraceae	SENVU	-	-	+	-	-	+
Setaria verticillata (L.) P. Beauv.	Poaceae	SETVE	-	-	-	+	+	+
Sinapis arvensis L.	Brassicaceae	SINAR	-	-	+	-	-	-
Solanum nigrum L.	Solanaceae	SOLNI	-	-	-	-	-	+
Sonchus asper subsp. *glaucescens* (Jord.) Ball	Asteraceae	SONAR	-	-	-	-	-	+
Sonchus oleraceus L.	Asteraceae	SONOL	-	-	-	-	-	+
Stellaria media (L.) Vill.	Caryophyllaceae	STEME	-	-	-	-	-	+
Triticum spp.	Poaceae	-	-	-	+	-	-	-
Urtica dioica L.	Urticaceae	URTDI	-	+	+	-	-	-
Veronica peregrina L.	Plantaginaceae	VERPG	-	-	-	-	-	+
Total richness (No. species)			12	12	28	14	13	20

A principal component analysis (PCA) was carried out to evaluate the effect of the ASC on the development, quantity, and distribution of the weed community. With respect to soil management data analysis, Component 1 explained 20.97% of the total variability, while Component 2 explained 15.71% (Table 4). According to the PCA results relative to the spring and autumn analysis, as shown in Figure 1A,B for the spring stage, there were no significant differences in terms of distribution between the plots analyzed. Regarding the distribution of the spontaneous flora community in the inter-row with different soil management (ZT and TI) (Figure 1A), weed species appeared divided into four main groups (Figure 1A) characterizing the community: perennial species (namely CONAR, CYNDA, and CYPRO), AMARE, POROL, and DACGL (group 1) were negatively correlated to POLAV, URTDI, BETVU, FUMOF, and LACSE (group 2), and PAPRH (group 3), whereas two completely independent grass species appeared, AVEST and LOLPE (group 4). Despite this, PCA did not show clear differences in terms of abundance and distribution. On the other hand, the zero-tillage community was characterized by the presence of AVEST and LOLPE, whereas BETVU (BEAVX) and URTDI showed a higher relationship with minimum tillage (Figure 2A,C). At this stage (spring), the intra-rows with sage, lemongrass curry plant, and thyme living mulch and control all presented a weed community where all the specimens had an average distribution, with some peak presence of AMARE in sage mulch rows and of SETVE in the control row (Figure 2B,D). These records are an overview of 1 year of the field trial and still need to be re-evaluated in the long term management of the orchard. Similar results were obtained for the second assessment in autumn (not shown).

Table 4. (**A**,**B**) Principal component analysis (PCA) eigenvalues and percentage variance of the studied samples from experimental trial in relation to the inter-row (**A**) and intra-row (**B**) management in spring.

A			B		
PC	**Eigenvalue**	**% Variance**	**PC**	**Eigenvalue**	**% Variance**
1	3.36	20.97	1	3.02	11.17
2	2.51	15.71	2	2.38	8.80
3	2.41	15.08	3	2.15	7.95
4	1.78	11.13	4	1.87	6.93
5	1.30	8.15	5	1.53	5.67
6	1.25	7.80	6	1.49	5.50
7	1.19	7.41	7	1.39	5.17
8	0.68	4.24	8	1.26	4.67
9	0.55	3.42	9	1.13	4.17
10	0.44	2.77	10	1.12	4.14
11	0.26	1.60	11	1.04	3.86
12	0.20	1.23	12	1.01	3.76
13	0.05	0.29	13	0.94	3.49
14	0.02	0.13	14	0.89	3.30
15	0.01	0.07	15	0.85	3.16
			16	0.77	2.86
			17	0.71	2.62
			18	0.65	2.39
			19	0.58	2.14
			20	0.52	1.93
			21	0.38	1.42
			22	0.35	1.30
			23	0.26	0.97
			24	0.23	0.84
			25	0.19	0.69
			26	0.16	0.58
			27	0.13	0.50

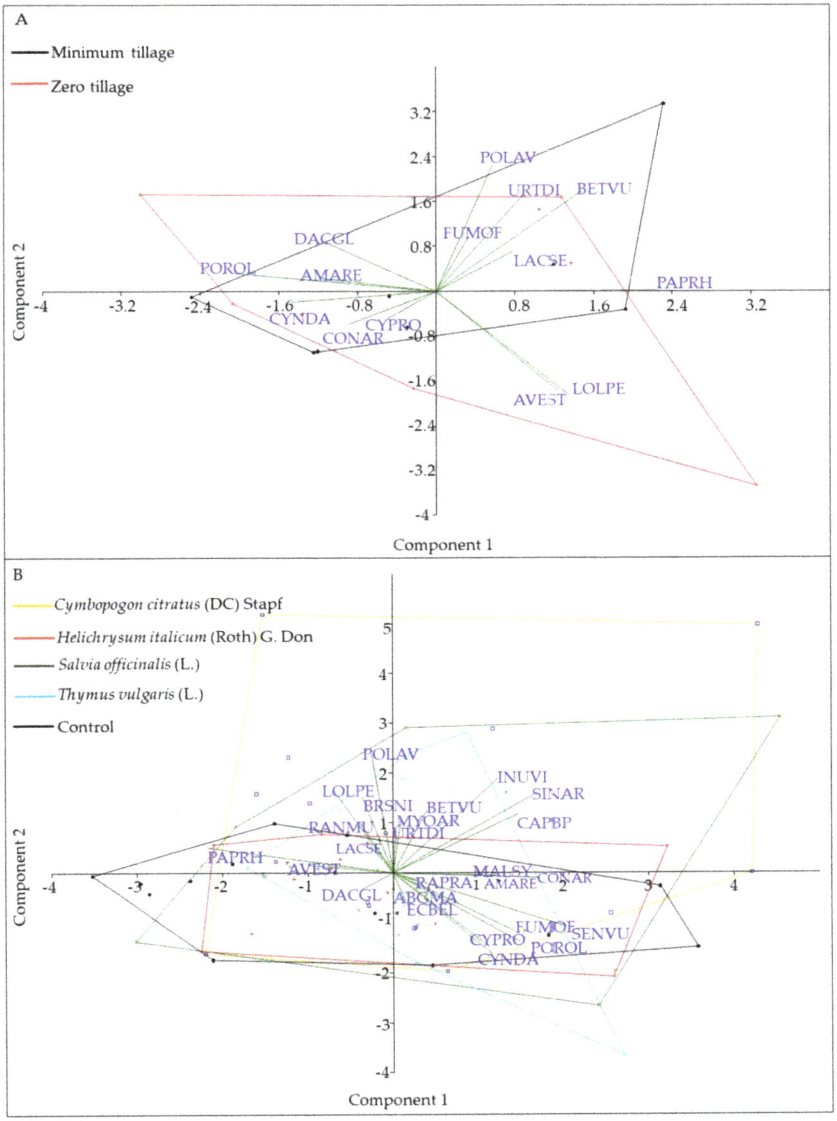

Figure 1. (**A**,**B**) Principal component analysis (PCA) ordination diagram (biplot) depicting the localization of the studied samples from the experimental trial in relation to the inter-row (**A**) and intra-row (**B**) management.

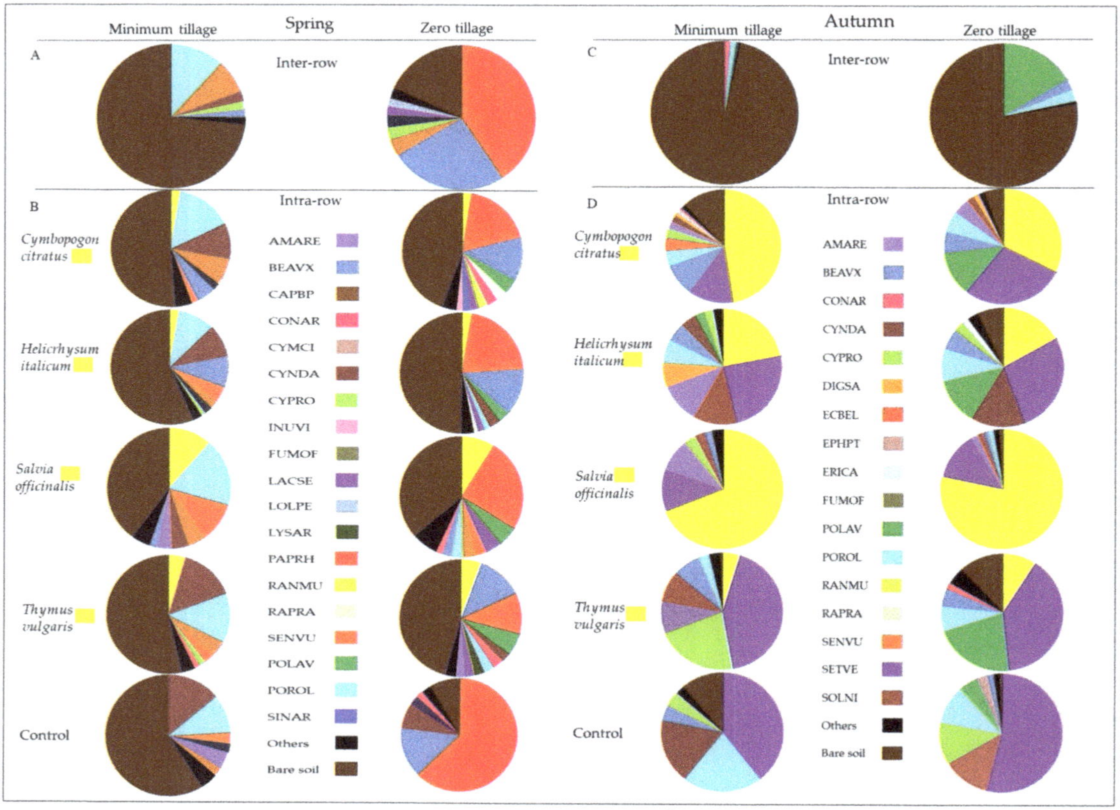

Figure 2. Spontaneous flora species covering percentage in spring (**A**) and autumn (**C**) over the inter-row and along the intra-row (**B**,**D**) of the experimental field 'long-term trial on organic olive (BiOlea)'.

2.3. Plant Growth Analysis

In terms of the produced biomass removed with winter pruning (in February), the most abundant quantity was recorded for the NE cultivar in both soil treatments. In September, the quantity of emitted material (suckers and shoots removed from the trunk) was the highest in NE-MT (Figure 3). Concerning the shoot growth monitoring, despite the absence of significant differences among treatments, a better performance for NE in both soil treatments was observed. In general, the growth rate was about 10–12 cm between day of the year (DOY) 145 and 180, about 8–10 cm between DOY 180 and 210, 2–3 cm between DOY 210 and 239, and 2–3 cm between DOY 239 and 272 (Figure 4). The plant growth response to the applied soil management is reported in Table 5. The trunk cross-sectional area reached the highest growth for both cultivars in the zero-tillage soil management. The canopy height increase (approximately 30%) was similar among treatments, although NE-ZT showed the highest growth. The trunk cross-sectional area (TCSA) showed more variable results, with NE-ZT and NB-ZT showing the highest growth (+105% and 96%, respectively), while NB-TI showed an expansion of about 48% and NE-TI of just 17%. According to the data presented in Figure 4, all variables had the same rate of growth, with an increase of about 10–12 cm between DOY 145 and 180, 8–10 cm between DOY 180 and 210, 2–3 cm between DOY 210 and 239, and 2–3 cm between DOY 239 and 272. This trend is in accordance with the normal development of the olive trees during their young phase, as well as with the climatic data and water intake registered during the trial.

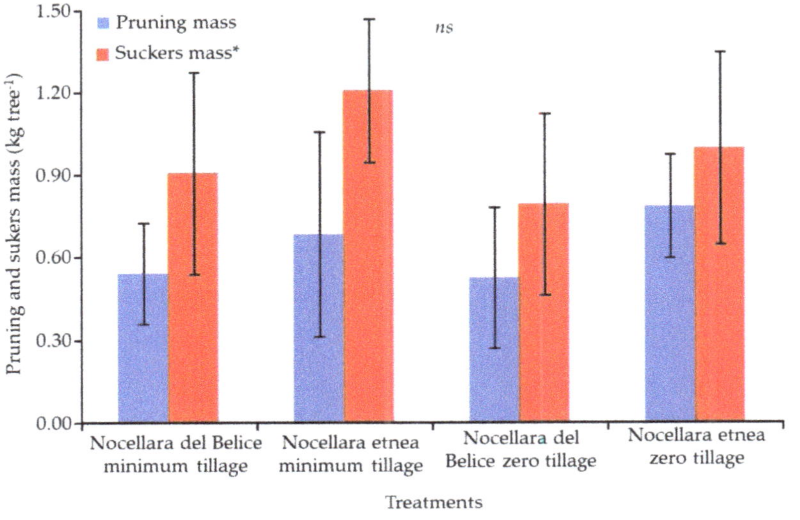

Figure 3. Influence of soil management strategy on winter pruning and sucker mass produced (ns = not significant within each parameter; bars indicate standard deviation) according to Tukey's HSD test, for each treatment and parameter. * Comprehensive record of the shoots weight grown from the ground level to the branch insertion.

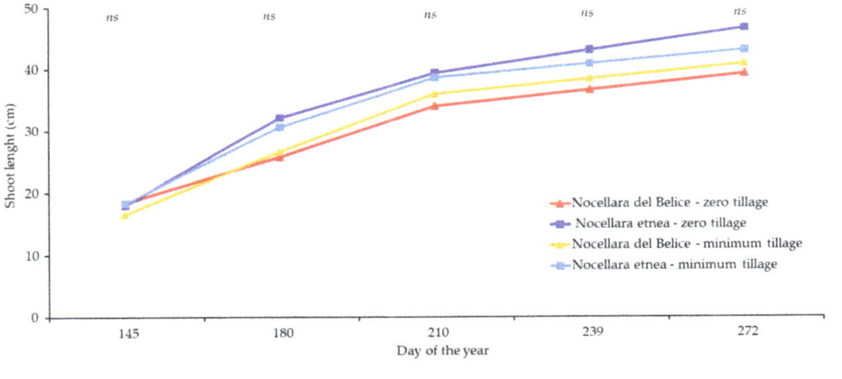

Figure 4. Influence of soil management strategy on mixed shoot growth (ns = not significant) according to Tukey's HSD test, for each treatment and parameter.

Table 5. Influence of soil management strategy on olive tree growth in pre-growing season on 15 December 2020 as compared with plant growth in autumn on 15 October 2021 and percentage increase. Means indicated by different letters are significantly different (lowercase $p \leq 0.05$, ±standard deviation) according to Tukey's HSD test, for each treatment and parameter. ^{ns} = not significant.

	15 December 2020			15 October 2021			Percentage Increase (Δ%)		
Treatment	Trunk Cross-Sectional Area (cm^2)	Canopy Height (cm)	Canopy Volume (m^3)	Trunk Cross-Sectional Area (cm^2)	Canopy Height (cm)	Canopy Volume (m^3)	Trunk Cross-Sectional Area (Δ%)	Canopy Height (Δ%)	Canopy Volume (Δ%)
Nocellara etnea—minimum tillage	6.32 ± 2.4 ^{ab}	103.9 ± 22.06 ^{a}	0.29 ± 0.11 ^{ns}	13.7 ± 2.69 ^{b}	152.6 ± 23.43 ^{ns}	1.55 ± 0.38 ^{a}	117	146	542
Nocellara del Belice—minimum tillage	4.99 ± 2.09 ^{b}	72.5 ± 18.65 ^{b}	0.21 ± 0.09 ^{ns}	12.4 ± 2.74 ^{b}	105.8 ± 22.31 ^{ns}	0.73 ± 0.19 ^{b}	148	145	339
Nocellara etnea—zero tillage	8.87 ± 2.03 ^{a}	82.6 ± 31.68 ^{ab}	0.32 ± 0.12 ^{ns}	18.2 ± 2.23 ^{a}	143.1 ± 29.29 ^{ns}	1.23 ± 0.22 ^{a}	205	173	387
Nocellara del Belice—zero tillage	4.13 ± 2.09 ^{ab}	82.13 ± 21.65 ^{ab}	0.34 ± 0.10 ^{ns}	8.1 ± 3.87 ^{b}	120.5 ± 28.20 ^{ns}	1.05 ± 0.19 ^{b}	196	146	438

3. Discussion

This study focused on three key indicators in agro-ecosystems: (1) the insect community, (2) the spontaneous flora diversity, and (3) the young olive response in terms of vegetative growth. Therefore, in our study, the entire soil–plant–atmosphere *continuum* (SPAC) was analyzed.

The entomological study was performed in terms of both pollinators and natural enemies. The research was conducted in an olive orchard located on a farm in a district with high relevance for citrus and other fruit crops. The collected Apoidea were observed on 23 species of wild plants, comprising a total of 23 plant genera within 16 plant families (Tables 1 and 2). The Asteraceae family was that frequented by the greatest number of pollinators (15 species), followed by Brassicaceae (12 spp.) and Ranunculaceae (five spp.) (Table 3). On the consociated plants, 39 species of pollinators were observed, 25 on *Thymus vulgaris*, 12 on *Salvia officinalis* (Lamiaceae), and nine on *Helichrysum italicum* (Asteraceae).

Currently, 686 species of bees are known in Sicily [39]. The comparison of bee fauna in the Palazzelli agro-ecosystem evidenced a total of 33 species (4.8% of the species known for the Sicilian fauna) belonging to Colletidae (one species) Andrenidae (seven spp.), Halictidae (four spp.), Megachilidae (five spp.), and Apidae (16 spp.) families.

The order Lepidoptera, the second most important group, was present with 23 species, comprising 16 butterflies and eight moths.

In terms of wild bees, it is significant to note that 72.72% (24 species) of the overall species nest in the ground, and their existence depends on the typology of soil management. In recent years, various regional surveys have focused on the biodiversity of these populations and the agroecological role of these two groups of insects [40–42] or as specific pollinators of crops [43–46].

In order to maintain Apoidea biodiversity, management practices should take into account that most species of wild bees nest in the ground [47], and different agronomic practices, including tillage of the land, usually render crops an unsuitable habitat for wild bees, especially in intensive management [48]. In particular, deep tillage and total removal of spontaneous vegetation represent a serious problem for the foraging and nesting of these pollinators [49]. Therefore, in agricultural environments, wild bees need semi-natural habitats for nesting, obtaining the floral resources, and overwintering. The elements of the landscape, in the field and around the field, also have the function of habitat for fauna in general and, in this context, of ecological corridors in intensely cultivated and biodiversity conservation areas [50,51]. It is also necessary to consider how useful effects are particularly important in Mediterranean agro-ecosystems subject to desertification [52–56].

The consociated plants in the intra-row were visited by 62.3% (43 species) of collected insects, 62.2% of all pollinators and 62.5% of all predators. Overall, 15.9% (11 species) of all reported insects were found only on consociated plants, 16.3% of pollinators and 12% of predators.

In our trial, conservative models were also proposed to increase soil fertility and biodiversity (insects and spontaneous flora in the inter-row), reducing the costs for soil management and improving the spontaneous flora control along the row. Our findings evidence small differences between the two soil management strategies. In particular, minimum tillage showed a higher reduction in weed presence at both sampling times (spring and autumn) as confirmed by the higher bare soil cover than in the zero-tillage system (Figure 3). This result evidence how single tillage is an efficient weed management strategy. On the other hand, ZT showed a higher weed cover than MT and a higher richness (data not shown). Nevertheless, ZT in spring showed the selection of perennial species (namely, CONAR, BEAVX, CYPRO, and LOLPE; Figure 2A,C) and a higher characterization of some grass-like species (AVEST and LOLPE; Figure 2B,D). This result is in line with previous findings on zero tillage as a filter to shift the community toward grassy annual and perennial species [57,58], representing a risk in terms of competition with young orchards.

The living mulches realized along the row showed different effects according to the adopted species. In spring, only sage covered the main portion of the soil, due to its habitus. In autumn, 6 months after planting, the sage showed a complete hedgerow, and the consociated flora was observed just at the ground level under the plants. Similarly, lemongrass, despite forming an almost dense hedgerow, completely prevented weed growth under the plants thanks to its strong tillering ability, while allowing growth between plants. Therefore, these species contributed to creating a wide soil cover before the winter season and improved the soil performance [59]. Thyme and curry plant recorded the lowest growth and showed reduced power for competition with the spontaneous flora. However, in these cases, the spontaneous flora had a role in the preservation of the essences during summer since they covered the little plants and permitted them to survive during this season. Perhaps, for these essences, two growing seasons are required to reach a complete hedgerow. Therefore, in the inter-row, lemongrass and sage reduced the need for further soil management. The adopted living mulches reduced the propagation of weeds without reducing the vigor and growth of olive trees. It is possible to assume that the distance from the trunk of the young olive trees to the plants of living mulch was about 40 cm, and it did not significantly affect the olive growth. It is important to highlight that the irrigation lines played a strong role for both the olives trees and the consociated species. Since the olive trees were young, full irrigation was useful to reach high growth rates as shown by the increase registered in morphological parameters (Figures 3 and 4, and Table 5). Among these, the canopy volume exhibited strong growth. According to our findings, it is possible to hypothesize two drip lines for differentiated irrigation between olive trees and living mulch species. From a practical point of view, in areas with hot and dry summers, planting in the field is possible in autumn or in spring. One plant every 50 cm is enough to boost the growth of the living mulch along the row, but it is important to consider that, after 6 months, the removal of the lines from the row is very difficult; therefore, positioning above the ground level is preferred.

In general, the obtained hedgerows could represent an integrative crop for a secondary income for the farmer, such as food, feed, or industrial products, increasing the resilience of the system to pest incidence and market volatility [60].

4. Materials and Methods

4.1. Site Description, Experimental Design, and Treatments

The study was carried out between June 2019 and October 2021, in the 'long-term trial on organic olive (BiOlea)', of the experimental farm of the Council for Agricultural Research and Economics (CREA), Research Center for Olive, Tree Fruit, and Citrus located at Palazzelli (Lentini district, Syracuse), Sicily, Italy, (latitude 37.17″ N, longitude 14.50″

E, elevation 45 m a.s.l.). The experiment focused on a young olive orchard, planted with two Sicilian main double aptitude olive cultivars 'Nocellara del Belice' (NB) and 'Nocellara etnea' (NE), grafted onto seedling rootstocks. Trees were planted in May 2019, in north–south-oriented rows, at a spacing of 6 m between rows and 5 m within the row. The adopted training system, since the first winter pruning season (February 2020), was the polyconic vase, aiming to maintain three main branches. Trees were drip-irrigated early in the morning three times per week, from June to September. Irrigation volume scheduling was based on the FAO-56 Penman–Monteith (P–M) approach [61,62], adjusted by the variable crop coefficient (kc) from 0.15 in the first growing season to 0.34 in the second one [63]. Each of the four drippers per tree emitted 2 $L{\cdot}h^{-1}$, for a total of 8 $L{\cdot}h^{-1}$, with an operational pressure of 1 bar. Plants were fully irrigated, corresponding to 95–98% of crop evapotranspiration, ET_c. The electrical conductivity of the water (at 25 °C) was 2.02 $dS{\cdot}m^{-1}$ and the pH was 7.30. Only organic fertilization was applied at the plantation.

The trial was designed as a split-plot system with four blocks of 10 rows with five plants each (Figure 5). The main plot was assigned to soil management practice comparing two systems: (1) minimum tillage (MT) consisting of one tillage (15 cm depth) performed at the end of the winter (first week of March) and (2) zero tillage (ZT) consisting of soil managed only through mechanical shredding, performed twice per year: at the end of the winter, in the same period of MT (first week of March), and at the beginning of summer (four week of June). The sub-plot was assigned to the variety alternating a row with NB and a row with NE, so that compared treatments were (1) Nocellara del Belice—minimum tillage (NB-MT), (2) Nocellara del Belice—zero tillage (NB-ZT), (3) Nocellara etnea—minimum tillage (NE-MT), and (4) Nocellara etnea—zero tillage (NE-ZT).

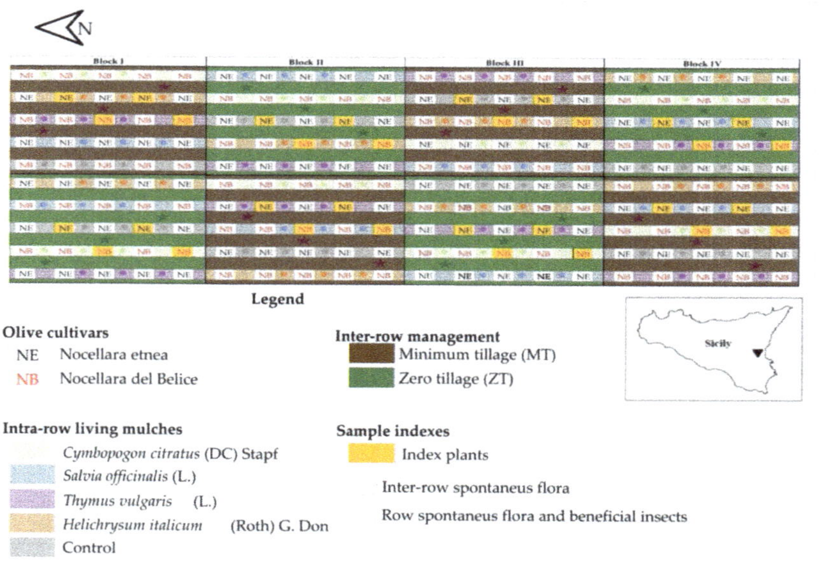

Figure 5. 'Long-term trial on organic olive (BiOlea)' experimental field design within the experimental farm of the CREA, Research Center for Olive, Tree Fruit, and Citrus located at Palazzelli, Sicily, Italy (latitude 37.17″ N, longitude 14.50″ E, elevation 45 m a.s.l.), with indications of the index plants and the samples points.

For the specific activity of this study, on 15 March 2021, a living mulch system was set down along the row using four officinal species as agro-ecological service crops (ASCs) planted at a distance of 0.5 m: (1) sage (*Salvia officinalis* L.), (2) thyme (*Thymus vulgaris* L.),

(3) curry plant (*Helichrysum italicum* (Roth) G. Don), and (4) lemongrass (*Cymbopogon citratus* (DC) Stapf). No living mulch between trees along the row was used as control (C), but the spontaneous flora was maintained. Inter-row soil management was used as a factor for field spontaneous flora assessment and for plant growth monitoring in both cultivars. The soil management and the living mulch interactions along the row were used both for the spontaneous flora and for the entomological assessments.

4.2. Soil Analysis and Climatic Data

At planting, soil characteristics were analyzed at 20–40 cm depth by three samplings per plot. Soil physical and chemical characteristics are reported in Table 6. Regarding physical characteristics, the quantity and distribution of sand, clay, and silt was obtained by particle-size analysis using the "micro-pipette" method [64]. In terms of chemical properties, total nitrogen (N), organic matter (OM), soil extractable phosphorus (mg/kg), soil exchangeable potassium (meq/100 g), cation exchange capacity, pH, and electrical conductivity (EC) determinations were determined as described in [65–71]. Total nitrogen was measured by Kjeldahl digestion using a Buchi Labortechnik GmbH N analyzer, and organic matter (OM) was measured by quantifying total organic carbon (TOC, mg·kg^{-1}). TOC was analyzed by means of elemental analyzer LECO (RC-612; St. Joseph, MI, USA) using a dry combustion method. Soil exchangeable potassium (meq/100 g) was determined in a solution of barium chloride and triethanolamine at pH 8.2 (2 g of soil: 25 mL). Cationic exchange capacity was analyzed by the $BaCl_2$ compulsive exchange method. The pH and EC determinations were carried out on a HI 9813 portable EC meter (Hanna Instruments, Woonsocket, RI, USA) and an AB 15 pH meter (Thermo Fisher Scientific, Waltham, MA, USA), respectively. Inductively coupled plasma optical emission spectrometry, ICP-OES, was conducted using an Optima 2000 DV, PerkinElmer Inc. Shelton, CT, USA). According to the United States Department of Agriculture (USDA) scheme, the olive-grove soil is classified as loamy sand [72]. The soil pH is subalkaline, and electrical conductivity is considered low [73].

Table 6. Main soil physical and chemical properties at the experimental field 'long-term trial on organic olive (BiOlea)'.

Parameter	Unit Measure	Value
Sand	%	60
Silt	%	21
Clay	%	19
pH		7.8
Electrical conductivity (1:2.5)	dS/m	0.26
Organic matter	%	2.69
Total nitrogen (N)	‰	0.140
Exchangeable phosphorus (P)	ppm P	53
Exchangeable potassium (K)	ppm K	3628
Cation exchange capacity (CEC)	meq/100 g	64.98

Climatic data, namely, monthly minimum, mean, and maximum air temperature, global solar radiation, rainfall, reference evapotranspiration (ET_0), cultural evapotranspiration (ET_c), and vapor pressure deficit, registered at the experimental field, were collected from an agro-meteorological station located in the experimental farm (Figure 6). The climate of the region is typical Mediterranean, with hot and dry summers. According to the available meteorological data (30 years, not shown), annual mean reference rainfall is about 550 mm, and the maximum temperature in summer during daytime often reaches 38–40 °C [74]. During the trial, the site's climate was characterized by mild and wet winters, while the summers were semiarid (first and second) and dry (third) in which no rainfall was recorded from May to August. The annual average temperature was

18.29 °C. The lowest minimum temperatures were recorded in January and February. Mean temperature values were always above 22 °C from April to November.

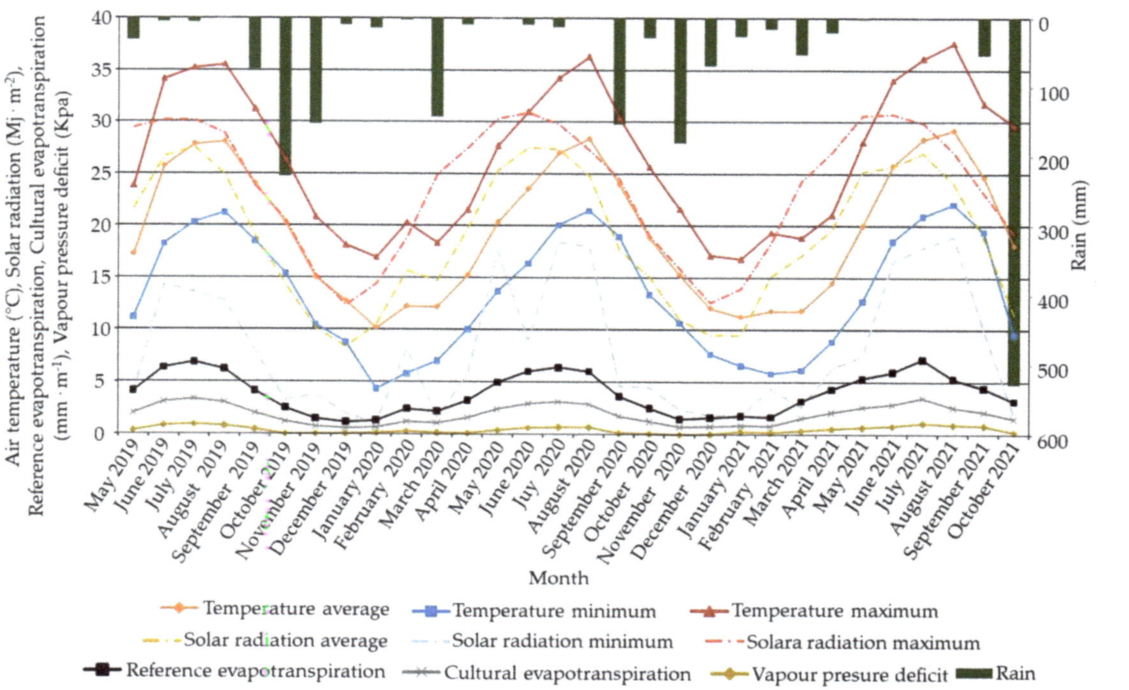

Figure 6. Monthly minimum, average, and maximum air temperature and solar radiation, rainfall, reference and cultural evapotranspiration, and vapor pressure deficit registered in the experimental field 'long-term trial on organic olive (BiOlea)'.

4.3. Entomological Samplings and Analysis

Entomological studies, regarding pollinators (Hymenoptera Apoidea, Lepidoptera, and Diptera Syrphidae) and predator insects (Neuroptera and Coleoptera Coccinellidae), were carried out twice per month, from March 2020 to October 2021. In particular, from 1 March 2020 to 28 February 2021, insects were collected from 2500 m^2 for each of the two soil management areas (125 m^2 each inter-row × 5 rows × 4 blocks = 2500 m^2) for a total of 5000 m^2. From 1 March 2021 to 31 October 2021, a defined linear transect of 25 m each in eight replicates (25.8 = 200 m) was used for the assessments of the beneficial insects along the row.

Specimens were collected with the net technique, from 10:00 a.m. to 4:00 p.m., on flowers (pollinators) and vegetative organs (predators) of the spontaneous and planted (intercropping) plant species. All specimens were transferred in the laboratory, dry prepared, and identified, when necessary, through the observation of sexual structures. The month of collection, number of specimens, and visited plants are given for all species. Specimens of wild bees were identified using the taxonomic keys in [75–77], as Lepidoptera [78], Diptera Syrphidae [38], Coleoptera Coccinellidae [79,80], and Neuroptera [81]. The classification followed Michener [47] for supra-specific taxa, and their nomenclature was according to [82,83]. The examined specimens were preserved in the collections of the authors and in the entomological collection of CREA-OFA of Acireale.

4.4. Spontaneous Flora Assessment and Analysis

Weed abundance and community composition and diversity were evaluated and monitored twice during the experiment: at the start of spring on 25 March 2021 and in autumn on 6 October 2021 at day of the year (DOY) 141 and 255, respectively, corresponding to the stages of maximum development of the natural cover (i.e., spring and autumn). At each sampling stage, weed cover (i.e., the percentage of the surface area of the quadrat covered by weeds) was evaluated at a species level by randomly placing three 1.0 m^2 quadrats within each block per soil management in the inter-row (3 squares $\times$ 4 subplots $\times$ 2 soil managements = 24) and three 1.0 m^2 quadrats for each intercropping species in each intra-row, in all blocks for each soil management (3 squares $\times$ 5 consociated species or control $\times$ 4 subplots $\times$ 2 soil managements = 120). Density was evaluated by placing two 0.60 $\times$ 0.60 m^2 quadrats in the intra-row space and four 0.25 $\times$ 0.25 m^2 quadrats in each soil management system per block. Cover and density assessment allowed providing the total cover (%) and the total density of the community.

4.5. Tree Growth Monitoring

Biometrical measurements of the young olive trees were conducted on 15 December 2020 and on 15 October 2021, and the relative increments were calculated. Measurements regarded the total height of the tree, the widths of the canopy (in two perpendicular directions from the projection on the ground at noon), and the canopy height, measured from the first primary branch insertion point to the top. The canopy volume was calculated assuming an elliptical shape [84]. The trunk cross-sectional area (TCSA) was calculated from the trunk circumference measured at 20 cm from the ground.

Pruning was realized on 15 February 2020, and the weight of the removed material was recorded, while the weight per tree of new emitted suckers was recorded in October 2021.

Moreover, the total vegetative growth was obtained by measuring the length improvement from the beginning of the vegetative growth (15 April 2021) to the end of the experiment (31 October 2021) of two 1 year old mixed shoots per plants, randomly selected and labeled around the canopy of the trees at 1.0–1.2 m height from the ground.

4.6. Statistical Analysis

Analysis of variance (ANOVA) was performed with Jamovi 2.0.0 statistical software (The jamovi project, 2021). One-way analysis of variance (ANOVA) was carried out on the differences among the canopy treatments. A post hoc analysis based on Tukey's HSD test (Tukey's honestly significant difference) was performed at a significance level (p-value) of 0.05, 0.01, and 0.001, respectively. Principal component analysis (PCA) was performed with Past 4.03 statistical software (Oyvind Hammer), to assess the effect of the ASC along the row, as well as the role of tillage used in the inter-row soil management in the development, abundance, and distribution of the weed community in spring and in autumn.

5. Conclusions

The obtained results, even if preliminary, evidence the role of diversification strategies in recovering rather than halting the loss of wild biodiversity in agricultural fields. In particular, the agronomical techniques proposed for the young organic olive, have been shown to be an evaluable option for promoting the presence of pollinators and, thus, supporting the potential production. The inter-row management resulted in a diversified spontaneous flora community, more service provider than competitor. In addition, the wild plants on the row had a sheltering effect on the living mulch species during the hot period, demonstrating a flow of services between the components of the agroecosystem. Among the studied living mulch species, sage and lemongrass were able to create an almost continuous hedge along the row and a semi-full soil cover, thus reducing the need for weed management in the intra-row soil strip and improving the beneficial insects without influencing the plant growth.

In a nutshell, current results indicated that the agroecological practices adopted increase the richness of the biota and, hence, the complexity of the Arthropod fauna in terms of number of species and taxonomic complexity. The knowledge of the two groups of insects investigated is of primary importance for evaluating the local populations of pollinators and predators of wild and cultivated plants.

Author Contributions: Conceptualization, F.F., C.C. and S.B.; methodology, F.F., E.M.L., C.C. and V.I.; validation, G.L.C., R.C. and E.N.; formal analysis, F.F., S.B., C.C., G.L.C. and E.N.; investigation, B.T., R.C. and A.G.; resources, F.F., G.L.C. and S.B.; data curation, V.I., A.G. and R.C.; writing—original draft preparation, F.F., S.B., E.M.L., C.C., G.L.C., A.G. and V.I.; writing—review and editing, F.F., S.B., C.C., G.L.C., E.M.L. and E.N.; visualization, F.F., B.T. and V.I.; supervision, F.F. and C.C.; project administration, F.F., E.M.L. and C.C.; funding acquisition, F.F. All authors have read and agreed to the published version of the manuscript.

Funding: This study was carried out within the project "Promozione e rafforzamento de dispositivi di lungo periodo in agricoltura biologica—PERILBIO" (Decree n. 3400, 20 December 2018; n. 92011 20 December 2018) funded by the Italian Ministry of Agriculture (MiPAAF).

Institutional Review Board Statement: Not applicable.

Informed Consent Statement: Not applicable.

Data Availability Statement: All data are available via email request to the corresponding authors.

Acknowledgments: The authors thank "G.S. S.p.A. Group Italia" for the economic support in publishing this study through the agreement "Azioni di studio e divulgazione finalizzate alla riduzione e ottimizzazione dell'uso di agrofarmaci in coltivazioni di pesche e nettarine, e all'individuazione di buone pratiche agronomiche al fine di preservare l'ambiente e le api (GS PES-NET21)" subscribed on 13 November 2020. We also wish to thank Vittorio Nobile (Ragusa, Italy) for confirming the determination of some species of Hymenoptera Apoidea.

Conflicts of Interest: The authors declare no conflict of interest. The funders had no role in the design of the study; in the collection, analyses, or interpretation of data; in the writing of the manuscript; or in the decision to publish the results.

References

1. Montanarella, L.; Panagos, P. The relevance of sustainable soil management within the European Green Deal. *Land Use Policy* **2021**, *100*, 104950. [CrossRef]
2. Alexandratos, N.; Bruinsma, J. *World Agriculture towards 2030/2050: The 2012 Revision*; ESAWorking Paper No. 12-03; FAO: Rome, Italy, 2012.
3. Food and Agriculture Organization. 2050: A Third More Mouths to Feed. 2009. Available online: http://www.fao.org/news/story/en/item/35571/icode/ (accessed on 1 September 2021).
4. Ciaccia, C.; Testani, E.; Fiore, A.; Iocola, I.; Di Pierro, M.; Mele, G.; Ferlito, F.; Cutuli, M.; Montemurro, F.; Farina, R.; et al. Organic Agroforestry Long-Term Field Experiment Designing Trough Actors' Knowledge towards Food System Sustainability. *Sustainability* **2021**, *13*, 5532. [CrossRef]
5. Michalopoulos, G.; Kasapi, K.A.; Koubouris, G.; Psarras, G.; Arampatzis, G.; Hatzigiannakis, E.; Kavvadias, V.; Xiloyannis, C.; Montanaro, G.; Malliaraki, S.; et al. Adaptation of Mediterranean Olive Groves to Climate Change through Sustainable Cultivation Practices. *Climate* **2020**, *8*, 54. [CrossRef]
6. Gómez, J.A.; Amato, M.; Celano, G.; Koubouris, G.C. Organic olive orchards on sloping land: More than a specialty niche production system? *J. Environ. Manag.* **2008**, *89*, 99–109. [CrossRef] [PubMed]
7. Stroosnijder, L.; Mansinho, M.I.; Palese, A.M. OLIVERO: The project analysing the future of olive production systems on sloping land in the Mediterranean basin. *J. Environ. Manag.* **2008**, *89*, 75–85. [CrossRef] [PubMed]
8. Beaufoy, G. *EU Policies for Olive Farming: Unsustainable on All Counts*; WWF Europe/Birdlife International: Brussels, Belgium, 2001; p. 16.
9. Connor, D.J.; Gomez-del-Campo, M.; Rousseaux, M.C.; Searles, P.S. Structure, management and productivity of hedgerow olive orchards: A review. *Sci. Hortic.* **2014**, *169*, 71–93. [CrossRef]
10. Tous, J. Olive production systems and mechanization. *Acta Hortic.* **2011**, *924*, 169–184. [CrossRef]
11. Caruso, G.; Palai, G.; Marra, F.P.; Caruso, T. High-Resolution UAV Imagery for Field Olive (*Olea europaea* L.) Phenotyping. *Horticulturae* **2021**, *7*, 258. [CrossRef]
12. Assirelli, A.; Romano, E.; Bisaglia, C.; Lodolini, E.M.; Neri, D.; Brambilla, M. Canopy Index Evaluation for Precision Management in an Intensive Olive Orchard. *Sustainability* **2021**, *13*, 8266. [CrossRef]

13. Neri, D.; Cioccolanti, T.; Zuccherelli, G.; Navacchi, O.; Giorgi, V.; Lodolini, E.M. Micropropagation effects on juvenile traits, flower differentiation, and tree architecture in young olive trees. *Agronomy* **2020**, *10*, 1742. [CrossRef]
14. The International Olive Oil Council. 2016. Available online: http://www.internationaloliveoil.org (accessed on 15 September 2021).
15. Eurostat. *Agriculture, Forestry and Fishery Statistics*, 2014 ed.; Eurostat: Luxembourg, 2015.
16. De Graaff, J.; Eppink, L.A.A.J. Olive oil production and soil conservation in southern Spain, in relation to EU subsidy policies. *Land Use Policy* **1999**, *16*, 259–267. [CrossRef]
17. Potts, S.G.; Petanidou, T.; Roberts, S.; O'Toole, C.; Hulbert, A.; Willmer, P. Plant-pollinator biodiversity and pollination services in a complex Mediterranean landscape. *Biol. Conserv.* **2006**, *129*, 519–529. [CrossRef]
18. Palese, A.M.; Vignozzi, N.; Celano, G.; Agnelli, A.E.; Pagliai, M.; Xiloyannis, C. Influence of soil management on soil physical characteristics and water storage in a mature rainfed olive orchard. *Soil Tillage Res.* **2014**, *144*, 96–109. [CrossRef]
19. Wagner, D.L. Insect Declines in the Anthropocene. *Annu. Rev. Entomol.* **2020**, *65*, 457–480. [CrossRef] [PubMed]
20. Nieto, A.; Roberts, S.P.M.; Kemp, J.; Rasmont, P.; Kuhlmann, M.; García Criado, M.; Biesmeijer, J.C.; Bogusch, P.; Dathe, H.H.; De la Rúa, P.; et al. *European Red List of Bees*; Publication Office of the European Union: Luxembourg, 2014; p. 84p.
21. Palm, C.; Blanco-Canqui, H.; DeClerck, F.; Gatere, L.; Grace, P. Conservation agriculture and ecosystem services: An overview. *Agric. Ecosyst. Environ.* **2014**, *187*, 87–105. [CrossRef]
22. Ciaccia, C.; Testani, E.; Roccuzzo, G.; Canali, S. The role of agrobiodiversity in sustainable food systems design and management. In *Genetic Diversity in Horticultural Plants*; Springer: Cham, Switzerland, 2019; pp. 245–271.
23. Landis, D.A. Designing agricultural landscapes for biodiversity-based ecosystem services. *Basic Appl. Ecol.* **2017**, *18*, 1–12. [CrossRef]
24. Wood, S.A.; Karp, D.S.; DeClerck, F.; Kremen, C.; Naeem, S.; Palm, C.A. Functional traits in agriculture: Agrobiodiversity and ecosystem services. *Trends Ecol. Evol.* **2015**, *30*, 531–539. [CrossRef]
25. Canali, S.; Diacono, M.; Campanelli, G.; Montemurro, F. Organic No-Till with Roller Crimpers: Agro-ecosystem Services and Applications in Organic Mediterranean Vegetable Productions. *Sustain. Agric. Res.* **2015**, *4*, 70–79. [CrossRef]
26. Bommarco, R.; Kleijn, D.; Potts, S.G. Ecological intensification: Harnessing ecosystem services for food security. *Trends Ecol. Evol.* **2013**, *28*, 230–238. [CrossRef]
27. Ramos, M.E.; Benítez, E.; García, P.A.; Robles, A.B. Cover crops under different managements vs. frequent tillage in almond orchards in semiarid conditions: Effects on soil quality. *Appl. Soil. Ecol.* **2010**, *44*, 6–14. [CrossRef]
28. Nieto, O.M.; Castro, J.; Fernandez-Ondono, E. Conventional tillage versus cover crops in relation to carbon fixation in Mediterranean olive cultivation. *Plant Soil* **2013**, *365*, 321–335. [CrossRef]
29. Mourugàn-Coronado, A.; Linares, C.; Gòmez-Lòpeez, M.D.; Faz, A.; Zornova, R. The impact of intercropping, tillage and fertilizer type on soil and crop yield in fruit orchards under Mediterranean conditions: A meta-analysis of field studies. *Agric. Syst.* **2020**, *178*, 102736. [CrossRef]
30. Sestini, A. *Il Paesaggio*; Club Italiano: Milano, Italy, 1963.
31. Paris, P.; Camilli, F.; Rosati, A.; Mantino, A.; Mezzalira, G.; Dalla Valle, C.; Franca, A.; Seddaiu, G.; Pisanelli, A.; Lauteri, M.; et al. What is the future for agroforestry in Italy? *Agroforest Syst.* **2019**, *93*, 2243–2256. [CrossRef]
32. Nicolosi, E.; Ferlito, F.; Allegra, M.; Cicala, A.; Trovato, F.; La Malfa, S. Influences of aspect and tillage on two winegrape cultivars on Mount Etna. *N. Z. J. Crop Hortic. Sci.* **2016**, *44*, 83–102. [CrossRef]
33. Ferlito, F.; Allegra, M.; Torrisi, B.; Pappalardo, H.; Gentile, A.; La Malfa, S.; Continella, A.; Stagno, F.; Nicolosi, E. Early defoliation effect on water status, fruit yield and must quality of 'Nerello mascalese' grapes. *Sci. Agric.* **2020**, *77*, 1–10. [CrossRef]
34. Nicolosi, E.; Iovino, V.; Distefano, G.; Di Guardo, M.; La Malfa, S.; Gentile, A.; Palliotti, A.; Las Casas, G.; Ferlito, F. Mid-Term Effects of Conservative Soil Management and Fruit-Zone Early Leaf Removal Treatments on the Performance of Nerello Mascalese (*Vitis vinifera* L.) Grapes on Mount Etna (Southern Italy). *Agronomy* **2021**, *11*, 1070. [CrossRef]
35. Bennici, S.; Las Casas, G.; Distefano, G.; Di Guardo, M.; Continella, A.; Ferlito, F.; Gentile, A.; La Malfa, S. Elucidating the contribution of wild related species on autochthonous pear germplasm: A case study from Mount Etna. *PLoS ONE* **2018**, *13*, e0198512. [CrossRef]
36. Russo, C.; Cappelletti, G.M.; Nicoletti, G.M.; Di Noia, A.E.; Michalopoulos, G. Comparison of European Olive Production Systems. *Sustainability* **2016**, *8*, 825. [CrossRef]
37. Magdoff, F. Ecological agriculture: Principles, practices, and constraints. *Renew. Agric. Food Syst.* **2007**, *22*, 109–117. [CrossRef]
38. Burgio, G.; Sommaggio, D.; Birtele, D. I Sirfidi (Ditteri): Biodiversità e conservazione. In *Manuali e Linee Guida*; ISPRA: Roma, Italy, 2015; Volume 128, p. 182.
39. Comba, M. Hymenoptera: Apoidea: Anthophila of Italy. 2021. Available online: http://digilander.libero.it/mario.comba (accessed on 27 July 2021).
40. Mazzeo, G.; Bella, S.; Seminara, A.R.; Longo, S. Bumblebees in natural and agro-ecosystems at different altitudes from Mount Etna, Sicily (Hymenoptera Apidae Bombinae): Long-term faunistic and ecological observations. *Redia* **2015**, *XCVIII*, 123–131.
41. Mazzeo, G.; Longo, S.; Seminara, A.R.; Bella, S. Faunistic and ecological studies on Apidae (Hymenoptera, Apoidea) in natural and cultivated ecosystems in Sicily. *Redia* **2019**, *102*, 153–162. [CrossRef]
42. Bella, S.; Catania, R.; Nobile, V.; Mazzeo, G. New or little known bees (Hymenoptera, Apoidea) from Sicily. *Fragm. Entomol.* **2020**, *52*, 113–117. [CrossRef]

43. Incalcaterra, G.; Iapichino, G.; D'Anna, F.; Sinacori, A. Influences of different Pollinators on Winter Melon Grown under Polyethylene Tunnel. *Acta Hort.* **2003**, *614*, 297–299. [CrossRef]
44. Lo Verde, G.; La Mantia, T. The role of native flower visitors in pollinating *Opuntia ficus-indica* (L.) Mill. naturalized in Sicily. *Acta Oecol.* **2011**, *37*, 413–417. [CrossRef]
45. Mazzeo, G.; Scavo, A.; Lo Monaco, A.; Longo, S.; Mauromicale, G. Insect pollinators improve seed production in globe artichoke (*Cynara cardunculus* var. *scolymus). Ann. Appl. Biol.* **2020**, *176*, 241–248. [CrossRef]
46. Bella, S.; Catania, R.; Baviera, C. First record of the genus *Serangium* Blackburn, 1889 (Coleoptera Coccinellidae) in Italy. *Redia* **2021**, *104*, 185–191. [CrossRef]
47. Michener, C.D. *The Bees of the World*, 2nd ed.; John Hopkins University Press: Baltimore, MD, USA, 2007; p. 953.
48. Holzschuh, A.; Steffan-Dewenter, I.; Kleijn, D.; Tscharntke, T. Diversity of flower-visiting bees in cereal fields: Effects of farming system, landscape composition and regional context. *J. Appl. Ecol.* **2007**, *44*, 41–49. [CrossRef]
49. Scheper, J.A. *Promoting wild bees in European agricultural landscapes. The Role of Floral Resources in Driving and Mitigating Wild Bee Decline*; Alterra, Wageningen University & Research Centre: Wageningen, The Netherlands, 2015; 175p.
50. Wezel, A.; Casagrande, M.; Celette, F.; Vian, J.-F.; Ferrer, A.; Peigné, J. Agroecological practices for sustainable agriculture. A review. *Agron. Sustain. Dev.* **2014**, *34*, 1–20. [CrossRef]
51. Baviera, C.; Bellavista, M.; Altadonna, G.; Turrisi, G.F.; Bella, S.; Muscarella, C.; Sparacio, I. The Cerambycidae (Coleoptera: Chrysomeloidea) of Sicily: Recent records and updated checklist. *AAPP Atti Accad. Peloritana dei Pericolanti Cl. Sci. Fis. Mat. Nat.* **2017**, *95*, 2.
52. Ramos, M.E.; Robles, A.B.; Sánchez-Navarro, A.; González-Rebollar, J.L. Soil responses to different management practices in rainfed orchards in semiarid environments. *Soil Tillage Res.* **2011**, *112*, 85–91. [CrossRef]
53. Koch, J.B.; Lozier, J.; Strange, J.P.; Ikerd, H.; Griswold, T.; Cordes, N.; Solter, L.; Stewart, I.; Cameron, S.A. USBombus, a database of contemporary survey data for North American Bumble Bees (Hymenoptera, Apidae, Bombus) distributed in the United States. *Biodivers. Data J.* **2015**, 3, e6833. [CrossRef] [PubMed]
54. Taguas, E.V.; Gómez, J.A. Vulnerability of olive orchards under the current CAP (Common Agricultural Policy) regulations on soil erosion: A study case in Southern Spain. *Land Use Policy* **2015**, *42*, 683–694. [CrossRef]
55. Almagro, M.; de Vente, J.; Boix-Fayós, C.; García-Franco, N.; de Aguilar, J.M.; González, D.; Solé-Benet, A.; Martínez-Mena, M. Sustainable land management practices as providers of several ecosystem services under rainfed Mediterranean agroecosystems. *Mitig. Adapt. Strateg. Glob. Chang.* **2016**, *21*, 1029–1043. [CrossRef]
56. Chamizo, S.; Serrano-Ortiz, P.; López-Ballesteros, A.; Sánchez-Cañete, E.P.; Vicente-Vicente, J.L.; Kowalski, A.S. Net ecosystem CO2 exchange in an irrigated olive orchard of SE Spain: Influence of weed cover. *Agric. Ecosyst. Environ.* **2017**, *239*, 51–64. [CrossRef]
57. Armengot, L.; Berner, A.; Blanco-Moreno, J.M.; Mäder, P.; Sans, F.X. Long-term feasibility of reduced tillage in organic farming. *Agron. Sustain. Develop.* **2015**, *35*, 339–346. [CrossRef]
58. Nichols, V.; Verhulst, N.; Cox, R.; Govaerts, B. Weed dynamics and conservation agriculture principles: A review. *Field Crops Res.* **2015**, *183*, 56–68. [CrossRef]
59. Mia, M.J.; Massetani, F.; Murri, G.; Neri, D. Sustainable alternatives to chemicals for weed control in the orchard a review. *Hortic. Sci.* **2020**, *47*, 1–12. [CrossRef]
60. Bateni, S.; Vosoughifar, H.; Ek, M.; Xu, T. Estimation of Daily Reference Evapotranspiration from Limited Climatic Variables in Coastal Regions. In Proceedings of the AGU Fall Meeting, San Francisco, CA, USA, 9–13 December 2019; American Geophysical Union: Washington, DC, USA, 2019.
61. Allen, R.G.; Pereira, L.S.; Raes, D.; Smith, M. Crop Evapotranspiration. In *Guidelines for Computing Crop Water Requirements*; FAO Irrigation and Drainage Paper; FAO: Rome, Italy, 1998; Volume 56, p. 300.
62. Saitta, D.; Vanella, D.; Ramírez-Cuesta, J.M.; Longo-Minnolo, G.; Ferlito, F.; Consoli, S. Comparison of Orange Orchard Evapotranspiration by Eddy Covariance, Sap Flow, and FAO-56 Methods under Different Irrigation Strategies. *J. Irrig. Drain.* **2020**, *146*, 05020002. [CrossRef]
63. Lo Cicero, L.; Puglisi, I.; Nicolosi, E.; Gentile, A.; Ferlito, F.; Continella, A.; Lo Piero, A.R. Anthocyanin levels and expression analysis of biosynthesis-related genes during ripening of sicilian and international grape berries subjected to leaf removal and water deficit. *J. Agric. Sci.* **2016**, *18*, 1333–1344.
64. Miller, W.P.; Miller, D.M. A micro-pipette method for soil mechanical analysis. *Commun. Soil Sci. Plant Anal.* **1987**, *18*, 1–15. [CrossRef]
65. Ferlito, F.; Torrisi, B.; Allegra, M.; Stagno, F.; Caruso, P.; Fascella, G. Evaluation of Conifer Wood Biochar as Growing Media Component for Citrus Nursery. *Appl. Sci.* **2020**, *10*, 1618. [CrossRef]
66. Torrisi, B.; Allegra, M.; Amenta, M.; Gentile, F.; Rapisarda, P.; Fabroni, S.; Ferlito, F. Physico-chemical and multielemental traits of anaerobic digestate from Mediterranean agro-industrial wastes and assessment as fertiliser for citrus nurseries. *Waste Manag.* **2021**, *131*, 201–213. [CrossRef] [PubMed]
67. Ferlito, F.; Distefano, G.; Gentile, A.; Allegra, M.; Lakso, A.N.; Nicolosi, E. Scion–rootstock interactions influence the growth and behaviour of the grapevine root system in a heavy clay soil. *Aust. J. Grape Wine Res.* **2020**, *26*, 68–78. [CrossRef]
68. Olsen, S.; Cole, C.; Watanabe, F.; Dean, L. *Estimation of Available Phosphorus in Soils by Extraction with Sodium Bicarbonate (Circular No. 939)*; United States Department of Agriculture: Washington, DC, USA, 1954.

69. Sumner, M.E.; Miller, W.P. Cation exchange capacity and exchange coefficients. In *Methods of Soil Analysis*; Sparks, D.L., Ed.; Part 3, Chemical Methods; Book Series No. 5; Soil Science Society of America: Madison, WI, USA, 1996.
70. Springer, U.; Klee, J. Prufung der Leistungsfahigkeit von einigen wichtigeren Verfahren zur Bestimming des Kohlemstoffs mittels Chromschwefelsaure sowie Vorschlag einer neuen Schnellmethode. *Z. Pflanz. Düngung Bodenkd.* **1954**, *64*, 1–8. [CrossRef]
71. Leonardi, M.; Caruso, G.M.; Carroccio, S.C.; Boninelli, S.; Curcuruto, G.; Zimbone, M.; Allegra, M.; Torrisi, B.; Ferlito, F.; Miritello, M. Smart nanocomposites of chitosan/alginate nanoparticles loaded with copper oxide as alternative nanofertilizers. *Environ. Sci. Nano* **2021**, *8*, 174–187. [CrossRef]
72. Klingebiel, A.A.; Montgomery, P.H. *Land Capability Classification*; Agriculture USDA: Washington, DC, USA, 1961.
73. Saitta, D.; Consoli, S.; Ferlito, F.; Torrisi, B.; Allegra, M.; Longo-Minnolo, G.; Ramírez-Cuesta, J.M.; Vanella, D. Adaptation of citrus orchards to deficit irrigation strategies. *Agric. Water Manag.* **2021**, *247*, 106734. [CrossRef]
74. Ciaccia, C.; La Torre, A.; Ferlito, F.; Testani, E.; Battaglia, V.; Salvati, L.; Roccuzzo, G. Agroecological Practices and Agrobiodiversity: A Case Study on Organic Orange in Southern Italy. *Agronomy* **2019**, *9*, 85. [CrossRef]
75. Schmid-Egger, C.; Scheuchl, E. Illustrierte Bestimmungstabellen der Wildbienen Deutschlands und Österreichs Unter Berücksichtigung der Arten der Schweiz. In *Band III, Andrenidae*; Eigenverlag: Velden, Austria, 1997; p. 180.
76. Scheuchl, E. Illustrierte Bestimmungstabellen der Wildbienen Deutschlands und Österreichs. In *Band I: Anthophoridae. 2. Erweiterte Auflage*; Eigenverlag: Berlin, Germany, 2006; 158p.
77. Scheuchl, E. Illustrierte Bestimmungstabellen der Wildbienen Deutschlands und Österreichs Für Osmia s.1. Unter Berücksichtigung der Arten der Schweiz, Norditaliens, Ungarns, Sloweniens und der Slowakei. In *Band II, Schlüssel der Arten der Familien Megachilidae und Melittidae*, 2nd ed.; Apollo Books: Tsim Sha Tsui, Japan, 2006; p. 192.
78. Mazzei, P.; Morel, D.; Panfili, R. Moths and Buttes of Europe and North Africa. 2021. Available online: http://www.leps.eu (accessed on 2 November 2021).
79. Fürsch, H. Übersicht über die Genera und Subgenera der Scymnini mit besonderer Berücksichtigung der Westpalaearktis (Insecta, Coleoptera, Coccinellidae). *Entomol. Abh. (Dresd.)* **1987**, *51*, 57–74.
80. El-Saeady, A.A.; Hafez, S.F.; Abied, M.K.; Bedewy, M.M.M. Taxonomy of the Tribe Scymnini (Coleoptera: Coccinellidae: Scymninae). *Egypt. Bull. Ent. Soc.* **2016**, *93*, 117–150.
81. Brooks, S.J. A taxonomic review of the common green lacewing genus *Chrysoperla* (Neuroptera: Chrysopidae). *Bull. Br. Nat. Hist. (Entomol.)* **1994**, *63*, 137–210.
82. Polaszek, A. Fauna Europaea: Apidae. In *Fauna Europaea: Hymenoptera—Apocryta. Fauna Europaea Version 2.6*; Mitroiu, M.D., Noyes, J., Cetkovic, A., Nonveiller, G., Radchenko, A., Polaszek, A., Ronquist, F., Forshage, M., Pagliano, G., Gusenleitner, J., et al., Eds.; Fauna Europaea Secretariat: Berlin, Germany, 2013. Available online: http://www.fauna-eu.org (accessed on 27 July 2021).
83. Turrisi, G.F.; Bella, S.; Catania, R.; La Greca, P.; Nobile, V.; D'urso, V. Bee diversity in fragmented areas of Volcano Etna (Sicily, Italy) at different degrees of anthropic disturbance (Hymenoptera: Apoidea, Anthophila). *J. Entomol. Acarol. Res.* **2021**, *53*, 10362. [CrossRef]
84. Lodolini, E.M.; Paoletti, A.; Nolasco, A.; Ferlito, F.; Cutuli, M.; Torrisi, B.F.; Santilli, E.; Zaffina, F.; Desando, M.; Rosti, A.; et al. Biomass recovery from olive rejuvenation pruning in different varieties. In Proceedings of the 29th European Biomass Conference and Exhibition, Marseille, France, 26–29 April 2021; pp. 273–278. [CrossRef]

plants

Article

The Impact of Different Weed Management Systems on Weed Flora and Dry Biomass Production of Barley Grown under Various Barley-Based Cropping Systems

Muhammad Naeem [1], Shahid Farooq [2] and Mubshar Hussain [1,3,*]

[1] Department of Agronomy, Bahauddin Zakariya University, Multan 60800, Pakistan; naeem_agrarian@yahoo.com
[2] Department of Plant Protection, Faculty of Agriculture, Harran University, Şanlıurfa 63250, Turkey; csfa2006@gmail.com
[3] School of Veterinary and Life Sciences, Murdoch University, 90 South Street, Murdoch, WA 6150, Australia
* Correspondence: mubashiragr@gmail.com; Tel.: +92-301-7164879

Abstract: Weeds are among the major issues responsible for reduction in yield and profit in any crop production system. Herbicides are the easiest and quickest solution of weeds; however, their frequent use exert negative consequences on environment, human health, and results in the evolution of herbicide-resistant weed species. Due to these reasons, alternative weed management methods that are less harmful to environment and human health are needed. This two-year study evaluated the impact of different weed management options, i.e., false seedbed (FS), allelopathic water extracts (AWE), chemical control (CC), weed-free (WF) weedy-check (WC) on weed spectrum in various barley-based cropping systems, i.e., fallow-barley (FB), maize-barley (MB), cotton-barley (CB), mungbean-barley (M*B), and sorghum-barley (SB). Data relating to density, diversity, and biomass production of weed species prevailing in the studied cropping systems were recorded. Interactive effect of weed management methods and barley-based cropping systems significantly altered weed diversity, and densities of individual, broadleaved, and grassy weeds. A total 13 weed species (ten broadleaved and three grass) were recorded during both years of study. The highest dry biomass, diversity, and density of individual, broadleaved, and grassy weeds were noted in WC treatment, whereas WF treatment resulted in the lowest values of these traits. Chemical control resulted in the highest suppression of weed flora and improved dry biomass production of barley followed by AWE. The SB cropping system with CC or AWE resulted in the least weed flora. The M*B cropping system with CC or AWE produced the highest dry biomass of barley. It is concluded that including sorghum crop in rotation and applying AWE could suppress weeds comparable to herbicides. Similarly, including mungbean in rotation and applying AWE could increase dry biomass production of barley. In conclusion, herbicides can be replaced with an eco-friendly approach, i.e., allelopathy and inclusion of sorghum crop could be helpful in suppressing weed flora.

Keywords: weeds; allelopathy; barley; false seedbed; cropping system

Citation: Naeem, M.; Farooq, S.; Hussain, M. The Impact of Different Weed Management Systems on Weed Flora and Dry Biomass Production of Barley Grown under Various Barley-Based Cropping Systems. *Plants* **2022**, *11*, 718. https://doi.org/10.3390/plants11060718

Academic Editors: Alessia Restuccia and Aurelio Scavo

Received: 17 February 2022
Accepted: 7 March 2022
Published: 8 March 2022

1. Introduction

Barley (*Hordeum vulgare*) is the fourth major cereal in terms of production globally after wheat, maize, and rice. Barley is grown for fodder, brewing, human food, and in the production of malt around the world [1,2]. Barley is cultivated in ~100 different countries [3]. It performs better in low rainfall areas where other crops fail to establish and can survive under adverse environmental and conditions [4,5]. However, it gives better production on moderately saline soils and higher salinity could obstruct its growth leading to reduced yield [6]. Barley is tolerant to several abiotic and biotic conditions; nonetheless, weed infestation can significantly reduce its yield and productivity [7–9].

Weeds exert negative impacts on quality and quantity of agricultural products; thus, reduce farmers' profits to a significant extent [10]. Weeds compete with crop plants either through competing for moisture, sunlight, nutrients, and space or through secreting allelochemicals, which adversely impact seed germination and growth of crop plants [11–13]. Nevertheless, weeds produce significant number of seeds, which are deposited to soil seed bank; thus, laying the foundation for future weed infestation [14–16]. Therefore, weeds must be controlled to reduce weed seed bank in soil and crop yield losses [17,18]. Several weed management methods, i.e., cultural, chemical, mechanical, and biological are opted to suppress the growth of weed flora [19,20]. Labor unavailability and high wages along with unreliable weed control are the main issues faced in manual/cultural weed control [21]. Mechanical weed control, on the other hand, is expensive because of sophisticated equipment required for each crop [22], and involves extra soil disturbance resulting in the disruption of soil structure and reduced soil fertility [23]. Similarly, frequent use of herbicide in chemical weed control results in the evolution of herbicide-resistant weeds, environmental contamination, and health hazards [20,24].

Because of the disadvantages associated with the prevalent weed management methods, alternative weed control methods with low environmental contamination, health hazards, and lesser herbicide resistance are needed [24]. Adoption of preventive weed control method like false or stale seedbeds provide effective weed control during crop growth with less labor cost [25–27]. However, the efficacy of such methods is strongly reliant on time available for the preparation of stale seedbeds, method used, and soil and climatic conditions [25,26]. Recently, plant-based natural products that could serve as alternatives to herbicides have been focused on weed management research globally [28]. Residues' incorporation of allelopathic crops, and inclusion of such crops in rotation could improve weed control [8,11,29]. The crops with high allelopathic potential include sunflower, rye, wheat, rice, barley, and sorghum, which have been shown to suppress weed flora in different crops [29–31]. The allelopathic compounds found in mulberry (tannins steroids and phenols), sunflower (phenolic compounds and terpenes), and sorghum (sorgoleone and phenolics) are responsible for suppression of weed flora [13,29,32–34].

Diversifying the crops to be sown on a particular area could suppress weed flora since it has the potential to inhibit weed growth [35,36]. Selection of similar crops for longer time periods results in the proliferation and establishment of particular weed species, which become established and are difficult to control [37]. The inclusion of allelopathic crops, i.e., sorghum in rotation could provide significant control over weeds compared to a rotation having no allelopathic crop [34]. Sorghum releases various allelopathic compounds from its grains, stems, and root hairs; thus, considered as an important candidate for crop rotation to suppress weed flora [30,34]. Several studies explored the allelopathic potential of sorghum as cover crop, mulch, and aqueous extracts on different weeds and concentration-dependent, selective, and species-specific allelopathic effects have been reported [7,8,20,38,39]. Therefore, inclusion of sorghum in barley-based cropping system could suppress weed flora.

This two-year field experiment evaluated the effect of different weed control methods on weed infestation and dry matter production of barley in different barley-based cropping systems. It was hypothesized that different weed control methods will differ in weed infestation level, density, and composition of weed flora. It was further hypothesized that barley-based cropping systems including an allelopathic crop would have lower weed infestation compared to those having no allelopathic crop. The results would help to improve the weed control in barley-based cropping systems and lower the adverse impacts of herbicides on environment and human health.

2. Results

2.1. Weed Flora

A total 13 weed species (ten broadleaved and three grass) were recorded from the study area during both years of the study. Of the recorded weed species, four were perennial, whereas the remaining nine had an annual life cycle. The weed species belonged to 11 pant families, of which Asteraceae and Leguminosae were represented by two species each, while the remaining families were represented by one species only (Table 1).

Table 1. Common and Latin names, family, and life cycle of different weed species recorded in barley crop during both years of the study.

Species	Common Name	Family	Life Cycle
	Broadleaved weed species		
Chenopodium murale L.	Fat hen	Amaranthaceae	Annual
Melilotus indicus (L.) All.	Yellow sweet clover	Leguminosae	Annual
Rumex obtusifolius L.	Bitter dock	Polygonaceae	Perennial
Anagallis arvensis L.	Blue pimpernel	Primulaceae	Annual
Chenopodium album L.	Common goosefoot	Amaranthaceae	Annual
Sonchus arvensis L.	Perennial sow thistle	Asteraceae	Perennial
Conyza stricta Willd.	Horseweed	Asteraceae	Annual
Convolvulus arvensis L.	Field bindweed	Convolvulaceae	Perennial
Medicago polymorpha L.	Yellow trefoil	Leguminosae	Annual
Coronopus didymus L. Sm.	Swine-cress	Brassicaceae	Annual
	Grassy weed species		
Polypogon monspeliensis L. Desf.	Winter grass	Poaceae	Annual
Spergula arvensis L.	Corn spurry	Caryophyllaceae	Annual
Bolboschoenus maritimus (L.) Palla	Salt marsh	Cyperaceae	Perennial

2.2. Weeds Diversity (Number of Weed Species)

The interaction between barley-based cropping systems and weed control methods (WCM) significantly altered weeds' diversity during both years (Figure 1). During the first year of study, the highest weed diversity was recorded in cotton-barley (CB) cropping system with weedy-check (WC) treatment, which was like mungbean-barley (M*B) system with WC condition. During the second year, M*B cropping system with WC and false seedbed (FS) treatments resulted in the highest weeds' diversity, which was similar to M*B system with allelopathic water extracts (AWE), and CB and fallow-barley (FB) systems with WC treatment (Figure 1). However, all cropping systems with chemical control (CC) during the first year and sorghum-barley (SB) cropping system with CC during the second year observed the lowest weed diversity (Figure 1).

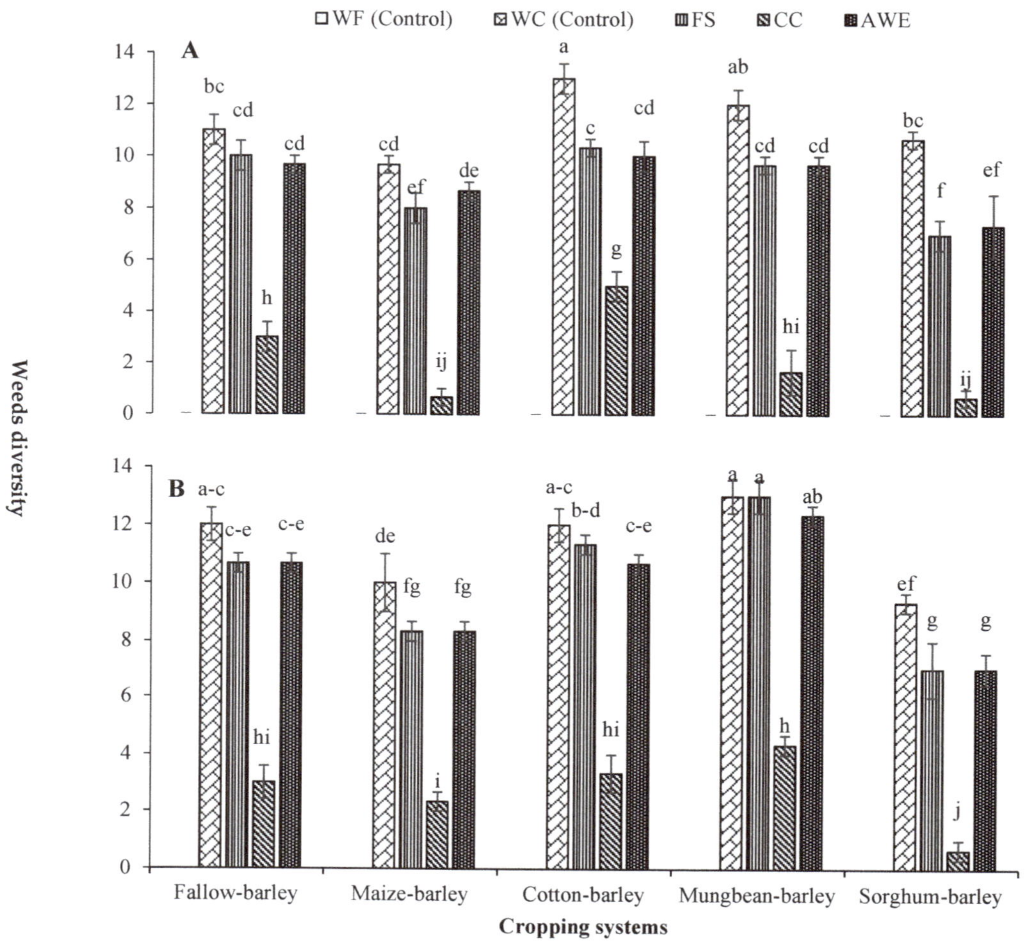

Figure 1. Influence of different barley-based cropping systems on weed diversity (number of weeds species) under various weed management methods during 2017–2018 (**A**) and 2018–2019 (**B**) ±S.E. In the legend, WF = weed free (control), WC = weedy check (control), FS = false seedbed, CC = chemical control, AWE = allelopathic water extracts. The means sharing the same letters do not differ significantly at $p \leq 0.05$ (LSD value at $p \leq 0.05$ for 2017–2018 = 1.41, 2018–2019 = 1.37).

2.3. Density of Broadleaved Weed Species

Barley-based cropping systems, WCM, and their interaction had significant effect on the density of broadleaved weed species. The highest and the lowest density of broadleaved weed species was noted for WC and weed-free (WF) treatments, respectively (Table 2). Chemical control resulted in higher reduction in the density of broadleaved weed species compared to FS and AWE during both years of study. The CB and M*B cropping systems observed the highest density of broadleaved weed species during the first and second years, respectively, while the lowest density of broadleaved weed species was recorded for SB cropping system during both years (Table 2). Regarding interaction, the highest density of broadleaved weeds was recorded in CB cropping system during the first year and M*B cropping system during the second year with WC treatment, whereas all cropping systems (CS) with WF and CC treatments had little or no infestation of broadleaved weed species during both years (Table 2).

Table 2. Influence of different barley-based cropping systems on the overall density (m^{-2}) of broadleaved and grassy weed species under various weed management methods during 2017–2018 and 2018–2019.

Cropping Systems	2017–2018					2018–2019				
	WC	FS	CC	AWE	Means (CS)	WC	FS	CC	AWE	Means (CS)
				Broadleaved weeds (m^{-2})						
FB	67.00 b	30.00 f	0.33 j	39.00 d	27.27 B	84.00 b	41.33 de	0.00 k	37.67 ef	32.60 B
MB	35.67 de	18.00 h	0.00 j	16.33 h	14.00 D	33.00 gh	14.67 j	0.00 k	12.67 j	12.06 D
CB	73.00 a	34.67 e	1.33 j	47.33 c	31.27 A	67.33 c	30.00 hi	0.00 k	26.00 i	24.67 C
M*B	70.00 ab	24.00 g	0.33 j	32.33 ef	25.33 C	98.67 a	44.00 c	3.67 k	36.00 fg	36.47 A
SB	32.67 ef	9.67 i	0.00 j	11.33 i	10.73 E	29.33 hi	10.33 j	0.00 k	10.33 j	10.00 E
Means (WCS)	55.67 A	23.27 C (58.20)	0.40 D (99.28)	29.27 B (47.42)		62.47 A	28.07 B (55.06)	0.73 D (98.83)	24.53 C (60.72)	
LSD value ($p < 0.05$)	WCS = 1.51, CS = 1.51, WCS × CS = 3.38					WCS = 1.98, CS = 1.98, WCS × CS = 4.44				
				Grassy weeds (m^{-2})						
FB	43.00 a	27.00 c	4.33 h–k	12.33 e	17.33 A	38.67 a	14.33 e–g	7.67 h	19.67 c	16.06 A
MB	9.33 f	2.67 j–m	0.67 m	4.00 i–l	3.33 D	16.67 de	6.67 h	3.00 ij	8.00 h	6.87 C
CB	39.00 b	18.00 d	5.67 hi	8.67 fg	14.27 B	37.67 a	12.00 g	5.67 hi	18.33 cd	14.73 B
M*B	29.00 c	8.33 fg	2.00 k–m	5.00 h–j	8.86 C	40.00 a	15.00 ef	3.33 ij	22.67 b	16.20 A
SB	6.67 gh	2.33 k–n	0.67 m	1.67 k–m	2.27 D	12.67 fg	3.00 ij	0.67 j	5.67 hi	4.40 D
Means (WCS)	25.40 A	11.67 B (54.05)	2.67 D (89.48)	6.33 C (75.07)		29.13 A	10.20 C (64.98)	4.07 D (86.02)	14.87 B (48.95)	
LSD value ($p < 0.05$)	WCS = 1.13, CS = 1.13, WCS × CS = 2.52					WCS = 1.31, CS = 1.31, WCS × CS = 2.92				

Means not having common letter for individual and interactive effects significantly vary from each other at $p \leq 0.05$. Here, WF = weed free (control), WC = weedy check (control), FS = false seedbed, CC = chemical control, AWE = allelopathic water extracts, FB = fallow-barley, MB = maize-barley, CB = cotton-barley, M*B = mungbean-barley, SB = sorghum-barley, WCS = weed control strategies, CS = cropping system, DAS = days after sowing. The values presented in brackets indicated the % decrease in the number of broadleaf weeds than WC (control).

2.4. Density of Grassy Weed Species

The individual and interactive effects of barley-based cropping systems and WCM had significant effect on the density of grassy weed species (Table 2). The highest and the lowest density of grassy weed species was noted for WC and CC treatments, during both years (Table 2). Regarding CS, FB cropping system had the highest and SB as well as maize-barley (MB) cropping systems recorded the lowest density of grassy weed species density during the first year. However, FB and M*B systems resulted in the highest density of grassy weed species, whereas SB system had the lowest density of grassy weed species during the second year (Table 2). Regarding interaction, FB cropping system with WC treatment had the highest density of grassy weeds, while SB and MB cropping systems with CC recorded the lowest density of grassy weed species during the first year of the study. Similarly, FB, CB, and M*B cropping systems with WC treatment recorded the highest density of grassy weed species, whereas SB cropping system with CC resulted in the lowest density of grassy weed species during the second year of the study (Table 2).

2.5. Density of Individual Weed Species

Different barley-based cropping systems, WCM and their interaction significantly altered density of individual weed species during both years (Tables 3–6). The highest and the lowest density of salt marsh (*Bolboschoenus maritimus* (L.) Palla) was recorded for WC and WF treatments, respectively, during both years (Table 4). The highest density of salt marsh was noted for FB cropping system during the first year, whereas the lowest density was recorded for SB and MB cropping systems during both years (Table 3). Regarding interaction, FB cropping system during 1st year and M*B cropping system during the second year with WC treatment had the highest density of salt marsh, whereas the lowest density was recorded in all cropping systems with CC treatment during both years (Table 3).

Table 3. Influence of different barley-based cropping systems on individual density (m^{-2}) of grassy weed species under various weed management methods during 2017–2018 and 2018–2019.

Cropping Systems	2017–2018					2018–2019				
	WC	FS	CC	AWE	Means	WC	FS	CC	AWE	Means
				Salt marsh						
FB	16.00 a	10.33 c	1.33 hi	6.00 de	6.73 A	14.00 b	4.00 g–i	2.67 ij	6.33 d–f	5.40 A
MB	4.67 ef	1.00 hi	0.00 i	1.33 hi	1.40 D	7.00 de	2.33 ij	1.33 jk	3.33 hi	2.80 C
CB	12.33 b	6.67 d	2.00 gh	3.00 g	4.80 B	11.67 c	3.33 hi	1.33 jk	5.33 e–g	4.33 B
M*B	9.33 c	3.33 fg	1.00 hi	1.00 hi	2.93 C	16.33 a	5.00 f–h	1.00 jk	8.00 d	6.07 A
SB	3.00 g	1.00 hi	0.00 i	0.67 hi	0.93 D	6.33 d–f	1.33 jk	0.00 k	2.67 ij	2.07 C
Means (WCS)	9.07 A	4.47 B (50.71)	0.87 D (90.40)	2.40 C (73.53)		11.07 A	3.20 C (71.09)	1.27 D (88.52)	5.13 B (53.65)	
LSD value ($p < 0.05$)	WCS = 0.68, CS = 0.68, WCS × CS = 1.53					WCS = 0.83, CS = 0.83, WCS × CS = 1.85				
				Corn spurry						
FB	3.33 b	1.33 c	0.00 d	1.00 c	1.13 B	7.67 a	3.33 c	1.67 d	4.00 c	3.33 A
MB	0.67 cd	0.00 d	0.00 d	0.67 cd	0.27 C	1.00 d–f	1.67 d	0.33 ef	0.00 f	0.60 C
CB	6.00 a	1.33 c	0.67 cd	0.67 cd	1.73 A	6.00 b	1.33 de	1.33 de	3.33 c	2.40 B
M*B	3.00 b	0.67 cd	0.00 d	0.00 d	0.73 B	0.00 f	1.00 d–f	0.00 f	2.00 d	0.60 C
SB	0.67 cd	0.00 d	0.00 d	0.00 d	0.13 C	0.00 f	0.00 f	0.00 f	0.00 f	0.00 D
Means (WCS)	2.73 A	0.67 B (75.45)	0.13 CD (95.23)	0.47 BC (82.78)		2.93 A	1.47 B (49.82)	0.67 C (77.13)	1.87 B (36.17)	
LSD value ($p < 0.05$)	WCS = 0.44, CS = 0.44, WCS × CS = 0.99					WCS = 0.45, CS = 0.45, WCS × CS = 1.01				
				Winter grass						
FB	10.67 b	5.00 d	1.00 ef	0.00 f	3.33 B	0.00	0.00	0.00	0.00	0.00
MB	0.00 f	0.00 f	0.00 f	0.00 f	0.00 D	0.00	0.00	0.00	0.00	0.00
CB	12.00 a	5.00 d	1.33 e	1.00 ef	3.87 A	0.00	0.00	0.00	0.00	0.00
M*B	9.00 c	0.00 f	0.00 f	1.33 e	2.07 C	0.00	0.00	0.00	0.00	0.00
SB	0.00 f	0.00 f	0.00 f	0.00 f	0.00 D	0.00	0.00	0.00	0.00	0.00
Means (WCS)	6.33 A	2.00 B (68.4)	0.47 C (92.57)	0.47 C (92.57)		0.00	0.00	0.00	0.00	
LSD value ($p < 0.05$)	WCS = 0.52, CS = 0.52, WCS × CS = 1.16					WCS = NS, CS = NS, WCS × CS = NS				

Means not having common letter for individual and interactive effects significantly vary from each other at $p \leq 0.05$. Here, WF = weed free (control), WC = weedy check (control), FS = false seedbed, CC= chemical control, AWE = allelopathic water extracts, FB= fallow-barley, MB = maize-barley, CB = cotton-barley, M*B = mungbean-barley, SB = sorghum-barley, WCS = weed control strategies, CS = cropping system, DAS = days after sowing, NS = Non-significant. The values presented in brackets indicated the % decrease in the number of winter grass plants than WC (control).

Table 4. Influence of different barley-based cropping systems on individual density (m^{-2}) of broadleaved weed species under various weed management methods during 2017–2018 and 2018–2019.

Cropping Systems	2017–2018					2018–2019				
	WC	FS	CC	AWE	MEANS	WC	FS	CC	AWE	MEANS
				Common goosefoot						
FB	7.33 b	4.67 cd	0.33 h	3.00 ef	3.06 A	9.00 b	2.67 e–g	0.00 i	4.33 cd	3.20 B
MB	3.33 d–f	1.33 gh	0.00 h	0.33 h	1.00 B	4.33 cd	1.00 hi	0.00 i	2.00 f–h	1.47 D
CB	8.00 b	5.67 c	0.67 h	4.00 de	3.67 A	7.67 b	1.33 g–i	0.00 i	3.00 d–f	2.40 C
M*B	10.33 a	2.33 fg	0.00 h	3.33 d–f	3.20 A	12.67 a	3.67 de	0.00 i	5.67 c	4.40 A
SB	3.00 ef	1.00 gh	0.00 h	1.00 gh	1.00 B	4.00 de	1.00 hi	0.00 i	1.33 g–i	1.27 D
Means (WCS)	6.40 A	3.00 B (53.12)	0.20 D (96.87)	2.33 C (63.59)		7.53 A	1.93 C (74.36)	0.00 D (100)	3.27 B (56.57)	
LSD value ($p < 0.05$)	WCS = 0.66, CS = 0.66, WCS × CS = 1.48					WCS = 0.68, CS = 0.68, WCS × CS = 1.53				
				Perennial sow thistle						
FB	4.33 a	2.33 b	0.00 e	1.33 b–d	1.60	7.67 b	2.67 e–h	0.00 j	3.67 c–e	2.80
MB	1.33 b–d	0.67 de	0.00 e	0.67 de	0.53	3.33 d–f	1.00 ij	0.00 j	1.67 g–i	1.20
CB	2.33 b	1.00 c–e	0.67 de	1.00 c–e	1.00	8.33 ab	2.33 e–i	0.00 j	5.00 c	3.13
M*B	5.00 a	2.00 bc	0.00 e	1.67 b–d	1.73	9.33 a	2.33 e–i	1.00 ij	4.67 cd	3.47
SB	1.67 b–d	1.00 c–e	0.00 e	0.67 de	0.67	3.00 e–g	1.33 h–j	0.00 j	2.00 f–i	1.27
Means (WCS)	2.93 A	1.40 B (52.21)	0.13 C (95.56)	1.07 B (63.48)		6.33 A	1.93 C (69.51)	0.20 D (96.84)	3.40 B (46.28)	
LSD value ($p < 0.05$)	WCS = 0.48, CS = NS, WCS × CS = 1.08					WCS = 0.68, CS = NS, WCS × CS = 1.51				

Table 4. *Cont.*

Cropping Systems	2017–2018					2018–2019				
	WC	FS	CC	AWE	MEANS	WC	FS	CC	AWE	MEANS
				Bitter dock						
FB	17.67 b	9.00 ef	0.00 j	14.33 c	8.20 B	24.33 b	15.00 e	0.00 m	8.33 hi	9.53 A
MB	9.00 ef	3.67 hi	0.00 j	4.67 h	3.47 D	10.33 g	7.00 i	0.00 m	3.00 kl	4.07 C
CB	20.33 a	10.67 de	0.00 j	16.33 b	9.47 A	20.00 c	12.67 f	0.00 m	6.33 j	7.80 B
M*B	13.33 c	6.67 g	0.00 j	11.00 d	6.20 C	26.33 a	17.67 d	0.00 m	7.00 ij	10.20 A
SB	6.00 fg	2.00 i	0.00 j	2.67 i	2.53 E	9.00 gh	4.33 k	0.00 m	1.67 lm	3.00 D
Means (WCS)	13.67 A	6.40 C (53.18)	0.00 D (100)	9.80 B (28.31)		18.00 A	11.33 B (37.05)	0.00 D (100)	5.27 C (70.72)	
LSD value ($p < 0.05$)	WCS = 0.84, CS = 0.84, WCS × CS = 1.88					WCS = 83, CS = 0.83, WCS × CS = 1.86				
				Fat hen						
FB	6.33 a	4.00 bc	0.00 h	4.00 bc	2.87 A	6.67 a	3.00 bc	0.00 g	2.67 cd	2.47 A
MB	2.00 d–f	2.00 d–f	0.00 h	1.67 e–g	1.13 C	1.67 d–f	1.00 e–g	0.00 g	1.33 ef	0.80 CD
CB	4.00 bc	2.33 b–f	0.00 h	2.67 de	1.80 B	4.00 b	2.00 c–e	0.00 g	1.67 d–f	1.53 B
M*B	5.00 b	1.67 e–g	0.00 h	3.00 cd	1.93 B	3.00 bc	1.33 ef	0.00 g	1.00 e–g	1.07 BC
SB	2.00 d–f	1.33 fg	0.00 h	0.67 gh	0.80 C	1.33 ef	0.67 fg	0.00 g	0.67 fg	0.53 D
Means (WCS)	3.87 A	2.27 B (41.34)	0.00 C (100)	2.40 B (37.98)		3.33 A	1.60 B (51.95)	0.00 C (100)	1.47 B (55.85)	
LSD value ($p < 0.05$)	WCS = 0.56, CS = 0.56, WCS × CS = 1.24					WCS = 0.49, CS = 0.49, WCS × CS = 1.10				

Means not having common letter for individual and interactive effects significantly vary from each other at $p \leq 0.05$. Here, WF = weed free (control), WC = weedy check (control), FS = false seedbed, CC = chemical control, AWE = allelopathic water extracts, FB = fallow-barley, MB = maize-barley, CB = cotton-barley, M*B = mungbean-barley, SB = sorghum-barley, WCS = weed control strategies, CS = cropping system, DAS = days after sowing. The values presented in brackets indicated the % decrease in the number of fat hen plants than WC (control).

Table 5. Influence of different barley-based cropping systems on individual weeds density (m^{-2}) under various weed management methods during 2017–2018 and 2018–2019.

Cropping Systems	2017–2018					2018–2019				
	WC	FS	CC	AWE	MEANS	WC	FS	CC	AWE	MEANS
				Field bindweed						
FB	0.00 f	0.00 f	0.00 f	0.00 f	0.00 NS	0.00 b	0.00 b	0.00 b	0.00 b	0.00 NS
MB	1.67 bc	0.67 d–f	0.00 f	1.00 c–e	0.67	0.00 b	0.00 b	0.00 b	0.00 b	0
CB	2.00 b	0.67 d–f	0.00 f	0.33 ef	0.6	0.00 b	0.00 b	0.00 b	0.00 b	0
M*B	4.00 a	1.67 bc	0.33 ef	1.33 b–d	1.46	0.00 b	1.33 a	0.00 b	0.00 b	0.27
SB	1.33 b–d	0.00 f	0.00 f	0.67 d–f	0.4	0.00 b	0.00 b	0.00 b	0.00 b	0
Means (WCS)	1.80 A	0.60 B (66.67)	0.07 C (96.11)	0.67 B (62.77)		0.00 B	0.27 A (100)	0.00 B (100)	0.00 B (100)	
LSD value ($p < 0.05$)	WCS = 0.38, CS = NS, WCS × CS = 0.85					WCS = 0.08, CS = NS, WCS × CS = 0.19				
				Yellow trefoil						
FB	13.00 b	7.33 de	0.00 i	7.67 d	5.60 A	13.67 a	6.67 cd	0.00 i	8.67 b	5.80 A
MB	10.33 c	5.00 ef	0.00 i	2.33 gh	3.53 B	5.00 de	1.00 hi	0.00 i	2.00 f–h	1.60 C
CB	12.67 b	7.00 d	0.00 i	5.00 ef	4.93 A	9.33 b	3.33 ef	0.00 i	5.33 d	3.60 B
M*B	18.33 a	6.33 de	0.00 i	4.00 fg	5.73 A	15.33 a	6.00 d	1.33 g–i	8.33 bc	6.20 A
SB	8.00 d	3.00 gh	0.00 i	1.33 hi	2.47 C	5.67 d	1.33 g–i	0.00 i	3.00 fg	2.00 C
Means (WCS)	12.47 A	5.73 B (54.05)	0.00 D (100)	4.07 C (67.36)		9.80 A	3.67 C (62.55)	0.27 D (97.24)	5.47 B (44.18)	
LSD value ($p < 0.05$)	WCS = 0.89, CS = 0.89, WCS × CS = 2.00					WCS = 0.78, CS = 0.78, WCS × CS = 1.75				
				Yellow sweet clover						
FB	14.33 c	2.67 ij	0.00 k	7.33 ef	4.87 B	16.00 b	9.00 d	0.00 l	6.67 ef	6.33 A
MB	8.00 e	4.67 g–i	0.00 k	5.33 f–h	3.60 C	6.33 e–g	3.00 i–k	0.00 l	1.67 j–l	2.20 C
CB	20.33 a	7.33 ef	0.00 k	16.67 b	8.87 A	12.33 c	5.67 fg	0.00 l	3.33 h–j	4.27 B
M*B	12.00 d	3.33 h–j	0.00 k	8.00 e	4.67 B	18.33 a	7.67 de	1.33 kl	4.67 g–i	6.40 A
SB	6.67 e–g	1.33 jk	0.00 k	4.00 hi	2.40 D	5.00 f–h	1.67 j–l	0.00 l	1.67 j–l	1.67 C
Means (WCS)	12.27 A	3.87 C (68.45)	0.00 D (100)	8.27 B (32.59)		11.60 A	5.40 B (53.44)	0.27 D (97.67)	3.60 C (68.96)	
LSD value ($p < 0.05$)	WCS = 0.90, CS = 0.90, WCS × CS = 2.01					WCS = 0.82, CS = 0.82, WCS × CS = 1.83				

Table 5. *Cont.*

Cropping Systems	2017–2018					2018–2019				
	WC	FS	CC	AWE	MEANS	WC	FS	CC	AWE	MEANS
				Swine cress						
FB	4.00 a	0.00 d	0.00 d	1.33 c	1.06 NS	2.33 c	1.00 ef	0.00 g	1.33 de	0.93 B
MB	0.00 d	0.00 d	0.00 d	0.33 d	0.06	1.00 ef	0.67 e–g	0.00 g	0.33 fg	0.40 C
CB	2.00 b	0.00 d	0.00 d	1.33 c	0.67	3.67 b	2.00 cd	0.00 g	1.33 de	1.40 A
M*B	2.00 b	0.00 d	0.00 d	0.00 d	0.4	5.33 a	2.33 c	0.00 g	1.00 ef	1.73 A
SB	0.00 d	0.00 d	0.00 d	0.00 d	0	0.67 e–g	0.00 g	0.00 g	0.00 g	0.13 C
Means (WCS)	1.60 A	0.00 C (100)	0.00 C (100)	0.60 B (62.5)		2.60 A	1.20 B (53.84)	0.00 D (100)	0.80 C (69.23)	
LSD value ($p < 0.05$)	WCS = 0.29, CS = 0.29, WCS × CS = 0.65					WCS = 0.39, CS = 0.39, WCS × CS = 0.88				

Means not having common letter for individual and interactive effects significantly vary from each other at $p \leq 0.05$. Here, WF = weed free (control), WC = weedy check (control), FS = false seedbed, CC = chemical control, AWE = allelopathic water extracts, FB = fallow-barley, MB = maize-barley, CB = cotton-barley, M*B = mungbean-barley, SB = sorghum-barley, WCS = weed control strategies, CS = cropping system, DAS = days after sowing, NS = Non-significant. The values presented in brackets indicated the % decrease in the number of winter grass plants than WC (control).

Table 6. Influence of different barley-based cropping systems on individual weeds density (m^{-2}) under various weed management methods during 2017–2018 and 2018–2019.

Cropping Systems	2017–2018					2018–2019				
	WC	FS	CC	AWE	MEANS	WC	FS	CC	AWE	MEAN
				Blue pimpernel						
FB	0.00	0.00	0.00	0.00	0.00	3.00 b	1.33 de	0.00 f	2.00 cd	1.27
MB	0.00	0.00	0.00	0.00	0.00	1.00 e	0.00 f	0.00 f	0.67 ef	0.33
CB	0.00	0.00	0.00	0.00	0.00	0.00 f	0.00 f	0.00 f	0.00 f	0.00
M*B	0.00	0.00	0.00	0.00	0.00	4.33 a	1.33 de	0.00 f	2.33 bc	1.60
SB	0.00	0.00	0.00	0.00	0.00	0.67 ef	0.00 f	0.00 f	0.00 f	0.13
Means (WCS)	0.00	0.00	0.00	0.00		1.80 A	0.53 C (70.55)	0.00 D (100)	1.00 B (44.44)	
LSD value ($p < 0.05$)	WCS = NS, CS = NS, WCS × CS = NS					WCS = 0.39, CS = NS, WCS × CS = 0.88				
				Horseweed						
FB	0.00	0.00	0.00	0.00	0.00	1.33 c	0.00 d	0.00 d	0.00 d	0.27
MB	0.00	0.00	0.00	0.00	0.00	0.00 d	0.00 d	0.00 d	0.00 d	0.00
CB	0.00	0.00	0.00	0.00	0.00	2.00 b	0.00 d	0.00 d	0.00 d	0.40
M*B	0.00	0.00	0.00	0.00	0.00	2.67 a	0.00 d	0.00 d	1.33 c	0.80
SB	0.00	0.00	0.00	0.00	0.00	0.00 d	0.00 d	0.00 d	0.00 d	0.00
Means (WCS)	0.00	0.00	0.00	0.00		1.20 A	0.00 C	0.00 C	0.27 B	
LSD value ($p < 0.05$)	WCS = NS, CS = NS, WCS × CS = NS					WCS = 0.25, CS = NS, WCS × CS = 0.56				

Means not having common letter for individual and interactive effects significantly vary from each other at $p \leq 0.05$. Here, WF = weed free (control), WC = weedy check (control), FS = false seedbed, CC = chemical control, AWE = allelopathic water extracts, FB = fallow-barley, MB = maize-barley, CB = cotton-barley, M*B = mungbean-barley, SB = sorghum-barley, WCS = weed control strategies, CS = cropping system, DAS = days after sowing, NS = Non-significant. The values presented in brackets indicated the % decrease in the number of winter grass plants than WC (control).

The highest density of corn spurry (*Spergula arvensis* L.) was found in WC treatment, while the lowest density was noted for CC and WF treatments during both years (Table 3). The CB cropping system recorded the highest corn spurry density, while the lowest was recorded for SB and MB cropping systems during the first year (Table 3). However, FB cropping system noted the highest corn spurry density during the second year, while no infestation was noted in SB system during the second year (Table 3). Regarding interaction, CB cropping system with WC treatment observed the highest density of corn spurry during the first year. Similarly, the highest corn spurry density was noted in FB cropping system with WC treatment during the second year. However, there was little, or no infestation recorded for all cropping systems with CC and WF treatments during the second year (Table 3).

Winter grass (*Polypogon monspeliensis* L. Desf.) was only recorded during the first year, while no infestation of this weed was noted during the second year (Table 3). The WC treatment had the highest density of winter grass, while CC and AWE treatments had the lowest infestation like WF treatment (Table 3). The CB system recorded the highest density of winter grass, while no infestation of this weed was recorded in MB and SB cropping systems. Regarding interaction, the highest density of winter grass was observed in CB cropping system with WC treatment, whereas all cropping systems with CC and AWE had low or no infestation like WF treatment (Table 3).

The highest common goosefoot (*Chenopodium album* L.) density was noted for WC treatment, while the lowest infestation was recorded for CC and WF treatments during both years (Table 4). The FB, CB, and M*B cropping systems recorded the highest, while MB and SB cropping systems had the lowest common goosefoot infestation during the first year (Table 4). However, M*B cropping system recorded the highest, while SB and MB cropping systems observed the lowest common goosefoot density during the second year (Table 4). Regarding interaction, M*B cropping system with WC treatment had the highest common goosefoot density, while all cropping systems with CC and WF treatments recorded no infestation of this weed during both years (Table 4).

Weedy-check treatment recorded the highest perennial sow thistle (*Sonchus arvensis* L.) density, while the lowest density was noted in CC and WF treatments during both years (Table 4). Cropping systems had non-significant effect on perennial sow thistle density during both years (Table 4). Regarding interaction, the highest perennial sow thistle infestation was noted in M*B and FB cropping systems with WC treatment during the first year. The M*B cropping system with WC treatment recorded the highest infestation of perennial sow thistle, which was statistically at par with CB cropping system under WC treatment during the second year (Table 4). However, all cropping systems with CC and WF treatments observed little or no infestation of this weed during both years (Table 4).

The highest bitter dock (*Rumex obtusifolius* L.) density was recorded in WC treatment, while no infestation of this weed was found in CC and WF treatments during both years (Table 4). The CB cropping system during the first year and M*B as well as FB cropping systems during the second year recorded the highest bitter dock infestation, while the lowest infestation was noted in SB cropping system during both years (Table 4). Regarding interaction, the highest density of bitter dock was noted in CB and M*B cropping systems with WC treatments during the first and second year, respectively. No infestation of this weed was noted in all cropping systems with CC and WF treatments during both years (Table 4).

Weedy-check treatment had the highest and CC as well as WF treatments recorded the lowest fat hen (*Chenopodium murale* L.) density during both years (Table 4). The FB cropping system recorded the highest density of fat hen, while the lowest density was recorded for SB and MB cropping systems during both years (Table 4). The FB cropping system with WC resulted in the highest fat hen infestation, while the lowest infestation was noted in all cropping system with CC and WF treatments during both years (Table 4).

The highest density of field bindweed (*Convolvulus arvensis* L.) was recorded for WC treatment, whereas the lowest was noted for CC and WF treatments during the first year. However, all WCM except FS recorded no infestation during the second year (Table 5). All cropping systems had non-significant effect on field bindweed density during both years (Table 5). Regarding interaction, the highest infestation of field bindweed was recorded in M*B cropping system with WC treatment, while little or no infestation was recorded in all cropping systems with CC and WF treatments (Table 5).

Weedy-check treatment resulted in the highest density of yellow trefoil (*Medicago polymorpha* L.), while the lowest density was recorded for CC and WF treatments during both years (Table 5). The FB, CB, and M*B cropping systems observed the highest infestation of yellow trefoil density, while the lowest infestation was recorded in SB cropping system during the first year (Table 5). Nonetheless, the highest density of yellow trefoil was noted in M*B and FB cropping systems, while MB and SB cropping systems recorded the lowest density during the second year (Table 5). The M*B cropping system with WC treatment had the highest infestation of yellow trefoil during the first year. During the second year, M*B and FB cropping systems with WC treatment observed the highest density of yellow trefoil (Table 5). The lowest infestation of yellow trefoil was observed in all cropping systems with CC and WF treatments during both years (Table 5).

The highest and the lowest density of yellow sweet clover (*Melilotus indicus* (L.) All.) was recorded for WC treatment, and CC and WF treatments, respectively, during both years (Table 5). The CB cropping system had the highest and SB cropping system recorded the lowest density of yellow sweet clover during the first year. The M*B and FB cropping systems noted the highest, while SB and MB cropping systems recorded the lowest infestation of yellow sweet clover during the second year (Table 5). The CB cropping system with WC treatment had the highest yellow sweet clover density during the first year; however, M*B cropping system with WC treatment had the highest yellow sweet clover density during the second year. The lowest density of yellow sweet clover was noted in all cropping systems under CC and WF treatments during both years (Table 5).

Weedy-check treatment had the highest infestation of swine cress (*Coronopus didymus* L. Sm.), while CC, AWE, and WF treatments recorded no infestation during the first year (Table 5). During the second year, WC treatment observed the highest swine cress density, while CC and WF treatments recorded no infestation (Table 5). Cropping systems had non-significant effect on swine cress density during the first year; however, M*B and CB cropping systems observed the highest, while SB and MB cropping systems resulted in the lowest infestation of swine cress during the second year (Table 5). Regarding interaction, FB cropping system with WC treatment resulted in the highest swine cress density, while no infestation was noted in all cropping systems with CC, FS, and WF treatments (Table 5). The M*B cropping system with WC treatment recorded the highest swine cress density, while no infestation was noted in all cropping systems under CC and WF treatments during the second year (Table 5).

All cropping systems with all WCM recorded no infestation of blue pimpernel (*Anagallis arvensis* L.) during the first year (Table 6). However, WC treatment had the highest, while CC and WF treatments recorded no infestation during the second year (Table 6).

No horseweed (*Conyza stricta* Willd.) infestation was recorded in all cropping systems under all WCM during the first year (Table 6). However, WC treatment observed the highest horseweed density, while CC, FS, and WF treatments recorded no infestation during the second year (Table 6). There was non-significant effect of cropping systems during the second year (Table 6). Regarding interaction, M*B cropping system with WC treatment recorded the highest horseweed density, while all cropping systems under CC, FS, and WF treatments, all cropping systems except M*B system under AWE, MB, and SB cropping systems under WC treatment did not have any of this weed during the second year (Table 6).

Dry biomass yield of barley was significantly ($p < 0.05$) influenced by WCM and CS; however, their interactive effect was non-significant during both years (Table 7). The FB cropping system had the lowest, whereas M*B system recorded the highest dry biomass yield, which was at par with MB cropping systems during the first year. However, FB system recorded the lowest dry biomass yield, which was statistically similar to SB system, while the highest was recorded in M*B system during the second year (Table 7). In case of WCM, WF treatment produced the highest dry biomass yield of barley, which was at par with CC treatment, while the lowest was recorded in FS treatment, which was similar to AWE treatment during the first year of the experiment. Nonetheless, WF treatment recorded the highest, and FS, as well as AWE treatment, had the lowest dry biomass yield during the second year (Table 7).

Table 7. Influence of different barley-based cropping systems on dry biomass yield (g m^{-2}) under various weed management methods.

Cropping Systems	2017–2018					
	WF	WC	FS	CC	AWE	Means (CS)
FB	332.99	248.14	316.97	330.07	324.63	310.56 C
MB	349.35	273.39	329.00	343.35	336.53	326.32 AB
CB	345.79	265.06	328.92	341.49	332.83	322.82 B
M*B	369.85	259.24	342.94	346.29	341.36	331.94 A
SB	338.56	258.07	328.65	337.08	327.66	318.00 BC
Means (WCS)	347.31 A	260.78 D	329.30 C	339.65 AB	332.60 BC	
LSD at $p \leq 0.05$	WCS = 9.03, CS = 9.03, WCS × CS = NS					
	2018–2019					
FB	334.33	256.28	316.50	330.40	322.85	312.07 C
MB	346.86	277.38	325.03	343.37	331.20	324.77 B
CB	351.25	270.65	326.08	339.24	328.52	323.15 B
M*B	372.25	267.95	343.48	346.43	340.27	334.07 A
SB	340.15	266.45	326.05	336.65	326.11	319.08 BC
Means (WCS)	348.97 A	267.74 D	327.43 C	339.22 B	329.79 C	
LSD at $p \leq 0.05$	WCS = 8.90, CS = 8.90, WCS × CS = NS					

Means not having common letter for individual and interactive effects significantly vary from each other at $p \leq 0.05$. Here, WF = weed free (control), WC = weedy check (control), FS = false seedbed, CC = chemical control, AWE = allelopathic water extracts, FB = fallow-barley, MB = maize-barley, CB = cotton-barley, M*B = mungbean-barley, SB = sorghum-barley, WCS = weed control strategies, CS = cropping system, DAS = days after sowing, NS = Non-significant.

3. Discussion

Weed flora, including weed diversity and densities of broadleaved, grassy, and individual weed species were significantly altered by different barley-based CS, WCM, and their interaction (Figure 1, Tables 1–6). This supported our hypothesis that barley-based cropping systems and WCM would differ for weed flora and dry biomass production of barley. The highest density and diversity were recorded in WC treatment, while the lowest were noted in WF treatment. The weeds were efficiently controlled by CC as compared to the rest of WCM used in the study. Interestingly, AWE provided sufficient control over weeds after CC indicating that this method could be used to mitigate the adverse effects associated with CC. The highest weeds' diversity and density were recorded in CB and M*B cropping systems during the first and second year, respectively. The lowest was noted in SB cropping system (Figure 1, Tables 2–6).

Weed diversity and density were significantly reduced by herbicides compared to weedy-check treatment in wheat crop [20]. However, unnecessary and overuse of herbicides has increased the evolution of herbicide resistance in several weed species [40]. Similarly, widespread use of herbicides causes anxiety in the public regarding the adverse effects on the environment and human health [41]. Therefore, it is essential to use some integrated weed management approaches for efficient weed control [42]. False or stale seedbed is regarded as an efficient integrated weed management method as it significantly reduced weeds density and dry biomass compared to WC [20,25,26]. Some weeds like *Eclipta prostrata* L., *Fimbristylis miliacea* (L.) Vahl, *Cyperus iria* L., *Leptochloa chinensis* (L.) Nees, and *Cyperus difformis* L. are comparatively more influenced by FS treatment due to their inability to emerge from a depth >1 cm and low seed dormancy [43]. In this experiment, weed density and diversity were significantly reduced by AWE due to phytotoxic effects. However, FS failed to suppress weed flora. The possible reasons of FS failure are unavailability of sufficient moisture before true seedbed, which reduced seed germination of weed species. Allelopathic water extracts inhibit photosynthesis, cell division, thickness of seminal roots, protein synthesis, and respiration by reducing nutrients and water uptake through roots, which negatively affect weed growth [44]. Sorghum is a renowned allelopathic crop that has the ability to control weed growth owing to the release of sorgoleone from roots [34]. Members of the Brassicaceae family release glucosinolate that gets decomposed into many biologically active compounds, including isothiocyanate [45]. Isothiocyanate effectively suppresses weed growth [46]. Weeds are also controlled by the allelochemicals of sunflower (terpenes and phenolic compounds) [32] and mulberry (steroids, phenols, and tannins) [33]. Therefore, weeds could be controlled by the combination of sorghum, sunflower, eucalyptus, and mulberry alleopathic water extracts.

The CB and M*B cropping systems had the highest weed diversity and density, while SB cropping system recorded the lowest in this regard (Figure 1, Tables 2–6). Similar results were reported by Shehzad et al. [20], where fallow-wheat and rice-wheat rotations favored different weed flora, while sorghum–wheat rotation reduced weed growth. In the current study, CB and M*B cropping systems favored the infestation of common goosefoot, bitter dock, yellow trefoil, yellow sweet clover, and swine cress. Similarly, FB cropping system observed the highest density of yellow sweet clover, yellow trefoil, fat hen, bitter dock, common goosefoot, corn spurry, and salt marsh, while SB cropping system recorded the lowest weed density. Rotating crops that have different cultivation practices or life cycles is an efficient cultural practice for controlling problematic weed species through disturbing their life cycles [47]. It is an effective approach to control weeds; however, it is more efficient when combined with any other weed management practice [48]. Similarly, weeds are suppressed by different management method and the inclusion allelopathic crops in rotation [49]. Different experiments showed that the growth of cultivated crops is significantly affected by allelochemicals exuded from sorghum roots [13,49]. Therefore, the lowest weeds population was recorded in SB cropping system during both years in the current study.

The highest dry biomass yield of barley was noted in WF treatment, while the lowest was recorded in WC treatment (Table 7). The M*B cropping system had the highest dry biomass yield, while the lowest was recorded in FB cropping system (Table 7). The FB cropping system had more weed infestation, which reduced yield-related traits of barley [8]. Weeds negatively affect crop growth by competing for nutrients and other essential resources [50]. However, crops can perform better in the absence of weeds [20].

The M*B cropping system improved dry biomass yield due to better soil condition resulting in better allometric traits and root growth. Therefore, plants dry biomass yield was improved by absorbing more water and nutrients from soil. It has been described by Zhao et al. [51] that the soil fertility and crop productivity can be efficiently increased by practicing legume-based crop rotation. Crop diversification with legumes had significant effect on soil fertility as it improves the status of phosphorus nitrogen, carbon, and soil organic carbon depending upon the soil type [52]. Similar results were reported in the current study.

4. Materials and Methods

4.1. Experimental Site and Soil

This experiment was conducted during 2017–2018 and 2018–2019 at the Agronomic Research Area, Department of Agronomy, Bahauddin Zakariya University, Multan (30.2° N, 71.43° E and 122 m above sea level), Pakistan. The study area had an arid to semi-arid climate. Weather data of the experimental site during study period are given in Table 8. The study site has loamy soil with pH values of 8.20–8.25, ECe 2.78–2.80 mS cm^{-1}, 0.60–0.63% organic matter content, 0.03% total nitrogen, 7.25–7.18 mg kg^{-1} available phosphorus and 240–230 mg kg^{-1} available potash during the first and second year of the study, respectively.

Table 8. Weather data for the period of research at the experimental site.

Months	2017–2018				2018–2019			
	Mean Temperature (°C)	Mean Relative Humidity (%)	Mean Daily Sunshine (h)	Total Monthly Rainfall (mm)	Mean Temperature (°C)	Mean Relative Humidity (%)	Mean Daily Sunshine (h)	Total Monthly Rainfall (mm)
May	34.00	63.05	4.80	0.10	32.90	52.60	10.30	0.00
June	33.10	74.90	4.50	45.60	34.60	64.70	3.50	0.00
July	33.65	73.00	7.20	4.90	33.20	71.20	5.50	0.00
August	31.80	85.20	7.70	30.00	32.40	75.10	4.30	0.00
September	30.60	77.10	8.00	10.00	29.80	77.10	6.80	0.00
October	27.00	77.60	7.40	4.20	23.00	75.10	5.50	0.00
November	18.00	81.40	3.70	16.00	18.90	82.25	4.40	0.00
December	14.65	75.00	5.20	16.00	14.25	85.00	5.90	0.00
January	13.65	83.10	4.40	0.00	12.20	86.35	4.30	11.00
February	17.50	75.40	4.90	6.80	14.45	80.60	6.70	25.10
March	23.50	70.90	7.20	0.00	19.55	75.95	7.30	21.00
April	29.45	56.70	5.40	3.00	28.60	73.15	7.70	12.70

4.2. Experiment Description

Barley was cultivated in five different cropping systems, i.e., fallow-barley (FB), maize-barley (MB), cotton-barley (CB), mungbean-barley (M*B), and sorghum-barley (SB). Similarly, five different weed control methods, i.e., weed-free (control; WF), weedy-check (control; WC), false seedbeds (FS), chemical control (CC), and allelopathic water extracts (AWE) were used to test their impact on weed flora and biomass production of weeds and barley. Regarding WF treatment, the weeds were completely removed from the experimental plots during the entire growth period of barley crop, whereas weeds were retained for the whole cropping period in WC treatment. In case of FS treatment, experimental field was tilled and kept fallow for seven days to allow weeds' growth. Afterwards, the emerged weeds were removed by cultivating the field and seedbed was prepared. For CC treatment, 'Bromoxynil + MCPA' (60% EC) was sprayed @1.25 L ha^{-1} after one week of 1st irrigation. In AWE treatment, water extracts of mulberry, sorghum, eucalyptus, and sunflower were prepared and mixed in equal ratio. Afterwards these were sprayed @12 L ha^{-1} after one week of 1st irrigation. The leaves and branches of all crops were taken, chopped into small pieces, and dried under sun for the preparation of AWE. The dried materials were then soaked in distilled water (1:20 ratio), separately for 24 h. The solutions were filtered after 24 h to obtain the extracts. The resulting extracts were then mixed in a 1:1:1:1 ratio, diluted by 10 times, and sprayed. Each treatment was replicated three times and net subplot size was 2.7 × 5 m. The study was carried out according to randomized complete block design (RCBD) with factorial arrangement. Barley-based cropping systems were the main factor, whereas weed control methods were considered as a sub-factor.

4.3. Crop Husbandry

Before sowing of all crops, 10 cm irrigation was applied to whole field during both years of study. Afterwards, seedbeds of all crops were prepared once the soil attained feasible moisture level. All crops were sown according to their recommended production technology as given in Table 9. All crops were irrigated by surface irrigation method to fulfill their moisture requirements. All agronomic and plant protection measures were adopted to ensure healthy crop and to avoid pest and diseases. Finally, all crops were harvested at their harvest maturity.

Table 9. Crop husbandry of different crops included in barley-based cropping systems of the study.

Crops	Sowing Time	Cultivars	Seed Rate (kg ha^{-1})	Fertilizer NPK (kg ha^{-1})	P–P (cm)	R–R (cm)	Harvest Date
Year 2017 and 2018 (Summer Season)							
Cotton	15 May	IUB-2013	25	250–200–0	20	75	28 October
Sorghum	10 June	YS-16	10	100–60–0	15	60	29 October
Mungbean	15 June	NIAB-Mung 2011	20	20–60–0	10	30	27 September
Maize	25 July	YH-1898	25	200–150–0	22	75	30 October
Year 2017–2018 and 2018–2019 (Winter Season)							
Barley	10 November	Haider-93	80	50–25–0		25	7 and 10 April

P–P = Plant spacing; R–R = Row spacing.

4.4. Weeds Data Collection

Data relating to weeds' diversity (number of weed species), density of broad-leaved and grassy weeds, and density of all individual weeds were recorded at 60 DAS during both years of study. Data were collected from three randomly selected locations in each experimental plot with the help of 1 m^2 quadrate [8,37]. Weed diversity was recorded by observing all species in 1 m^2 at three random places in each experimental unit and averaged. Total number of weed species present in each quadrate were noted, identified, and averaged to record the weeds' diversity. The densities of broadleaved, grassy, and individual weeds were recorded by randomly placing the quadrate at three different places in each experimental unit. The observed weed species for density were separated into broadleaved, grassy, and individual weeds.

4.5. Biomass Yield

Two central rows of barley from each experimental unit were harvested at 105 DAS. The barley plants were manually harvested at ground level to observe biomass production. The harvested samples were sun-dried for three days and then oven-dried at 75 °C until constant weight. Dry weight of these samples was recorded by using a digital weighing balance.

4.6. Statistical Analysis

The data were tested for differences among experimental years, which indicated that years' effect was significant. Therefore, data of each year were analyzed and interpreted separately. Collected data for both years statistically analyzed by analysis of variance (ANOVA) [53] according to general linear model procedure. Treatments means were compared by least significance difference (LSD) test at 5% probability level, where ANOVA indicated significant differences.

5. Conclusions

Different barley-based cropping systems and weed control methods significantly altered weed flora during both years of the current study. Chemical control resulted in the highest suppression of weed flora and improved dry biomass production of barley followed by allelopathic crop water extracts. The SB cropping system with chemical control or allelopathic crop water extracts resulted in the lowest weed infestation. The M*B cropping system with chemical control, or allelopathic crop water extracts produced the highest dry biomass of barley. It is concluded that including sorghum crop in rotation and applying allelopathic extracts could suppress weeds comparable to herbicides. Similarly, including mungbean in rotation and applying allelopathic extracts could increase dry biomass production of barley. In conclusion, herbicides can be replaced with an eco-friendly approach, i.e., allelopathy and inclusion of sorghum crop could be helpful in suppressing weed flora.

Author Contributions: Conceptualization, M.H.; Methodology, M.N.; Software, M.N.; Validation, M.N., S.F. and M.H.; Formal analysis, M.N.; Investigation, M.N.; Resources, M.H.; Data curation, M.N.; Writing—original draft preparation, M.N.; Writing—review and editing, S.F.; Visualization, M.H.; Supervision, M.H.; Project administration, M.H.; Funding acquisition, M.N. All authors have read and agreed to the published version of the manuscript.

Funding: Funding was provided by Higher Education Commission, Islamabad, Pakistan, to complete the doctoral studies through grant number 17-6/HEC/HRD/IS-II-5/2019.

Institutional Review Board Statement: Not applicable.

Informed Consent Statement: Not applicable.

Data Availability Statement: All data are within the manuscript.

Acknowledgments: The first author acknowledges the financial grant from Higher Education Commission, Islamabad, Pakistan.

Conflicts of Interest: The authors declare no conflict of interest

References

1. Meints, B.; Hayes, P.M. Breeding Naked Barley for Food, Feed, and Malt. In *Plant Breeding Reviews*; Wiley: Hoboken, NJ, USA, 2019; pp. 95–119, ISBN 9781119616801.
2. Visioni, A.; Al-Abdallat, A.; Elenien, J.A.; Verma, R.P.S.; Gyawali, S.; Baum, M. Genomics and molecular breeding for improving tolerance to abiotic stress in barley (*Hordeum vulgare* L.). In *Genomics Assisted Breeding of Crops for Abiotic Stress Tolerance*; Springer: Berlin/Heidelberg, Germany, 2019; Volume II, pp. 49–68.
3. Mwando, E.; Angessa, T.T.; Han, Y.; Li, C. Salinity tolerance in barley during germination—Homologs and potential genes. *J. Zhejiang Univ. B* **2020**, *21*, 93–121. [CrossRef] [PubMed]
4. Cammarano, D.; Ceccarelli, S.; Grando, S.; Romagosa, I.; Benbelkacem, A.; Akar, T.; Al-Yassin, A.; Pecchioni, N.; Francia, E.; Ronga, D. The impact of climate change on barley yield in the Mediterranean basin. *Eur. J. Agron.* **2019**, *106*, 1–11. [CrossRef]
5. Ikram ul Haq, M.; Maqbool, M.M.; Ali, A.; Farooq, S.; Khan, S.; Saddiq, M.S.; Khan, K.A.; Ali, S.; Ifnan Khan, M.; Hussain, A.; et al. Optimizing planting geometry for barley-Egyptian clover intercropping system in semi-arid sub-tropical climate. *PLoS ONE* **2020**, *15*, e0233171. [CrossRef] [PubMed]
6. Zeeshan, M.; Lu, M.; Sehar, S.; Holford, P.; Wu, F. Comparison of biochemical, anatomical, morphological, and physiological responses to salinity stress in wheat and barley genotypes deferring in salinity tolerance. *Agronomy* **2020**, *10*, 127. [CrossRef]
7. Naeem, M.; Mehboob, N.; Farooq, M.; Farooq, S.; Hussain, S.; Ali, H.M.; Hussain, M. Impact of Different Barley-Based Cropping Systems on Soil Physicochemical Properties and Barley Growth under Conventional and Conservation Tillage Systems. *Agronomy* **2021**, *11*, 8. [CrossRef]
8. Naeem, M.; Hussain, M.; Farooq, M.; Farooq, S. Weed flora composition of different barley-based cropping systems under conventional and conservation tillage practices. *Phytoparasitica* **2021**, *49*, 751–769. [CrossRef]
9. Diyanat, M.; Baziar, S. The effect of weed control on growth, yield and physicochemical properties of valerian (*Valeriana officinalis* L.). *J. Res. Weed Sci.* **2020**, *3*, 420–435. [CrossRef]
10. Zimdahl, R.L. *Fundamentals of Weed Science*; Academic Press: Cambridge, MA, USA, 2018; ISBN 0128111445.
11. Riaz Marral, M.W.; Khan, M.B.; Ahmad, F.; Farooq, S.; Hussain, M. The influence of transgenic (Bt) and non-transgenic (non-Bt) cotton mulches on weed dynamics, soil properties and productivity of different winter crops. *PLoS ONE* **2020**, *15*, e0238716. [CrossRef]
12. Farooq, M.; Nawaz, A.; Saharawat, Y.S.; Reeves, T.; Siddique, K. Crop and cropping systems management practices and benefits in Conservation Agriculture systems. In *Advances in Conservation Agriculture: Volume 2: Practice and Benefits*; Burleigh Dodds Science Publishing Limited: Cambridge, UK, 2020.
13. Farooq, N.; Abbas, T.; Tanveer, A.; Jabran, K. Allelopathy for weed management. In *Co-Evolution of Secondary Metabolites*; Springer: Cham, Switzerland, 2020; pp. 505–519. [CrossRef]
14. Farooq, S.; Onen, H.; Tad, S.; Ozaslan, C.; Mahmoud, S.F.; Brestic, M.; Zivcak, M.; Skalicky, M.; El-Shehawi, A.M. The influence of environmental factors on seed germination of *Polygonum perfoliatum* L.: Implications for management. *Agronomy* **2021**, *11*, 1123. [CrossRef]
15. Farooq, S.; Onen, H.; Ozaslan, C.; Baskin, C.C.; Gunal, H. Seed germination niche for common ragweed (*Ambrosia artemisiifolia* L.) populations naturalized in Turkey. *S. Afr. J. Bot.* **2019**, *123*, 361–371. [CrossRef]
16. Önen, H.; Farooq, S.; Tad, S.; Özaslan, C.; Gunal, H.; Chauhan, B.S. The Influence of Environmental Factors on Germination of Burcucumber (*Sicyos angulatus*) Seeds: Implications for Range Expansion and Management. *Weed Sci.* **2018**, *66*, 494–501. [CrossRef]
17. Chauhan, B.S.; Matloob, A.; Mahajan, G.; Aslam, F.; Florentine, S.K.; Jha, P. Emerging challenges and opportunities for education and research in weed science. *Front. Plant Sci.* **2017**, *8*, 1537. [CrossRef] [PubMed]

18. van der Meulen, A.; Chauhan, B.S. A review of weed management in wheat using crop competition. *Crop Prot.* **2017**, *95*, 38–44 [CrossRef]
19. Maqsood, Q.; Abbas, R.N.; Iqbal, M.A.; Serap, K.A.; Iqbal, A.; El Sabagh, A. Overviewing of weed management practices to reduce weed seed bank and to increase maize yield. *Planta Daninha* **2020**, *38*. [CrossRef]
20. Shahzad, M.; Jabran, K.; Hussain, M.; Raza, M.A.S.; Wijaya, L.; El-Sheikh, M.A.; Alyemeni, M.N. The impact of different weed management strategies on weed flora of wheat-based cropping systems. *PLoS ONE* **2021**, *16*, e0247137. [CrossRef]
21. Carballido, J.; Rodríguez-Lizana, A.; Agüera, J.; Pérez-Ruiz, M. Field sprayer for inter and intra-row weed control: Performance and labor savings. *Span. J. Agric. Res.* **2013**, *11*, 642. [CrossRef]
22. Bond, W.; Grundy, A.C. Non-chemical weed management in organic farming systems. *Weed Res.* **2001**, *41*, 383–405. [CrossRef]
23. Smith, R.G.; Ryan, M.R.; Menalled, F.D.; Hatfield, J.L.; Sauer, T.J. Direct and indirect impacts of weed management practices on soil quality. *Soil Manag. Build. Stable Base Agric.* **2011**, 275–286. [CrossRef]
24. Annett, R.; Habibi, H.R.; Hontela, A. Impact of glyphosate and glyphosate-based herbicides on the freshwater environment. *J. Appl. Toxicol.* **2014**, *34*, 458–479. [CrossRef]
25. Kanatas, P.J.J.; Travlos, I.S.S.; Gazoulis, J.; Antonopoulos, N.; Tsekoura, A.; Tataridas, A.; Zannopoulos, S. The combined effects of false seedbed technique, post-emergence chemical control and cultivar on weed management and yield of barley in Greece. *Phytoparasitica* **2020**, *48*, 131–143. [CrossRef]
26. Kanatas, P.; Travlos, I.; Papastylianou, P.; Gazoulis, I.; Kakabouki, I.; Tsekoura, A. Yield, quality and weed control in soybean crop as affected by several cultural and weed management practices. *Not. Bot. Horti Agrobot. Cluj-Napoca* **2020**, *48*, 329–341. [CrossRef]
27. Merfield, C.N. False and Stale Seedbeds: The most effective non-chemical weed management tools for cropping and pasture establishment. *FFC Bull.* **2015**, *2015*, 25.
28. Hussain, M.I.; Reigosa, M.J. Evaluation of photosynthetic performance and carbon isotope discrimination in perennial ryegrass (*Lolium perenne* L.) under allelochemicals stress. *Ecotoxicology* **2017**, *26*, 613–624. [CrossRef] [PubMed]
29. Abbas, T.; Ahmad, A.; Kamal, A.; Nawaz, M.Y.; Jamil, M.A.; Saeed, T.; Abid, M.A.; Ali, H.H.; Ateeq, M. Ways to Use Allelopathic Potential for Weed Management: A Review. *Int. J. Food Sci. Agric.* **2021**, *5*, 492–498. [CrossRef]
30. Jabran, K. Sunflower allelopathy for weed control. In *Manipulation of Allelopathic Crops for Weed Control*; Springer: Berlin/Heidelberg, Germany, 2017; pp. 65–75.
31. Jabran, K. Allelopathy: Introduction and concepts. In *Manipulation of Allelopathic Crops for Weed Control*; Springer: Berlin/Heidelberg, Germany, 2017; pp. 1–12.
32. Weston, L.A.; Alsaadawi, I.S.; Baerson, S.R. Sorghum Allelopathy—From Ecosystem to Molecule. *J. Chem. Ecol.* **2013**, *39*, 142–153. [CrossRef] [PubMed]
33. Díaz Solares, M.; Cazaña Martínez, Y.; Pérez Hernández, Y.; Valdivia Ávila, A.; Prieto Abreu, M.; Lugo Morales, Y. Qualitative evaluation of secondary metabolites in extracts of *Morus alba* L. (Mulberry) varieties and hybrids. *Rev. Cuba. Plantas Med.* **2015**, *20*, 358–366.
34. Hussain, M.I.; Danish, S.; Sánchez-Moreiras, A.M.; Vicente, Ó.; Jabran, K.; Chaudhry, U.K.; Branca, F.; Reigosa, M.J. Unraveling Sorghum Allelopathy in Agriculture: Concepts and Implications. *Plants* **2021**, *10*, 1795. [CrossRef]
35. Graziani, F.; Onofri, A.; Pannacci, E.; Tei, F.; Guiducci, M. Size and composition of weed seedbank in long-term organic and conventional low-input cropping systems. *Eur. J. Agron.* **2012**, *39*, 52–61. [CrossRef]
36. Hosseini, P.; Karimi, H.; Babaei, S.; Mashhadi, H.R.; Oveisi, M. Weed seed bank as affected by crop rotation and disturbance. *Crop Prot.* **2014**, *64*, 1–6. [CrossRef]
37. Shahzad, M.; Farooq, M.; Jabran, K.; Hussain, M. Impact of different crop rotations and tillage systems on weed infestation and productivity of bread wheat. *Crop Prot.* **2016**, *89*, 161–169. [CrossRef]
38. Mahmood, A.; Cheema, Z.A.; Mushtaq, M.N.; Farooq, M. Maize–Sorghum intercropping systems for purple nutsedge management. *Arch. Agron. Soil Sci.* **2013**, *59*, 1279–1288. [CrossRef]
39. Cheema, Z.A.; Khaliq, A. Use of sorghum allelopathic properties to control weeds in irrigated wheat in a semi arid region of Punjab. *Agric. Ecosyst. Environ.* **2000**, *79*, 105–112. [CrossRef]
40. Heap, I. The International Herbicide-Resistant Weed Database. Available online: http://www.weedscience.org/Home.aspx (accessed on 21 December 2021).
41. Mehdizadeh, M.; Mushtaq, W.; Anusha Siddiqui, S.; Ayadi, S.; Kaur, P.; Yeboah, S.; Mazraedoost, S.; AL-Taey, D.K.A.; Tampubolon, K. Herbicide Residues in Agroecosystems: Fate, Detection, and Effect on Non-Target Plants. *Rev. Agric. Sci.* **2021**, *9*, 157–167. [CrossRef]
42. Perotti, V.E.; Larran, A.S.; Palmieri, V.E.; Martinatto, A.K.; Permingeat, H.R. Herbicide resistant weeds: A call to integrate conventional agricultural practices, molecular biology knowledge and new technologies. *Plant Sci.* **2020**, *290*, 110255. [CrossRef]
43. Chauhan, B.S.; Johnson, D.E. The role of seed ecology in improving weed management strategies in the tropics. In *Advances in Agronomy*; Academic Press: Cambridge, MA, USA, 2010; Volume 105, pp. 221–262, ISBN 0065-2113.
44. Jafariehyazdi, E.; Javidfar, F. Comparison of allelopathic effects of some brassica species in two growth stages on germination and growth of sunflower. *Plant Soil Environ.* **2011**, *57*, 52–56. [CrossRef]
45. Weston, L.A.; Duke, S.O. Weed and crop allelopathy. *CRC. Crit. Rev. Plant Sci.* **2003**, *22*, 367–389. [CrossRef]
46. Bangarwa, S.K.; Norsworthy, J.K. Brassicaceae Cover-Crop Effects on Weed Management in Plasticulture Tomato. *J. Crop Improv.* **2014**, *28*, 145–158. [CrossRef]

47. Chhokar, R.S.; Sharma, R.K.; Singh, R.K.; Gill, S.C. Herbicide resistance in littleseed canarygrass (Phalaris minor) and its management. In Proceedings of the 14th Australian Agronomy Conference, Adelaide, Australia, 21–25 September 2008; Australian Society of Agronomy: Adelaide, Australia, 2008; p. 106.
48. Garrison, A.J.; Miller, A.D.; Ryan, M.R.; Roxburgh, S.H.; Shea, K. Stacked Crop Rotations Exploit Weed-Weed Competition for Sustainable Weed Management. *Weed Sci.* **2014**, *62*, 166–176. [CrossRef]
49. Farooq, M.; Nawaz, A.; Ahmad, E.; Nadeem, F.; Hussain, M.; Siddique, K.H.M. Using Sorghum to suppress weeds in dry seeded aerobic and puddled transplanted rice. *Field Crops Res.* **2017**, *214*, 211–218. [CrossRef]
50. Zhang, J.; Wu, L.F. Impact of tillage and crop residue management on the weed community and wheat yield in a wheat–maize double cropping system. *Agriculture* **2021**, *11*, 265. [CrossRef]
51. Zhao, J.; Yang, Y.; Zhang, K.; Jeong, J.; Zeng, Z.; Zang, H. Does crop rotation yield more in China? A meta-analysis. *Field Crops Res.* **2020**, *245*, 107659. [CrossRef]
52. Sauvadet, M.; Trap, J.; Damour, G.; Plassard, C.; Van den Meersche, K.; Achard, R.; Allinne, C.; Autfray, P.; Bertrand, I.; Blanchart, E.; et al. Agroecosystem diversification with legumes or non-legumes improves differently soil fertility according to soil type. *Sci. Total Environ.* **2021**, *795*, 148934. [CrossRef] [PubMed]
53. Steel, R.; Torrei, J.; Dickey, D. *Principles and Procedures of Statistics A Biometrical Approach*; McGraw-Hill: New York, NY, USA, 1997.

plants

MDPI

Article

Effect of Row Spacing and Seeding Rate on Russian Thistle (*Salsola tragus*) in Spring Barley and Spring Wheat

Judit Barroso * and Nicholas G. Genna

Columbia Basin Agricultural Research Center, Oregon State University Adams, OR 97810, USA; nicholas.genna@oregonstate.edu
* Correspondence: judit.barroso@oregonstate.edu; Tel.: +1-541-278-4394

Abstract: Russian thistle (*Salsola tragus* L.) is a persistent post-harvest issue in the Pacific Northwest (PNW). Farmers need more integrated management strategies to control it. Russian thistle emergence, mortality, plant biomass, seed production, and crop yield were evaluated in spring wheat and spring barley planted in 18- or 36-cm row spacing and seeded at 73 or 140 kg ha^{-1} in Pendleton and Moro, Oregon, during 2018 and 2019. Russian thistle emergence was lower and mortality was higher in spring barley than in spring wheat. However, little to no effect of row spacing or seeding rate was observed on Russian thistle emergence or mortality. Russian thistle seed production and plant biomass followed crop productivity; higher crop yield produced higher Russian thistle biomass and seed production and lower crop yield produced lower weed biomass and seed production. Crop yield with Russian thistle pressure was improved in 2018 with 18-cm rows or by seeding at 140 kg ha^{-1} while no effect was observed in 2019. Increasing seeding rates or planting spring crops in narrow rows may be effective at increasing yield in low rainfall years of the PNW, such as in 2018. No effect may be observed in years with higher rainfall than normal, such as in 2019.

Keywords: crop competition; cultural management; rainfall; rain-fed agriculture; seed production; weed suppression; weed density

Citation: Barroso, J.; Genna, N.G. Effect of Row Spacing and Seeding Rate on Russian Thistle (*Salsola tragus*) in Spring Barley and Spring Wheat. *Plants* **2021**, *10*, 126. https://doi.org/10.3390/plants10010126

Received: 11 December 2020
Accepted: 6 January 2021
Published: 9 January 2021

1. Introduction

Russian thistle (*Salsola tragus* L.) is a summer annual broadleaf weed that is widely distributed throughout the western United States [1]. A healthy, well-established winter wheat crop is competitive with Russian thistle, whereas spring crops such as spring wheat (*Triticum aestivum*) or spring barley (*Hordeum vulgare*) suffer from competition-associated yield loss that is exacerbated during dry years [2]. In addition to potential crop yield loss, Russian thistle can reduce the yield of the subsequent crop through rapid postharvest regrowth and soil water depletion [3,4]. Individual Russian thistle plants were previously shown to use 70 L of soil water during spring wheat development and an additional 170 L of soil water following harvest [3].

A majority of wheat produced in the inland Pacific Northwest (PNW) relies on non-selective herbicides or tillage to control weeds following harvest and during fallow. The ubiquitous glyphosate use in this region is selecting for glyphosate-resistance in major agronomic weeds [5]. Russian thistle is one weed that is a persistent issue in low and intermediate precipitation zones of the PNW that has recently been identified as glyphosate-resistant in Oregon, Montana, and Washington [6,7]. Farmers in the PNW need more integrated strategies to control Russian thistle and other weeds to prolong the usefulness of herbicides and the sustainability of wheat production systems in the region. Diversifying the common winter wheat–fallow cropping system in the inland PNW with the introduction of a spring crop could help reduce herbicide pressure and provide opportunities to control winter annuals, in addition to other benefits [8,9]. However, spring crops could facilitate the increase in summer annuals such as Russian thistle.

Reducing crop row spacing or increasing crop seeding rates are two cultural management practices that can increase crop competition and weed suppression in cereals [10–12]. For example, increasing spring barley planting density from approximately 63 to 161 plants m^{-2} increased yield and decreased rigid ryegrass (*Lolium rigidum*) tiller number [13]. Furthermore, for example, decreasing durum winter wheat inter-row spacing from 25 to 5 cm provided significantly greater control of various weed species in two cultivars, with a significant increase in grain yield in one cultivar [14].

However, a higher seeding rate or reduced row spacing is not always beneficial and may have no effect on weed control. Increasing durum winter wheat seeding density from 190 to 570 seeds m^{-2} did not significantly reduce weed biomass or increase yield [14]. Similarly, Kolb et al. [15,16] found similar spring wheat and spring barley yields between different crop densities ranging from 200 to 600 plants m^{-2} and crop row widths ranging from 11 to 23 cm with *Sinapis alba* weed pressure.

Modifying row spacing or seeding rate is an effective cultural management practice with most crops and weed species, although there have been reported cases where it was not effective. Various reasons are employed in instances where weed control is insufficient, including soil quality, year-to-year climate variation, crop cultivar performance, or weed species of interest. Research is necessary to evaluate the efficacy of reducing row spacing and increasing seeding rates to reduce Russian thistle density in spring crops. The objective of this research was to determine how seeding rate and/or row spacing affect Russian thistle emergence, mortality, plant biomass, and seed production within spring barley and spring wheat crops.

2. Results and Discussion

Total precipitation received at the Pendleton site from seeding to harvest was 111 and 150 mm in 2018 and 2019, respectively, and 63 and 83 mm in Moro during 2018 and 2019, respectively. The average total rainfall for Pendleton in 2018 and 2019 was 401 and 412 mm, respectively, while the average total rainfall for Moro in 2018 and 2019 was 215 and 241 mm, respectively (Figure 1).

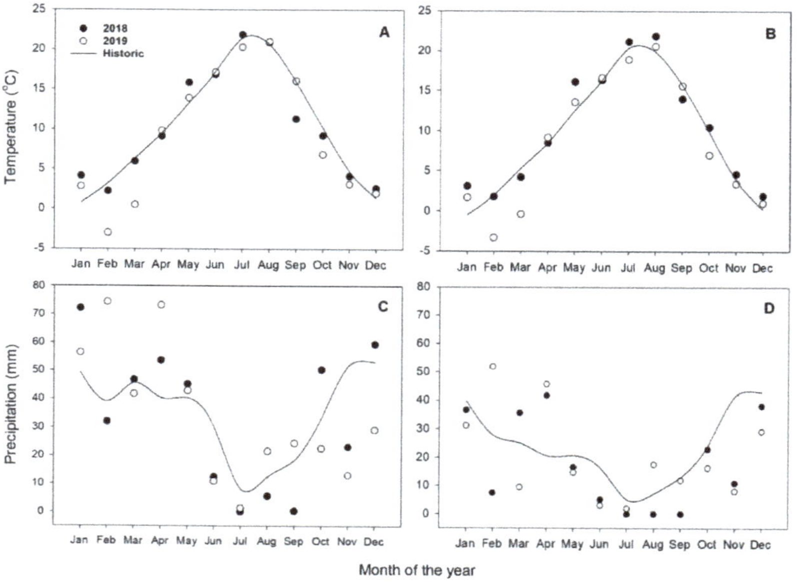

Figure 1. Annual temperature and precipitation for Pendleton (**A**,**C**) and Moro (**B**,**D**) in 2018 (black circles) and 2019 (open circles). Historic data (solid line) date back to 1932 in Pendleton and 1897 in Moro.

2.1. Russian Thistle Emergence and Mortality

Russian thistle emergence differed between years, sites, crops, and their interaction (Table 1). Pendleton, in 2018, had the highest emergence (55.5 plants m^{-2}), and Moro, in 2019, had the lowest emergence (9.1 plants m^{-2}). Higher rainfall was likely a primary driver of higher emergence in Pendleton compared to Moro. In Lind, WA, an area that receives approximately 248 mm $year^{-1}$ of rainfall, Russian thistle density in spring wheat was 2.2 times lower in two years that received an average rainfall of 215 mm compared to a year that received 356 mm when averaged across treatments [2].

Table 1. Effect of year, site, crop, and their interactions on Russian thistle emergence, mortality, seed production, plant biomass, and crop yield (mean ± standard deviation). All means were compared with Tukey's method.

Factor	Emergence (plants m^{-2})	Mortality (%)	Seed Production (seeds $plant^{-1}$)	Plant Biomass (g)	Crop Yield (kg ha^{-1})
Year	*	ns	ns	ns	***
2018	37.8 (±23.9) a	49.2 (±27.9)	789 (±528)	17.7 (±11.2)	2997 (±1957) b
2019	15.0 (±9.3) b	35.2 (±26.8)	838 (±911)	22.8 (±21.0)	3505 (±1220) a
Site	***	ns	**	ns	***
Pendleton	38.2 (±23.0) a	38.7 (±26.4)	715 (±536) b	13.9 (±7.9)	3811 (±1348) a
Moro	15.1 (±10.6) b	46.1 (±29.6)	910 (±897) a	26.6 (±21.0)	2664 (±1738) b
Crop	*	*	ns	*	ns
SB	24.4 (±20.0) b	47.7 (±30.4) a	604 (±467)	14.4 (±10.2) b	3789 (±1825)
SW	29.4 (±22.5) a	37.0 (±24.7) b	1024 (±897)	26.0 (±20.4) a	2706 (±1228)
Year × Site	***	*	***	*	***
Pendleton_18	55.5 (±20.4) a	51.2 (± 24.9) a	1054 (±540) a	17.1 (±8.3) b	4776 (±1095) a
Pendleton_19	21.4 (±7.7) b	26.2 (±21.7) b	376 (±234) b	10.6 (±6.3) b	2846 (±758) b
Moro_18	20.7 (±10.4) b	47.2 (±30.9) a	516 (±355) b	18.3 (±13.6) b	1218 (±230) c
Moro_19	9.1 (±6.2) c	44.9 (±28.6) a	1331 (±1097) a	34.9 (±24.5) a	4207 (±1235) a
Year × Crop	ns	*	***	**	**
SB_18	34.0 (±23.2)	59.2 (±26.7) a	698 (±528) cb	14.1 (±9.2) b	3493 (±2216) b
SB_19	14.8 (±9.3)	36.1 (±29.8) b	511 (±382) c	14.7 (±11.2) b	4086 (±1296) a
SW_18	41.7 (±24.3)	39.1 (±25.8) b	872 (±522) b	21.2 (±12.0) b	2501 (±1538) c
SW_19	16.2 (±9.5)	34.9 (±23.6) b	1243 (±1161) a	30.8 (±26.1) a	2911 (±752) c
Year × Site × Crop	ns	ns	**	*	***
Pendleton_18_SB	50.2 (±21.0)	64.7 (±20.8)	993 (±544) b	17.7 (±8.4) bc	5611 (±734) a
Pendleton_18_SW	59.8 (±20.3)	37.6 (±21.4)	1114 (±549) b	16.5 (±7.8) bc	3941 (±671) c
Pendleton_19_SB	21.6 (±6.7)	26.7 (±24.1)	345 (±208) c	6.9 (±6.4) c	3308 (±709) c
Pendleton_19_SW	21.2 (±8.9)	25.7 (±19.6)	407 (±261) c	14.4 (±6.2) bc	2384 (±482) d
Moro_18_SB	17.7 (±10.6)	53.7 (±31.1)	402 (±309) c	10.6 (±10.0) bc	1375 (±203) e
Moro_18_SW	23.6 (±10.6)	40.7 (±30.2)	629 (±370) bc	26.0 (±15.1) b	1060 (±124) e
Moro_19_SB	8.0 (±5.8)	45.6 (±32.6)	677 (±447) bc	22.6 (±12.6) bc	4864 (±1294) b
Moro_19_SW	10.5 (±6.6)	44.1 (±24.5)	2078 (±1152) a	46.8 (±25.8) a	3439 (±577) c

Significance indicated by ns $p > 0.05$, * $p \leq 0.05$, ** $p \leq 0.01$, and *** $p \leq 0.001$ for main effects and letters a, b, c, d, and e ($\alpha = 0.05$) for means separation. SB: spring barley; SW: spring wheat.

Russian thistle density decreased in all plots during the growing season. Plant mortality differed between crops and the interaction between year × site or year × crop (Table 1). Russian thistle mortality was higher in spring barley (48%) compared to spring wheat (37%). Similarly, Borger et al. [17] found that barley suppressed *L. rigidum* plant density and biomass to a greater extent compared to wheat in two of three sites in one year. These authors also demonstrated that photosynthetically active radiation was lower in barley in the inter-row space, which may have contributed to greater *L. rigidum* suppression. Barley may be more competitive than wheat due to greater early biomass, higher leaf area index throughout development, and potential allelopathic activity [18–20]. This research supports the broad notion of barley's competitiveness and offers novel support for greater competitiveness with Russian thistle compared to spring wheat.

Russian thistle mortality was highest in spring barley in 2018 (59%) and lowest in spring wheat in 2019 (35%). Mortality was similar among Pendleton 2018 (51%), Moro

2018 (47%), and Moro 2019 (45%), while mortality was lowest in Pendleton 2019 (26%). Russian thistle mortality rate was at an average of 37% in spring wheat across sites and years, which was much higher than the 8% mortality reported by Young [4] in spring wheat across two years. It is not currently known how Russian thistle mortality is related to shading, emergence time, or emergence position (i.e., inter-row vs. in-furrow). Further research is necessary to address this knowledge gap.

Higher crop seeding density reduced Russian thistle emergence in spring barley in Pendleton 2019 and increased mortality in spring wheat in Moro 2018. Russian thistle emergence did not differ between row spacing treatments in spring barley in either year, whereas emergence was different in spring wheat in 2019 at both sites with a location-dependent effect (Table 2). Narrow row spacing decreased Russian thistle emergence in Moro 2019 when the crop was competitive due to higher rainfall. To the contrary, wide row spacing produced lower Russian thistle emergence in Pendleton 2019 when the crop had a significant infestation of netseed lambsquarter (*Chenopodium berlandieri*) that was removed in mid-June. It is possible that the presence of netseed lambsquarter in the trial area could have reduced Russian thistle emergence due to higher competition in the wide row spacing treatment. Higher soil disturbance in narrow rows may have also favored Russian thistle emergence. Russian thistle emergence has been observed to increase in disturbed soil compared with no tilled soil [21]. Row spacing did not affect Russian thistle mortality.

2.2. Effects on Russian Thistle Seed Production and Plant Biomass

There was a significant three-way interaction between year, site, and crop with respect to Russian thistle seed production (Table 1). Russian thistle plants produced the greatest number of seeds in spring wheat in Moro 2019, averaging 2078 seeds plant^{-1}. Russian thistle plants produced the fewest seeds in spring barley and spring wheat in Pendleton during 2019 and in spring barley in Moro during 2018, averaging approximately 385 seeds plant^{-1}.

Plant biomass followed a similar pattern to seed production with a significant three-way interaction between year, site, and crop (Table 1). Russian thistle plants were largest in spring wheat in Moro 2019 (46.8 g) and smallest in spring barley in Pendleton 2019 (6.9 g). Russian thistle seed production increased linearly with plant biomass and was similar among sites and years (analysis of covariance (ANCOVA) $F_{3,119} = 0.655$. $p = 0.581$; Figure 2). This indicates that larger Russian thistle plants have a greater likelihood of producing more seeds irrespective of location and year. Russian thistle grows rapidly following harvest, and final plant size may depend on size at crop harvest [3,4]. Young [4] demonstrated that Russian thistle growing in winter wheat accumulated less biomass by harvest compared to Russian thistle growing in spring wheat. In this study, biomass was not measured until approximately two months after harvest, but Russian thistle biomass was significantly lower in spring barley (14 g) compared to spring wheat (26 g) (Table 1).

Row spacing and seeding rate did not affect Russian thistle seed production or plant biomass in either year (Table 2). The lack of a broad effect of row spacing or seeding rate on Russian thistle seed production and plant biomass mirrored the weak effects observed on Russian thistle emergence and mortality. In contrast, Paynter and Hills [13] demonstrated a linear decrease in *L. rigidum* tiller number with increasing spring barley planting density. Borger et al. [22] also demonstrated declining *L. rigidum* seed number m^{-2} in various crops, including barley and wheat, as row spacing treatments declined from 36 to 9 cm. Unlike *L. rigidum*, however, Russian thistle does not senesce by harvest, continuing to grow until a killing frost [3].

Table 2. Effect of row spacing (Row) and crop seeding rate (Density) on Russian thistle emergence, mortality, seed production, plant biomass, and crop yield (mean ±standard deviation) in two locations (Pendleton and Moro) over two years (2018 and 2019) with spring barley (SB) and spring wheat (SW). All means were compared with Tukey's method.

Factor	Emergence (plants m^{-2})	Mortality (%)	Seed Production (seeds plant^{-1})	Plant Biomass (g)	Crop Yield (kg ha^{-1})	Factor	Emergence (plants m^{-2})	Mortality (%)	Seed Production (seeds plant^{-1})	Plant Biomass (g)	Crop Yield (kg ha^{-1})
	Pendleton 2018		SB						SW		
Density						Density					
Low	27.6 (±8.4)	67.0 (±22.9)	1130 (±615)	16.5 (±8.4)	5298 (±919) *	Low	29.8 (±9.2)	34.5 (±23.2)	1240 (±563)	21.4 (±8.7)	3615 (±826) *
High	22.7 (±12.3)	62.3 (±19.8)	856 (±460)	13.8 (±8.8)	5923 (±294) *	High	30.0 (±11.7)	40.7 (±20.6)	988 (±539)	16.6 (±6.6)	4265 (±202) *
Row						Row					
Narrow	19.7 (±7.3)	68.0 (±25.2)	1153 (±361)	17.0 (±7.5)	5652 (±434)	Narrow	26.3 (±8.3)	43.8 (±17.5)	1186 (±582)	20.1 (±9.4)	4165 (±333)
Wide	30.5 (±10.7)	61.4 (±16.3)	834 (±667)	13.3 (±9.5)	5569 (±981)	Wide	33.5 (±11.1)	31.5 (±24.3)	1043 (±542)	17.9 (±6.4)	3715 (±858)
	Moro 2018		SB						SW		
Density						Density					
Low	8.2 (±4.6)	38.9 (±29.3)	479 (±380)	15.8 (±12.5)	1367 (±137)	Low	9.5 (±4.3)	22.0 (±28.5) **	770 (±346)	30.6 (±13.8)	1074 (±156)
High	9.5 (±6.2)	68.5 (±26.9)	326 (±215)	10.5 (±6.5)	1382 (±262)	High	14.2 (±5.4)	59.3 (±18.9) **	488 (±358)	16.3 (±13.3)	1046 (±90)
Row						Row					
Narrow	10.8 (±6.4)	59.9 (±27.7)	384 (±270)	13.3 (±9.7)	1277 (±213) *	Narrow	10.9 (±5.5)	45.2 (±33.9)	591 (±308)	23.7 (±17.6)	1115 (±134)
Wide	6.9 (±3.2)	47.5 (±35.1)	421 (±361)	12.9 (±11.0)	1472 (±144) *	Wide	12.8 (±5.3)	36.2 (±27.6)	668 (±442)	23.3 (±13.2)	1005 (±89)
	Pendleton 2019		SB						SW		
Density						Density					
Low	24.9 (±9.6) *	33.5 (±21.7)	363 (±288)	11.4 (±7.9)	2941 (±316) *	Low	24.0 (±9.6)	31.5 (±21.7)	474 (±288)	13.9 (±7.0)	2487 (±316)
High	18.4 (±7.6) *	19.8 (±16.7)	328 (±228)	7.4 (±3.8)	3674 (±610) *	High	18.5 (±7.6)	19.8 (±16.7)	340 (±228)	9.9 (±4.9)	2281 (±610)
Row						Row					
Narrow	22.9 (±7.8)	22.6 (±22.5)	310 (±229)	8.5 (±6.2)	3485 (±769)	Narrow	26.6 (±9.0) *	22.2 (±19.6)	295 (±150)	9.1 (±3.7)	2417 (±485)
Wide	20.4 (±5.5)	30.8 (±26.5)	380 (±193)	10.3 (±6.7)	3129 (±643)	Wide	15.8 (±4.6) *	29.1 (±20.3)	519 (±306)	14.7 (±7.2)	2350 (±508)
	Moro 2019		SB						SW		
Density						Density					
Low	8.6 (±6.2)	44.2 (±34.4)	739 (±373)	20.5 (±9.4)	5373 (±804) *	Low	11.7 (±6.7)	44.2 (±31.7)	1894 (±1226)	49.9 (±32.7)	3339 (±726) **
High	7.4 (±5.7)	47.0 (±33.0)	614 (±528)	19.6 (±15.0)	4355 (±1533) *	High	9.3 (±6.9)	43.9 (±17.3)	2263 (±1136)	49.1 (±19.2)	3574 (±403) **
Row						Row					
Narrow	6.2 (±5.3)	34.9 (±33.8)	643 (±551)	18.0 (±13.9)	4590 (±1690)	Narrow	5.4 (±4.3) *	32.3 (±27.4)	1858 (±1238)	42.4 (±17.2)	3138 (±651)
Wide	9.8 (±6.1)	56.2 (±29.6)	711 (±347)	22.1 (±11.6)	5137 (±749)	Wide	15.6 (±3.7) *	55.8 (±15.1)	2298 (±1108)	56.6 (±32.0)	3775 (±248)

Significance indicated by $p \leq 0.1$, * $p \leq 0.05$, and ** $p \leq 0.01$.

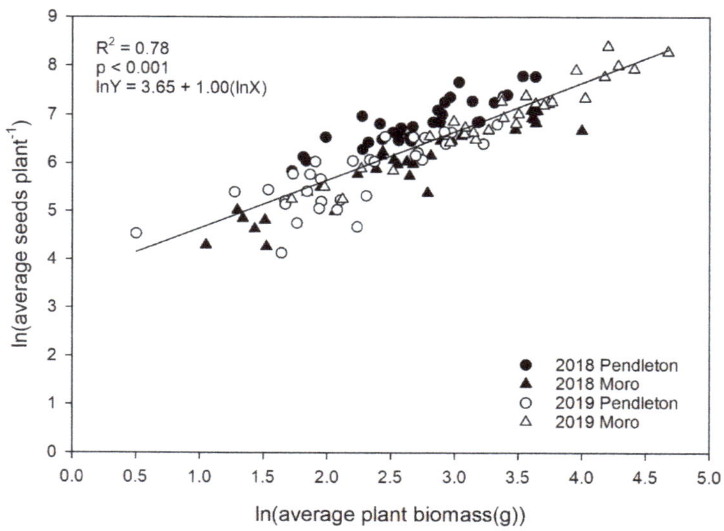

Figure 2. Relationship between average Russian thistle plant biomass (g) and Russian thistle seed production (seeds plant^{-1}) across sites and years. Solid line represents the average across sites and years.

2.3. *Effects on Crop Yield*

The combination of year, site, and crop type produced a significant three-way interaction when yield with seeded Russian thistle was examined (Table 1). In 2018, the wheat yield in Pendleton was 3.7 times higher than the wheat yield in Moro at 3941 and 1060 kg ha^{-1}, respectively. Similarly, the barley yield in Pendleton was 4.1 times higher than the barley yield in Moro during 2018 at 5611 and 1375 kg ha^{-1}, respectively. The relationship reversed in 2019 for both crops with 1.5 and 1.6 times higher wheat and barley yields, respectively, in Moro compared to Pendleton. Pendleton is typically higher yielding than Moro due to its location in a higher rainfall zone. The lower yield in Pendleton in 2019 may have been due to the trial following winter wheat instead of fallow, as occurred in 2018. The trial in Moro 2019 also followed spring barley instead of winter wheat, as occurred in 2018. This combination of events likely contributed to the higher yield in Moro during 2019.

Spring barley yield was 1.4 times higher than spring wheat yield in both years (Table 1). Barley yields significantly more produce than wheat in the PNW and in other areas due to greater above-ground dry mass [19,23]. Spring barley yield was also significantly higher in 2019 than in 2018, whereas spring wheat yield was similar in 2018 and 2019. Higher spring barley yield in 2019 was likely driven by higher spring precipitation that year.

Narrow row spacing did not produce higher yields than wider row spacing treatments. For example, spring barley planted in wider rows produced higher yields than the narrow row spacing treatment in Moro 2018. In contrast, yield tended to increase with higher crop seeding density in spring barley and spring wheat, except for Moro 2019 where spring barley yield was higher in the lower seeding density treatment (Table 2). Previous research has demonstrated that yield increases when crops are sown in narrower rows or at higher seeding rates with weed pressure since greater crop competition suppresses weed growth and biomass [24,25]. In this study, however, higher yield with a higher crop density was independent of weed presence, since yield in Russian-thistle-free sub-sub-plots was similar to infested sub-sub-plots in 2019 (Table 3). This demonstrates that a Russian thistle density of 15 plants m^{-2} or lower may not have a significant impact on spring wheat or spring barley yield in a year with higher growing season rainfall than average. Similarly, Young [2] demonstrated no effect of Russian thistle on spring wheat yield in one year with high early growing season rainfall, while two other years with less rainfall were affected by Russian thistle pressure. Unfortunately, Russian-thistle-free yield data were not obtained in 2018, precluding broad conclusions about effects of Russian thistle on yield in a lower rainfall year, except for observations that emergence was higher in 2018.

Table 3. Comparison of crop yield with and without Russian thistle pressure for each location and crop in 2019.

Site	Crop	*p*-Value
Pendleton	Spring wheat	0.374
	Spring barley	0.820
Moro	Spring wheat	0.700
	Spring barley	0.219

3. Materials and Methods

3.1. Location

A two-year field experiment was established in Umatilla and Sherman counties located in Oregon during 2018 and 2019. The Umatilla site was located at the Columbia Basin Agricultural Research Center (CBARC) in Adams, Oregon, hereafter referred to as the Pendleton site. The Sherman site was located at CBARC in Moro, Oregon, hereafter referred to as the Moro site. Both sites are rain-fed sites. The soil at the Pendleton site was a Walla Walla silt loam (8% clay, 27% sand, and 65% silt) with 2.3% organic matter and a pH of 5.4. The soil at the Moro site was a Walla Walla silt loam (7% clay, 30% sand, and 63% silt) with 1.2% organic matter and a pH of 6.6. The Pendleton site is located in an intermediate precipitation zone while the Moro site is located in a low rainfall zone of the PNW. Long-term average precipitation at the Pendleton site is 421 mm year^{-1} while the Moro site receives 287 mm year^{-1}. Fertilization was applied in all sites following standard recommendations for the region [26]. All sites were managed following conventional tillage practices, except for the site in Moro 2018 which was a no-till site. The Pendleton site used in 2018 was fallow in 2017 and the Pendleton site used in 2019 had winter wheat in 2018. The Moro site used in 2018 had winter wheat in 2017 and the Moro site used in 2019 had spring barley in 2018.

3.2. Experimental Design

The experimental design in 2018 was a split-plot randomized complete block design with four replications. Row spacing (18 or 36 cm) was the main plot factor and seeding rate (73 or 140 kg ha^{-1}) was the sub-plot factor. The main plot was doubled to accommodate two crop types (spring wheat "WB6341" or spring barley "Champion"). Both crops were included in the same experimental area to guard against a soil type or field topography effect on the crop and Russian thistle development. The main plot was 1.7 m × 36.6 m while the sub-plot was 1.7 m × 9.15 m. Each main plot had then four sub-plots (spring wheat at 73 and 140 kg ha^{-1} and spring barley at 73 and 140 kg ha^{-1}). Crop type and seeding rate were randomized within main plots. Each main plot was separated by a 1.7-m alley. Russian thistle seed was spread by hand at a rate of 0.43 g m^{-2} before seeding the crops with a Hege 9-row drill on April 19 in the Pendleton site and April 20 in the Moro site.

The experimental design in 2019 was a split-split-plot randomized complete block design with four replications. Main plot size increased to 1.7 m × 48.8 m and sub-plot size increased to 1.7 m × 12.2 m. The first 6.1 m of the sub-plot were hand-seeded with Russian thistle while the second 6.1 m of the sub-plot remained Russian-thistle-free. Crop type, row spacing, and seeding rate treatments were the same as in 2018. Experiments were seeded on 19 and 22 April 2019 at the Pendleton and Moro sites, respectively. In the Pendleton site, sub-sub-plots without seeded Russian thistle were sprayed with pyrasulfotole + bromoxynil herbicide (Huskie®, Bayer CropScience, Chesterfield, MO, USA) and sub-plots seeded with Russian thistle were hand weeded on 14 June 2019 to control a weed infestation of netseed lambsquarter that occurred in 2019 at that site. The lambsquarter was fully controlled in a few days after its detection. For all the other sites, the weed control before the crop seeding and Russian thistle seed spreading (by tillage or by a non-residual, non-selective herbicide) was enough to maintain the plots weed-free except for the Russian thistle infestation.

3.3. Data Collection

Russian thistle evaluations were conducted by placing a sampling frame (0.5 m × 0.5 m = 0.25 m^2) in six random locations within each sub-plot in 2018. In 2019, larger sampling frames (1 m × 0.5 m = 0.5 m^2) were used to sample six random locations due to a lower Russian thistle infestation that year. Post-seeding evaluations (hereafter referred to as emergence evaluations) at each location were conducted on 22–23 May in 2018 and 25–26 June in 2019. Spring wheat and spring barley were harvested on 25 and 31 July 2018 at the Moro and Pendleton sites, respectively. Harvest in 2019 was on 20 and 21 August in Pendleton and Moro sites, respectively. Yield data were determined per sub-plot in 2018 and sub-sub-plot in 2019.

Russian thistle plants were harvested on 15 and 19 October 2018 and 25 October and 1 November 2019 at the Pendleton and Moro sites, respectively. Five plants were randomly removed from each sub-plot in 2018 and sub-sub-plot in 2019, placed inside paper bags, and moved to a greenhouse to be processed at a later date. Russian thistle plants were first weighed to determine dry biomass and then hand-threshed. The processed material was passed through a series of sieves to obtain a seed and chaff mixture. The seed and chaff mixture was weighed and seed number was determined from an approximate 0.5-g sample of the mixture. Seed number per plant was determined by dividing the seed number in each sample by 0.5 and then multiplying the result by the total weight of the seed and chaff.

3.4. Statistical Analyses

Linear mixed models (LMMs) were used to study the effect of year, site, crop type, row spacing, and seeding rate on Russian thistle emergence, mortality, seed production, plant biomass, and crop yield. Due to the high number of independent variables, two analyses were conducted separately. The first analysis included year, site, and crop type effects on Russian thistle emergence, mortality, plant biomass, seed production, and crop yield. The second analysis included row spacing and seeding density effects on the aforementioned variables. Fixed effects in the first analysis were year, site, and crop, while seeding density and row spacing were the fixed variables in the second analysis. Replications in the first analysis were the random effect, while replications nested within row spacing were the random effect in the second analysis. RStudio v.1.2 software (RStudio Team, Boston, MA, USA) was used for all analyses. The least-square means function (lsmeanS) in R studio was used for mean separation in the LMMs, specifying Tukey's test.

Linear regression was used to evaluate the relationship between Russian thistle plant biomass and seed production among sites and years. A natural logarithm transformation was applied to both axes to satisfy normality. Differences in slopes between regression models were assessed with analysis of covariance (ANCOVA), specifying the ANOVA_test function in the rstatix package of R studio.

Mean crop yield with and without Russian thistle pressure for each location and crop in 2019 was assessed with a one-way analysis of variance and means were compared with Tukey's test. All data were checked graphically for normality assumptions before conducting the analyses. All figures were created in SigmaPlot v.14 (Systat Software, San Jose, CA, USA).

4. Conclusions

The absence of a broad effect of row spacing, seeding rate, or their interaction on Russian thistle emergence, mortality, plant biomass, or seed production in this research indicates that these cultural management practices had a minor effect on Russian thistle suppression. These findings contrast previous research with different weed species in spring barley [13] or spring wheat [25,27,28]. This research does not support using high crop density or narrow row spacing as a broad prescription for Russian thistle suppression. However, spring barley was more competitive with Russian thistle than spring wheat. Including spring barley over spring wheat in a winter wheat–fallow rotation in the PNW is recommended to suppress Russian thistle.

Author Contributions: Conceptualization, J.B.; methodology, J.B.; formal analysis, J.B. and N.G.G.; writing—original draft preparation, J.B. and N.G.G.; writing—review and editing, J.B. and N.G.G.; project administration, J.B.; funding acquisition, J.B. Both authors have read and agreed to the published version of the manuscript.

Funding: This research was supported by the Oregon Wheat Commission.

Institutional Review Board Statement: Not applicable.

Informed Consent Statement: Not applicable.

Data Availability Statement: The data presented in this study are openly available at the ScholarArchive@OSU (https://ir.library.oregonstate.edu/concern/datasets/r781wp412).

Acknowledgments: Authors thank the Oregon Wheat Commission for funding this research and Jennifer Gourlie for data collection.

Conflicts of Interest: No conflict of interest have been declared.

References

1. Beckie, H.J.; Francis, A. The biology of Canadian weeds. 65. *Salsola tragus* L. (updated). *Can. J. Plant Sci.* **2009**, *89*, 775–789. [CrossRef]
2. Young, F.L. Effect of Russian thistle (*Salsola iberica*) interference on spring wheat (*Triticum aestivum*). *Weed Sci.* **1988**, *36*, 594–598. [CrossRef]
3. Schillinger, W.F.; Young, F.L. Soil water use and growth of Russian thistle after wheat harvest. *Agron. J.* **2000**, *92*, 167–172. [CrossRef]
4. Young, F.L. Russian thistle (*Salsola iberica*) growth and development in wheat (*Triticum aestivum*). *Weed Sci.* **1986**, *34*, 901–905. [CrossRef]
5. Nakka, S.; Jugulam, M.; Peterson, D.; Asif, M. Herbicide resistance: Development of wheat production systems and current status of resistant weeds in wheat cropping systems. *Crop J.* **2019**, *7*, 750–760. [CrossRef]
6. Barroso, J.; Gourlie, J.A.; Lutcher, L.K.; Liu, M.; Mallory-Smith, C.A. Identification of glyphosate resistance in *Salsola tragus* in north-eastern Oregon. *Pest Manag. Sci.* **2018**, *74*, 1089–1093. [CrossRef]
7. Kumar, V.; Spring, J.F.; Jha, P.; Lyon, D.J.; Burke, I.C. Glyphosate-resistant Russian-thistle (*Salsola tragus*) identified in Montana and Washington. *Weed Technol.* **2017**, *31*, 238–251. [CrossRef]
8. Davis, A.S.; Hill, J.D.; Chase, C.A.; Johanns, A.M.; Liebman, M. Increasing cropping system diversity balances productivity, profitability and environmental health. *PLoS ONE* **2012**, *7*, e47149. [CrossRef]
9. Kinkegaard, J.A.; Chrisen, O.; Krupinsky, J.; Layzell, D. Break crop benefits in temperate wheat production. *Field Crop. Res.* **2008**, *107*, 185–195. [CrossRef]
10. Van der Meulen, A.; Chauhan, B.S. A review of weed management in wheat using crop competition. *Crop Prot.* **2017**, *95*, 38–44. [CrossRef]
11. Scott, B.J.; Martin, P.; Riethmuller, G.P. *Graham Centre Monograph No. 3: Row Spacing of Winter Crops in Broad Scale Agriculture in Southern Australia*; Nugent, T., Nicholls, C., Eds.; NSW Department of Primary Industries: Orange, NSW, Australia, 2013; p. 94.
12. Solie, J.B.; Solomon, S.G., Jr.; Self, K.P.; Peeper, T.F.; Koscelny, J.A. Reduced row spacing for improved wheat yields in weed-free and weed-infested fields. *Trans. ASABE* **1991**, *34*, 1654–1660. [CrossRef]
13. Paynter, B.H.; Hills, A.L. Barley and rigid ryegrass (*Lolium rigidum*) competition is influenced by crop cultivar and density. *Weed Technol.* **2009**, *23*, 40–48. [CrossRef]
14. De Vita, P.; Colecchia, S.A.; Pecorella, I.; Saia, S. Reduced inter-row distance improves yield and competition against weeds in a semi-dwarf durum wheat variety. *Eur. J. Agron.* **2017**, *85*, 69–77. [CrossRef]
15. Kolb, L.N.; Gallandt, E.R.; Molloy, T. Improving weed management in organic spring barley: Physical weed control vs. interspecific competition. *Weed Res.* **2010**, *50*, 597–605. [CrossRef]
16. Kolb, L.N.; Gallandt, E.T.; Mallory, E.B. Impact of spring wheat planting density, row spacing, and mechanical weed control on yield, grain protein, and economic return in Maine. *Weed Sci.* **2012**, *60*, 244–253. [CrossRef]
17. Borger, C.P.D.; Hashem, A.; Powles, S.C. Manipulating crop row orientation and crop density to suppress *Lolium rigidum*. *Weed Res.* **2015**, *56*, 22–30. [CrossRef]

18. Bertholdsson, N.-O. Early vigour and allelopathy—Two useful traits for enhanced barley and wheat competitiveness against weeds. *Weed Res.* **2005**, *45*, 94–102. [CrossRef]
19. López-Castañeda, C.; Richards, R.A. Variation in temperature cereals in rainfed environments II. Phasic development and growth *Field Crop. Res.* **1994**, *37*, 63–75. [CrossRef]
20. Woldeamlak, A.; Bastiaans, L.; Struik, P.C. Competition and niche differentiation in barley (*Hordeum vulgare*) and wheat (*Triticum aestivum*) mixtures under rainfed conditions in the Central Highlands of Eritrea. *Neth. J. Agric.* **2001**, *49*, 95–112. [CrossRef]
21. Genna, N.G.; Wuest, S.; Barroso, J. Russian-Thistle Ecology in Dryland Wheat Cropping Systems of the PNW. Status: Unpublished; article in preparation.
22. Borger, C.P.D.; Riethmuller, G.; D'Antuono, M. Eleven years of integrated weed management: Long-term impacts of row spacing and harvest weed seed destruction on *Lolium rigidum* control. *Weed Res.* **2016**, *56*, 335–406. [CrossRef]
23. López-Castañeda, C.; Richards, R.A. Variation in temperate cereals in rainfed environments I. Grain yield, biomass and agronomic characteristics. *Field Crop. Res.* **1994**, *37*, 51–62.
24. Sardana, V.; Gulshan, M.; Jabran, K.; Chauhan, B.S. Role of competition in managing weeds: An introduction to the special issue. *Crop Prot.* **2017**, *95*, 1–7. [CrossRef]
25. Kristensen, L.; Olsen, J.; Weiner, J. Crop density, sowing pattern, and nitrogen fertilization effects on weed suppression and yield in spring wheat. *Weed Sci.* **2008**, *56*, 97–102. [CrossRef]
26. Wysocki, D.J.; Lutcher, L.K.; Horneck, D.A.; Hart, J.M.; Petrie, S.E. Fertilizer guide: Winter wheat and spring grains in continuous cropping systems. (Low precipitation zone). In *Oregon State University Extension Service*; Oregon State University: Corvallis, OR, USA, 2007; p. 81.
27. Olsen, J.; Kristensen, L.; Weiner, J. Influence of sowing density and spatial pattern of spring wheat (*Triticum aestivum*) on the suppression of different weed species. *Weed Biol. Manag.* **2006**, *6*, 165–175. [CrossRef]
28. Champion, G.T.; Froud-Williams, R.J.; Holland, J.M. Interactions between wheat (*Triticum aestivum* L.) cultivar, row spacing and density and the effect on weed suppression and crop yield. *Ann. Appl. Biol.* **1998**, *133*, 443–453. [CrossRef]

plants MDPI

Communication

Winter Wheat (*Triticum aestivum* L.) Tolerance to Mulch

Matthew R. Ryan [1,*], Sandra Wayman [1], Christopher J. Pelzer [1], Caitlin A. Peterson [2], Uriel D. Menalled [1] and Terry J. Rose [3]

[1] Section of Soil and Crop Sciences, School of Integrative Plant Science, Cornell University, Ithaca, NY 14850, USA; sw783@cornell.edu (S.W.); cjp254@cornell.edu (C.J.P.); udm3@cornell.edu (U.D.M.)
[2] Water Policy Center, Public Policy Institute of California, San Francisco, CA 94111, USA; peterson@ppic.org
[3] Faculty of Science and Engineering, Southern Cross Plant Science, Southern Cross University, Lismore, NSW 2480, Australia; terry.rose@scu.edu.au
[*] Correspondence: mryan@cornell.edu; Tel.: +1-(607)-255-4964

Abstract: Mulch from cover crops can effectively suppress weeds in organic corn (*Zea mays* L.) and soybean (*Glycine max* L.) as part of cover crop-based rotational no-till systems, but little is known about the feasibility of using mulch to suppress weeds in organic winter small grain crops. A field experiment was conducted in central NY, USA, to quantify winter wheat (*Triticum aestivum* L.) seedling emergence, weed and crop biomass production, and wheat grain yield across a gradient of mulch biomass. Winter wheat seedling density showed an asymptotic relationship with mulch biomass, with no effect at low rates and a gradual decrease from moderate to high rates of mulch. Selective suppression of weed biomass but not wheat biomass was observed, and wheat grain yield was not reduced at the highest level of mulch (9000 kg ha^{-1}). Results indicate that organic winter wheat can be no-till planted in systems that use mulch for weed suppression. Future research should explore wheat tolerance to mulch under different conditions, and the potential of no-till planting wheat directly into rolled-crimped cover crops.

Keywords: cover crop; weed management; organic; no-till

Citation: Ryan, M.R.; Wayman, S.; Pelzer, C.J.; Peterson, C.A.; Menalled, U.D.; Rose, T.J. Winter Wheat (*Triticum aestivum* L.) Tolerance to Mulch. *Plants* **2021**, *10*, 2047. https://doi.org/10.3390/plants10102047

Academic Editors: Alessia Restuccia and Aurelio Scavo

Received: 26 August 2021
Accepted: 27 September 2021
Published: 29 September 2021

1. Introduction

No-till crop production has received widespread attention over the past several decades as a strategy to conserve topsoil and improve soil health while reducing fuel and labor inputs. In 2017, no-till was practiced on 26% of cropland in the United States [1]. The adoption of no-till practices since the 1980s was enabled by synthetic herbicides and improved planting equipment [2]. However, in organic production systems where synthetic herbicides are prohibited, soil tillage and cultivation are commonly used for weed management [3]. Weeds can also be suppressed by surface mulch from cover crops that are mechanically terminated with roller-crimpers, and researchers have demonstrated success with this approach for organic no-till corn (*Zea mays* L.) and soybean (*Glycine max* L.) [4]. Although cover crop-based, organic no-till production has the potential to provide some soil health benefits [5], soil tillage is typically used for establishing cover crops and small grain crops in the crop rotation. This rotational no-till approach limits the soil health benefits that manifest over a longer period (e.g., increased soil organic matter, enhanced water infiltration from preferential flow channels, etc.) [6]. Thus, research is needed to explore the potential of other crops beyond corn and soybean that can be no-till planted into mulch, which could allow for extended sequences without soil tillage in organic cropping systems.

Selective suppression of weed seedlings but not crop seedlings by mulch is important for successful cover crop-based, organic no-till production. Previous research has shown that weed suppression from mulch is a function of physical impedance and light deprivation [7] as well as chemical inhibition from allelochemicals [8]. Whereas cooler soils and reduced light transmittance from mulch tend to reduce weed emergence, mulch can also conserve soil moisture and increase weed seedling emergence, especially at low levels of mulch biomass [9]. Cover crop biomass is an important driver of weed suppression from mulch [10] and past work has suggested that 8000 kg ha^{-1} is the minimum biomass required to achieve consistent weed suppression [11]. However, this threshold likely varies by weed community as well as crop variety, environment, and management practices (i.e., G × E × M).

Weed tolerance to mulch is often correlated with seed size, where species with larger seeds are less likely to be suppressed by mulch [12]. Crop species with small seeds and light requirements for germination are generally more susceptible to suppression by mulch due to smaller nutrient reserves, among other considerations [13]. Seed size may also be the main lever that provides selective weed but not crop suppression in cover crop-based organic no-till [4,13]. Wheat seeds are smaller than corn and soybean (e.g., 0.05, 0.17 and 0.30 g seed^{-1} for wheat, soybean, and corn, respectively) [14], but still larger than many weed species [15]. As wheat is generally weed suppressive in the study region compared with corn and soybean, a lower rate of cover crop biomass might be adequate for consistent weed suppression, while avoiding suppression of wheat seedlings with lower nutrient reserves than corn and soybean. A field experiment was conducted in central New York, United States to evaluate weed suppression, wheat emergence, and wheat grain yield of two winter wheat varieties across a biomass gradient of grass-clover hay mulch. We hypothesized that weed biomass but also wheat seedling density, wheat biomass, and wheat grain yield would decrease with increasing mulch biomass. We also hypothesized that seedling emergence would vary by wheat variety and be greater in the variety marketed for its superior performance in high-residue environments.

2. Results and Discussion

In general, winter wheat performed well at all mulch rates, with neither wheat biomass nor grain yield suppressed by increasing amounts of mulch. Wheat seedling emergence was tolerant to mulch rates at or below 3000 kg ha^{-1} and remained at roughly 80% of the no-mulch control at mulch rates of 6000 kg ha^{-1}. In contrast, weed biomass was suppressed at mulch rates above 6000 kg ha^{-1}, which is lower than the mulch rates recommended for the summer annual crops (e.g., corn, soybean) where high-residue production is more commonly implemented [11]. These results suggest that minor management adjustments such as increased seeding rates could ensure acceptable winter wheat crop stands while effectively suppressing weeds in a no-till system.

2.1. Wheat Emergence

Wheat seedling emergence was not affected by wheat variety. When data were pooled across the two varieties, wheat seedling emergence was consistently high at low to moderate mulch rates (3000 kg ha^{-1}), after which it gradually decreased to 55% of the emergence rate in the no-mulch control at the highest mulch rate (9000 kg ha^{-1}; Figure 1). In addition to the lack of support for our second hypothesis about wheat varieties, the functional form of the relationship between wheat density and mulch rate was unexpected. In contrast to our results, previous experiments with broadleaf weeds spanning mulch rates like those used in our experiment have reported an exponential decline with weed seedling emergence in response to increasing mulch rate. For example, Teasdale and Mohler [7] found an exponential decline in *Amaranthus retroflexus* (redroot pigweed) emergence at increasing mulch rates.

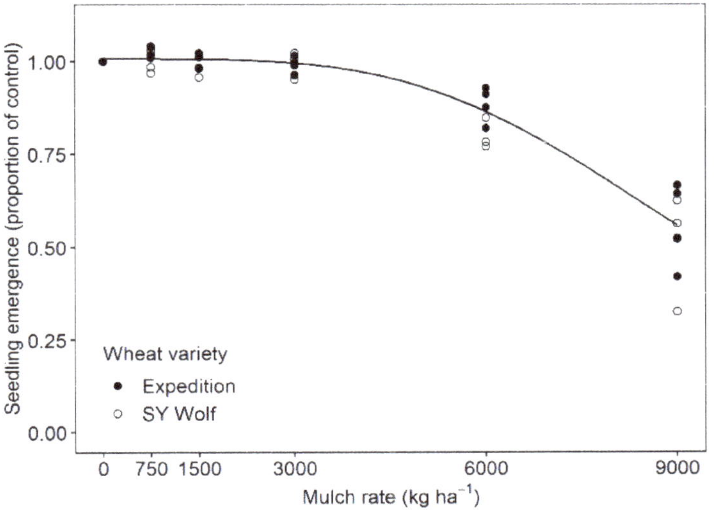

Figure 1. Winter wheat seedling emergence (*E*) of two winter wheat varieties, "Expedition" and "SY Wolf", under increasing grass-clover hay mulch rates (*M*), expressed as a proportion of the emergence rate in the no-mulch control. The log-logistic response was estimated as $E = 1.006/(1 + \exp^{(3.9(\log(M) - \log(9217)))})$. All coefficients were significant at the $\alpha = 0.05$ level, indicating a significant log-logistic response of wheat emergence to mulch rate. The response curve did not vary significantly by wheat variety ($p = 0.26$).

2.2. Weed and Wheat Biomass

Wheat biomass was unaffected by mulch rate ($p = 0.06$), averaging 11,265 ± 183 kg ha^{-1} dry matter across mulch rates (stem and seed head; Figure 2A). This trend did not differ between wheat varieties ($p = 0.18$) and there was no interaction between mulch rate and wheat variety ($p = 0.98$).

In contrast to the non-effect of mulch on winter wheat biomass, we observed stimulation of weed growth at low mulch rates. Weed biomass increased from an average of 245 ± 41 with no mulch to 518 ± 122 kg ha^{-1} of biomass at a mulch rate of 1500 kg ha^{-1}, but subsequently declined to a low of 89 ± 36 kg ha^{-1} biomass at the highest mulch rate of 9000 kg ha^{-1} (Figure 2B). The stimulation of weed growth at lower mulch rates was indicated by a positive mulch stimulation parameter (*a* in Equation (2); $p < 0.001$). However, at mulch rates above 2000 kg ha^{-1} weed biomass decreased (Figure 2B). Weed suppression at higher mulch biomass rates was likely due to attenuation of germination cues that reduced weed seedling density as well as light deprivation, physical interference, phytotoxin inhibition that reduced weed seedling growth [7,11]. Promotion of plant emergence at low mulch rates has long been described in the turfgrass literature, and light mulching is often used to help establish lawns [16,17]. Our results suggest that certain weed species may behave similarly to turfgrasses, with light mulch improving soil microclimate (i.e., higher and more consistent moisture, reduced temperature) in ways that promote seedling emergence and growth.

A noteworthy difference between our results and previous work on cover cropping for weed suppression in summer annual crops is that weed suppression in winter wheat was achieved at much lower biomass levels than usually recommended. A synthesis of cover cropping work in soybean and corn suggested that mulch rates of 8000 kg ha^{-1} or more are required to achieve consistent weed suppression [11]. In our study, weed biomass was not a limiting factor on wheat yield at any of the mulch rates, including rates below 8000 kg ha^{-1}. Instead, wheat biomass production appeared optimal at rates of about 5000 kg ha^{-1}, when the stimulatory effect of low mulch rates on weed emergence was surpassed, but wheat emergence was not yet affected. This result suggests that one of the common limitations of using cover crops for weed suppression—production of sufficient biomass—could be a smaller obstacle in a winter wheat-based system.

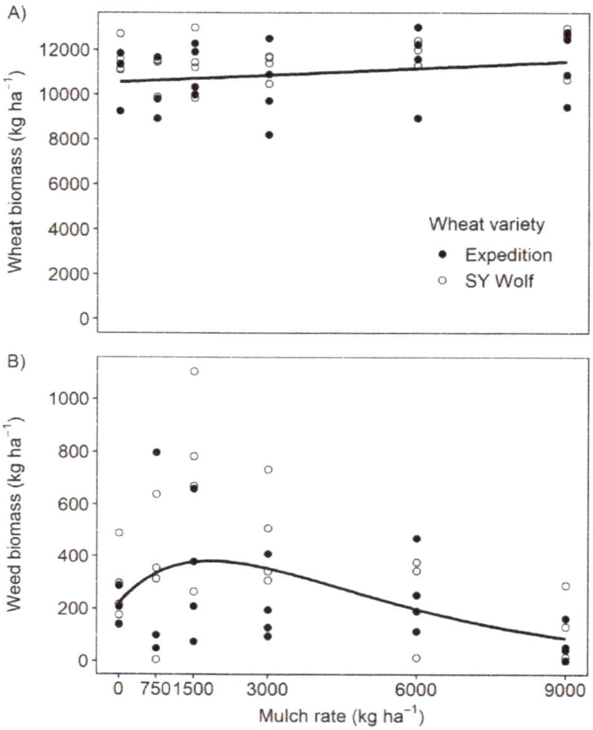

Figure 2. (**A**) Winter wheat biomass and (**B**) weed biomass (W_b) collected from plots of two varieties of winter wheat at different grass-clover hay mulch rates (M). There was no significant effect of mulch rate on winter wheat biomass production (p = 0.06). Weed biomass response to mulch rate was estimated as $W_b = 222(1 + 0.0014M)(\exp^{-0.0004M})$. All coefficients were significant at the α = 0.05 level. There was no difference in the response curves between the two varieties for either response variable (wheat biomass p = 0.1; weed biomass p = 0.1).

2.3. Winter Wheat Yield

Mulch rate and wheat variety did not interact to affect wheat yield (p = 0.34). However, there was a slight positive trend in the response of wheat grain yield to mulch rate ($p < 0.05$; Figure 3). Across mulch rates, wheat yields averaged 4684 ± 84 kg ha^{-1}, which is approximately 11% greater than average wheat yields in the region [18]. Contrary to our hypothesis, yield response to mulch rates did not differ significantly between varieties, despite the marketing of the SY Wolf variety for superior performance in high-residue conditions.

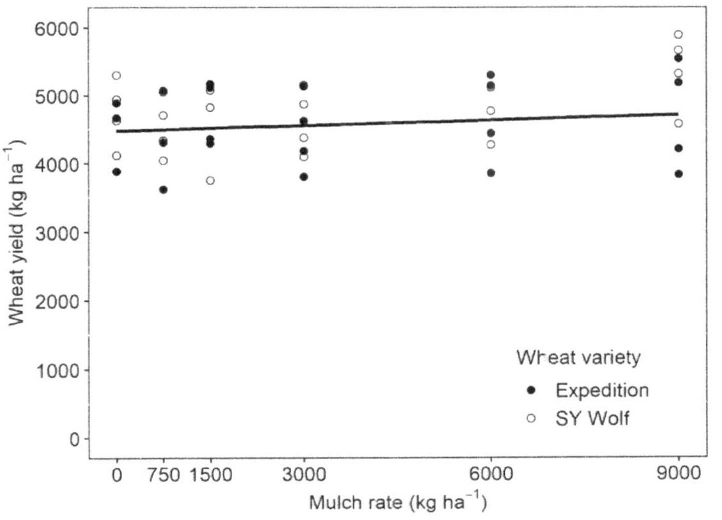

Figure 3. Wheat yield as affected by grass-clover hay mulch rate. There was a small positive effect of mulch rate on wheat yield ($p < 0.05$) and no difference in yield responses between wheat varieties ($p = 0.72$).

3. Materials and Methods

3.1. Site Description

A field experiment was conducted in Ithaca, NY (42.45° N, 76.46° W) to quantify the effects of mulch mass on wheat emergence, weed suppression, and wheat yield. Soils at the site are very fine, sandy loams in the Williamson series, Typic Fragiochrept (Table 1). Prior to the experiment, the field was planted with intermediate wheatgrass (*Thinopyrum intermedium*, (Host) Barkworth & DR Dewey) in the fall of 2018 and then disked and cultipacked in fall 2019. Temperatures in 2019 and 2020 generally trended with the 30-yr normal, but precipitation in 2019 was generally higher than both the 30-yr normal and 2020 (Figure 4).

Table 1. Selected soil properties from the field site at the Caldwell Farm at Cornell University in Ithaca, NY, USA. Soil samples (20 cm depth) were analyzed by Dairy One Agronomy Services, Ithaca, NY, USA. Cations were measured using the Morgan method.

Soil Property	Value
pH	6.1
Organic matter (%)	2.6
Phosphorus (kg ha^{-1})	6.7
Potassium (kg ha^{-1})	141
Calcium (kg ha^{-1})	2180
Magnesium (kg ha^{-1})	278
Iron (kg ha^{-1})	26
Manganese (kg ha^{-1})	23
Zing (kg ha^{-1})	0.45
Aluminium (kg ha^{-1})	88

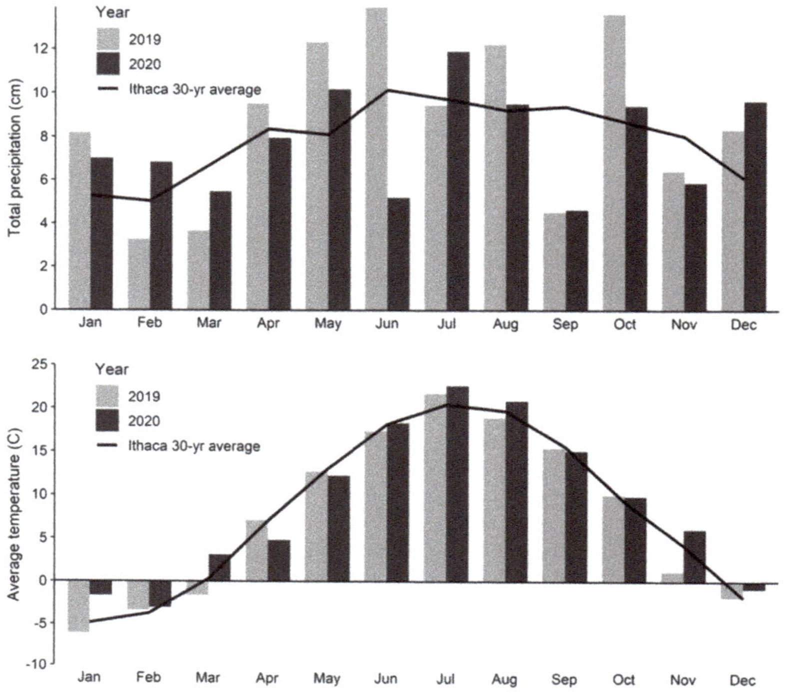

Figure 4. Total precipitation (cm) and average temperature (°C) by month for 2019–2020 at the Cornell Caldwell Farm (Ithaca, NY, USA). Precipitation and air temperature data were collected from an on-site weather station.

3.2. Experimental Design

Treatments included six mulch rates (0, 750, 1500, 3000, 6000 and 9000 kg ha^{-1} dry weight) and two hard red winter wheat varieties: (1) 'SY Wolf' and (2) 'Expedition'. Expedition is commonly used by organic farmers in the northeast USA, whereas SY Wolf is a new variety marketed for good performance in heavy residue [19,20]. All wheat seed was untreated. The experimental design was a spatially balanced, split-plot randomized complete block with four blocks. Wheat variety was the main-plot factor and mulch rate was the sub-plot factor. Sub-plots measured 1 m wide × 4 m long.

3.3. Soil and Crop Management

The field was moldboard plowed, disked, and cultipacked several weeks prior to planting. Wheat seed was drilled to a depth of 2.5 cm into bare soil on 5 October 2019, using an Almaco heavy duty grain drill with 15 cm row spacing. Following planting, all plots were smoothed with a garden rake. Within 24 h of planting, dry grass-clover hay was weighed in the field and evenly distributed on the soil surface by hand at the six target mulch biomass rates. The grass-clover hay used as mulch was harvested from a nearby farm and consisted primarily of perennial species including orchardgrass (*Dactylis glomerata* L.), timothy (*Phleum pratense* L.), and red clover (*Trifolium pratense* L.). No amendments or fertilizer was applied to the field over the course of the experiment.

3.4. Data Collection

Wheat seedling emergence was documented 33 days after planting on 27 November 2019, by counting seedlings in the center four rows of each plot within a 0.5 m^2 quadrat. On 26 July 2020, subplots were sampled by hand to determine wheat yield. In each variety by mulch rate sub-plot a 0.25 m^2 quadrat centered over the middle 4 rows of wheat was harvested, bagged, and dried to a constant weight. Dry weights were measured for total wheat biomass as well as stem and seed head biomass. Yields are reported as threshed grain dry weights. Weeds in each quadrat were bulk harvested and dried to constant weight to determine total weed biomass, and dominant weed species in each quadrat were recorded.

3.5. Statistical Analysis

All data were analyzed using R version 4.0.3 [21] with packages 'lme4' [22], 'nlme' [23], and 'stats' [21]. We used nonlinear mixed-effects models [23] to estimate wheat seedling density and weed biomass responses to wheat variety and mulch rates. Linear mixed-effects models were used to estimate wheat biomass and wheat yield responses to mulch rate. In all models, random effects accounted for the split plot design of the experiment by nesting wheat variety within the field block. Before modeling, a Grubbs test was used to identify and remove one weed biomass and one wheat yield outlier ($p < 0.001$). Furthermore, wheat seedling density data was modeled as a proportion of seedling density in the no-mulch control (0 kg ha^{-1}) within each block.

The effect of wheat variety on wheat seedling density and weed biomass was elucidated by comparing a reduced model, where data was pooled across the two wheat varieties, and a full model, where the response variables differed as a function of wheat variety. The reduced and full nonlinear models were compared using a log-likelihood ratio test. If no significant difference ($p > 0.05$) was detected between log-likelihood ratios, wheat variety had a negligible effect on the response variable. Conversely, a significant log-likelihood ratio indicated a better fit for the full model, wheat variety was considered to be a significant predictor of seedling emergence or weed biomass. If evaluation of the fitted estimates suggested violation of the assumption of homoscedacity, we fit a separate model allowing for unequal variances and compared the two fits using the same log-likelihood test. In cases where multiple nonlinear equations could reasonably be used to describe the response curve, the equation with the best fit was determined using the Akaike Information Criterion (AIC).

Wheat seedling emergence (E) response to mulch rate (M) was modeled with a 3-parameter log-logistic equation Hill equation [24]:

$$E = \frac{d}{1 + exp^{b(\log(M) - \log(e))}} \quad (1)$$

where d is the upper asymptote of seeding emergence, b is the slope at the inflection point, and e is the mulch rate halfway between the upper and lower asymptote of seedling emergence.

The change in weed biomass (W_b) across the mulch biomass (M) gradient was modeled with a right-skewed, hump-shaped equation devised by Teasdale and Mohler (2000):

$$W_b = W_0(1 + aM)\left(exp^{-bM}\right) \quad (2)$$

where W_0 represents the intercept, a represents the mulch stimulation effect, and b represents the mulch suppression effect.

For the linear mixed-effects models, mulch rate, wheat variety, and the interaction of these two predictor variables were fixed effects. Weed biomass did not affect wheat yield or biomass ($p \geq 0.50$) and was not included as a predictor variable in the linear mixed models. Denominator degrees of freedom were calculated using the Kenward-Roger method and the linear mixed effects models were assessed through type III ANOVA tests.

4. Conclusions

We compared the effects of increasing mulch biomass on weed suppression and winter wheat performance and our results suggest wheat is relatively tolerant to mulch. The wheat seedlings in our experiment were subject to more extreme emergence restrictions than usual because mulch was placed on top of wheat seeds after seeding into bare ground, thus reducing any aid to emergence created by coulters slicing through the mulch during planting. Although wheat seedling density was reduced at high levels of mulch, wheat biomass and grain yield were tolerant to increasing mulch. On the other hand, weed biomass was stimulated at low levels and suppressed at high levels of mulch.

Given the differential tolerance to mulch, wheat may be a viable candidate for no-till planting into rolled-crimped cover crops. More research is needed across different environments, wheat varieties, and types of mulch to understand the full potential of organic no-till wheat production and how multiple organic no-till cash crops can be combined in a rotation. Such extended sequences of no-till production could provide enhanced soil health benefits.

Author Contributions: Conceptualization, M.R.R. and T.J.R.; methodology, M.R.R., S.W., C.J.P.; software, C.A.P., U.D.M.; validation, M.R.R., C.J.P.; formal analysis, M.R.R., C.A.P., U.D.M.; investigation, S.W., C.J.P.; resources, M.R.R.; data curation, S.W.; writing—original draft preparation, M.R.R., C.A.P., S.W.; writing—review and editing, M.R.R., S.W., C.J.P., U.D.M.; visualization, C.A.P., M.R.R., S.W.; supervision, M.R.R.; project administration, S.W., C.J.P.; funding acquisition, M.R.R. and T.J.R. All authors have read and agreed to the published version of the manuscript.

Funding: This research was supported by the USDA National Institute of Food and Agriculture, Organic Agriculture Research and Extension Initiative project 2020-51300-32183; the Cornell University Agricultural Experiment Station (Hatch funds) and Cornell Cooperative Extension (Smith Lever funds) project 1023863; and the New York State Environmental Protection Fund for the New York Soil Health Initiative, administered through the New York State Department of Agriculture and Markets Contract No. C00178GS-3000000. TR acknowledges the receipt of a fellowship from the OECD Co-operative Research Programme: Biological Resource Management for Sustainable Agricultural Systems in 2019.

Institutional Review Board Statement: Not applicable.

Informed Consent Statement: Not applicable.

Data Availability Statement: The datasets generated during and/or analyzed during the current study are available from the corresponding author on reasonable request.

Acknowledgments: Thanks to James Cagle and Dylan Rodgers for assistance processing samples. The experiment reported in this article was conducted on the traditional homelands of the Gayogohó:nǫ' (the Cayuga Nation). The Gayogohó:nǫ' are members of the Haudenosaunee Confederacy, an alliance of six sovereign Nations with a historic and contemporary presence on this land. The Confederacy precedes the establishment of Cornell University, New York state, and the United States of America. We acknowledge the painful history of Gayogohó:nǫ' dispossession, and honor the ongoing connection of Gayogohó:nǫ' people, past and present, to these lands and waters.

Conflicts of Interest: The authors declare no conflict of interest. The funders had no role in the design of the study; in the collection, analyses, or interpretation of data; in the writing of the manuscript, or in the decision to publish the results.

References

1. National Agricultural Statistics Service. *2017 Census of Agriculture: United States Summary and State Data*; U.S. Department of Agriculture: Washington, DC, USA, 2019; p. 128.
2. Triplett, G.B.; Dick, W.A. No-tillage crop production: A revolution in agriculture! *Agron. J.* **2008**, *100*, S-153–S-165. [CrossRef]
3. Baker, B.P.; Mohler, C.L. Weed management by upstate New York organic farmers: Strategies, techniques and research priorities. *Renew. Agric. Food Syst.* **2015**, *30*, 418–427. [CrossRef]
4. Vincent-Caboud, L.; Vereecke, L.; Silva, E.; Peigné, J. Cover crop effectiveness varies in cover crop-based rotational tillage organic soybean systems depending on species and environment. *Agronomy* **2019**, *9*, 319. [CrossRef]
5. Crowley, K.A.; Van Es, H.M.; Gómez, M.I.; Ryan, M.R. Trade-offs in cereal rye management strategies prior to organically managed soybean. *Agron. J.* **2018**, *110*, 1492. [CrossRef]
6. Aziz, I.; Mahmood, T.; Islam, K.R. Effect of long term no-till and conventional tillage practices on soil quality. *Soil Tillage Res.* **2013**, *131*, 28–35. [CrossRef]

7. Teasdale, J.R.; Mohler, C.L. The quantitative relationship between weed emergence and the physical properties of mulches. *Weed Sci.* **2000**, *48*, 385–392. [CrossRef]
8. Petersen, J.; Belz, R.; Walker, F.; Hurle, K. Weed suppression by release of *Isothiocyanates* from turnip-rape mulch. *Agron. J.* **2001**, *93*, 37–43. [CrossRef]
9. Mohler, C.L.; Teasdale, J.R. Light transmittance, soil temperature, and soil moisture under residue of hairy vetch and rye. *Agron. J.* **1993**, *85*, 673–680.
10. Teasdale, J.R. Contribution of cover crops to weed management in sustainable agricultural systems. *J. Prod. Agric.* **1996**, *9*, 475–479. [CrossRef]
11. Mirsky, S.B.; Ryan, M.R.; Teasdale, J.R.; Curran, W.S.; Reberg-Horton, C.S.; Spargo, J.T.; Wells, M.S.; Keene, C.L.; Moyer, J.W. Overcoming weed management challenges in cover crop–based organic rotational no-till soybean production in the Eastern United States. *Weed Technol.* **2013**, *27*, 193–203. [CrossRef]
12. Liebman, M.; Sundberg, D.N. Seed mass affects the susceptibility of weed and crop species to phytotoxins extracted from red clover shoots. *Weed Sci.* **2006**, *54*, 340–345. [CrossRef]
13. Mohler, C.L. Ecological bases for the cultural control of annual weeds. *J. Prod. Agric.* **1996**, *9*, 468–474. [CrossRef]
14. Deivasigamani, S.; Swaminathan, C. Evaluation of seed test weight on major field crops. *Int. J. Res. Stud. Agric. Sci.* **2018**. [CrossRef]
15. Uva, R.; Neal, J.; DiTomaso, J. *Weeds of the Northeast*; Comstock Publishing Associates: Ithaca, NY, USA, 1997.
16. Dudeck, A.E.; Swanson, N.P.; Mielke, L.N.; Dedrick, A.R. Mulches for grass establishment on fill slopes. *Agron. J.* **1970**, *62*, 810–812. [CrossRef]
17. Patchan, G.; Payne, K.; Smith, T. *Turf Tips for the Homeowner, Extension Bulletin E-1491, Cooperative Extension Service*; Michigan State University: East Lansing, MI, USA, 1981.
18. Anonymous. Northwest New York Dairy, Livestock & Field Crops. Cornell Cooperative Extension. 2021. Available online: https://nwnyteam.cce.cornell.edu/grains.php (accessed on 19 August 2021).
19. AgriPro SY Wolf: The Heavy Residue Champion. Available online: https://agriprowheat.com/variety/sy-wolf (accessed on 4 March 2020).
20. Albert Lea Seed SY Wolf Hard Red Winter Wheat. Available online: https://www.alseed.com/product/sy-wolf/ (accessed on 11 March 2020).
21. R Core Team. *R: A Language and Environment for Statistical Computing*; R Foundation for Statistical Computing: Vienna, Austria, 2020.
22. Bates, D.; Maechler, M.; Bolker, B.; Walker, S. Fitting linear mixed-effects models using lme4. *J. Stat. Softw.* **2015**, *67*, 1–48. [CrossRef]
23. Pinheiro, J.; Bates, D.; DebRoy, S.; Sarkar, D.; R Development Core Team. nlme: Linear and Nonlinear Mixed Effects Models. Available online: https://cran.r-project.org/web/packages/nlme/index.html (accessed on 25 August 2021).
24. Archontoulis, S.V.; Miguez, F.E. Nonlinear regression models and applications in agricultural research. *Agron. J.* **2015**, *107*, 786–798. [CrossRef]

plants

MDPI

Article

Differentiated Weed-Suppressive Ability of Modern and Old Durum Wheat Cultivars after Long-Term Cultivation under Semi-Arid Climate

Aurelio Scavo, Alessia Restuccia *, Mario Bannò and Giovanni Mauromicale

Department of Agriculture, Food and Environment (Di3A), University of Catania, 95123 Catania, Italy
* Correspondence: a.restuccia@unict.it

Abstract: Durum wheat (*Triticum turgidum* spp. *durum*) is one of the most important grain crops cultivated across the Mediterranean Basin, where a strong return to local landraces cultivation is occurring to meet the market demand for high-quality food and low-input cropping systems. A characterisation of the long-term effect (10 years) of durum wheat landraces and modern cultivars on the potential and real weed flora is still lacking. Hence, a multilocation trial over 10 farms in Central-Eastern Sicily was carried out to investigate the repeated cultivation of several old landraces (OLD) and modern cultivars (MOD) on the abundance and diversity of weed flora. Overall, OLD was associated with a 47% reduction of the soil seedbank size and to −64% of the aboveground weed biomass compared to MOD. In addition, diversity indices pointed out a high similarity between MOD and OLD farm groups for the soil seedbank, while a lower diversity was found in OLD for aboveground weed communities. From the principal component analysis emerged that the species compositions of MOD and OLD were quite separated for both soil seedbank and real flora, with the latter showing few specific associations with major weeds. These findings demonstrated the indirect effect of durum wheat landraces in sustainably reducing weed pressure without the adoption of chemical weed control.

Keywords: *Triticum durum*; weed management; soil seedbank; species diversity; weed communities; old landraces; multivariate statistics

Citation: Scavo, A.; Restuccia, A.; Bannò, M.; Mauromicale, G. Differentiated Weed-Suppressive Ability of Modern and Old Durum Wheat Cultivars after Long-Term Cultivation under Semi-Arid Climate. *Plants* **2022**, *11*, 3368. https://doi.org/10.3390/plants11233368

Academic Editor: Sheng Qiang

Received: 3 November 2022
Accepted: 1 December 2022
Published: 4 December 2022

1. Introduction

Durum wheat (*Triticum turgidum* subsp. *durum* (Desf.) Husn., $2n = 4x = 28$, AABB), although grown on just 8–10% of the global land surface, is one of the most important cereal crops in semi-arid zones, especially in the Mediterranean Basin, where more than 80% of the total European-harvested production take place [1]. In Europe, Italy is the first country of economic importance with 3.8×10^6 Mg of harvested production obtained from 1.2×10^6 ha [2], primarily concentrated in the southern regions. Mediterranean durum wheat germplasm is characterised by the largest biodiversity, as demonstrated by the high number of local landraces adapted to numerous pedoclimatic conditions [3]. However, many of these old landraces are no longer cropped or are underutilised due to the spread of modern genetically improved and high-yielding cultivars, thus causing a serious genetic erosion [4,5].

Aside from major abiotic constrains such as drought, nitrogen supply, high temperatures and soil properties, weeds are the most important biotic threat reducing the yields and quality characteristics of durum wheat [6,7], especially in the Mediterranean Basin. To control weeds, over the last decades, durum wheat has been subjected to a considerable chemical weed control that caused a number of negative effects, mainly the development of highly resistant weed populations and the persistence of herbicides in the environment and in the food chain. Considering the adverse effects determined by the irrational chemical weed control, on the one hand, and the raising of organic cropping systems where synthetic herbicides are banned, on the other hand, the exploitation of eco-friendly weed management practices in durum wheat agroecosystems has an outstanding relevance [8]. In recent years, several sustainable weed management practices involving the manipulation of allelopathy, such as the use of plant water extracts [9], intercropping [10] and mulching [11], have been proposed in durum wheat [12]. Within the integrated weed management (IWM) systems, a key role is played by prevention, including those strategies or agronomic choices aimed at preventing weed adaptation, and it is based on the reduction of the soil seedbank and the improvement of the crop competitiveness against weeds [8]. The former is the most important and challenging aspect, given that the soil seedbank is the primary source of new infestations and that the real weed flora derives almost exclusively from the potential weed population communities [13]. The main goal is getting its control below 20 million weed seeds ha^{-1} in order to simplify and reduce the direct weed control methods. The second strategy, which is closely connected to the soil seedbank control, can be addressed by choosing weed-competitive cultivars with high root development, early vigour, faster seedling emergence, high growth rates, wide leaf areas and an allelopathic ability [14]. In this regard, several findings suggest a higher weed-suppressive ability of old durum wheat landraces than modern cultivars due to a combined competitive–allelopathic effect [15,16]. Fields of old durum wheat landraces, in fact, generally show lower weed densities than modern cultivars, according to our experience. This aspect, together with the increasingly importance of low-input agricultural systems (especially in the European Union) and a greater market demand for high-quality food, is determining a reawakened interest in durum wheat local landraces. They are particularly appreciated by the market by virtue of their high-quality flours, especially for the production of pasta, pizza and bread.

Based on these considerations, the present research started from the hypothesis that durum wheat landraces may have a higher weed-suppressive ability than modern cultivars and that the repeated cultivation of local landraces may reduce the soil seedbank (potential weed flora) and the real weed flora pressure. A scientific verification directly on a field scale has never been done. Hence, the goals were to evaluate the effect derived from the long-term rotation, including some old durum wheat landraces, compared to modern cultivars, on the abundance and diversity of potential and real weed flora in Central-Eastern Sicily, a semi-arid environment representing an important production centre of durum wheat.

2. Results

2.1. Potential Weed Flora (Soil Seedbank)

2.1.1. Weed Abundance

Pooling over the five farms belonging to the MOD and OLD groups (Figure 1), it is clearly visible that the repeated cultivation of old durum wheat landraces was associated with a 46.8% reduced seedbank size compared to modern cultivars in the studied area (1760.0 vs. 3306.7 seeds m^{-2}). From the analysis of variance (ANOVA), it emerged that the soil seedbank varied significantly across the ten farms under study (Figure 1). In detail, the highest seedbank size was found at the Bannò Farm (6266.7 seeds m^{-2}), where the cv. Anco Marzio is cultivated from many years in rotation with vetch, followed by the Spitaleri Farm (3933.3 seeds m^{-2}), which sows an Iride–Simeto–Core mix in rotation with vetch and fava beans. The lowest seedbank size was detected at the Mocciaro Farm (666.7 seeds m^{-2}), despite it sowing modern cv. Core alternately with a fodder mix composed of vetch, clover, Sulla, ryegrass and oat, followed by Minio (866.7 seeds m^{-2}), which sows the old cv. Perciasacchi. Delizia, which also cultivates the cv. Perciasacchi, and which is the only farm adopting the stale seedbed, showed the highest seedbank size (3600.0 seeds m^{-2}) within the OLD group.

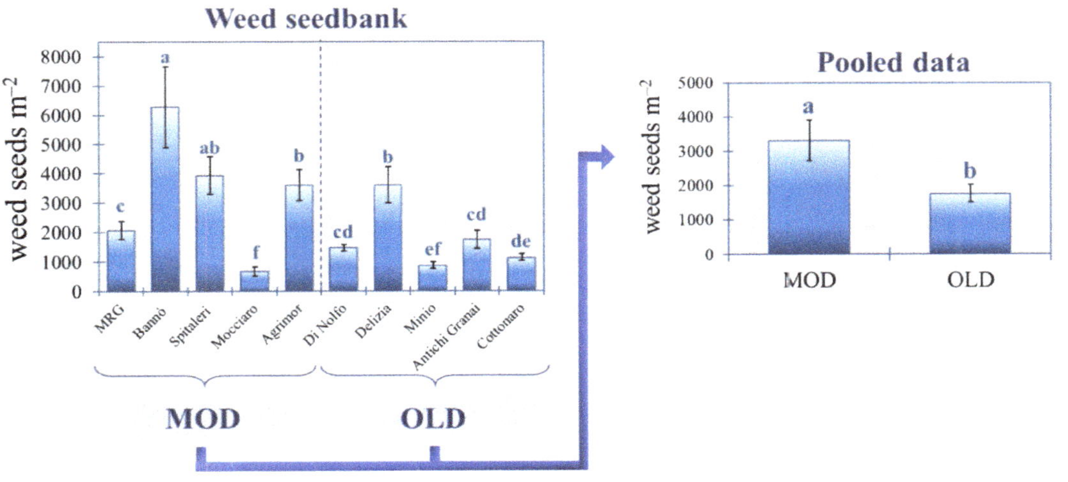

Figure 1. Size of the weed soil seedbank (0–15 cm) across the 10 farms under study. Bars are the standard deviation (n = 3). Different letters indicate statistical significance by applying one-way analysis of variance with Tukey's HSD test at $p \leq 0.05$. MOD: wheat modern varieties group; OLD: wheat old landraces group.

2.1.2. Weed Diversity

Throughout the ten farms, the total 0–15 cm soil seedbank consisted of 13 weed species or genera belonging to 11 botanical families (Table 1). All detected taxa were annual therophytes, excluding the biennial hemicryptophyte *Stellaria media* (L.) Vill. The soil seedbank was dominated by six major weeds (i.e., with a RD $\geq$ 5%): in decreasing order, *Euphorbia helioscopia* L., *Anagallis arvensis* L., *Helminthotheca echioides* (L.) Holub, *S. media*, *Sinapis arvensis* L. and *Fallopia convolvulus* (L.) Á. Löve. Their sum accounted for 86% of the total weed seedbank density. The ANOVA of the species richness and RAIs did not show significant differences among the MOD and OLD for each species (data not shown). However, Table 1 highlights some interesting findings. About the major weeds, *E. helioscopia* and *S. arvensis* were more abundant at the Delizia Farm (0.54 and 0.22, respectively), *A. arvensis* at Antichi granai (0.72), *E. echoiides* at MRG (0.38), *S. media* at Di Nolfo (0.45) and *F. convolvulus* at Minio (0.28). Moreover, the weeds *Glebionis coronaria* (L.) Cass. ex Spach, *Portulaca oleracea* L. and *Veronica* sp. were detected only in MOD, whereas *Galium aparine* L. and *S. media* were exclusive of OLD. Interestingly, *A. arvensis, E. helioscopia, F. convolvulus* and *Fumaria* sp. had higher mean RAIs in the MOD farm group, while the RAIs of *Amaranthus retroflexus* L., *E. echoiides* and *S. arvensis* were higher in farms belonging to the OLD group.

Taking into account the α-diversity, no significant differences were observed for the Margalef's (D_{MG}), Shannon–Wiener (H) and Pielou's (J) indices between the MOD and OLD (Table 2). Within the MOD, Agrimor, cultivating cv. Core in rotation with vetch, showed the highest D_{MG} (2.08) and H (1.48) values, indicating a higher biodiversity compared to the other MOD farms. On the contrary, Mocciaro, which was the farm with the lowest seedbank size, had the highest J value (0.94), indicating the presence of a few dominant species, namely *A. arvensis*, *E. helioscopia* and *P. oleracea* (Table 1). Within OLD, similarly to the Mocciaro Farm, Minio showed the highest J (0.91), while Antichi granai had the lowest, α-diversity indices. Concerning β-diversity (Table 2), a high similarity between the MOD and OLD groups was found both in terms of presence/absence (Sørensen's, 76.2%) and abundance (Steinahus's, 54.7%).

Table 1. Mean relative abundance values and mean relative densities (RD) of weed species in the total seedbank (0–15 cm) across the 10 farms under study. Weeds are grouped by botanical family, life cycle and biological group (BG).

Binomial Name	Botanical Family	Life Cycle	BG †	MOD Farm Group					OLD Farm Group					RD (%) ‡
				MRG	Bannò	Spitaleri	Mocciaro	Agrimor	Di Nolfo	Delizia	Minio	Antichi Granai	Cottonaro	
Amaranthus retroflexus L.	Amaranthaceae	annual	T	-	0.12	-	-	-	0.07	-	-	-	-	1.3
Anagallis arvensis L.	Primulaceae	annual	T	0.27	0.09	0.45	0.29	0.32	-	-	-	0.72	0.27	28.1
Euphorbia falcata L.	Euphorbiaceae	annual	T	-	-	-	-	0.04	-	0.06	-	0.12	-	0.8
Euphorbia helioscopia L.	Euphorbiaceae	annual	T	0.17	0.45	0.05	0.47	0.21	-	0.54	0.38	0.04	0.30	31.7
Fallopia convolvulus (L.) Á. Löve	Polygonaceae	annual	T	-	0.09	-	-	-	-	-	0.28	-	0.21	5.1
Fumaria sp.	Fumariacee	annual	T	0.06	-	0.06	-	0.08	0.24	0.06	0.10	-	0.07	4.4
Galium aparine L.	Rubiaceae	annual	T	-	-	-	-	-	0.24	0.06	-	-	-	2.0
Glebionis coronaria (L.) Cass. ex Spach	Asteraceae	annual	T	-	0.04	0.13	-	-	-	-	-	-	-	1.0
Helminthotheca echioides (L.) Holub	Asteraceae	annual	T	0.38	-	0.12	-	0.08	-	-	0.24	0.12	-	8.6
Portulaca oleracea L.	Portulacaceae	annual	T	-	0.22	-	0.24	-	-	-	-	-	-	4.0
Sinapis arvensis L.	Brassicaceae	annual	T	0.12	-	0.20	-	0.18	-	0.22	-	-	0.07	5.9
Stellaria media (L.) Vill.	Caryophyllaceae	biennial	H	-	-	-	-	-	0.45	0.06	-	-	0.07	6.7
Veronica sp.	Plantaginaceae	annual	T	-	-	-	-	0,09	-	-	-	-	-	0.6

† T: therophytes; H: hemicryptophytes; ‡ Averaged over all ten farms under study.

Table 2. α- and β-diversity indices of weed species in the total seedbank (0–15 cm) across the 10 farms under study.

	α-Diversity				β-Diversity	
	Margalef	Shannon-Weiner	Pielou	Whittaker	Sørensen ‡	Steinhaus ‡
MOD farm group †	**1.72 ± 0.23 A**	**1.13 ± 0.20 A**	**0.70 ± 0.15 A**	**2.6 ± 0.71 A**	76.2%	54.7%
MRG	1.71 b	0.99 b	0.61 b	2.6 b		
Bannò	1.45 b	1.02 b	0.57 b	2.2 c		
Spitaleri	1.68 b	1.15 b	0.64 b	2.2 c		
Mocciaro	1.66 b	1.03 b	0.94 a	4.3 a		
Agrimor	2.08 a	1.48 a	0.76 b	1.9 d		
OLD farm group †	**1.91 ± 0.60 A**	**1.05 ± 0.38 A**	**0.68 ± 0.23 A**	**2.9 ± 0.65 A**		
Di Nolfo	1.51 cd	1.07 b	0.77 a	3.3 a		
Delizia	1.73 c	0.77 c	0.43 b	2.2 b		
Minio	2.05 b	1.27 ab	0.91 a	3.3 a		
Antichi granai	1.39 d	0.59 c	0.43 b	3.5 a		
Cottonaro	2.88 a	1.53 a	0.86 a	2.2 b		

Different capital letters indicate significant differences between the MOD and OLD groups at $p \leq 0.05$ (Tukey's HSD). Different lowercase letters indicate significant differences within groups at $p \leq 0.05$ (Tukey's HSD test). † Data are the mean ± standard deviation. For within groups values, the standard deviation is always 0.1. ‡ Similarity between the MOD and OLD groups.

2.2. Real Weed Flora

2.2.1. Weed Abundance

Averaged over the MOD and OLD (Figure 2), emerged durum wheat landraces may have determined a 64.4% reduction of the real weed flora abundance in the studied area with respect to modern cultivars (2.1 vs. 12.7 g DW m^{-2}). Similar to the soil seedbank, ANOVA showed that the real weed flora abundance was significantly different for the ten farms (Figure 2). The Spitaleri Farm, in particular, had the highest weed aboveground biomass (180.1 g DW m^{-2}), followed by MRG (91.6 g DW m^{-2}) and Agrimor (80.4 g DW m^{-2}). Within OLD, the highest weed aboveground biomass was found at Antichi granai (73.0 g DW m^{-2}), which is the only farm performing fertilisation, and Delizia (68.6 g DW m^{-2}), the farm with the highest seedbank size. At Cottonaro, which carries out a long-term rotation durum wheat cv. Senatore Cappelli with a leguminous mix (vetch, clover and Sulla) and which is the only OLD farm performing chemical weed control, no emerged weeds were detected.

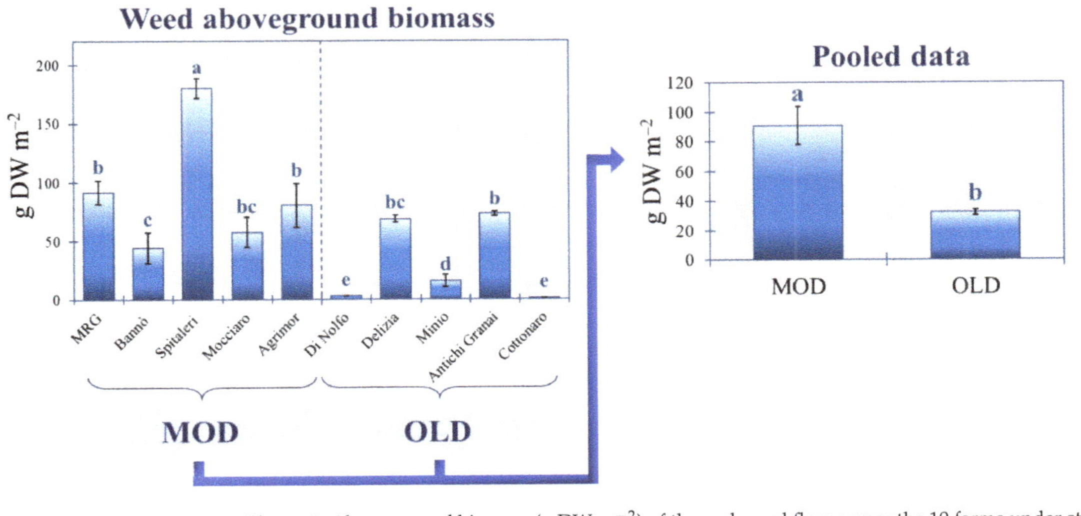

Figure 2. Aboveground biomass (g DW m^{-2}) of the real weed flora across the 10 farms under study. Bars are the standard deviation (n = 3). Different letters indicate statistical significance by applying one-way analysis of variance with Tukey's HSD test at $p \leq 0.05$. MOD: wheat modern varieties group; OLD: wheat old landraces group.

2.2.2. Weed Diversity

Nineteen weed species or genera were recorded throughout the study in the real weed flora, of which 74% were annuals, 16% perennials and just one biennial hemicryptophyte, namely *S. media* (Table 3). Among the 19 detected taxa, 39% belong to Asteraceae, 23% to Poaceae and 15% to Brassicaceae. Seven weeds or genera had a RD $\geq$ 5% and thus dominated the real weed flora: *Avena fatua* L., *S. media*, *G. aparine*, *G. coronaria*, *Lolium* sp., *Centaurea* sp. and *Phalaris paradoxa* L., which, altogether, accounted for 71.2% of the total density. As observed for the soil seedbank, although the ANOVA of the species richness and RAI was not significant, the following statements could be highlighted: *A. fatua* and *Lolium* sp. were more abundant at the Mocciaro Farm (0.54 and 0.31, respectively), *S. media* and *G. aparine* at Di Nolfo (0.27 and 0.54, respectively), *G. coronaria* at Spitaleri (0.27), *Centaurea* sp. at MRG (0.18) and *P. paradoxa* at Agrimor (0.23). Furthermore, *Artemisia vulgaris* L., *Inula helenium* L., *Papaver rhoeas* L., *Polygonum aviculare* L. and *S. arvensis* were recorded only at the MOD farms, whereas *Convolvulus arvensis* L., *Diplotaxis erucoides* (L.) DC., *E. helioscopia*, *Erodium cicutarium* (L.) L'Hér. and *G. aparine* only at the OLD. From Table 3, it is also possible to see that *Centaurea* sp., *G. coronaria*, *P. paradoxa* and *Sonchus* sp. had a higher RAI in MOD, while *D. carota*, *Lolium* sp. and *S. media* showed a higher RAI in OLD.

In contrast with the soil seedbank, for the real weed flora, the α-diversity was significantly higher in the MOD farm group than the OLD for the D_{MG}, H and J (Table 4). In detail, MRG had the highest α-diversity (D_{MG} = 3.6; H = 1.9; J = 0.8) across the MOD farm group, while Di Nolfo showed the lowest values (D_{MG} = 1.2; H = 0.8; J = 0.7) across the OLD farms. Despite significant α-diversity differences, the MOD and OLD showed a medium-high qualitative β-diversity (Sørensen's = 64.3%) but a Steinahus's coefficient < 50%.

2.3. *Species Composition of Potential and Real Weed Flora*

The associations between major weeds and farms were analysed by PCA on the correlation matrix of standardised weed densities. The eigen analysis showed that, for both potential and real weed flora, the first three PCs gave eigenvalues greater than one and accounted for most of the variance (Table 5). Interrelationships among major weeds and farms were observed graphically through ordination biplots constructed with the first two components explaining the maximum variance. For the soil seedbank, *A. arvensis*, *H. echioides* and *S. media* captured 67.4% of the variance in PC1, and *E. helioscopia* and *F. convolvulus* added a 59.8% variance in PC2, while *S. arvensis* had the highest weight on PC3. In addition, PC1 was positively correlated to *A. arvensis*, *E. helioscopia*, *H. echioides* and *S. arvensis*, thus positioning them on the right side of the biplot (Figure 3), while a negative correlation (left side) was found with *F. convolvulus* and *S. media*. PC2 correlated positively (top of the biplot) with *A. arvensis*, *H. echioides*, *S. arvensis* and *S. media* and negatively (bottom) with *E. helioscopia* and *F. convolvulus*. For the real weed flora, *A. fatua*, *Lolium* sp. and *S. media* accounted for 59.7% of the PC1 variance, *Centaurea* sp. and *G. aparine* for 54.1% in PC2 and *G. coronaria* and *P. paradoxa* for 39.5% in PC3 (Table 5). Moreover, the weeds *Centaurea* sp., *G. aparine*, *G. coronaria* and *S. media* showed a positive correlation with PC1, whereas *A. fatua*, *Lolium* sp. and *P. paradoxa* correlated negatively. Except for *G. aparine*, *Lolium* sp. and *S. media*, all weeds correlated negatively with PC2 (bottom of the biplot). The ordination biplots show that the farms discriminated mainly along PC1 for the soil seedbank and along PC2 for the real weed flora (Figure 3). In particular, except for Antichi granai, the OLD farms were positioned on the left side of the soil seedbank biplot; about real flora, all the OLD farms, excluding Minio, were positioned on the top of the biplot. Therefore, the MOD and OLD were quite separated for both soil seedbank and real flora, with OLD farms that showed few specific associations with the major weeds. Di Nolfo, in particular, was not associated with any species.

Table 3. Mean relative abundance values and mean relative densities (RD) of weed species in the real flora across the 10 farms under study. Weeds are grouped by botanical family, life cycle and biological group (BG).

Binomial Name	Botanical Family	Life Cycle	BG †	MOD Farm Group					OLD Farm Group					RD (%) ‡
				MRG	Bannò	Spitaleri	Mocciaro	Agrimor	Di Nolfo	Delizia	Minio	Antichi Granai	Cottonaro	
Anagallis arvensis L.	Primulaceae	annual	T	-	0.17	-	-	-	-	-	0.09	-	-	2.2
Artemisia vulgaris L.	Asteraceae	perennial	H	0.24	-	-	-	-	-	-	-	-	-	2.4
Avena fatua L.	Poaceae	annual	T	0.09	0.26	0.31	0.54	0.26	-	0.37	-	-	-	21.7
Centaurea sp.	Asteraceae	annual	T	0.18	-	0.12	-	0.15	-	0.05	0.14	0.10	-	5.9
Convolvulus arvensis L.	Convolvulaceae	perennial	G	-	-	-	-	-	-	-	0.13	0.26	-	4.4
Daucus carota L.	Apiaceae	biennal	H	0.07	-	-	-	-	0.11	0.05	0.13	0.07	-	3.2
Diplotaxis erucoides (L.) DC.	Brassicaceae	annual	T	-	-	-	-	-	-	-	0.03	-	-	0.1
Euphorbia helioscopia L.	Euphorbiaceae	annual	T	-	-	-	-	-	-	-	-	0.25	-	2.9
Erodium cicutarium (L.) L'Hér.	Geraniaceae	annual	T	-	-	-	-	-	0.09	0.06	-	0.11	-	1.8
Galium aparine L.	Rubiaceae	annual	T	-	-	-	-	-	0.54	0.19	-	-	-	8.0
Glebionis coronaria (L.) Cass. ex Spach	Asteraceae	annual	T	0.09	0.14	0.27	-	-	-	0.06	0.12	-	-	6.5
Inula helenium L.	Asteraceae	perennial	H	-	-	-	-	0.05	-	-	-	-	-	0.3
Lolium sp.	Poaceae	annual	T	0.07	-	-	0.31	0.05	-	0.22	-	-	-	5.8
Papaver rhoeas L.	Papaveraceae	annual	T	0.11	-	-	-	-	-	-	-	-	-	1.4
Phalaris paradoxa L.	Poaceae	annual	T	-	-	0.06	0.15	0.23	-	-	-	0.10	-	5.0
Polygonum aviculare L.	Polygonaceae	annual	T	-	0.24	-	-	-	-	-	-	-	-	2.6
Sinapis arvensis L.	Brassicaceae	annual	T	-	0.06	-	-	0.20	-	-	-	-	-	3.2
Sonchus sp.	Asteraceae	annual	T	0.15	0.14	-	-	-	-	-	0.12	0.07	-	4.4
Stellaria media (L.) Vill.	Caryophyllaceae	biennal	H	-	-	0.23	-	0.07	0.27	-	0.24	0.05	-	8.4

† T: therophytes; H: hemicryptophytes; G: geophytes. ‡ Averaged over all ten farms under study.

Table 4. α- and β-diversity indices of weed species in the real flora across the 10 farms under study.

	α-Diversity				β-Diversity	
	Margalef	**Shannon-Weiner**	**Pielou**	**Whittaker**	**Sørensen ‡**	**Steinhaus ‡**
MOD farm group †	**2.45 ± 0.68 A**	**1.49 ± 0.36 A**	**0.88 ± 0.04 A**	**3.8 ± 1.5 A**	64.3%	42.3%
MRG	3.60 a	1.95 a	0.94 a	2.4 d		
Bannò	2.45 b	1.65 b	0.92 a	3.2 c		
Spitaleri	1.92 c	1.39 b	0.87 a	3.8 b		
Mocciaro	2.36 b	0.96 c	0.87 a	6.3 a		
Agrimor	1.93 c	1.48 b	0.83 a	3.2 c		
OLD farm group †	**1.55 ± 0.93 B**	**1.19 ± 0.76 B**	**0.65 ± 0.39 B**	**2.7 ± 1.6 A**		
Di Nolfo	1.34 b	0.94 c	0.68 c	4.8 a		
Delizia	2.20 a	1.40 b	0.72 bc	2.7 c		
Minio	2.06 a	1.85 a	0.89 ab	2.4 d		
Antichi granai	2.14 a	1.77 a	0.99 a	3.2 b		
Cottonaro	-	-	-	-		

Different capital letters indicate significant differences between the MOD and OLD groups at $p \leq 0.05$ (Tukey's HSD). Different lowercase letters indicate significant differences within the groups at $p \leq 0.05$ (Tukey's HSD test). † Data are the mean ± standard deviation. For within-group values, the standard deviation is always 0.1. ‡ Similarity between the MOD and OLD groups.

Table 5. Eigenvectors and eigen analysis of the first three PCs of 12 variables (6 and 7 major weeds for the soil seedbank and real flora, respectively) from PCA on the correlation matrix. Variables with the largest influence for each principal component are in bold.

Variable	Weed Communities					
	Soil Seedbank			**Real Flora**		
	PC1	**PC2**	**PC3**	**PC1**	**PC2**	**PC3**
ANGAR	**0.514**	0.070	0.081	-	-	-
AVEFA	-	-	-	**−0.504**	−0.171	−0.474
CENXX	-	-	-	0.216	**−0.545**	0.274
EUPHE	0.288	**−0.577**	−0.500	-	-	-
FALCO	−0.144	**−0.687**	0.026	-	-	-
GALAP	-	-	-	0.188	**0.618**	−0.202
GLECO	-	-	-	0.206	−0.502	**−0.594**
HELMEC	0.488	0.181	0.313	-	-	-
LOXX	-	-	-	**−0.507**	0.118	−0.278
PHALPHA	-	-	-	−0.383	−0.157	0.389
SINAR	0.311	0.308	**−0.731**	-	-	-
STELME	**−0.544**	0.292	−0.333	**0.467**	0.039	−0.279
Eigenvalue	2.346	1.491	1.038	2.505	1.693	1.178
% Variance	39.2	24.8	17.3	35.8	24.2	16.8
% Cumulative variance	39.1	64.0	81.3	35.8	60.0	76.8

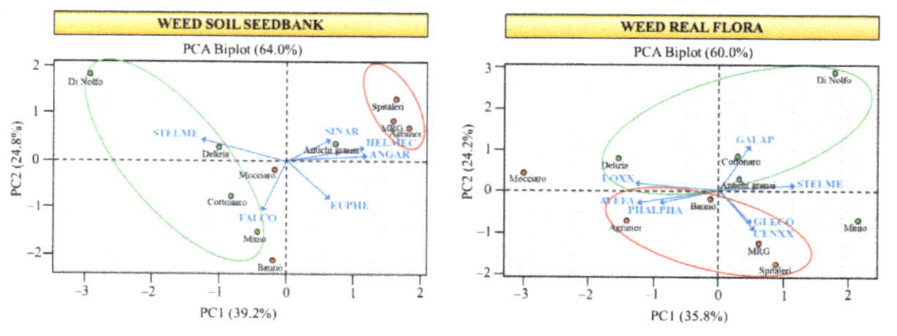

(a) (b)

Figure 3. Principal components analysis ordination biplot from the correlation matrix with the 6 major weeds for the soil seedbank (**a**) and with the 7 major weeds for the real weed flora (**b**) across the 10 farms under study. Farms belonging to the MOD group are labelled green, while farms from the ANT group are shown with red circles. Arrows highlight the discrimination of weeds along the principal components. Groups: MOD: wheat modern varieties; OLD: wheat old landraces. Weeds: ANGAR (*Anagallis arvensis*), AVEFA (*Avena fatua*), CENXX (*Centaurea* sp.), EUPHE (*Euphorbia helioscopia*), FALCO (*Fallopia convolvulus*), GALAP (*Galium aparine*); GLECO (*Glebionis coronaria*), HELMEC (*Helminthotheca echioides*), LOXX (*Lolium* sp.), PHALPHA (*Phalaris paradoxa*), SINAR (*Sinapis arvensis*) and STEMLE (*Stellaria media*).

3. Discussion

In this research, the weed-suppressive ability of the modern and old durum wheat cultivars was evaluated on both the soil seedbank and real weed flora under a semi-arid climate, namely Central-Eastern Sicily, an important Mediterranean centre production of such a crop. Following Travlos et al. [17] and Nkoa et al. [18], weed abundance and diversity were considered. Though the obtained results were farm-specific, pooling over farms emerged clearly that old landraces, compared to modern cultivars, were associated with a high decrease of soil seedbank, seed emergence and weed growth, as indicated by the lower aboveground biomass weight, despite four of the five farms cultivating old landraces not performing chemical weed control. Nevertheless, four of the five farms belonging to OLD group showed a seedbank size < 20 million seeds ha^{-1}, which is an important goal to reduce the direct weed control methods within IWM strategies. This significantly higher weed-suppressive ability of old landraces could be attributed to their greater competitive traits and allelopathic properties, as well as to the differences in the management practices performed. The crop competitive ability is conferred by a number of morphophysiological traits, such as fast seedling emergence, early vigour, high growth rates and root development, plant height, leaf area index, etc. [8–14]. Mwendwa et al. [19], for instance, reported a higher capacity in suppressing weed establishment by those bread wheat (*T. aestivum*) cultivars showing early vigour and early canopy closure, high biomass production and height. Lemerle et al. [15], screening several Australian wheat genotypes for their competitiveness against weeds, found that durum wheats were less competitive than *T. aestivum* and that old landraces suppressed the weeds more than all modern cultivars. Giambalvo et al. [20], after comparing one durum wheat landrace and two modern cultivars for their nitrogen use efficiency under induced interspecific competition, reported a higher competitive ability of the landrace Russello, likely due to its capacity in reducing the N availability to a competitor, a factor that increased with the increasing plant stature. About wheat allelopathy, the literature refers that the concentration of wheat allelochemicals, mainly belonging to benzoxazinoids, phenolic acids and short-chain fatty acids [20,21], varies considerably based on the cultivar choice [22]. In this regard, recently, Scavo et al. [16] found that three durum wheat old landraces (Timilia, Russello and Perciasacchi) were able to reduce seed germination and increase the mean germination time of the weeds *P. oleracea* and *S. media* more than the modern cultivar Mongibello. The authors supposed that the improved phytotoxicity of old landraces might be caused by their higher total polyphenol and total flavonoid contents. Lo Bianco et al. [5] indicated a specific and genotype-dependent pattern of phenolics concentration among ten Sicilian durum wheat landraces and three genetically improved cultivars, with coumarin, vanillic acid, luteolin and apigenin conjugates that were more abundant in local landraces. These phenols are recognised as well-known allelochemicals against several weeds [23]. In addition, Di Loreto et al. [24] reported a twofold greater content of vanillin, *p*-coumaric acid and 4-hydroxybenzaldehyde, as well as a 1.6-times greater amount of ferulic acid and syringaldehyde in old landraces than in modern cultivars. Here, it is likely that, in addition to their highly competitive traits, the repeated cultivation of durum wheat old landraces caused a build-up of allelochemicals into the rhizosphere through root exudation and plant residue decomposition. To reinforce this hypothesis, Belz and Hurle [25], after screening 146 cultivars of four Triticeae species, including durum wheat, demonstrated a high cultivar dependence in benzoxazinoids exudation, with hexaploid species that accumulate preferentially DIMBOA and only low levels of DIBOA, while the tetraploid *T. durum* accumulates substantial levels of both glucosides. Once in the rhizosphere, these allelochemicals interact with the complex of soil physical, chemical and biological characteristics that affect their bioavailability and phytotoxic level [26].

About species diversity, in this study, the potential and the real weed flora showed some common aspects, taking into account that emerged weeds are derived, to a large extent, from the soil seedbank. Both types of flora were largely composed of therophytes and annual weed species, as common under semi-arid climates [27], and were dominated by the few species present at high density. Moreover, in both cases, no significant differences were observed in terms of the species richness. Indeed, the diversity indices pointed out a high similarity between the MOD and OLD farm groups for the soil seedbank in terms of both the α- and β-diversity. On the contrary, in the real weed flora, the α-diversity was significantly higher in the MOD than the OLD farm group, and the two groups showed a quantitative β-diversity below 50%, the limit below which a dissimilarity can be interpreted. The values obtained here are in line with Hyvonen et al. [28], who registered values ranging from 50% to 80% in weed communities of cereal crops under a temperate climate. To better visualise the species compositions among the MOD and OLD farm groups, a PCA was carried out on the major weeds of both potential and real weed flora. The biplots showed that the magnitude of changes in the weed community composition varied between the potential and real weed flora, thus indicating a low correspondence between the below and aboveground weed communities, in accordance with Cardina and Sparrow [29]. The values of the seedbank size and aboveground biomass, in fact, did not always correspond with each other. Davis et al. [30] also reported little predictive value between the weed seedbanks and weed biomass within a long-term corn–soybean–wheat crop sequence under conventional and no tillage systems. In contrast with our findings, Ghersa and Ghersa-Martinez [31] indicated a strong predictive capacity of potential flora for aboveground weed communities in no tillage systems, because the shallow depth placement of seeds leads to greater proportional recruitment. Therefore, it is likely that the low correspondence between the soil seedbank and aboveground weed communities detected here is due to tillage systems performed in the studied area. In both the soil seedbank and real weed flora, however, the MOD and OLD farm groups were quite separated, suggesting different shifts and patters of weed communities between the modern cultivars and old durum wheat landraces. Moreover, the OLD farm group showed a few specific associations with major weeds, meaning that the local landraces were associated, to a lesser extent, with the major weeds than the modern cultivars.

4. Materials and Methods

4.1. Description of Survey Area

The present research was performed across 10 wheat farms located in Central-Eastern Sicily, an area devoted to cereal cultivation with a long tradition of durum wheat production. The climate of the zone is semi-arid Mediterranean, characterised by dry, long summers and mild, wet winters. The average annual rainfall in the last 30 years was 623 mm, mainly distributed over the autumn–winter period (Figure 4). Mean monthly air temperature is 15.1 °C, with July and January as the months with, respectively, the highest (24.8 °C) and the lowest (6.4 °C) temperatures (Figure 4). According to the USDA classification [32], the soils are Regosoils (Typic Xerorthensis or Xerochrepts) and Alluvial (Typic Vertic Xerofluvents), with moderately clayey texture.

4.2. Agronomic Management

In order to study the long-term effects derived from the repeated cultivation of some old landraces and modern durum wheat genotypes on weed flora, 10 cereal farms were selected for their long-term cultivation of durum wheats. Half of them grow modern durum wheat varieties, and the other half cultivates old landraces. Tables 6 and 7 show the geographical coordinates and the agronomic management of the ten farms. The modern durum wheat cultivars under study, characterised by early or early-medium maturity, were Antalis; Anco Marzio; Core and a mix composed of Iride, Simeto and Core. Old durum wheat landrace, showing a medium-late cycle, were Perciasacchi, Timilia and Senatore Cappelli. In the studied zone, they are usually sown in late autumn, with a seeding rate ranging from 160 to 300 seeds m^{-2}, and harvested in late spring or at the beginning of summertime [33]. All farms cultivating modern cultivars controlled the weeds chemically and applied a mineral fertilisation. On the contrary, among the farms cultivating old landraces, Antichi granai was the only one that fertilised the crop and Cottonaro the only farm performing weed chemical control. In addition, only the Delizia Farm carried out the stale seedbed. Long-term crop sequences of the ten farms are shown in Table 8.

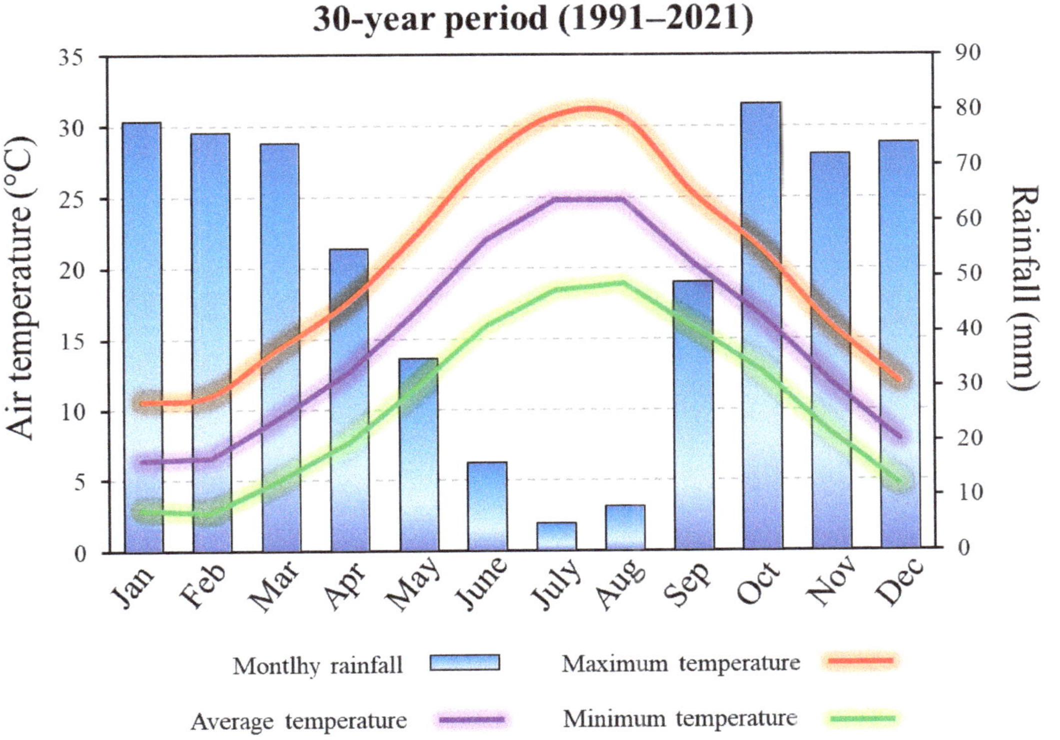

Figure 4. Long-term period of total monthly rainfall and monthly air temperatures (maximum, average and minimum) in the survey area (Central-Eastern Sicily, South Italy).

4.3. Weed Flora Analysis

Both potential (soil seedbank) and real weed flora (aboveground species) were analysed by considering the abundance and diversity. Prior to sampling, the study sites were monitored with a field scouting to visualise weed distribution and locate the sampling units, excluding the borders of each plot and the nonrepresentative areas. Due to the variation between and within the study sites, a stratified random sampling was adopted [18], which consisted of dividing each sampling zone into homogeneous strata. In detail, a 1000 m^2 area was selected in each farm—within which, three sampling zones were located. Soil and aboveground weed samples were collected on June 2nd and 3rd 2022, just before wheat harvest.

Table 6. Geographical coordinates and agronomic management of the 5 farms belonging to the MOD group.

Farm	Geographical Coordinates	Wheat Genotype	Seeding Density (kg ha^{-1})	Tillage	Fertiliser Type and Fertilisation Time	Fertiliser Amount (kg ha^{-1})	Active Principle for Weed Chemical Control
MRG	37°35′45″ N 14°28′11″ E	Antalis	200	Hoeing (late August-early September) Deep ploughing (−25 cm) Light ploughing (−15 cm)	Ammonium nitrate (34%) in post-emergence between February and March	120	Thifensulfuron-methyl, Tribenuron-methyl (Amedeus Top); Clodinafop-propargyl, Cloquintocet-mexyl, Pinoxaden (Traxos Pronto 60)
Bannò	37°35′20″ N 14°27′45″ E	Anco Marzio	260	Disc ploughing Deep ploughing (−25 cm) Light ploughing (−15 cm)	Urea (46%) in post-emergence between February and March	150	Mefenpir-diethyl, Mesosulfuron-methyl (half-dosed Atlantis); 2,4-D.
Spitaleri	37°35′28″ N 14°28′00″ E	Iride, Simeto and Core	300	Hoeing (late August-early September) Disc ploughing Ploughing cultivator Rolling after seeding	Urea (46%) in post-emergence between February and March	250	Clodinafop-propargyl, Cloquintocet-mexyl, Pinoxaden (Traxos Pronto 60)
Mocciaro	37°41′09″ N 14°23′53″ E	Core	240	Subsoiling (September) Deep ploughing (−25 cm) Light ploughing (−15 cm)	Urea (46%) in post-emergence between February and March	150	Mefenpir-diethyl, Mesosulfuron-methyl (Atlantis)
Agrimor	37°35′26″ N 14°27′01″ E	Core	230–280	Hoeing (April–May) Deep ploughing (−25 cm) on August-September Light ploughing (−15 cm) on October	Diammonium phosphate (18% N, 46% P_2O_5) at seeding Urea (46%) in post-emergence between February and March	115–140 120	Clodinafop-propargyl, Cloquintocet-mexyl, Pinoxaden (Traxos Pronto 60)

Table 7. Geographical coordinates and agronomic management of the 5 farms belonging to the OLD group.

Farm	Geographical Coordinates	Wheat Genotype	Seeding Density (kg ha^{-1})	Tillage	Fertiliser Type and Fertilisation Time	Fertiliser Amount (kg ha^{-1})	Active Principle for Weed Chemical Control
Di Nolfo	37°35′22″ N 14°32′52″ E	Perciasacchi	220–230	Subsoiling Deep ploughing (−25 cm) Light ploughing (−10–15 cm)	-	-	-
Delizia	37°31′55.2″ N 14°12′52.7″ E	Perciasacchi	200	Disc ploughing (after wheat harvest) Subsoiling (September) Deep ploughing (October) Light ploughing (−10–15 cm) Stale seedbed with precision seeder	-	-	-
Minio	37°35′29″ N 14°30′48″ E	Perciasacchi	200–220	Hoeing (late August–early September) Deep ploughing (−25 cm) Light ploughing (−10–15 cm)	-	-	-
Antichi granai	37°36′08.5″ N 14°34′53.7″ E	Timilia	200	Subsoiling Deep ploughing (−25 cm) Light ploughing (−10–15 cm) Pre-seeding ploughing	Organic N (8.5%); organic C (28%)-(AMMINO-BIO) in post-emergence in April	20	-
Cottonaro	37°35′48″ N 14°19′03″ E	Senatore Cappelli	160	Subsoiling (September) Deep ploughing (−25 cm) Light ploughing (−10 cm)	-	-	Fluroxipir meptyl-eptyl ester, Clopiralid pure ethylammonium salt, MCPA pure potassium salt (Ariane II); Pyroxsulam, Florasulam, Cloquintocet mexyl (Floramix)

Table 8. Crop sequence of the ten farms under study over the last ten years.

Farm	Crop Sequence Alternating with Wheat
MOD farm group	
MRG	vetch-clover and vetch-fava bean mix
Bannò	vetch and leguminous mix
Spitaleri	fava bean and vetch
Mocciaro	vetch-clover-sulla-ryegrass-oat mix
Agrimor	vetch
OLD farm group	
Di Nolfo	vecth-sulla-clover mix
Delizia	vetch-clover-sulla-ryegrass-oat mix
Minio	sulla, fava bean and vecth
Antichi granai	fallow, chickpea and lentil
Cottonaro	vetch-clover-sulla mix

4.3.1. Soil Seedbank (Potential Weed Flora)

Soil samples were taken with a core sampler 10–15 cm deep along the diagonals of the central part of each sampling zone [34]. A sample was obtained by pooling 5 randomly distributed subsamples, each 0.75 dm^3, for a total of 150 soil cores (10 farms × 3 replicates × 5 subsamples) collected. Following Scavo et al. [13], the soil samples were freed from inert fraction (stones, pebbles and dead debris), and the seeds were extracted from the soil by using a metal tube (Karcher, K 3500 model, Winnenden, Germany) equipped with a removable cap consisting of a 250 μm steel mesh. For high clayey soils, a pre-treatment with 5 g of sodium hexametaphosphate solution for 20 min was necessary to disperse the colloid matrix. The count and identification of weeds was performed with a MS5 Leica stereomicroscope (Leica Microsystems, Wetzlar, Germany) in Petri dishes after 24 h of air-drying. The seedbank size was calculated as the number of seeds per square metre of surface area for each plot.

4.3.2. Aboveground Species (Real Weed Flora)

The analysis of the real weed flora was carried out over three randomly placed quadrats (each of 1 m^2) per sampling zone [27]. For the total aboveground weed biomass, weeds were clipped at the soil surface, dried to constant mass and weighed (pooled weight at the quadrat level was considered). In each quadrat, weeds were sorted by species or genera, together with the number of individual plants per species.

4.3.3. Weed Abundance

Aside from the seedbank size for potential flora and aboveground biomass for real flora, weed species abundance was calculated under the relative density (RD), relative frequency (RF) and relative abundance index (RAI), in accordance with Scavo et al. [34]:

$$\mathrm{RD}\ (\%) = \left(\frac{\sum \mathrm{Y_i}}{\mathrm{S}}\right) \times 100 \quad (1)$$

$$\mathrm{RF}\ (\%) = \left(\frac{\mathrm{F_i}}{\sum \mathrm{F}}\right) \times 100 \quad (2)$$

$$\mathrm{RAI} = \frac{\mathrm{RD} + \mathrm{RF}}{2} \quad (3)$$

where: $\sum Y_i$ = sum of the number of individuals or seeds for a weed species, S = species richness within the plot, F_i = number of sampling units in which the species *i* occurred and $\sum F$ = sum of the absolute frequencies of all species. The RAI is a valuable parameter in characterising weed communities, since it takes into account both the weed density and evenness, thus overcoming the problems caused by a nonuniform weed distribution [35].

4.3.4. Floristic Composition and Species Diversity of Weed Communities

After seed or individual plant identification, the floristic composition was assessed based on Conti et al. [36], grouping weed species or genera by botanical family, life cycle and biological group. Species diversity was explored as richness and evenness. The former is the total number of weed species present in a community, and the latter provides information about the abundance of each species in a community [17]. Considering the variability across the study sites, the species diversity was estimated within the community (α-diversity) and between communities (β-diversity). Margalef's (D_{MG}), Shannon–Wiener (H) and Pielou's (J) indices were computed for the α-diversity:

$$D_{MG} = (S - 1)/\ln(N) \tag{4}$$

$$H = -\sum p_i \ln p_i \tag{5}$$

$$J = H/\ln S \tag{6}$$

where N = total number of seeds or individuals of all species in the community, and p_i = proportional abundance of the ith species. Margalef's diversity index is a rapid method to measure the gross species diversity only based on richness, but it is very sensitive to the sample size. In contrast, J measures only the evenness, and H includes both species richness and evenness.

For the β-diversity, three common diversity indices were chosen, i.e., Whittaker's statistics (W), Sørensen's (S_S) and Steinahus's (S_A) coefficients, in accordance with Ramírez et al. [37] and Restuccia et al. [27]:

$$W = \frac{Y}{S} \tag{7}$$

$$S_S = \left[\frac{2J}{(a+b)}\right] \times 100 \tag{8}$$

$$S_A = \left(\frac{2W}{A+B}\right) \times 100 \tag{9}$$

where Υ = total number of all species in the entire study area, J = number of common species to each community; a + b = sum of the total number of species in each community, W = sum of the lower of the two abundances of each species in the community, A = total number of individuals in population A and B = total number of individuals in population B. Whittaker's statistics measures the rate of species turnover, and S_S measures the binary similarity in terms of presence/absence, while S_A accounts for the differences in abundance.

4.4. Statistical Analysis

All data were subjected to ANOVA by applying a generalised linear model (GLM) with the protected Tukey's HSD means separation test at α = 0.05. Prior to ANOVA, the normality and homogeneity of variance were respectively assessed with the Shapiro–Wilk's and Bartlett's tests. To meet the ANOVA assumptions, the seedbank and aboveground biomass data were $\log_{(x+1)}$-transformed, the RAI data needed an arcsine–square root transformation and the H and J data were respectively square root- and logit-transformed, while the species richness showed a homogeneous variance distribution [34]. For an easier interpretation of the results, data from the five farms cultivating modern cultivars and from the other five farms cultivating old landraces were respectively pooled and presented into two groups: MOD and OLD. Multivariate statistics was used to analyse the weed community composition. Following Restuccia et al. [27], a principal component analysis (PCA) on the correlation matrix of the standardised major weeds (those with RD $\geq$ 5%) was performed on both the potential and real weed flora. The results were graphically presented on "distance" biplots derived from the first two principal components explaining the maximum variance [38]. Biplots allowed visualising the relations between the variables (major weeds) and wheat genotypes. Minitab® version 16 (Minitab Inc., State College, PA, USA) statistical software was employed for all analyses.

5. Conclusions

The results obtained here, as a whole, demonstrate that the old landraces of durum wheat may possess a stronger weed-suppressive ability than modern cultivars in terms of seedbank and aboveground biomass reduction. Within the studied area, landraces were also able to reduce the aboveground weed species diversity and to cause shifts in weed populations. The importance of these findings is even greater if considering that four of the five farms cultivating old landraces did not perform chemical weed control. Therefore, this research provided the scientific basis for the increased interest that consumers, government policies and scientists have moved toward durum wheat landraces by virtue of their sustainable cultivation, high product quality and remuneration. On the other hand, although a multilocation trial involving ten different farms was carried out, our study considered just one growing season. Furthermore, it should be considered that the high heterogeneity in terms of the management practices across the farms under study may have affected the obtained results. In future steps, we aim to perform an economic analysis of these findings, as we noted that old landraces can be grown with lower inputs, and their products can be sold at a higher price than modern cultivars.

Author Contributions: Conceptualisation, A.S. and G.M.; methodology, A.S., A.R., M.B. and G.M.; software, A.S.; formal analysis, A.S., A.R. and M.B.; investigation, A.S., A.R., M.B. and G.M.; data curation, A.S.; writing—original draft preparation, A.S.; writing—review and editing, A.S., A.R. and G.M. and supervision, A.S. and G.M. All authors have read and agreed to the published version of the manuscript.

Funding: This research received no external funding.

Institutional Review Board Statement: Not applicable.

Informed Consent Statement: Not applicable.

Data Availability Statement: All data are available via email request to the corresponding author.

Acknowledgments: The authors gratefully acknowledge all ten farms involved in the research for hosting the field experiments.

Conflicts of Interest: The authors declare no conflict of interest.

References

1. Sall, A.T.; Chiari, T.; Legesse, W.; Seid-Ahmed, K.; Ortiz, R.; van Ginkel, M.; Bassi, F.M. Durum Wheat (*Triticum durum* Desf.): Origin, Cultivation and Potential Expansion in Sub-Saharan Africa. *Agronomy* **2019**, *9*, 263. [CrossRef]
2. Eurostat. European Commission. 2021. Available online: https://ec.europa.eu/eurostat/data/database (accessed on 10 June 2022).
3. Kabbaj, H.; Sall, A.T.; Al-Abdallat, A.; Geleta, M.; Amri, A.; Filali-Maltouf, A.; Belkadi, B.; Ortiz, R.; Bassi, F.M. Genetic Diversity within a Global Panel of Durum Wheat (*Triticum durum*) Landraces and Modern Germplasm Reveals the History of Alleles Exchange. *Front. Plant Sci.* **2017**, *8*, 1277. [CrossRef] [PubMed]
4. Visioli, G.; Giannelli, G.; Agrimonti, G.; Spina, A.; Pasini, G. Traceability of Sicilian Durum Wheat Landraces and Historical Varieties by High Molecular Weight Glutenins Footprint. *Agronomy* **2021**, *11*, 143. [CrossRef]
5. Lo Bianco, M.; Siracusa, L.; Dattilo, S.; Venora, G.; Ruberto, G. Phenolic fingerprint of Sicilian modern varieties and old durum wheat landraces: A tool to assess with diversity. *Cereal Chem.* **2017**, *94*, 1045–1051. [CrossRef]
6. Beres, B.L.; Rahmani, E.; Clarke, J.M.; Grassini, P.; Pozniak, C.J.; Geddes, C.M.; Porker, K.D.; May, W.E.; Ransom, J.K. A Systematic Review of Durum Wheat: Enhancing Production Systems by Exploring Genotype, Environment, and Management (G × E × M) Synergies. *Front. Plant Sci.* **2020**, *11*, 568657. [CrossRef]
7. Jabran, K.; Mahmood, K.; Melander, B.; Bajwa, A.A.; Kudsk, P. Weed Dynamics and Management in Wheat. *Adv. Agron.* **2017**, *145*, 97–166. [CrossRef]
8. Scavo, A.; Mauromicale, G. Integrated Weed Management in Herbaceous Field Crops. *Agronomy* **2020**, *10*, 466. [CrossRef]
9. Carrubba, A.; Labruzzo, A.; Comparato, A.; Muccilli, S.; Spina, A. Use of Plant Water Extracts for Weed Control in Durum Wheat (*Triticum turgidum* L. Subsp. *durum* Desf.). *Agronomy* **2020**, *10*, 364. [CrossRef]

10. Campiglia, E.; Mancinelli, R.; Radicetti, E. Evaluating spatial arrangement for durum wheat (*Triticum durum* Desf.) and subclover (*Trifolium subterraneum* L.) intercropping systems. *Field Crop. Res.* **2014**, *169*, 49–57. [CrossRef]
11. Stagnari, F.; Galieni, A.; Speca, S.; Cafiero, G.; Pisante, M. Effects of straw mulch on growth and yield of durum wheat during transition to conservation agriculture in Mediterranean environment. *Field Crops Res.* **2014**, *167*, 51–63. [CrossRef]
12. Scavo, A.; Mauromicale, G. Crop Allelopathy for Sustainable Weed Management in Agroecosystems: Knowing the Present with a View to the Future. *Agronomy* **2021**, *11*, 2104. [CrossRef]
13. Scavo, A.; Restuccia, A.; Abbate, C.; Mauromicale, G. Seeming field allelopathic activity of *Cynara cardunculus* L. reduces the soil weed seed bank. *Agron. Sustain. Dev.* **2019**, *39*, 41. [CrossRef]
14. Bàrberi, P. Weed management in organic agriculture: Are we addressing the right issues? *Weed Res.* **2002**, *42*, 177–193. [CrossRef]
15. Lemerle, D.; Verbeek, B.; Cousens, R.D.; Coombes, N.E. The potential for selecting wheat varieties strongly competitive against weeds. *Weed Res.* **1996**, *36*, 505–513. [CrossRef]
16. Scavo, A.; Pandino, G.; Restuccia, A.; Caruso, P.; Lombardo, S.; Mauromicale, G. Allelopathy in Durum Wheat Landraces as Affected by Genotype and Plant Part. *Agronomy* **2022**, *11*, 1021. [CrossRef]
17. Travlos, I.S.; Cheimona, N.; Roussis, I.; Bilalis, D.J. Weed-species abundance and diversity indices in relation to tillage systems and fertilization. *Front. Environ. Sci.* **2018**, *6*, 11. [CrossRef]
18. Nkoa, R.; Owen, M.D.K.; Swanton, C.J. Weed Abundance, Distribution, Diversity, and Community Analyses. *Weed Sci.* **2015**, *63*, 64–90. [CrossRef]
19. Mwendwa, J.M.; Brown, W.B.; Weidenhamer, J.D.; Weston, P.A.; Quinn, J.C.; Wu, H.; Weston, L.A. Evaluation of commercial wheat cultivars for canopy architecture, early vigour, weed suppression, and yield. *Agronomy* **2020**, *10*, 983. [CrossRef]
20. Giambalvo, D.; Ruisi, P.; Di Miceli, G.; Frenda, A.S.; Amato, G. Nitrogen use efficiency and nitrogen fertilizer recovery of durum wheat genotypes as affected by interspecific competition. *Agron. J.* **2010**, *102*, 707–715. [CrossRef]
21. Aslam, F.; Khaliq, A.; Matloob, A.; Tanveer, A.; Hussain, S.; Zahir, Z.A. Allelopathy in agro-ecosystems: A critical review of wheat allelopathy-concepts and implications. *Chemoecology* **2017**, *27*, 1–24. [CrossRef]
22. Ma, Y. Allelopathic studies of common wheat (*Triticum aestivum* L.). *Weed Biol. Manag.* **2005**, *5*, 93–104. [CrossRef]
23. Scavo, A.; Pandino, G.; Restuccia, A.; Mauromicale, G. Leaf extracts of cultivated cardoon as potential bioherbicide. *Sci. Hortic.* **2020**, *261*, 109024. [CrossRef]
24. Di Loreto, A.; Bosi, S.; Montero, L.; Bregola, V.; Marotti, I.; Sferrazza, R.E.; Dinelli, G.; Herrero, M.; Cifuentes, A. Determination of phenolic compounds in ancient and modern durum wheat genotypes. *Electrophoresis* **2018**, *39*, 2001–2010. [CrossRef]
25. Belz, R.G.; Hurle, K. Differential exudation of two benzoxazinoids one of the determining factors for seedling allelopathy of Triticeae species. *J. Agric. Food Chem.* **2005**, *53*, 250–261. [CrossRef] [PubMed]
26. Scavo, A.; Abbate, C.; Mauromicale, G. Plant allelochemicals: Agronomic, nutritional and ecological relevance in the soil system. *Plant Soil* **2019**, *442*, 23–48. [CrossRef]
27. Restuccia, A.; Scavo, A.; Lombardo, S.; Pandino, G.; Fontanazza, S.; Anastasi, U.; Abbate, C.; Mauromicale, G. Long-term effect of cover crops on species abundance and diversity of weed flora. *Plants* **2020**, *9*, 1506. [CrossRef]
28. Hyvonen, T.; Ketoja, E.; Salonen, J.; Jalli, H.; Tiainen, J. Weed species diversity and community composition in organic and conventional cropping of spring cereals. *Agric. Ecosyst. Environ.* **2003**, *97*, 131–149. [CrossRef]
29. Cardina, J.; Sparrow, D.H. A comparison of methods to predict weed seedling populations from the soil seedbank. *Weed Sci.* **1996**, *44*, 46–51. [CrossRef]
30. Davis, A.S.; Renner, K.A.; Gross, K.L. Weed seedbank and community shifts in a long-term cropping systems experiment. *Weed Sci.* **2005**, *53*, 296–306. [CrossRef]
31. Ghersa, C.M.; Ghersa-Martinez, M.A. Ecological correlates of weed seed size and persistence in the soil under different tilling systems: Implications for weed management. *Field Crops Res.* **2000**, *67*, 141–148. [CrossRef]
32. Soil Survey Staff. *Soil Taxonomy: A Basic System of Soil Classification for Making and Interpreting Soil Surveys*, 2nd ed.; U.S. Government Publishing Office: Washington, DC, USA, 1999.
33. Pandino, G.; Mattiolo, E.; Lombardo, S.; Lombardo, G.M.; Mauromicale, G. Organic Cropping System Affects Grain Chemical Composition, Rheological and Agronomic Performance of Durum Wheat. *Agriculture* **2020**, *10*, 46. [CrossRef]
34. Scavo, A.; Restuccia, A.; Abbate, C.; Lombardo, S.; Fontanazza, S.; Pandino, G.; Anastasi, U.; Mauromicale, G. *Trifolium subterraneum* cover cropping enhances soil fertility and weed seedbank dynamics in a Mediterranean apricot orchard. *Agron. Sustain. Dev.* **2021**, *41*, 70. [CrossRef]
35. Derksen, D.A.; Lafond, G.P.; Thomas, A.G.; Loeppky, H.A.; Swanton, C.J. Impact of agronomic practices on weed communities: Tillage systems. *Weed Sci.* **1993**, *41*, 409–417. [CrossRef]
36. Conti, F.; Abbate, G.; Alessandrini, A.; Blasi, C. *An Annotated Checklist of the Italian Vascular Flora*; Palombi Editore: Rome, Italy, 2005.
37. Ramírez, J.S.; Hoyos, C.V.; Plaza, T.G. Phytosociology of weeds associated with rice crops in the department of Tolima, Colombia. *Agron. Colomb.* **2015**, *33*, 64–73. [CrossRef]
38. Legendre, P.; Legendre, L. *Numerical Ecology*; Elsevier: Amsterdam, The Netherlands, 2012; pp. 429–463.

plants

MDPI

Article

Weed Management and Crop Establishment Methods in Rice (*Oryza sativa* L.) Influence the Soil Microbial and Enzymatic Activity in Sub-Tropical Environment

Sarthak Pattanayak [1], Satyananda Jena [1], Priyanka Das [2], Sagar Maitra [3,*], Tanmoy Shankar [3], Subhashisa Praharaj [3], Prasannajit Mishra [1], Santanu Mohanty [4], Madhusmita Pradhan [4], Deepak Kumar Swain [5], Biswajit Pramanick [6], Ahmed Gaber [7] and Akbar Hossain [8,*]

1 Department of Agronomy, College of Agriculture, Odisha University of Agriculture and Technology, Bhubaneswar 751003, Odisha, India; imanagronomist@gmail.com (S.P.); snjena2008@gmail.com (S.J.); prasannajitmishra71@gmail.com (P.M.)
2 Department of Agronomy, Palli Siksha Bhavana, Visva-Bharati, Sriniketan 731204, West Bengal, India; pdaspriyanka06@gmail.com
3 Department of Agronomy, Centurion University of Technology and Management, Paralakhemundi 761211, Odisha, India; tanmoy@cutm.ac.in (T.S.); subhashisa.praharaj@cutm.ac.in (S.P.)
4 Department of Soil Science and Agriculture Chemistry, College of Agriculture, Odisha University of Agriculture and Technology, Bhubaneswar 751003, Odisha, India; santanu.madhu1960@gmail.com (S.M.); madhusmita.pradhan47@gmail.com (M.P.)
5 Department of Agricultural Statistics, Institute of Agricultural Science, Siksha-o-Anusandhan Deemed to be University, Bhubaneswar 751030, Odisha, India; deepakkswain@soa.ac.in
6 Department of Agronomy, Dr. Rajendra Prasad Central Agricultural University, Pusa, Samastipur 848125, Bihar, India; biswajit@rpcau.ac.in
7 Department of Biology, College of Science, Taif University, P.O. Box 11099, Taif 21944, Saudi Arabia; a.gaber@tu.edu.sa
8 Department of Agronomy, Bangladesh Wheat and Maize Research Institute, Dinajpur 5200, Bangladesh
* Correspondence: sagar.maitra@cutm.ac.in (S.M.); akbar.hossain@bwmri.gov.bd (A.H.)

Citation: Pattanayak, S.; Jena, S.; Das, P.; Maitra, S.; Shankar, T.; Praharaj, S.; Mishra, P.; Mohanty, S.; Pradhan, M.; Swain, D.K.; et al. Weed Management and Crop Establishment Methods in Rice (*Oryza sativa* L.) Influence the Soil Microbial and Enzymatic Activity in Sub-Tropical Environment. *Plants* **2022**, *11*, 1071. https://doi.org/10.3390/plants11081071

Academic Editors: Alessia Restuccia and Aurelio Scavo

Received: 14 March 2022
Accepted: 10 April 2022
Published: 14 April 2022

Abstract: Weed management has become the most important and inevitable aspect of crop management for achieving a higher rice yield. Nowadays, chemical herbicide application has become a popular practice for managing weeds in different rice cultures. However, herbicide application can have qualitative and quantitative impacts on soil microorganisms and soil enzymes, particularly in the case of new herbicide molecules and their indiscriminate use for a longer period. Further, different rice establishment methods also play a significant role in soil microbial population dynamics as well as soil biological properties. Keeping these in view, a field experiment was conducted at the Agronomy Main Research Farm, Orissa University of Agriculture and Technology (OUAT), India, during the *kharif* season of 2016 and 2017, on the impact of crop establishment methods and weed management practices on soil microbial and enzymatic status. The field experiment was laid out in a split-plot design with three replications with four crop establishment methods in the main plot, viz., M_1, Direct Seeded Rice (DSR); M_2, Wet Seeded Rice (WSR); M_3,Unpuddled Transplanted Rice (NPTR); M_4, Puddled Transplanted Rice (PTR), and six weed management practices in the sub-plot, viz., W_1, Weedy check; W_2, Bensulfuron methyl 0.6% + Pretilachlor 6% (pre-emergence (PE)) 0.660 kg ha^{-1} + Hand weeding (HW) at 30 days after sowing/transplanting (days after sowing/transplanting (DAS/T)); W_3, Bensulfuron methyl 0.6% + Pretilachlor 6% (PE) 0.495 kg ha^{-1} + HW at 30 DAS/T; W_4, Bensulfuron methyl 0.6% + Pretilachlor 6% (PE) 0.495 kg ha^{-1} + Bispyribac-Sodium (post-emergence(POE)) 0.025 kg ha^{-1} at 15 DAS/T; W_5, Cono weeding (CW) at 15 DAS/T + hand weeding 30 DAS/T, and W_6, Brown manuring/Green manuring. The initial decline in the microbial population was observed due to herbicide application in NPTR and PTR up to 7 DAS/T and then it increased up to 28 DAS/T. There was a reduction in soil microbial and enzymatic status after the application of herbicides Bensulfuron methyl 0.6% + Pretilachlor 6% (PE) and Bispyribac-Sodium (POE) that again followed an upward graph with crop age. Significant variation in enzymatic activity and the microbial count was also observed among treatments involving crop establishment methods. The study revealed that improved microbial population and enzyme activity were noted in unpuddled transplanted rice under organic weed management due to favorable conditions, and chemical weed control initially affected microbial population and activities.

Keywords: rice; cultural methods; herbicides; impacts; soil microorganisms; soil enzymes

1. Introduction

Rice (*Oryza sativa* L.) is the most widely consumed staple food for more than 50% of the world's human population. Around 90% of the global rice production is being grown and consumed in Asia where rice is synonymous with the livelihood of people [1,2]. Globally, rice is grown in 120 countries in an area of 164.1 million hectares with a production of 756.74 million tons and productivity of 4.6 t ha^{-1} [3]. India is the second-largest rice-producing country in the world where rice is the main staple food crop. In India, rice occupies 4.4 million hectares with a production of 118 million tons and productivity of 2.7 t ha^{-1} [4]. Odisha, one of the leading rice-growing states of India, grows rice in 3.89 Mha with a production of 8.04 million tons and productivity of 2.1 t ha^{-1}, which is low compared to global and national average productivity [4].

Achieving self-sufficiency in rice production and maintaining price stability are important objectives in low-income countries [5]. Kharif rice is grown in different ecosystems in Odisha such as rainfed upland, rainfed lowland, deep water, and tidal wetland thanks to the high adaptability of rice to a wide range of geohydrological situations. Challenges to rice cultivation have increased manyfold due to the impact of climate change, shrinking landmass, labor, and water scarcity. This has triggered the need to revisit crop management strategies, more importantly, crop establishment methods, and to formulate better weed management strategies for higher rice production.

Life and biodiversity of soil may be prone to destruction due to the intensive use of plant protection chemicals [6]. Plant protection materials are chemically derived compounds containing one or more than one active ingredients with mutually complementary properties but with different modes of action [7] and their application is exceptionally beneficial for agriculture. However, we have to realize the threats attached to soil microbial diversity [8–10]. The application of plant protection chemicals exceeding the recommended dose may impact the growth and development of microbial assemblages [11], plants [12], animals [13], and people [14]. It has been experimentally demonstrated by many researchers that only a small percentage of applied plant protection chemicals are actually involved in controlling target organisms while a huge percentage pervades soil, water, and living organisms [15,16].

Climatic condition with soil physicochemical properties affects the persistence of these chemicals in soil and their movement to water and agroecosystem [17]. Various effects of plant protection chemicals can be seen on soil microorganisms as well as detrimental impacts on some microorganisms [18,19]. Microorganisms are an excellent source of carbon and energy and so they play multifaceted roles in the agroecosystem inclusive of enhancement of plant nutrients and bioremediation of soil from harmful abiotic factors [17,20,21]. Herbicide application results in better control of weeds with higher weed control efficiency and a lower labor requirement. Initially, herbicide application was mostly confined to plantation crops, but nowadays it is mostly used in field crops where wheat and rice constitute around 42% and 32% of total herbicide consumption, respectively [22].

However, herbicide application can have a qualitative and quantitative impact on the alternation of soil microbial population and soil enzymatic activity [23]. Soil microorganisms and plant roots are an important source of soil enzymes, therefore, the effect of herbicides on soil microorganism invariably affect soil enzymatic activity [16,18]. Moreover, a change in the soil microflora has been listed as one of the possible causes of productivity decline in rice-based cropping systems [24]. Soil microbial biomass is considered an active nutrient supply pool to plants. Herbicides are considered to have no major or long-term effect on soil microbial counts when applied at the standard field recommended dose. However, indiscriminate use of herbicides and new molecules of herbicide may affect soil health, particularly soil microorganisms and soil enzymatic activity, and may lead to long term accumulation.

The Bensulfuron methyl, methyl ester of bensulfuron, an acetolactase synthatse inhibitor and pretilachlor [2–chloro–2,6–diethyl–N–(2–propoxyethyl) acetanilide] is a chloroacetanilide herbicide. Bensulfuron methyl + pretilachlor is a new herbicide combination reported to be used as a pre-emergence herbicide providing effective control of grasses, sedges, and broad-leaved weeds in rice without any phytotoxic symptoms in the crop [25]. Bispyribac sodium is popularly used as a post-emergence herbicide in rice [26,27], it acts on the enzyme acetolactase synthetase (ALS), which in turn inhibits the production of amino acids such as valine, isoleucine, and leucine [28] that inhibit protein synthesis in plants, causing their death. Due to the rapid photochemical transformation and low vitalization potential, its environmental impact is considered to be negligible and is moderate to highly persistent in anaerobic flooded rice soil [29]. Detailed studies on the impact of new molecules are lacking, demanding an experiment to establish the cause and effect.

On the other hand, brown and green manuring practices are effective in managing the weed population in rice and improving soil health [30,31]. Knowledge of the effect of herbicides on soil biological activity and crop yield is very helpful in developing a better crop management strategy. Further, rice establishment methods have a significant impact on the growth and productivity of crops inclusive of microbial activities [32,33]. Variation in water regime and associated cultural operations may result in varying microbial count and enzymatic activity. The variation in microbial activity was earlier recorded under different rice establishment methods by Latha and Gopal [34] and Ramalakshmi et al. [35].

Considering the above facts, this study was planned to evaluate microbial population dynamics, namely, total bacteria, fungi, and actinomycetes, and soil enzymatic activity of soil enzymes (viz., urease, alkaline phosphatase, and dehydrogenase) in *Kharif* rice under various weed management and crop establishment methods. This will help to evaluate the effectiveness of new herbicide molecules along with the impact of different weed management practices through organic and mechanical methods on soil health and soil biological properties, and will also enable researchers, policy makers, and farmers to evaluate better treatment combinations for higher yield by maintaining soil health.

2. Materials and Methods

2.1. Site Description

The experiment was conducted during two Kharif seasons in 2016 and 2017 at the Agronomy Main Research Farm, Department of Agronomy, Odisha University of Agriculture and Technology, Bhubaneswar, Odisha, India which lies at 20°15′N latitude and 85°52′E longitude, respectively, with an altitude 25.9 m above the mean sea level. The station falls under the Agro-Climatic Zone of the East and Southeastern Plain Zone of Odisha, India with a moisture deficit index (MDI) of −0 to −20 and a growing season length of 180 to 210 days. The climate of the area is warm and moist with hot humid summer and mild winter. Broadly, the climate falls in the category of moist hot [36]. The Rainfall code of Bhubaneswar, Odisha, India is D_1A_2 $(B_1A_2B_1)$ $C_1D_1E_2$.

Representative soil samples were collected from a depth of 0–10 cm in a zig-zag manner using a soil auger before the beginning of the experiment to analyze and study the soil textural classes and soil physiochemical and biological properties; each sample was a composite of three locations within a plot. The experimental soil was slightly acidic and sandy loam in texture with a medium organic matter content with low soil available nitrogen, high phosphorus, and medium available potassium. The data on soil textural class and soil chemical and biological characteristics are given in Tables 1–3.

Table 1. Textural class of soil of the experimental area.

Sl. No	Constituents	0–10 cm	Method Followed
1	Sand (%)	83.7	Bouycous Hydrometer method [37]
2	Silt (%)	6.8	
3	Clay (%)	10.4	
4	Textural classes	Sandy loam	

Table 2. Chemical composition of the soil.

Parameters	0–10 cm		Method Adopted
	Values	Remarks	
pH	5.9	Acidic	Digital electronic pH meter with 1:2.5, soil: water [38]
Organic carbon (%)	0.53	Medium	Walkely and Black's rapid titration method [38]
Available Nitrogen (kg ha^{-1})	226.4	Low	Alkaline potassium permanganate method [39]
Available Phosphorus (kg ha^{-1})	32.6	High	Bray's-1 method [38]
Available Potassium (kg ha^{-1})	132.6	Medium	Ammonium acetate flame photometer method [38]

Table 3. Initial microbiological properties of experimental soil.

Parameters	Values	Methods
Total bacterial population	41.9 ($\times 10^6$ CFU/g soil)	Serial dilution and spread plate technique [40]
Total fungal population	11.9 ($\times 10^4$ CFU/g soil)	Serial dilution and spread plate technique [40]
Total actinomycetes population	34.2 ($\times 10^3$ CFU/g soil)	Serial dilution and spread plate technique [40]
Urease	26.7 (µg NH3 released/g soil/h)	Tabatabai [41]
Alkaline phosphatase	178.5 (µg p-nitrophenol/g soil/h)	Tabatabai and Bremner [42]
Dehydrogenase	84.1 (µg TPF/g soil/24 h)	Tabatabai [41]

2.2. Meteorological Conditions

Detailed weather parameters of the growing seasons of both years are given in Figure 1. The maximum temperature was recorded in May (38.8 °C) followed by 34.8 °C in June, while the lowest of 17.4 °C was in November 2016. Similarly, the maximum temperature of 38.8 °C was recorded in May followed by 35.2 °C in June while November recorded the lowest at 18.7 °C in the year 2017. Rainfall received from May to November in 2016 and 2017 was 1241 mm (93 rainy days) and 1492.9 mm (81 rainy days), respectively (Department of Agrometeorology, College of Agriculture, OUAT, Bhubaneswar, Odisha, India).

2.3. Treatment and Layout

The field experiment was laid out in a split-plot design with a standard gross plot size of 5.8 m × 4.5 m with three replications by taking twenty-four treatment combinations with four crop establishment methods in the main plot and six weed management practices in the sub plots (Table 4).

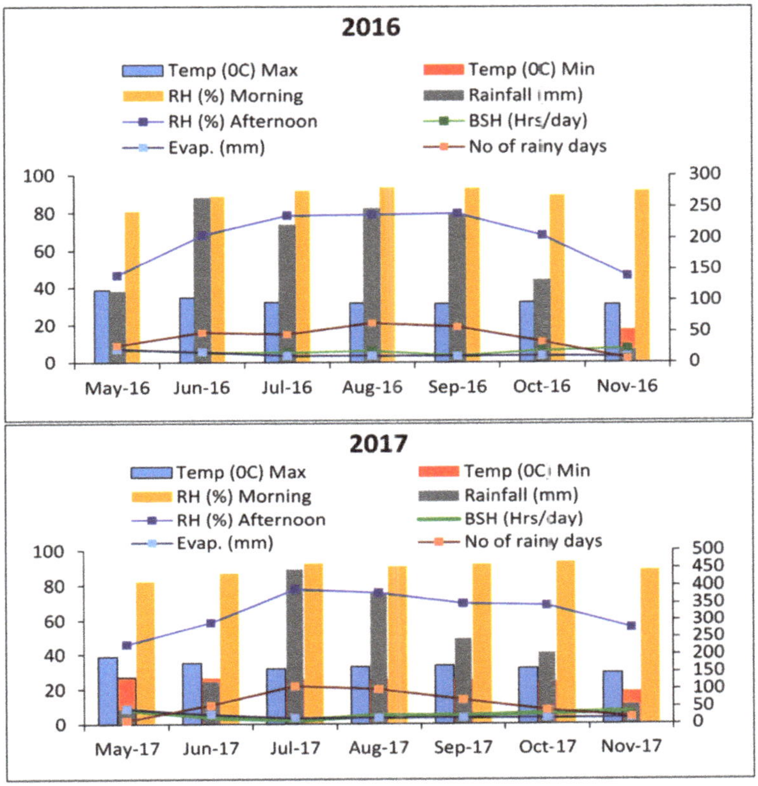

Figure 1. Monthly meteorological data during crop growing season (*kharif*) during 2016 and 2017 (Department of Agrometeorology, College of Agriculture, OUAT, Bhubaneswar, Odisha, India).

Table 4. Treatment details of the experiment.

Main Plots	Crop Establishment Methods
M_1	Direct Seeded Rice (DSR)
M_2	Wet seeded Rice (WSR)
M_3	Non-Puddled Transplanted Rice (NPTR)
M_4	Puddled Transplanted Rice (PTR)
Sub Plots	**Weed Management Practices**
W_1	Weedy Check
W_2	Bensulfuron methyl 0.6% + Pretilachlor 6% (PE) 0.660 kg ha^{-1} + Hand weeding (HW) at 30 DAS/T
W_3	Bensulfuron methyl 0.6% + Pretilachlor 6% (PE) 0.495 kg ha^{-1} + HW at 30 DAS/T
W_4	Bensulfuron methyl 0.6% + Pretilachlor 6% (PE) 0.495 kg ha^{-1} + Bispyribac-Sodium (POE) 0.025 kg ha^{-1} at 15 DAS/T,
W_5	Cono weeding at 15 DAS/T + hand weeding 30 DAS/T
W_6	Brown manuring/Green manuring

2.4. Crop Culture

Medium (115–130 days) duration rice variety Naveen (parentage- CR-749-20-2/IET-14461) with medium bold variety was taken as the test crop.

2.4.1. Rice Establishment Methods

As stated earlier, the following are the details of the crop establishment methods:

Direct Seeding (M_1): Seeds were directly sown in solid rows at 20 cm row-to-row spacing on well-pulverized soil.

Wet Seeding (M_2): The seeds were soaked for 24 h in water and were incubated for 48 h for sprouting; sprouted seeds were directly sown in solid rows at a spacing of 20 cm row-to-row on puddled soil.

Unpuddled Transplanted Rice (M3): 21 days old seedlings raised in a dry nursery were transplanted at two seedlings per hill at a spacing of 20 cm × 10 cm in well-cultivated non-puddled soil. Irrigation was applied to moisten the soil and allowed to settle for 12–24 h before transplanting in unpuddled transplanted treatment.

Puddled Transplanted Rice (M4): 21 days old seedlings raised in a dry nursery were transplanted at two seedlings/hill at a spacing of 20 cm × 10 cm in puddled transplanted soil.

Well decomposed Farm Yard Manure (FYM) at the rate of 5 t ha^{-1} was incorporated into the soil at final land preparation and inorganic fertilizers at 80-40-40 kg ha^{-1} N, P_2O_5, K_2O were applied to all the plots irrespective of treatments imposed. The full dose of P and K and 25% of N were applied at final plowing/puddling as basal dose. However, the rest of N was applied in a 2:1 ratio at tillering and panicle initiation stages, respectively, through Urea (46-0-0). The fertilizers used in the study were urea, diammonium phosphate (18-46-0), and muriate of potash (0-0-60).

2.4.2. Weed Management Practices

The application of the treatments related to weed management practices tested in the study has been described below:

Brown manuring/Green manuring: Sesbania seed was sown at 25 kg ha^{-1} together with rice. After 25–30 days of growth, when Sesbania was 30–40 cm tall, it was killed with 2, 4-D ester at 0.5 kg ha^{-1}.

Cono weeding and hand weeding: Cono weeding with a cono weeder, a mechanical farm implement, was operated in the inter-row at 15 DAS/T and hand weeding was operated with manual hand weeder at 30 DAS/T in W_5 plots while hand weeding operation in W_2 and W_3 plots was carried out at 30 DAS/T.

Herbicide application: Granular herbicides were uniformly applied to plots as per treatments after mixing with sand while liquid herbicide was applied by a knapsack sprayer using a flat fan nozzle as per treatments so as to spray the fluid uniformly throughout the targeted area. Bensulfuron Methyl (0.6%) + Pretilachlor (6%) (Ready-mix) was applied as pre-emergence herbicide 3 days after sowing/transplanting (DAS/T) at two doses at 0.660 kg ha^{-1} (W_2) and 0.495 kg ha^{-1} (W_3) while Bispyribac-Sodium (POE) 0.025 kg ha^{-1} was applied as post-emergence herbicide at 15 DAS/T in (W_3). For further details, refer to Table 3.

2.5. Microbial Analysis

2.5.1. Enumeration of Soil Microbial Population

Serial dilution and spread plate techniques were adopted to determine the soil microbial population, where one gram of the soil samples was added to ten tubes containing 9 mL of distilled water, serially diluted, and spread over nutrient agar and potato dextrose agar for enumeration of total bacteria and actinomycetes and fungi, respectively. The plates were incubated at 30 °C for 24 h for bacterial isolation and 48 h for actinomycetes and fungi. The following formula was considered for the calculation of soil microbial population.

$$\text{CFU/mL} = \frac{\text{No. of colony} \times \text{inverse of dilution taken}}{\text{Volume of inoculum taken}} \tag{1}$$

Here, CFU denoted the colony-forming unit of microbes, and ml represented milliliter.

2.5.2. Urease Activity

The method adopted to evaluate the urease activity of soil was essentially the same as adopted by Pancholy and Rice [43], except that the ammonia liberated due to hydrolysis used in the reaction mixture was determined by nesslerization as described by Jackson [38]. Ten grams of each freshly collected soil sample was placed in 100 mL capacity Erlenmeyer flasks to which 1 mL toluene was added and allowed to stand for 15 min to permit complete penetration into soils. Each of these flasks then had 10 mL of phosphate buffer (pH 6.7) and 10 mL of 10% urea solution added. For control flasks, the urea solution was replaced by an equal quantity of distilled water. The contents of the flasks were well shaken for 5 min and incubated at 30 °C for 24 h. The contents of the flask were filtrated through Whatman No. 42-filter paper after incubation. The remaining soil in the flask was mixed with 15 mL of 1 N KCl solution, shaken for 5 min, and filtered. The volume of the total filtrate was increased to 100 mL in a volumetric flask using distilled water.

The ammonia present in the filtrate was determined by nesslerization. A total of 1 mL filtrate of each sample was transferred to a 20 mL volumetric flask to which 2 mL tartrate solution and 0.5 mL Nessler's reagent were added. The volume was increased to 20 mL with distilled water. The yellow color developed after 30 min was measured at 410 nm using a Graphicord Shimadzu UV-visible spectrophotometer (model UV 240) against the reagent blank.

2.5.3. Soil Phosphatase (Acid and Alkaline) Activity

Soil phosphatase activity involved colorimetric estimation of the p-nitrophenol released by phosphatase activity when soil is incubated with buffered (pH 6·5 and 11) sodium p-nitrophenyl phosphate solution and toluene at 37 °C for 1 h [42]. First, 1 g of fresh soil was weighed. To it, 0.2 mL of toluene was added. Then, 4 mL acid phosphatase MUB buffer (pH 6.5 or pH 11) was added, respectively, for acid phosphatase or alkaline phosphatase. Then, 1 mL of p-nitrophenol phosphate solution was mixed and placed in an incubator at 37 °C for 1 h. After 1 h, 1 mL of 0.5(M) $CaCl_2$ and 4 mL of 0.5(M) NaOH was added and shaken for 1 min. The soil suspension was filtered. The yellow color intensity was measured at 420 nm by spectrophotometer.

2.5.4. Soil Dehydrogenase Activity

The soil dehydrogenase activity was measured following the method of reduction of 2, 3, 5- triphenyl tetrazolium chloride (TTC) to the creaming red-colored triphenyl formazan (TPF). First, 3 g of fresh soil was weighed. To it, 0.03 g $CaCO_3$ and 1 mL of TTC were added. Then, 2.5 mL of distilled water was mixed and placed in an incubator at 37 °C for 24 h. After 24 h, 10 mL methanol was added and shaken for 1 min. The soil suspension was filtered. The filtered solution volume was increased to 100 mL with methanol. The red color intensity was measured at 485 nm by spectrophotometer [41,44].

2.6. Statistical Analysis

The experimental data were analyzed by adopting Fisher's method of analysis of variance as outlined by Gomez and Gomez [45] to draw a valid conclusion. The variations in the treatment mean were tested by using critical difference (CD) values at a 5% level of significance. The ANOVA model adopted for the above experiment was as follows:

$$Y_{ijk} = \mu + \rho_i + \tau_j + \delta_{ij} + \beta_k + (\tau\beta)_{jk} + \varepsilon_{ijk}$$

where i = 1, 2, ..., r, j = 1, 2, ..., a, k = 1, 2, ..., b; Y_{ijk} = observation corresponding to the kth level of sub-plot factor(W), jth level of main plot factor (M), and ith replication. μ = general mean, ρ_i = ith block effect, τ_j = jth main plot effect β_k = kth sub-plot effect, $(\tau\beta)_{jk}$ = interaction between the jth level of main plot treatment and the kth level of sub-plot treatment.

The error components δ_{ij} and ε_{ijk} are independently and normally distributed with means zero and respective variances σ_{δ}^2 and σ_{ε}^2.

Correlation and coefficient analysis was made as per the statistical methods (Cochran and Cox [46].

3. Results

3.1. Effect of Crop Establishment Methods and Weed Management Practices on Soil Microbial Population

The total bacterial, fungal, and actinomycetes population was significantly influenced by the type of weed management practices, types of herbicides, their concentration, and the days after herbicide application.

3.1.1. Effect of Crop Establishment Methods and Weed Management Practices on Total Bacterial Population

Crop establishment methods and weed management practices in rice significantly influenced the total bacterial population and data on different days after sowing/transplanting are presented in Table 5.

Table 5. Total bacterial population ($\times 10^6$ CFU g^{-1} soil) in rice soil as influenced by crop establishment methods and weed management practices.

Treatments	Total Bacterial Population ($\times 10^6$ CFU g^{-1} Soil)							
	7 DAS/T		14 DAS/T		21 DAS/T		28 DAS/T	
	2016	2017	2016	2017	2016	2017	2016	2017
				Establishment Methods				
M_1	37.57 ^{c}	38.48 ^{c}	40.83 ^{c}	41.75 ^{c}	42.97 ^{c}	43.73 ^{c}	44.68 ^{b}	46.40 ^{b}
M_2	38.53 ^{b}	39.38 ^{b}	41.87 ^{b}	43.33 ^{b}	44.08 ^{b}	44.70 ^{b}	45.33 ^{b}	45.47 ^{b}
M_3	39.47 ^{a}	40.50 ^{a}	43.82 ^{a}	44.18 ^{a}	45.67 ^{a}	45.83 ^{a}	49.57 ^{a}	48.85 ^{a}
M_4	35.93 ^{d}	37.78 ^{d}	40.65 ^{c}	40.88 ^{d}	43.08 ^{c}	42.03 ^{d}	45.32 ^{b}	46.45 ^{ab}
SEm (±)	0.081	0.063	0.067	0.064	0.049	0.066	0.199	0.348
CD (0.05)	0.282	0.218	0.23	0.221	0.17	0.23	0.688	1.204
				Weed Management Practice				
W_1	42.00 ^{b}	43.48 ^{a}	42.93 ^{c}	44.58 ^{c}	43.88 ^{c}	45.98 ^{c}	44.23 ^{c}	47.50 ^{b}
W_2	31.93 ^{d}	32.93 ^{c}	38.53 ^{d}	40.38 ^{d}	42.78 ^{d}	45.45 ^{d}	44.95 ^{c}	47.93 ^{ab}
W_3	33.25 ^{c}	35.30 ^{b}	35.83 ^{e}	39.40 ^{e}	38.35 ^{e}	44.68 ^{e}	40.75 ^{d}	46.53 ^{b}
W_4	33.25 ^{c}	35.30 ^{b}	35.83 ^{e}	39.40 ^{e}	33.98 ^{f}	32.55 ^{f}	38.43 ^{e}	41.45 ^{c}
W_5	41.98 ^{b}	43.48 ^{a}	44.05 ^{b}	45.20 ^{b}	46.25 ^{b}	47.10 ^{b}	53.75 ^{b}	49.73 ^{a}
W_6	44.85 ^{a}	43.75 ^{a}	53.60 ^{a}	46.28 ^{a}	58.48 ^{a}	48.70 ^{a}	55.25 ^{a}	47.63 ^{ab}
SEm (±)	0.074	0.067	0.01	0.019	0.014	0.011	0.234	0.428
CD (0.05)	0.215	0.195	0.03	0.056	0.042	0.031	0.684	1.249

M_1—Direct Seeded Rice (DSR), M_2— Wet Seeded Rice (WSR), M_3—Unpuddled Transplanted Rice (NPTR), M_4—Puddled Transplanted Rice (PTR), W_1—Weedy check, W_2—Bensulfuron methyl 0.6% + Pretilachlor 6% (PE) 0.660 kgha^{-1} + Hand weeding (HW) at 30 DAS/T, W_3—Bensulfuron methyl 0.6% + Pretilachlor 6% (PE) 0.495 kgha^{-1} + HW at 30 DAS/T, W_4—Bensulfuron methyl 0.6% + Pretilachlor 6% (PE) 0.495 kgha^{-1} + Bispyribac-Sodium (POE) 0.025 kg ha^{-1} at 15 DAS/T, W_5—Cono weeding at 15 DAS/T + hand weeding 30 DAS/T, W_6—Brown manuring/Green manuring. SE, standard Error, CD, Critical difference at 5% level of probability. Different superscript letters indicate a significant difference between the mean.

A declining trend was seen in the total bacterial population at 7 DAS/T in all establishment methods from the initial total bacterial population of 41.9 × 10^6 CFU/g soil. Among all establishment methods, there was a higher decline in the bacterial population in PTR (35.93 and 37.78), followed by WSR (38.53 and 39.38) and DSR (37.57 and 38.48) in both years, respectively. While the lowest decline in the total bacterial population at 7 DAS/T was recorded in NPTR treatment (39.47 and 40.50, respectively) during both years. Thereafter, a continuous increase in the bacterial population was seen at 14 DAS/T, 21 DAS/T, and 28 DAS/T. Among the four different establishment methods taken, significantly higher values of the bacterial population were recorded in unpuddled transplanted rice (NPTR) viz. at 14 DAST/T (43.82 and 44.18), 21 DAST/T (45.67 and 45.83), and 28 DAS/T (49.57 and 48.85) during both years of the experiment, respectively, which was followed by wet seeded rice (WSR) viz. 14 DAS/T (41.87 and 43.33), 21 DAS/T (44.08 and 44.70), and at 28 DAS/T (45.33 and 45.47), respectively, during both years (Table 5).

Among the six different weed management strategies taken, the total bacterial population increased in brown/green manuring, weedy check, and CW+HW treatments where no chemicals were applied, while the total bacterial population declined in treatments where chemical herbicides were applied viz., Bensulfuron methyl 0.6% + Pretilachlor 6% at 0.495 kg ha^{-1} and Bensulfuron methyl 0.6% + Pretilachlor 6% (PE) at 0.660 kg ha^{-1} at initial 7 DAS/T. Thereafter, there was a continuous increase in total bacterial population in all treatments until 28 DAS/T, except for the treatment Bensulfuron methyl 0.6% + Pretilachlor 6% (PE) at 0.495 kg ha^{-1} + Bispyribac-Sodium at 0.025 kg ha^{-1} POE where there was a decline at 21 DAS/T (33.98 732.55) from 14 DAS/T (35.83 and 39.40) which may be due to Post-emergence application of Bispyribac-Sodium at 0.025 kg ha^{-1} at 15 DAS/T, but the bacterial population also increased in same treatment at 28 DAS/T (38.43 and 41.45, respectively) during both years. Bensulfuron methyl 0.6% + Pretilachlor 6% (PE) at 0.660 kg ha^{-1} and Bensulfuron methyl 0.6% + Pretilachlor 6% (PE) at 0.495 kg ha^{-1} recorded a 22.67% and 18.13% decrease, respectively, in total bacterial population over the initial value at 7 days after sowing, 3 days after herbicide application.

Application of Bensulfuron methyl 0.6% + Pretilachlor 6% (PE) at 0.660 kg ha^{-1} + HW at 30 DAS/T, Bensulfuron methyl 0.6% + Pretilachlor 6% (PE) at 0.495 kg ha^{-1} + HW at 30 DAS/T and Bensulfuron methyl 0.6% + Pretilachlor 6% (PE) at 0.495 kg ha^{-1} + Bispyribac-Sodium at 0.025 kg ha^{-1} POE reduced the total bacterial population by 1.8%, 7.6%, and 25.8% while brown/green manuring resulted in a 16.2% increase over weedy check at 21 DAS/T.

3.1.2. Effect of Crop Establishment Methods and Weed Management Practices on Total Fungi Population

The reduction in total fungal population from the initial values of 11.9 ($\times 10^4$ CFU g^{-1} soil) was seen at 7 DAS/T in all establishment methods taken, while the highest decline was seen in Direct seeded rice (9.95 and11.43) which was significantly at par with the fungal population of puddled transplanted rice (9.87 and11.62, respectively) during both years (Table 6).

The lowest decline was recorded in unpuddled transplanted rice (11.28) during the first year while the lowest decline in the second year was recorded in Wet seeded rice (12.48). Thereafter, a constant increase in the total fungal population was seen during the stage of data recording. Significantly, higher values of the total fungal population were recorded in NPTR at 14 DAS/T (17.35 and 21.47), 21 DAS/T (20.80 and 23.47), and 28 DAS/T (28.92 and 31.98, respectively) during both years among all other crop establishment methods taken (Table 6).

Brown manuring/green manuring, CW + HW, and weedy check recorded a constant increase in total fungal population, while Bensulfuron methyl 0.6% + Pretilachlor 6% (PE) at 0. 660 kg ha^{-1} and Bensulfuron methyl 0.6% + Pretilachlor 6% (PE) at 0.495 kg ha^{-1} recorded a 31.93% and 20.12% decrease in total fungal population over the initial value at 7 days after sowing, 3 days after herbicide application. Brown/green manuring treatment recorded a higher total fungal population at 7 DAS/T (12.65 and 15.45), 14 DAS/T (19.68 and 21.85), and 21 DAS/T (24.78 and 27.65, respectively). CW + HW recorded higher values of total fungal population (28.00 and 30.40) which was significantly at par with treatment with Bensulfuron methyl 0.6% + Pretilachlor 6% (PE) at 0.660 kg ha^{-1} + HW at 30 DAS/T (27.75 and 30.63) and was followed by the application of Bensulfuron methyl 0.6% + Pretilachlor 6% (PE) at 0.495 kg ha^{-1} + HW at 30 DAS/T (26.88 and 29.70) at 28 DAS/T, respectively, during both years (Table 6).

Table 6. Total Fungi population ($\times 10^4$ CFU g^{-1} soil) in rice soil as influenced by crop establishment methods and weed management practices.

Treatments	Total Fungi Population ($\times 10^4$ CFU g^{-1} Soil)							
	7 DAS/T		14 DAS/T		21 DAS/T		28 DAS/T	
	2016	2017	2016	2017	2016	2017	2016	2017
	Establishment Methods							
M_1	9.95 ^{c}	11.43 ^{c}	15.08 ^{c}	17.95 ^{a}	18.38 ^{c}	20.22 ^{c}	24.35 ^{c}	27.38 ^{d}
M_2	10.62 ^{b}	12.48 ^{a}	16.60 ^{b}	20.35 ^{a}	19.70 ^{b}	22.17 ^{b}	25.07 ^{b}	28.37 ^{b}
M_3	11.28 ^{a}	12.07 ^{b}	17.35 ^{a}	21.47 ^{a}	20.80 ^{a}	23.47 ^{a}	28.92 ^{a}	31.98 ^{a}
M_4	9.87 ^{c}	11.62 ^{c}	14.82 ^{c}	17.13 ^{a}	17.22 ^{d}	20.08 ^{c}	24.98 ^{b}	27.92 ^{c}
SEm (±)	0.05	0.037	0.042	0.054	0.045	0.037	0.044	0.046
CD (0.05)	0.173	0.127	0.145	1.743	0.156	0.129	0.151	0.16
	Weed management practices							
W_1	12.15 ^{b}	13.73 ^{c}	15.60 ^{c}	18.28 ^{a}	19.65 ^{d}	20.10 ^{d}	24.40 ^{d}	27.88 ^{c}
W_2	7.63 ^{e}	8.58 ^{e}	16.45 ^{b}	20.20 ^{a}	21.90 ^{b}	23.95 ^{b}	27.75 ^{a}	30.63 ^{a}
W_3	9.05 ^{d}	9.88 ^{d}	14.23 ^{d}	18.13 ^{a}	17.90 ^{e}	22.38 ^{c}	26.88 ^{b}	29.70 ^{b}
W_4	9.05 ^{d}	9.88 ^{d}	14.23 ^{d}	18.23 ^{a}	9.88 ^{f}	12.43 ^{e}	22.10 ^{e}	27.18 ^{d}
W_5	12.05 ^{c}	13.90 ^{b}	15.60 ^{c}	18.68 ^{a}	20.05 ^{c}	22.40 ^{c}	28.00 ^{a}	30.40 ^{a}
W_6	12.65 a	15.45 a	19.68 ^{a}	21.85 ^{a}	24.78 ^{a}	27.65 ^{a}	25.85 ^{c}	27.70 ^{c}
SEm (±)	0.006	0.015	0.007	0.598	0.011	0.016	0.059	0.041
CD (0.05)	0.018	0.044	0.02	1.746	0.033	0.47	0.171	0.121

Table 4 may be referred to for treatment details. SE, standard Error, CD, Critical difference at 5% probability level. Different superscript letters indicate a significant difference between the mean.

3.1.3. Effect of Crop Establishment Methods and Weed Management Practices on Total Actinomycetes Population

The data recorded on the total count of actinomycetes recorded at various growth stages are presented in Table 6. Apart from the unpuddled transplanted rice, where there was a slight increase in total actinomycetes population (34.83 and 34.38), there was a decline in total actinomycetes population in all crop establishment methods at 7 DAS/T. Significantly higher values of total actinomycetes were seen in NPTR at 14 DAS/T (39.55 and 39.15), 21 DAS/T (42.65 and 42.45), 28 DAS/T (48.22 and 45.23, respectively) during both years (Table 7).

Brown/green manuring recorded higher values of total actinomycetes population from 7 DAS/T (40.58 and 39.18), 14 DAS/T (44.63 and 43.53), 21 DAS/T (48.08 and 47.35), while CW+ HW recorded higher values at 28 DAS/T (50.00 and 49.35, respectively) during both years. There was a continuous increase in total actinomycetes population in all weed management practices from 7 DAS/T to 28 DAS/T apart from the treatment of Bensulfuron methyl 0.6% + Pretilachlor 6% (PE) at 0.495 kg ha^{-1} + Bispyribac-Sodium at 0.025 kg ha^{-1} POE (W_4) where there was a decline in total actinomycetes population at 21 DAS/T (24.60 and 29.33) from 14 DAS/T (31.90 and 33.58) that again increased at 28 DAS/T (35.88 and 39.53) during both test years, respectively. Bensulfuron methyl 0.6% + Pretilachlor 6% (PE) at 0.495 kg ha^{-1} + Bispyribac-Sodium at 0.025 kg ha^{-1} POE treatment recorded 36.01% less actinomycetes population than weedy check at 21 DAS/T.

3.2. *Effect of Crop Establishment Methods and Weed Management Practices on Soil Enzymatic Activities*

3.2.1. Effect of Crop Establishment Methods and Weed Management Practices on Urease Activity

Data on urease activity in rice soil was significantly affected by crop establishment methods and weed management practices and is presented in Table 8. Brown/green manuring recorded higher values of total actinomycetes population from 7 DAS/T (40.58 and 39.18), 14 DAS/T (44.63 and 43.53), 21 DAS/T (48.08 and 47.35), while CW+ HW recorded higher values at 28 DAS/T (50.00 and 49.35, respectively) during both years. There was a continuous increase in total actinomycetes population in all weed management practices from 7 DAS/T to 28 DAS/T apart from the treatment of Bensulfuron methyl 0.6% + Pretilachlor 6% (PE) at 0.495 kg ha^{-1} + Bispyribac-Sodium at 0.025 kg ha^{-1} POE (W_4) where there was a decline in total actinomycetes population at 21 DAS/T (24.60 and 29.33) from 14 DAS/T (31.90 and 33.58) that again increased at 28 DAS/T (35.88 and 39.53) during both test years, respectively. Bensulfuron methyl 0.6% + Pretilachlor 6% (PE) at 0.495 kg ha^{-1} + Bispyribac-Sodium at 0.025 kg ha^{-1} POE treatment recorded 36.01% less actinomycetes population than weedy check at 21 DAS/T. Urease activity declined at 7 DAS/T from initial values and increased thereafter with days after sowing/transplanting. PTR recorded the highest urease activity at all stages of data recorded viz. 7 DAS/T (24.57 and 27.00), 14 DAS/T (26.57 and 30.58), 21 DAS/T (27.44 and 31.36), and 28 DAS/T (30.53 and 35.02). At 7 DAS/T, the trend of urease activity was PTR > WSR > NPTR > DSR and at 28 DAS/T the trend was PTR > WSR> DSR > NPTR. PTR, WSR, NPTR, and DSR recorded a 3.4%, 7.9%, 10.1%, and 15.4% decline in urease activity from the initial value of 26.7 µg NH3 released g^{-1} soil h^{-1} at 7 DAS/T. Bensulfuron methyl 0.6% + Pretilachlor 6% (PE) at 0.660 kg ha^{-1}, Bensulfuron methyl 0.6% + Pretilachlor 6% (PE) at 0.495 kg ha^{-1} recorded a 28.1% and 25.1% decrease in urease activity than the initial value, while weedy check, CW+HW, and brown/green manuring resulted a 7.05%, 7.0%, and 7.3% higher urease activity than initial values at 7 DAS/T. Brown/green manuring recorded higher urease activity at all stages of data recording, 7 DAS/T (27.15 and 30.50), 14 DAS/T (28.80 and 32.60), 21 DAS/T (33.28 and 34.49), and 28 DAS/T (33.30 and 34.10) (Table 8).

Table 7. Total Actinomycetes population ($\times 10^3$ CFU g^{-1} soil) in rice soil as influenced by crop establishment methods and weed management practices.

Treatments	Total Actinomycetes Population ($\times 10^3$ CFU g^{-1} Soil)							
	7 DAS/T		14 DAS/T		21 DAS/T		28 DAS/T	
	2016	2017	2016	2017	2016	2017	2016	2017
Establishment Methods								
M_1	31.97 ^{c}	32.59 ^{a}	36.22 ^{c}	36.97 ^{c}	39.73 ^{c}	41.02 ^{b}	44.20 ^{c}	43.81 ^{b}
M_2	32.55 ^{b}	33.35 ^{a}	38.13 ^{b}	38.15 ^{b}	41.13 ^{b}	42.23 ^{a}	45.23 ^{b}	44.87 ^{a}
M_3	34.83 ^{a}	34.38 ^{a}	39.55 a	39.15 ^{a}	42.65 ^{a}	42.45 ^{a}	48.22 ^{a}	45.23 ^{a}
M_4	31.17 ^{d}	31.68 ^{a}	34.78 ^{d}	36.27 ^{d}	38.02 ^{d}	39.60 ^{c}	43.42 ^{d}	42.75 ^{c}
SEm (±)	0.056	0.972	0.044	0.051	0.046	0.028	0.075	0.051
CD (0.05)	0.195	NS	0.153	0.178	0.159	0.096	0.261	0.176
Weed Management Practices								
W_1	38.30 ^{b}	36.48 ^{a}	40.23 ^{c}	38.38 ^{c}	42.33 ^{c}	42.10 ^{d}	46.05 ^{d}	43.85 ^{d}
W_2	24.55 ^{d}	26.84 ^{b}	33.00 ^{d}	34.83 ^{d}	41.88 ^{d}	42.35 ^{c}	47.58 ^{b}	45.95 ^{b}
W_3	27.03 ^{c}	28.78 ^{b}	31.90 ^{e}	33.58 ^{e}	39.23 ^{e}	40.88 ^{e}	45.08 ^{e}	45.03 c
W_4	27.03 ^{c}	28.78 ^{b}	31.90 ^{e}	33.58 ^{e}	24.60 ^{f}	29.33 ^{f}	35.88 ^{f}	39.53 ^{f}
W_5	38.30 ^{b}	37.98 ^{a}	41.38 ^{b}	41.93 ^{b}	46.20 ^{b}	45.95 ^{b}	50.00 ^{a}	49.35 ^{a}
W_6	40.58 ^{a}	39.18 ^{a}	44.63 ^{a}	43.53 ^{a}	48.08 ^{a}	47.35 ^{a}	47.03 ^{c}	41.29 ^{e}
SEm (±)	0.015	0.745	0.017	0.01	0.019	0.017	0.066	0.06
CD (0.05)	0.043	2.174	0.05	0.03	0.054	0.049	0.0192	0.174

For treatment details, Table 4 may be referred to. SE, standard Error, CD, Critical difference at 5% probability level. Different superscript letters indicate a significant difference between the mean.

Among the different chemicals, the herbicide application of Bensulfuron methyl 0.6% + Pretilachlor 6% (PE) at 0.660 kg ha^{-1} recorded a higher decline in urease activity than the application of Bensulfuron methyl 0.6% + Pretilachlor 6% (PE) at 0.495 kg ha^{-1} at 7 DAS/T. There was a continuous increase in urease activity while in treatment with the application of Bensulfuron methyl 0.6% + Pretilachlor 6% (PE) at 0.495 kg ha^{-1} + Bispyribac-Sodium at 0.025 kg ha^{-1} POE (W4) and there was a decline in urease activity at 21 DAS/T (17.92

and 22.23) from 7 DAS/T (23.15 and 24.30); this may be due to post-emergence application of Bispyribac-Sodium at 0.025 kg ha^{-1} (23.15 and 24.30) which increased thereafter at 28 DAS/T during both years of experiment.

Table 8. Crop establishment methods and weed management practices influence the activity of the urease enzyme (μg NH3 released g^{-1} soil h^{-1}) in rice soil.

Treatments	The Activity of Soil Enzyme Urease (μg NH3 Released g^{-1} Soil hr^{-1})							
	7 DAS/T		14 DAS/T		21 DAS/T		28 DAS/T	
	2016	2017	2016	2017	2016	2017	2016	2017
	Establishment Methods							
M_1	21.62 ^{c}	23.57 ^{d}	23.77 ^{b}	26.20 ^{d}	24.93 ^{b}	28.05 ^{b}	27.60 ^{c}	31.50 ^{c}
M_2	23.60 ^{b}	25.58 ^{b}	25.75 ^{a}	29.28 ^{b}	26.78 ^{b}	30.28 ^{a}	28.28 ^{b}	33.08 ^{b}
M_3	23.05 ^{b}	25.03 ^{c}	24.43 ^{b}	28.17 ^{c}	25.73 ^{b}	30.30 ^{a}	26.78 ^{d}	31.07 ^{d}
M_4	24.57 ^{a}	27.00 ^{a}	26.75 ^{a}	30.58 ^{a}	27.44 ^{a}	31.36 ^{a}	30.53 ^{a}	35.02 ^{a}
SEm (±)	0.481	0.039	0.118	0.118	0.341	0.185	0.045	0.049
CD (0.05)	1.664	0.136	0.408	0.407	1.181	0.639	0.156	0.153
	Weed Management Practices							
W_1	26.95 ^{a}	30.55 ^{a}	22.35 ^{c}	31.68 ^{a}	21.30 ^{d}	32.78 ^{b}	20.83 ^{f}	33.38 ^{d}
W_2	19.08 ^{b}	19.33 ^{d}	25.50 ^{b}	26.58 ^{b}	28.45 ^{b}	29.25 ^{c}	29.90 ^{c}	33.48 ^{c}
W_3	19.45 ^{b}	20.55 ^{c}	23.15 ^{c}	24.30 ^{c}	25.93 ^{c}	27.35 ^{d}	28.00 ^{d}	31.55 ^{e}
W_4	19.50 ^{b}	20.55 ^{c}	23.15 ^{c}	24.30 ^{c}	17.92 ^{e}	22.23 ^{e}	25.55 ^{e}	27.73 ^{f}
W_5	27.13 ^{a}	30.30 ^{b}	28.10 ^{a}	31.90 ^{a}	30.48 ^{b}	33.89 ^{b}	32.23 ^{b}	35.78 ^{a}
W_6	27.15 ^{a}	30.50 ^{a}	28.80 ^{a}	32.60 ^{a}	33.28 ^{a}	34.49 ^{a}	33.30 ^{a}	34.10 ^{b}
SEm (±)	0.592	0.023	0.155	0.173	0.434	0.197	0.013	0.016
CD (0.05)	1.728	0.066	0.453	0.505	1.267	0.575	0.037	0.045

For treatment details, Table 4 may be referred to. SE, standard Error, CD, Critical difference at 5% probability level. Different superscript letters indicate a significant difference between the mean.

3.2.2. Crop Establishment Methods and Weed Management Practices Influence the Alkaline Phosphatase Activity

Data on alkaline phosphatase activity in rice soil was significantly affected by crop establishment methods and weed management practices and is presented in Table 9. A perusal of data revealed that values of alkaline phosphatase declined at 7 DAS/T, irrespective of crop establishment methods from the initial values; however, the decline was lowest in unpuddled transplanted rice, followed by WSR. Thereafter, the values of alkaline phosphatase increased with the age of crop production. Significantly higher values of alkaline phosphatase were recorded in NPTR at all the stages of data recording viz., 7 DAS/T (171.32 and 175.68), 14 DAS/T (186.28 and 191.70), 21 DAS/T (192.17 and 197.42), and at 28 DAS/T (197.95 and 215.47), followed by WSR viz. 7 DAS/T (163.05 and 169.05), 14 DAS/T (179.93 and 179.53), 21 DAS/T (185.90 and 188.55), and 28 DAS/T (190.45 and 201.07, respectively) during both years. Alkaline phosphatase values declined after herbicide application or a higher concentration of herbicide applied, resulting in an additional decline in enzymatic activity. After 7 DAS/T, the alkaline phosphatase value was higher in brown/green manuring followed by CW + HW and weedy check.

There was an initial decline in values of alkaline phosphatase in treatments with chemical herbicide application. In W_2 treatment, the application of Bensulfuron methyl 0.6% + Pretilachlor 6% (PE) at 0.660 kg ha^{-1} had an initial decline of alkaline phosphatase values (141.83 and 146.58), while in W_3 and W_4, the values were 148.70 and 152.70, respectively, during both years. However, higher values were recorded in W_1, W_5, and W_6. After the initial decline, values of alkaline phosphatase increased with days of crop production, except in treatment W_4 (Bensulfuron methyl 0.6% + Pretilachlor 6% (PE) at 0.495 kg ha^{-1} + Bispyribac-Sodium at 0.025 kg ha^{-1} POE) at 21 (149.10 and 149.05) from 14 DAS/T (160.03 and 158.93), which may be due to the post-emergence application of

Bispyribac-Sodium at 0.025 kg ha^{-1} at 15 DAS/T, while the values again increased at 28 DAS/T (178.65 and 181.33), during both years, respectively.

Table 9. The activity of soil enzyme alkaline phosphatase in rice as influenced by crop establishment methods and weed management practices.

The Activity of Soil Enzyme Alkaline Phosphatase (µg p-nitro-phenol g^{-1} soil h^{-1})								
Treatments	**7 DAS/T**		**14 DAS/T**		**21 DAS/T**		**28 DAS/T**	
	2016	**2017**	**2016**	**2017**	**2016**	**2017**	**2016**	**2017**
Establishment Methods								
M_1	155.45 ^{c}	166.33 ^{b}	175.45 ^{b}	175.35 ^{bc}	180.95 ^{c}	183.28 ^{bc}	187.10 ^{b}	194.62 ^{c}
M_2	163.05 ^{b}	169.05 ^{b}	179.93 ^{ab}	179.53 ^{b}	185.90 ^{b}	188.55 ^{b}	190.45 ^{b}	201.07 ^{b}
M_3	171.32 ^{a}	175.68 ^{a}	186.28 ^{a}	191.70 ^{a}	192.17 ^{a}	197.42 ^{a}	197.95 ^{a}	215.47 ^{a}
M_4	161.52 ^{b}	165.13 ^{b}	174.00 ^{b}	171.32 ^{c}	180.17 ^{c}	178.67 ^{c}	189.95 ^{b}	189.08 ^{d}
SEm (±)	0.413	0.565	0.8	0.696	0.286	0.872	0.45	0.506
CD (0.05)	1.427	1.953	2.767	2.407	0.99	3.019	1.556	1.751
Weed Management Practices								
W_1	180.70 ^{a}	183.25 ^{c}	184.70 ^{b}	190.50 ^{c}	164.83 ^{e}	192.05 ^{c}	160.23 ^{e}	194.08 ^{c}
W_2	141.83 ^{e}	146.58 ^{e}	168.40 ^{c}	162.95 ^{d}	184.63 ^{c}	185.68 ^{c}	189.05 ^{b}	200.50 ^{b}
W_3	148.70 ^{d}	152.70 ^{d}	160.03 ^{d}	158.93 ^{d}	176.38 ^{d}	172.85 ^{d}	184.20 ^{c}	193.00 ^{c}
W_4	148.70 ^{d}	152.70 ^{d}	160.03 ^{d}	158.93 ^{d}	149.10 ^{f}	149.05 ^{e}	178.65 ^{d}	181.33 ^{d}
W_5	179.53 ^{b}	187.85 ^{b}	188.40 ^{b}	197.13 ^{b}	209.40 ^{b}	203.60 ^{b}	216.70 ^{a}	214.63 ^{a}
W_6	177.55 ^{c}	191.23 ^{a}	211.95 ^{a}	208.43 ^{a}	224.45 ^{a}	218.65 ^{a}	219.35 ^{a}	216.83 ^{a}
SEm (±)	0.104	0.227	0.66	1.132	0.39	1.335	0.707	1.118
CD (0.05)	0.304	0.662	1.926	3.304	1.138	3.896	2.063	3.262

For treatment details, Table 4 may be referred to. SE, standard Error, CD, Critical difference at 5% probability level. Different superscript letters indicate a significant difference between the mean.

3.2.3. Crop Establishment Methods and Weed Management Practices Influence the Dehydrogenase Activity

Data on dehydrogenase activity in rice soil was significantly affected by crop establishment methods and weed management practices and is presented in Table 10. From the study, it is revealed that there was a decline in soil enzymatic activity at 7 DAS/T, which is 4 days after application of pre-emergence herbicide in all crop establishment methods, apart from unpuddled transplanted rice, where a slight increase in dehydrogenase activity was recorded. Thereafter, from 7 DAS/T, soil enzyme dehydrogenase activity increased with crop age. The non-puddled transplanted rice (NPTR) condition recorded the highest dehydrogenase enzyme activity at all stages of data recording, viz. 7 DAS/T (84.43 and 86.07), 14 DAS/T (94.32 and 96.03), 21 DAS/T (98.92 and 107.43), and 28 DAS/T (103.65 and 112.95, respectively) during both years. Among the different chemical weed management practices adopted, the highest decline of initial dehydrogenase activity was recorded with the application of Bensulfuron methyl 0.6% + Pretilachlor 6% (PE) at 0.660 kg ha^{-1} (72.55 and 71.95), followed by the application of Bensulfuron methyl 0.6% + Pretilachlor 6% (PE) at 0.495 kg ha^{-1} (74.28 and 74.73, respectively) during both years (Table 10).

Among all the weed management practices adopted, higher dehydrogenase activity was recorded in brown/green manuring (97.53 and 98.00), which was followed by CW + HW (86.63 and 88.65) and weedy check (86.63 and 88.20, respectively) during both years. Thereafter, an increase in soil enzymatic activity was recorded with crop age in all treatments of weed management applied, apart from the treatment with application of Bensulfuron methyl 0.6% + Pretilachlor 6% (PE) at 0.495 kg ha^{-1} + Bispyribac-Sodium at 0.025 kg ha^{-1} POE, at 21 DAS/T (75.48 and 75.00) that again increased at 28 DAS/T (92.30 and 90.48, respectively) during both years.

Table 10. The activity of soil enzyme dehydrogenase in rice as influenced by crop establishment methods and weed management practices.

Treatments	The Activity of Soil Enzyme Dehydrogenase (μg TPF g^{-1} soil 24 h^{-1})							
	7DAS		14DAS		21DAS		28DAS	
	2016	2017	2016	2017	2016	2017	2016	2017
	Establishment Methods							
M_1	80.75 ^{b}	81.13 ^{b}	88.67 ^{b}	92.05 ^{b}	91.63 ^{b}	99.45 ^{b}	94.22 c	101.52 ^{c}
M_2	82.67 ^{ab}	81.63 ^{ab}	90.83 ^{ab}	92.25 ^{b}	94.32 ^{b}	99.95 ^{b}	97.20 ^{b}	105.17 ^{bc}
M_3	84.43 ^{a}	86.07 ^{a}	94.32 ^{a}	96.03 ^{a}	98.92 ^{a}	107.95 ^{a}	103.65 ^{a}	112.95 ^{a}
M_4	80.07 ^{b}	82.00 ^{ab}	89.08 ^{b}	91.62 ^{b}	92.15 ^{b}	99.80 ^{b}	97.13 ^{b}	106.50 ^{b}
Sem (±)	0.365	0.529	0.491	0.307	0.334	0.339	0.299	0.462
CD (0.05)	1.261	1.831	1.698	1.061	1.154	1.174	1.034	1.601
	Weed Management Practices							
W_1	86.63 ^{b}	88.20 ^{b}	80.70 ^{d}	92.50 ^{c}	74.00 ^{e}	99.45 ^{d}	69.00 ^{d}	104.28 ^{d}
W_2	72.55 ^{c}	71.95 ^{c}	93.78 ^{b}	89.88 ^{d}	106.23 ^{b}	109.50 ^{b}	108.38 ^{a}	113.25 ^{b}
W_3	74.28 ^{c}	74.73 ^{c}	83.68 ^{c}	87.08 ^{e}	92.15 ^{d}	102.73 ^{c}	103.68 ^{b}	107.50 ^{c}
W_4	74.28 ^{c}	74.73 ^{c}	83.68 c	87.08 ^{e}	75.48 ^{e}	75.00 ^{e}	92.30 ^{c}	90.48 ^{e}
W_5	86.63 ^{b}	88.65 ^{b}	94.55 ^{b}	95.98 ^{b}	102.18 ^{c}	109.53 ^{b}	106.70 ^{a}	115.60 ^{a}
W_6	97.53 ^{a}	98.00 ^{a}	107.98 ^{a}	105.43 ^{a}	115.50 ^{a}	114.53 ^{a}	108.25 ^{a}	108.10 ^{c}
SEm (±)	0.356	0.496	0.422	0.323	0.454	0.495	0.337	0.409
CD (0.05)	1.038	1.446	1.232	0.943	1.326	1.445	0.983	1.196

For treatment details, Table 4 may be referred to. SE, standard Error, CD, Critical difference at 5% probability level. Different superscript letters indicate a significant difference between the mean.

3.3. Correlation between Micro-Organisms and Soil Enzymes at Different Growth Stages

Multiple correlations between micro-organisms and soil enzymes (n = 24) at different stages recorded at 7, 14, 21, and 28 DAS in both years are presented in Tables 11 and 12. Data in both Tables revealed that there was a positive significant correlation between micro-organisms and soil enzymes recorded at 7, 14, 21, and 28 DAS/T in both years.

Table 11. Multiple correlations between micro-organisms and soil enzymes at different stages were recorded at 7 and 14 DAS in both years.

Parameters	7 DAS/T in the Year 2016				
	Bacteria	Fungi	Actinomycetes	Urease	Phosphatase
	1	2	3	4	5
Fungi	0.98 **				
Actinomycetes	0.99 **	0.99 **			
Urease	0.92 **	0.92 **	0.94 **		
Phosphatase	0.95 **	0.98 **	0.97 **	0.95 **	
Dehydrogenase	0.97 **	0.93 **	0.96 **	0.88 **	0.88 **
	7 DAS/T in the Year 2017				
Parameters	Bacteria	Fungi	Actinomycetes	Urease	Phosphatase
	1	2	3	4	5
Fungi	0.97 **				
Actinomycetes	0.99 **	0.99 **			
Urease	0.95 **	0.95 **	0.94 **		
Phosphatase	0.99 **	0.98 **	1.00 **	0.95 **	
Dehydrogenase	0.94 **	0.97 **	0.97 **	0.91 **	0.97 **

Table 11. *Cont.*

Parameters	Bacteria	Fungi	Actinomycetes	Urease	Phosphatase
	1	2	3	4	5
			14 DAS/T in the Year 2016		
Fungi	0.88 **				
Actinomycetes	0.94 **	0.76 **			
Urease	0.67 *	0.59 *	0.55 *		
Phosphatase	0.99 **	0.86 **	0.97 **	0.65 *	
Dehydrogenase	0.80 **	0.87 **	0.63 *	0.84 **	0.77 **
			14 DAS/T in the Year 2017		
Parameters	Bacteria	Fungi	Actinomycetes	Urease	Phosphatase
	1	2	3	4	5
Fungi	0.55				
Actinomycetes	0.97 **	0.53			
Urease	0.85 **	0.25	0.85 **		
Phosphatase	0.99 **	0.53	0.98 **	0.85 **	
Dehydrogenase	0.87 **	0.66*	0.94 **	0.75 **	0.92 **

* Correlation is significant at a 5% level of probability (1-tailed); ** Correlation is significant at a 1% level of probability (1-tailed).

Table 12. Multiple correlations between micro-organisms and soil enzymes at different stages were recorded at 21 and 28 DAS/T in both years.

Parameters	**Bacteria**	**Fungi**	**Actinomycetes**	**Urease**	**Phosphatase**
	1	**2**	**3**	**4**	**5**
			21 DAS/T in the Year 2016		
Fungi	0.85 **				
Actinomycetes	0.82 **	0.95 **			
Urease	0.79 **	0.81 **	0.81 **		
Phosphatase	0.89 **	0.82 **	0.84 **	0.94 **	
Dehydrogenase	0.73 **	0.77 **	0.69 *	0.93 **	0.90 **
			21 DAS/T in the year 2017		
Parameters	Bacteria	Fungi	Actinomycetes	Urease	Phosphatase
	1	2	3	4	5
Fungi	0.92 **				
Actinomycetes	0.98 **	0.92 **			
Urease	0.86 **	0.75 **	0.89 **		
Phosphatase	0.90 **	0.87 **	0.94 **	0.91 **	
Dehydrogenase	0.95 **	0.97 **	0.96 **	0.81 **	0.88 **
			28 DAS/T in the year 2016		
Parameters	Bacteria	Fungi	Actinomycetes	Urease	Phosphatase
	1	2	3	4	5
Fungi	0.55 *				
Actinomycetes	0.76 **	0.90 **			
Urease	0.62 *	0.40	0.60 *		
Phosphatase	0.83 **	0.51	0.68 *	0.90 **	
Dehydrogenase	0.43	0.60 *	0.80 **	0.86 **	0.81 **

Table 12. *Cont.*

Parameters	28 DAS/T in the year 2017				
	Bacteria	Fungi	Actinomycetes	Urease	Phosphatase
	1	2	3	4	5
Fungi	0.65 *				
Actinomycetes	0.75 **	0.71 *			
Urease	0.73 *	0.16	0.55 *		
Phosphatase	0.78 **	0.56 *	0.50	0.48	
Dehydrogenase	0.94 **	0.77 **	0.80 **	0.72 **	0.76 **

* Correlation is significant at a 5% level of probability (1-tailed); ** Correlation is significant at a 1% level of probability (1-tailed).

3.4. Effect of Crop Establishment Methods and Weed Management Practices on Species Wise Weed Count

Among grasses, *E. colona* and *Leptochloa chinensis* were prominent in DSR, while *E. crusgall* and *Paspalum distichum* was prominent in WSR and PTR (Figure 2). Among sedges, *Cyperus iria* was prominent in DSR while *C. difformis* and *Fimbristylis miliacea* were prominent in WSR. Among BLWs, *Eclipta alba*, *Ludwigia parviflora*, and *Spilanthes acmella* were prominent in DSR, *Aeschynomene indica*, *Alternanthera philoxeroids*, and *Ammannia baccifera* were prominent in WSR.

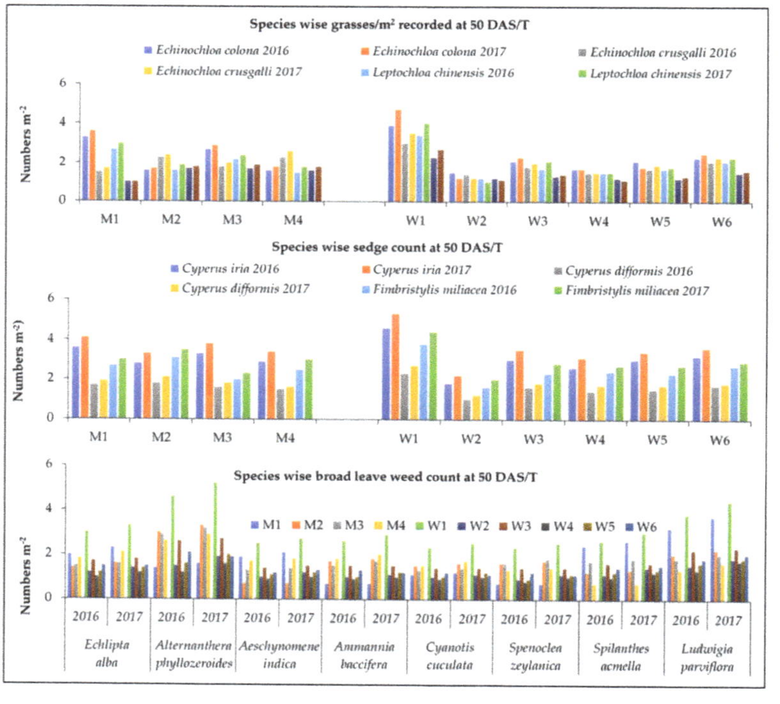

Figure 2. Effect of crop establishment methods and weed management practices on species-wise weed count at 50 DAS/T.

3.5. Effect of Crop Management Practices and Weed Management Practices on Grain Yield

Among the different crop establishment methods, PTR recorded the highest grain yield (4717 and 5033 kg ha^{-1}), followed by WSR (4579 and 4919 kg ha^{-1}), NPTR (4549 and 4816 kg ha^{-1}), and DSR (3257 and 3595 kg ha^{-1}). Among the different weed management practices (Figure 3), application of Bensulfuron methyl 0.6% + Pretilachlor 6% (PE) 0.660 kg ha^{-1} + Hand weeding (HW) at 30 DAS/T recorded the highest grain yield (5298 and 5675 kg ha^{-1}, respectively) during both years, followed by the application of Bensulfuron methyl 0.6% + Pretilachlor 6% (PE) 0.495 kg ha^{-1} + Bispyribac-Sodium (POE) 0.025 kg ha^{-1} at 15 DAS/T (5205 and 5565 kg ha^{-1}) and CW + HW (4659 and 4839 kg ha^{-1}), while the lowest was recorded in weedy check (1951 and 1790 kg ha^{-1}, respectively) during both years.

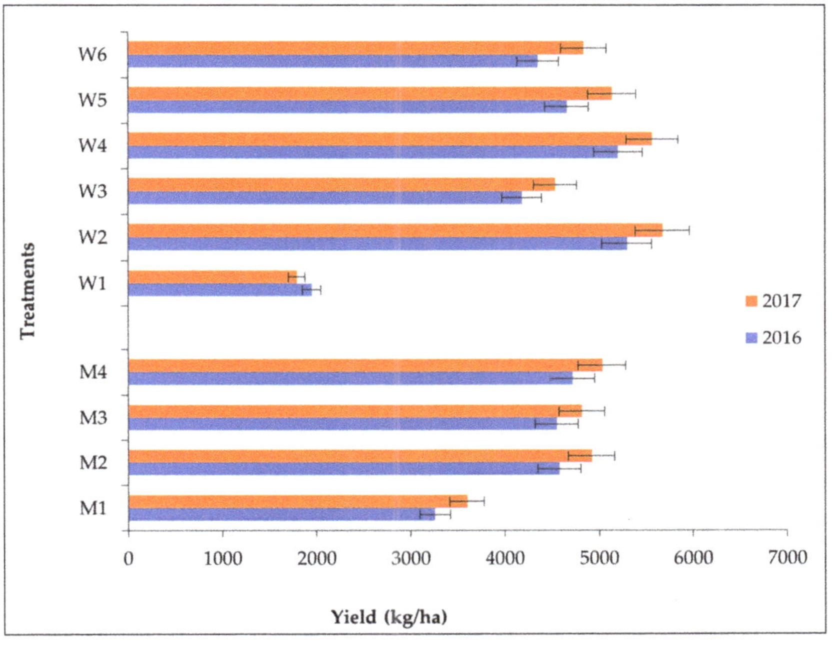

Figure 3. Effect of crop establishment methods and weed management practices on yield of rice.

4. Discussion

The impact of herbicides on soil microbiological and biochemical activity cannot be described by simple relationships as pesticides may be composed of one or more than one active ingredient that affects the soil microbial properties differently and may be toxic to soil microbes [47].

4.1. Effect of Crop Establishment Methods and Weed Management Practices -on Soil Microbial Activity

Significant inhibition of herbicides on microbial population exists that varies with the type of herbicides and dose of herbicide applied. Herbicide application causes a transient impact on microbial growth when applied at a recommended dose. There was an increasing trend of inhibition of microbial growth from initial application to 15 days after application, and no inhibition was found after 15 days after application (DAA) to harvest [48]. There coexists the ability of some microorganisms to degrade the herbicide, while some others are affected adversely by the type of herbicide and rate of application [49]. This has also been confirmed by the results of Zain et al. [50] who opined that the impact of herbicide may be stimulating or depressive on the growth of microorganisms, depending on the chemical types and dose, microbial species, and environmental factors. Microorganisms can degrade herbicides and utilize the products as a source of biogenic elements for their physiological activity, however, the herbicides have a toxic effect on microorganisms, reducing their abundance, activity, and diversity; while the toxic effect of herbicides is more severe immediately after application, at a later stage, microorganisms take part in the herbicides degradation process, converting it into carbon-rich substrates that in turn maximizes the microbial population in the rhizosphere [51].

Higher microbial activity (bacteria, fungi, and actinomycetes) was recorded under unpuddled transplanted rice conditions. While an initial decline in microbial activity was recorded immediately after both pre- and post-emergence herbicide application, that increased afterward, the rate of decline was dose-dependent, where a higher decrease was recorded with a higher concentration of herbicides. Similar results were also observed by Latha and Gopal [34] and Ramalakshmi et al. [35]. A decrease in microbial population was highest after the initial pre-emergence application of Bensulfuron methyl 0.6% + Pretilachlor 6% (PE) at 0.660 kg ha^{-1}, followed by application of Bensulfuron methyl 0.6% + Pretilachlor 6% (PE) at 0.495 kg ha^{-1}, while a continuous increase in microbial population was recorded where there were no chemical treatments applied. After the initial decline, there was an increase in microbial population unless there was no further additional chemical application. These results are also supported by Chauhan et al. [52] who revealed that the initial decrease followed by an increase in microbial population could also be due to microbial multiplication on the increased supply of nutrients available by the herbicide's degradation. The highest increase in microbial population and enzymatic activity was noticed in brown/green manuring plots. Jilani et al. [53] revealed that the organic amendments hold great promise as a source of multiple nutrients and the ability to improve soil characteristics.

4.2. *Effect of Crop Establishment Methods and Weed Management Practices on Soil Enzymatic Activity*

The activities of the soil enzymes urease, phosphatase, and dehydrogenase, as affected by different herbicides and weed management practices, have been discussed due to the significant role of these soil enzymes on soil biochemistry, nutrient availability, and plant growth. The urease enzyme catalyzes urea hydrolysis, which is susceptible to different soil-applied herbicides. It is a very useful soil quality indicator that is also widely used to analyze the xenobiotics' effect on different metabolic activities in the soil [54]. Higher urease activity was recorded under puddled transplanted conditions, followed by wet seeded rice, un-puddled transplanted rice, and dry seeded rice, where urease activity was higher under the flooded condition with a declining trend with the unflooded, dry condition among various weed management practices. Increasing activity of urease with the application of herbicides such as Alachlor, Butachlor, Propaquizafop, and Imazethapyr was recorded under flooded conditions [55,56]. From the study on herbicide, it was recorded that a higher dose of application of Bensulfuron methyl 0.6% + Pretilachlor 6% (PE) at 0.660 kg ha^{-1} recorded a greater decrease in initial urease activity than application at a dose of Bensulfuron methyl 0.6% + Pretilachlor 6% (PE) at 0.495 kg ha^{-1}, however, under both cases, the urease activity increased on a later stage, which was little higher at higher dose, that may be due to the degradation of the herbicide to biocarbon by microorganisms and its utilization that in turn may have increased the activity of soil urease, while again with the application of bispyribac sodium at 15 DAS/T at 0.025 kg ha^{-1} decreased the urease activity recorded at 21 DAS/T, that again increased at 28 DAS/T. A decrease in soil urease activity by increasing the concentration of various herbicides has also been reported by various authors [57–59]. Some authors have also found no effect of herbicides such as 2,4-dichlorofenoxy acetic acid, butachlor, and oxyfluorfen on soil urease activity [60].

The soil enzyme phosphatase plays an important role in converting organically bound phosphorous to an inorganic form, making it available for soil microorganisms as well as for plants. Both the stimulatory and inhibitory impact of herbicides on soil phosphatase activity has been reported by various researchers [54,61,62]. Higher activity of phosphatase was recorded in unpuddled transplanted conditions, while the lowest trend was recorded in puddled transplanted conditions. A similar report has been recorded by Rasool et al. [56] from their study on butachlor, which was also found to have an inhibitory impact on phosphatase activity under flooded conditions, while its impact was stimulatory under unflooded conditions. A steady increase in phosphatase activity was recorded in brown/green manured treatment and with the application of cono weeding and hand weeding treatments; better soil aeration and/or higher root activity may be the cause behind higher microbial and phosphatase activity, while the application of the pre-emergence herbicide Bensulfuron methyl 0.6% + Pretilachlor 6% (PE) at a higher dose of 0.660 kg ha^{-1} recorded a higher decrease in phosphatase activity at a dose of 0.495 kg ha^{-1}, while the increase in enzyme activity at a later stage was also higher where a higher dose of chemicals was applied (W_2). The stimulatory impact of the herbicide fomesafen has also been reported by Vladoiu et al. [63], Filimon et al. [64], and Ba'cmaga et al. [54], while the inhibitory effect of herbicides on phosphatase activity has been recorded by Filimon et al. [61] and Muñoz-Leoz et al. [64].

An activity study of the soil enzyme dehydrogenase has been considered one of the important methods of determining the effect of various chemical herbicides and pesticides on soil biochemistry. In the present study, an increase in dehydrogenase activity was recorded under brown/green manuring, which may be due to higher root activity, that may have increased the microbial population in various soil conditions, while at 7 DAS/T, a dose-dependent decline in dehydrogenase activity was recorded due to the pre-emergence application of herbicide Bensulfuron methyl 0.6% + Pretilachlor 6% (PE), that later increased at continuously after 14 DAS/T, and again, with the application of post-emergence herbicide bispyribac sodium, a declining trend was recorded in W_4 which increased at 28 DAS/T. A dose-dependent decrease in dehydrogenase activity due to herbicide S-metolachlor has also been reported by Filimon et al. [61] and Rasool et al. [56] whose studies found a detrimental effect of the herbicide butachlor on dehydrogenase activity under an unflooded soil condition while the impact was stimulatory under a flooded condition.

4.3. *Effect of Crop Establishment Methods and Weed Management Practices on Weed Count and Population*

It is imperative to know the weeds associated with any crop before recommending a control measure. Weeds present in any crop are mostly regulated by the growing season, cultural factors adopted by the farmer, and agro-ecological factors. A thorough study of weed taxonomy and biology is essential for success and efficient weed management. A critical study from the commencement of the crop (rice) until harvest indicated that as many as 15 different types of weeds existed in the crop field. Major weed flora in the experimental site enlisted of grasses, viz., *Echinochloa colona* (L.) Link, *Echinochloa crusgalli* (L.) Beauv, *Leptochloa chinensis* (L.) Nees, *Paspalum distichum* (L.), Sedges, *Cyperus iria* (L.), *Cyperus difformis* (L.), *Fimbristylis miliacea* (L.) Vahl, broad leave weeds, viz., *Eclipta alba* (L.) Hassk, *Alternanthera philoxeroids* (Mart.) Griseb, *Aeschynomene indica* (L.), *Ammannia baccifera* (L), *Cyanotis cucullata* (Roth) Kunth, *Spenoclea zeylanica*, *Spilanthes acmella* (L.), and *Ludwigia parviflora* (L.). Similar weed species in rice have been reported by Mahajan et al. [65] and Mohanty et al. [66].

4.4. *Effect of Crop Establishment Methods and Weed Management Practices on Grain Yield of Rice*

Higher yield in PTR might be due to higher yield attributing characters which may be the ultimate result of better availability and utilization of nutrients in properly spaced transplanted rice during the panicle growth period [67]. Similar results of higher yield were also observed by Jaiswal and Singh [68], Chauhan et al. [69], and Saharawat et al. [70] who recorded 0.45–0.61 t ha^{-1} lesser yield in both dry and wet direct-seeded rice than transplanted rice. Higher yield with the application of Bensulfuron methyl 0.6% + Pretilachlor 6% (PE) at 0.66 kg ha^{-1} + HW at 30 DAS/T might be the combined result of higher yield attributing characters and the lowest crop weed completion due to better weed control throughout the crop growth seasons, effective utilization of moisture, nutrients, light, and space as a whole. Similar results have also been observed by Sunil et al. [71], Reddy et al. [72], and Kumar et al. [73].

5. Conclusions

The present study revealed that microbial activity varies with different crop establishment methods and different weed management practices. A better microbial population with the microbial activity of dehydrogenase and phosphatase was recorded after an initial decline due to herbicide application in unpuddled transplanted conditions, providing a suitable environment for soil activity and soil health. The application of Bensulfuron methyl 0.6% + Pretilachlor 6% (PE) at 0.660 kg ha^{-1} recorded a greater decrease in initial urease, phosphatase, and dehydrogenase activity than the application of a dose of Bensulfuron methyl 0.6% + Pretilachlor 6% (PE) at 0.495 kg ha^{-1}. However, in both cases, the urease activity increased at a later stage, which was a little more than at a higher dose of herbicide, which might have increased the degradation of the herbicide to biocarbon by microorganisms, and its utilization, in turn, may have increased the activity of soil enzymes. Higher and better microbial activity with increased activity with crop duration was recorded in treatments where weed management was organically done. In treatments with chemical weed control, microbial activity suffered immediately after application but regained with time and the stage of the crop. Further study of different and new herbicides and their effect on soil microbial activity, soil enzymes, and bioassays will help the scientific community gain a better understanding of soil health.

Author Contributions: Conceptualization, S.P. (Sarthak Pattanayak), S.J., P.D., S.M. (Sagar Maitra), T.S., S.P. (Subhashisa Praharaj), P.M., M.P., S.M. (Santanu Mohanty) and D.K.S.; methodology and visualization, S.P.(Sarthak Pattanayak), S.J., P.D., S.M. (Sagar Maitra), T.S., S.P. (Sarthak Pattanayak), P.M., S.M. (Santanu Mohanty) and D.K.S.; software, S.P. (Subhashisa Praharaj), B.P. and A.H.; Validation, S.P. (Sarthak Pattanayak), S.J., P.D., S.M. (Sagar Maitra), T.S., S.P. (Subhashisa Praharaj), P.M., S.M. (Santanu Mohanty) and D.K.S.; formal analysis, S.P. (Sarthak Pattanayak), B.P. and A.H.; investigation, S.P. (Sarthak Pattanayak), S.J., P.D., S.M. (Sagar Maitra), S.P. (Sarthak Pattanayak), S.J. and P.D.; data curation, S.P. (Sarthak Pattanayak), B.P. and A.H.; writing—original draft preparation, S.P. (Sarthak Pattanayak), S.J., P.D., S.M. (Sagar Maitra), T.S., S.P. (Subhashisa Praharaj), P.M., M.P., S.M. (Santanu Mohanty) and D.K.S.; writing—review and editing, S.M. (Sagar Maitra), A.G. and A.H.; visualization, S.P. (Subhashisa Praharaj), S.J. and P.D.; supervision and project administration, S.J.; funding acquisition, A.H. and A.G. All authors have read and agreed to the published version of the manuscript.

Funding: The study was funded by the Department of Agronomy, College of Agriculture, Odisha University of Agriculture and Technology, Bhubaneswar 751003, Odisha, India. The current work was also partially funded by Taif University Researchers Supporting Project number (TURSP-2020/39), Taif University, Taif, Saudi Arabia.

Institutional Review Board Statement: Not applicable.

Informed Consent Statement: Not applicable.

Data Availability Statement: Data will be available after requesting to corresponding author.

Acknowledgments: The authors would like to give thanks to the Department of Agronomy, College of Agriculture, Odisha University of Agriculture and Technology, Bhubaneswar, Odisha, India 751003 for their valuable recommendation and direction during the tenure of the experiment. The authors also extend their appreciation to Taif University for supporting current work by Taif University Researchers Supporting Project number (TURSP-2020/39), Taif University, Taif, Saudi Arabia.

Conflicts of Interest: All authors declared no conflict of interest.

Disclaimer: We hereby declare that this manuscript contains no material which has been accepted for the award of any degree or diploma in any university, and that, to the best of our knowledge and belief, the manuscript contains no copy of any material previously published or written by another person except where due reference is made in the text. All authors in the manuscript have approved the submission in the journal.

References

1. Chauhan, B.S.; Mahajan, G.; Sardana, V.; Timsina, J.; Jat, M.L. Productivity and sustainability of the rice–wheat cropping system in the Indo-Gangetic Plains of the Indian subcontinent: Problems, opportunities, and strategies. *Adv. Agron.* **2012**, *117*, 315–369.
2. Shankar, T.; Malik, G.C.; Banerjee, M.; Dutta, S.; Maitra, S.; Praharaj, S.; Sairam, M.; Kumar, D.S.; Dessoky, E.S.; Hassan, M.M.; et al. Productivity and Nutrient Balance of an Intensive Rice–Rice Cropping System Are Influenced by Different Nutrient Management in the Red and Lateritic Belt of West Bengal, India. *Plants* **2021**, *10*, 1622. [CrossRef] [PubMed]
3. FAOSTAT. Data, Crops and Livestock Products. Food and Agriculture Organization of United Nations. 2022. Available online: https://www.fao.org/faostat/en/#data/QCL (accessed on 28 January 2022).
4. Government of India. *Agricultural Statistics at a Glance*; Directorate of Economics and Statistics, Ministry of Agriculture, Cooperation and Farmers Welfare, Government of India: New Delhi, India, 2020; pp. 48–49. Available online: https://foodprocessingindia.gov.in/uploads/publication/Agricultural-statistics-at-a-Glance-2020.pdf (accessed on 27 January 2022).
5. Ghosh, R.K.; Sharma, L.; Barman, S.; Dolai, A.K. System of rice intensification: The alternate approach for increasing production of field crops. *J. Crop Weed* **2009**, *5*, 63–67.
6. Pose-Juan, E.; Sánchez-Martín, M.J.; Herrero-Hernández, E.; Rodríguez-Cruz, M.S. Application of mesotrione at different doses in an amended soil: Dissipation and effect on the soil microbial biomass and activity. *Sci. Total Environ.* **2015**, *536*, 31–38. [CrossRef] [PubMed]
7. EPPO. *Guidelines of the European and Mediterranean Plant Protection Organization (EPPO) Concerning Research on the Efficiency of Plant Protection Chemicals*; Ministry of Agriculture and Rural Development: Paris, France, 2016.
8. Włodarczyk, M. Influence of formulation on mobility of metazachlor in soil. *Environ. Monit. Assess.* **2014**, *186*, 3503–3509. [CrossRef] [PubMed]
9. Kaczmarek, S.; Matysiak, K.; Kierzek, R. Weed control efficacy and selectivity of preemergence herbicides in *Sorghum vulgare* Perz. cultivation. *Prog. Plant Prot.* **2012**, *52*, 47–50.
10. Jones, D.L.; Edwards-Jones, G.; Murphy, D.V. Biochar mediated alterations in herbicide breakdown and leaching in soil. *Soil Biol. Biochem.* **2011**, *43*, 804–813. [CrossRef]
11. Bello, D.; Trasar-Cepeda, C.; Leirós, M.C.; Gil-Sotres, F. Modification of enzymatic activity in soils of contrasting pH contaminated with 2,4-dichlorophenol and 2,4,5-trichlorophenol. *Soil Biol. Biochem.* **2013**, *56*, 80–86. [CrossRef]
12. Tandon, S.; Pujari, A.; Sand, N.K. Degradation of Fentrazamide herbicide in soil under aerobic condition. *Bull. Environ. Contam. Toxicol.* **2012**, *89*, 312–315. [CrossRef]
13. Bro, E.; Devillers, J.; Millot, F.; Decors, A. Residues of plant protection products in grey partridge eggs in French cereal ecosystems. *Environ. Sci. Poll. Res.* **2016**, *23*, 9559–9573. [CrossRef]
14. Nikoloff, N.; Escobar, L.; Soloneski, S.; Larramendy, M.L. Comparative study of cytotoxic and genotoxic effects induced by herbicide S-metolachlor and its commercial formulation Twin Pack Gold® in human hepatoma (HepG2) cells. *Food Chem. Toxicol.* **2013**, *62*, 777–781. [CrossRef] [PubMed]
15. Lone, A.H.; Raverkar, K.P.; Pareek, N.; Chandra, R. Response of soil microbial communities to the selective herbicides: A microcosm approach. *J. Pure Appl. Microbiol.* **2014**, *8*, 1559–1567.
16. Kucharski, J.; Tomkiel, M.; Baćmaga, M.; Borowik, A.; Wyszkowska, J. Enzyme activity and microorganisms' diversity in soil contaminated with the boreal 58 WG herbicide. *J. Environ. Sci. Health B* **2016**, *51*, 446–454. [CrossRef] [PubMed]
17. Delgado-Moreno, L.; Peña, A. Compost and vermicompost of olive cake to bioremediate triazines-contaminated soil. *Sci. Total Environ.* **2009**, *407*, 1489–1495. [CrossRef]
18. Martins, P.F.; Carvalho, G.; Gratão, L.G.; Dourado, M.N.; Pileggi, M.; Araújo, W.L.; Azevedo, R.A. Effects of the herbicides acetochlor and metolachlor on antioxidant enzymes in soil bacteria. *Process Biochem.* **2011**, *46*, 1186–1195. [CrossRef]
19. Baćmaga, M.; Borowik, A.; Kucharski, J.; Tomkiel, M.; Wyszkowska, J. Microbial and enzymatic activity of soil contaminated with a mixture of diflufenican + mesosulfuron-methyl + iodosulfuron-methyl-sodium. *Environ. Sci. Poll. Res.* **2015**, *22*, 643–656. [CrossRef]
20. Midya, A.; Saren, B.K.; Dey, J.K.; Maitra, S.; Praharaj, S.; Gaikwad, D.J.; Gaber, A.; Alsanie, W.F.; Hossain, A. Crop Establishment Methods and Integrated Nutrient Management Improve: Part I. Crop Performance, Water Productivity and Profitability of Rice (*Oryza sativa* L.) in the Lower Indo-Gangetic Plain, India. *Agronomy* **2021**, *11*, 1860. [CrossRef]
21. Maitra, S.; Brestic, M.; Bhadra, P.; Shankar, T.; Praharaj, S.; Palai, J.B.; Shah, M.; Barek, V.; Ondrisik, P.; Skalický, M.; et al. Bioinoculants—Natural Biological Resources for Sustainable Plant Production. *Microorganisms* **2022**, *10*, 51. [CrossRef]
22. Yadaraju, N.T.; Mishra, J.S. Herbicides–boon or bane? *Pestology* **2002**, *26*, 43–46.
23. Rout, M.; Lodh, B.; Jena, S.N. Soil microbiological population in response to applied herbicide in green-gram. In Proceedings of the ISWS Golden Jubilee International Conference on Weeds and Society: Challenges and Opportunities, Jabalpur, India, 21–24 November 2018; p. 119.
24. Reichardt, W.; Dobermann, A.; George, T. Intensification of rice production systems: Opportunities and limits. In *Sustainability of Rice in the Global Food System*; Dowling, N.G., Greenfield, S.M., Fisher, K.S., Eds.; Pacific Basin Study Centre and IRRI Publ.: Manila, Philippines, 1998; pp. 127–144.
25. Sunil, C.M.; Shekara, B.G.; Kalyanamurthy, K.N.; Shankaralingappa, B.C. Growth and yield of aerobic rice as influenced by integrated weed management practices. *Ind. J. Weed Sci.* **2010**, *42*, 180–183.

26. Prakash, C.; Shivranand, R.K.; Koli, N.R. Bioefficacy of new herbicides on weed dynamics, yield attributes, yield and weed control efficiency in transplanted rice. *Trends. Biosci.* **2011**, *4*, 224–227.
27. Ramesh, M.; Yadav, D.B.; Ashok, Y.; Punia, S.S.; Malik, R.K.; Mehta, A. Weed control efficiency of bispyribac-sodium in transplanted and direct seeded rice and its residues in soil, rice grains and straw. *Environ. Ecol.* **2010**, *28*, 275–279.
28. Tranel, P.J.; Wright, T.R. Resistance of weeds to Al inhibiting herbicides: What have we learned? *Weed Sci.* **2002**, *50*, 700–712. [CrossRef]
29. EFSA (European Food Safety Authority). Conclusion on the peer review of the pesticide risk assessment of the active substance bispyribac (unless otherwise stated all data evaluated refer to the variant bispyribac-sodium). *EFSA J.* **2010**, *8*, 1692. [CrossRef]
30. Maitra, S.; Zaman, A. Brown manuring, an effective technique for yield sustainability and weed management of cereal crops: A review. *Int. J. Bioresour. Sci.* **2017**, *4*, 1–6. [CrossRef]
31. Maitra, S.; Zaman, A.; Mandal, T.K.; Palai, J.B. Green manures in agriculture: A review. *J. Pharm. Phytochem.* **2018**, *7*, 1319–1327.
32. Mohanta, S.; Banerjee, M.; Malik, G.C.; Shankar, T.; Maitra, S.; Ismail, I.A.; Dessoky, E.S.; Attia, A.O.; Hossain, A. Productivity and Profitability of Kharif Rice Are Influenced by Crop Establishment Methods and Nitrogen Management in the Lateritic Belt of the Subtropical Region. *Agronomy* **2021**, *11*, 1280. [CrossRef]
33. Midya, A.; Saren, B.K.; Dey, J.K.; Maitra, S.; Praharaj, S.; Gaikwad, D.J.; Gaber, A.; Alhomrani, M.; Hossain, A. Crop Establishment Methods and Integrated Nutrient Management Improve: Part II. Nutrient Uptake and Use Efficiency and Soil Health in Rice (*Oryza sativa* L.) Field in the Lower Indo-Gangetic Plain, India. *Agronomy* **2021**, *11*, 1894. [CrossRef]
34. Latha, S.; Gopal, H. Effect of Herbicides on Soil Microorganisms. *Ind. J. Weed Sci.* **2010**, *42*, 217–222.
35. Ramalakshmi, A.; Arthanari, P.M.; Chinnusamy, C. Effect of Pyrasosulfuron Ethyl, Bensulfuron Methyl, Pretilachlor and Bispyribac Sodium on Soil Microbial Community and Soil Enzymes under Rice-Rice Cropping System. *Int. J. Curr. Microbiol. App. Sci.* **2017**, *6*, 990–998. [CrossRef]
36. Lenka, D. Classification of rainfall zones and drought prone areas. In *Water Requirement of Crops in Orissa*; Publisher Directorate of Agriculture, Government of Orissa: Orissa, India, 1976; pp. 4–5.
37. Piper, C.S. *Soil and Plant Analysis*; A Monography from the Waite Agricultural Research Institute; The Waite Agricultural Research Institute: Urrbrae, Australia, 1950.
38. Jackson, M.L. *Soil Chemical Analysis*, 1st ed.; All India Coordinated Research Project on Weed Control; Prentice Hall of India Pvt. Ltd.: New Delhi, India, 1973.
39. Subbiah, B.V.; Asija, G.L. A Rapid Procedure for the Estimation of Available Nitrogen in Soils. *Curr. Sci.* **1956**, *25*, 259–260.
40. Hu, C.; Cao, S.Z. Activity of the Soil Microbial Biomass and Soil Enzyme Activity in Long Term Field Experiments. *World J. Agric. Sci.* **2007**, *3*, 63–70.
41. Tabatabai, M.A. Soil enzymes. In *Methods of Soil Analysis. Part 2. Chemical and Microbiological Properties*; Page, A.L., Miller, R.H., Keeney, D.R., Eds.; Agronomy No. 9.; ASA and SSSA: Madison, WI, USA, 1982; pp. 903–948.
42. Tabatabai, M.A.; Bremner, J.M. Use of p-nitrophenyl phosphate for assay of soil phosphatase activity. *Soil Biol. Biochem.* **1969**, *1*, 301–307. [CrossRef]
43. Pancholy, S.K.; Rice, E.L. Soil enzymes in relation to old field succession: Amylase, invertase, cellulose, dehydrogenase and urease. *Soil Sci. Soc. Am. J.* **1973**, *37*, 47–50. [CrossRef]
44. Casida, L.E.; Klein, D.A.; Santoro, T. Soil dehydrogenase activity. *Soil Sci.* **1964**, *98*, 371–376. [CrossRef]
45. Gomez, K.A.; Gomez, A.A. *Statistical Procedures for Agricultural Research*, 2nd ed.; John Wiley & Sons: Hoboken, NJ, USA, 1984; pp. 76–83.
46. Cochran, W.G.; Cox, G.M. Experimental Designs. John Wiley & Sons: New York, NY, USA, 1957; p. 611.
47. Tejada, M. Evolution of soil biological properties after addition of glyphosate, diflufenican and glyphosate + diflufenican herbicides. *Chemosphere* **2009**, *76*, 365–373. [CrossRef]
48. Adhikary, P.; Shil, S.; Patra, P.S. Effect of herbicides on soil microorganisms in transplanted chilli. *Glob. J. Biol. Agric. Health Sci.* **2014**, *3*, 36–238.
49. Sebiomo, A.; Ogundero, V.W.; Bankole, S.A. Effects of four herbicides on microbial population, soil organic matter and dehydrogenase activity. *Afr. J. Biotechnol.* **2011**, *10*, 770–778.
50. Zain, N.M.M.; Mohamad, R.B.; Sijam, K.; Morshed, M.M.; Awang, Y. Effects of selected herbicides on soil microbial populations in oil palm plantation of Malaysia: A microcosm experiment. *Afr. J. Microbiol. Res.* **2013**, *7*, 367–374.
51. Bera, S.; Ghosh, R.K. Soil Microflora and Weed Management as Influenced by Atrazine 50% WP in Sugarcane. *Univers. J. Agric. Res.* **2013**, *1*, 41–47. [CrossRef]
52. Chauhan, A.K.; Das, A.; Kharkwal, H.; Kharkwal, A.C.; Varma, A. Impact of micro-organisms on environment and health. In *Microbes: Health and Environment*; Chauhan, A.K., Varma, A., Eds.; IK International: New Delhi, India; New York, NY, USA, 2006; pp. 1–12.
53. Jilani, G.; Akram, A.; Ali, R.M.; Hafeez, F.Y.; Shamsi, I.H.; Chaudhry, A.N.; Chaudhry, A.G. Enhancing crop growth, nutrients availability, economics and beneficial rhizosphere microflora through organic and biofertilizers. *Ann. Microbiol.* **2007**, *57*, 177–183. [CrossRef]
54. Baćmaga, M.; Boros, E.; Kucharski, J.; Wyszkowska, J. Enzymatic activity in soil contaminated with the Aurora 40 WG herbicide. *Environ. Prot. Eng.* **2012**, *38*, 91–102.

55. Ramesh, A.; Joshi, O.P.; Billore, S.D. Effect of herbicides on soil dehydrogenase and urease activity in soybean (*Glycine max*). *Indian J. Agric. Sci.* **2000**, *70*, 218–219.
56. Rasool, N.; Reshi, Z.A.; Shah, M.A. Effect of butachlor (G) on soil enzyme activity. *Eur. J. Soil Biol.* **2014**, *61*, 94–100. [CrossRef]
57. Filimon, M.N.; Voia, S.O.; Popescu, R.; Bordean, D.M.; Vladoiu, D.L.; Mituletu, M.; Ostafe, V. The effect of chlorsulfurone and MCPB-Na on the enzymatic activity of microorganisms. *J. Serb. Chem. Soc.* **2014**, *79*, 1075–1084. [CrossRef]
58. Lupwayi, N.Z.; Hanson, K.G.; Harker, K.N.; Clayton, G.W.; Blackshaw, R.E.; O'Donovan, J.T.; Johnson, E.N.; Gan, Y.; Irvine, R.B.; Monreal, M.A. Soil microbial biomass, functional diversity and enzyme activity in glyphosate-resistant wheat canola rotations under low-disturbance direct seeding and conventional tillage. *Soil Biol. Biochem.* **2007**, *39*, 1418–1427. [CrossRef]
59. Kucharski, J.; Wyszkawska, J. Biological properties of soil contaminated with the herbicide Apyros 75 WG. *J. Elem.* **2008**, *13*, 57–371.
60. Baruah, M.; Mishra, R.R. Effect of herbicides butachlor, 2,4-D and oxyfluorfen on enzyme activities and CO_2 evolution in submerged paddy field soil. *Plant. Soil* **1986**, *96*, 287–291. [CrossRef]
61. Filimon, M.N.; Roman, D.L.; Caraba, I.V.; Isvoran, A. Assessment of the effect of application of the herbicide S-metolachlor on the activity of some enzymes found in soil. *Agriculture* **2021**, *11*, 469. [CrossRef]
62. Filimon, M.N.; Voia, S.O.; Vladoiu, D.L.; Isvoran, A.; Ostafe, V. Temperature dependent effect of difenoconazole on enzymatic activity from the soil. *J. Serb. Chem. Soc.* **2015**, *80*, 1127–1137. [CrossRef]
63. Vladoiu, D.L.; Filimon, M.N.; Ostafe, V.; Isvoran, A. Assessment of pesticides interactions with Bacillus pasteurii urease. A computational study. *Rom. J. Phys.* **2015**, *60*, 583–592.
64. Muñoz-Leoz, B.; Garbisu, C.; Charcosset, J.Y.; Sánchez-Pérez, J.M.; Antigüedad, I.; Ruiz-Romera, E. Non-target effects of three formulated pesticides on microbially-mediated processes in a clay-loam soil. *Sci. Total Environ.* **2013**, *449*, 345–354. [CrossRef] [PubMed]
65. Mahajan, G.; Chauhan, B.S.; Johnson, D.E. Weed management in aerobic rice in north western Indo-gangetic Plains. *J. Crop Improve.* **2009**, *23*, 366–382. [CrossRef]
66. Mohanty, S.K.; Behera, M.M.; Mishra, M.M.; Dash, R.; Mishra, P.J. Integrated weed management in rice-rice ropping system under east and southern eastern coastal plain zone of Orissa. In Proceedings of the 25th Asian Pacific Weed Society Conference on Weed Science for Sustainable Agriculture, Environment and Bio Diversity, Hyderabad, India, 13–16 October 2015.
67. Awan, T.H.; Ali, R.I.; Manzoor, Z.; Ahmad, M.; Akhtar, M. Effect of different nitrogen levels and row spacing on the performance of newly evolved medium grain rice variety, KSK-133. *J. Anim. Plant. Sci.* **2011**, *21*, 231–234.
68. Jaiswal, V.P.; Singh, G.R. Effect of planting methods, source and level of nitrogen on the growth and yield of rice (*Oryza sativa* L.) and on succeeding wheat (*Triticum aestivum*). *Indian J. Agron.* **2001**, *46*, 5–11.
69. Chauhan, B.S.; Ahmed, S.; Awan, T.H.; Manalil, S. Integrated weed management approach to improve weed control efficiencies for sustainable rice production in dry-seeded systems. *Crop Prot.* **2015**, *71*, 19–24. [CrossRef]
70. Sarawat, K.L. Fertility and organic matters in submerge rice soils. *Curr. Sci.* **2005**, *88*, 735–739.
71. Munda, S.; Saha, S.; Adak, T.; Jambhulkar, N. Weed management in aerobic rice: Role of establishment methods and herbicides. *Exp. Agric.* **2019**, *55*, 88–104. [CrossRef]
72. Reddy, B.G.M.; Ravishankar, G.; Balganvi, S.; Joshi, V.R.; Negalur, R.K. Efficacy of bensulfuron methyl plus pretilachlor for controlling weeds in transplanted rice. *Oryza Int. J. Rice* **2012**, *9*, 65–67.
73. Kumar, R.S.; Durairaj, S.N.; Daisy, M.; Archana, H.A. Studies on weed management practices in transplanted rice. *Trend. Biosci.* **2014**, *7*, 3882–3885.

plants

MDPI

Article

Effect of Herbicides on the Management of the Invasive Weed *Solanum rostratum* Dunal (Solanaceae)

Jackline Abu-Nassar and Maor Matzrafi *

Department of Plant Pathology and Weed Research, Newe Ya'ar Research Center, Agricultural Research Organization (ARO), Ramat Yishay 30095, Israel; jackline@volcani.agri.gov.il
* Correspondence: maorm@volcani.agri.gov.il

Abstract: *Solanum rostratum* Dunal is an invasive weed species that invaded Israel in the 1950s. The weed appears in several germination flashes, from early spring until late summer. Recently, an increase in its distribution range was observed, alongside the identification of new populations in the northern part of Israel. This study aimed to investigate the efficacy of herbicide application for the control of *S. rostratum* using two field populations originated from the Golan Heights and the Jezreel Valley. While minor differences in herbicide efficacy were recorded between populations, plant growth stage had a significant effect on herbicide response. Carfentrazone-ethyl was found to be highly effective in controlling plants at both early and late growth stages. Metribuzin, oxadiazon, oxyfluorfen and tembutrione showed reduced efficacy when applied at later growth stage (8–9 cm height), as compared to the application at an early growth stage (4–5 cm height). Tank mixes of oxadiazon and oxyfluorfen with different concentrations of surfactant improved later growth stage plant control. Taken together, our study highlights several herbicides that can improve weed control and may be used as chemical solutions alongside diversified crop rotation options. Thus, they may aid in preventing the spread and further buildup of *S. rostratum* field populations.

Keywords: alternative weed management; buffalobur; crop and herbicide rotation; herbicide efficacy; surfactant

Citation: Abu-Nassar, J.; Matzrafi, M. Effect of Herbicides on the Management of the Invasive Weed *Solanum rostratum* Dunal (Solanaceae). *Plants* **2021**, *10*, 284. https://doi.org/10.3390/plants10020284

Academic Editor: Alessia Restuccia
Received: 6 January 2021
Accepted: 31 January 2021
Published: 2 February 2021

1. Introduction

The family of *Solanaceae* counts many species widespread worldwide, either wild or cultivated [1]. These species are of international importance and numerous studies related to their biology and ecology as crops and weeds have been conducted. Several of these species, such as the subgenus of prickly nightshades (*Solanum* spp.: subgenus *Leptostemonum*) are considered as significant invasive species in many parts of the world [2,3]. Originated from of the southeastern US, these species include some of the most noxious weeds worldwide [3]. Two prickly nightshades, *S. elaeagnifolium* Cav. and *S. rostratum* Dunal, invaded Israel in the 1950s [4]. These troublesome weeds infest pastures, crops, roadsides and natural habitats. Their distribution range is quite different as *S. elaeagnifolium* has become widely spread, while *S. rostratum* remains of limited dispersal. However, in the last 10 years the distribution of *S. rostratum* increased, especially in agricultural habitats.

S. rostratum (common name buffalobur) is a native species of the Mexican highlands [5] which has invaded several countries worldwide, including Canada, China, Russia and Australia [6,7]. *S. rostratum* is a diploid, summer-annual, self-compatible weed species with prickles on both leaves and stems. It is a noxious weed, as it grows aggressively following habitat disturbance [6], and livestock abstains from grazing on vegetation where it grows as thorns cover the entire plant. In minor crops, such as watermelon (*Citrullus lanatus* Thunb.), onion (*Allium cepa* L.), chickpea (*Cicer arietinum* L.) and tomato (*Solanum lycopersicum* L.), for which chemical control options are poor and hand weeding is common practice, *S. rostratum* thorns may increase the difficulty of this task.

Although *S. rostratum* is highly abundant in China, it is usually found in open, disturbed habitats, such as roadsides, fallow fields and field margins [6]. In the US, this weed was reported as an agricultural pest in Oklahoma [8], Nebraska and Wyoming [9]. Field experiments conducted in cotton (*Gossypium hirsutum* L.) fields in Oklahoma showed that crop plant heights were decreased at greater densities of *S. rostratum* [8]. In the same study, yield reduction was also observed, with a damage of 480 kg ha^{-1} at a weed density of 64 plants/10 m row. In Israel, *S. rostratum* was first documented at the Jezreel Valley in 1953 [4]. Since then, several field populations were located in the Jordan Valley, the Golan Heights, the Hulla Valley and at the Mediterranean Sea coastline (Figure 1). *S. rostratum* may appear in several germination flashes, starting from spring and continuing throughout the summer; thus, both young and mature plants may coexist at the same field. Different sensitivity levels between young and mature *S. rostratum* plants can reduce the effectiveness of chemical weed control.

Figure 1. Distribution and characterization of *Solanum rostratum* at different habitats across Israel. (**a**) Geographical distribution of major *Solanum rostratum* populations. (**b**) View of mature *Solanum rostratum*. Representative photos of *S. rostratum* plants in onion (**c**) and corn (**d**) fields.

Studies relating to *S. rostratum* management approaches are scarce, warranting research into optimal weed control strategies, especially herbicide application. In two field experiments conducted at Nebraska and Wyoming, broadleaf weed control was tested as part of an overall strategy to reduce *S. rostratum* infestation in two summer cereal crops, proso (*Panicum miliaceum* L.) and foxtail millet (*Setaria italica* P. Beauv.) [9]. Treatments with carfentrazone-ethyl [protoporphyrinogen oxidase (PPO) inhibitor (group E), combined with either 2,4-D amine (auxin inhibitor; group O) or prosulfuron (acetolactate synthase inhibitor; group B), were highly effective in controlling *S. rostratum* plants. In Israel, PPO inhibitors such as oxyfluorfen and oxadiazon are registered for use in both onion and tomato, while carfentrazone-ethyl is registered for use in chickpea. Although the oxyfluorfen or oxadiazon herbicide labels do not require the addition of a surfactant, previous research suggested that the effectiveness of PPO inhibitors may be ameliorated by the addition of a nonionic surfactant to the tank mix [10,11].

Due to the recent increasing infestation of *S. rostratum* plants in several fields across Israel (Figure 1), this research aimed to evaluate the efficacy of herbicides with differing modes of action for the control of *S. rostratum* at different growth stages. In addition, it assessed the additive effect of surfactant to PPO inhibitors in *S. rostratum* control protocols.

2. Results

2.1. Herbicide Response of Ginegar (GO) Plants Treated at Different Growth Stages

The comparison of fresh final shoot weight indicated that plant response was significantly affected by both the specific herbicides and growth stage of the weed at the time of application. For plants that were sprayed at the 4–5 cm growth stage, both clomazone and rimsulfuron showed low efficacy in reducing plant final weight, with mean shoot fresh weights measuring 79.5% and 93.8%, respectively, of the untreated control (Figure 2a; Table 1). Clopyralid showed a more significant effect, as presented by a shoot fresh weight of 42.6% that of untreated plants. For plants treated at the 8–9 cm growth stage, low efficacy was recorded for almost all tested herbicides, with the exception of carfentrazone-ethyl, which reduced mean shoot fresh weight down to 0.2% of that of the control plants (Figure 2b; Table 1). Oxadiazon and oxyfluorfen applied at this same growth stage were significantly less effective in controlling GO *S. rostratum* plants, reducing final shoot fresh weight down to 39.7% and 33.1%, respectively, of untreated control plants. Metribuzin and tembutrione showed the same response as oxadiazon and oxyfluorfen and, while achieving high control at the 4–5 cm growth stage, at the 8–9 cm growth stage, the two herbicides provided poor weed control.

Table 1. Differences in mean shoot fresh weight among *Solanum rostratum* populations in response to various herbicides.

		Mean Shoot Fresh Weight (% of Control)		
Population [a]	**Herbicide** [b]	**4–5 cm** [c]	**8–9 cm**	***p*-Value**
GY	Carfentrazone-ethyl	0.57	0.32	0.2567
	Clomazone	36.34	87.56	<0.0001
	Clopyralid	60.09	98.00	<0.0001
	Metribuzin	0.89	47.75	<0.0001
	Oxadiazon	3.01	30.44	<0.0001
	Oxyfluorfen	4.77	48.75	<0.0001
	Rimsulfuron	87.84	56.00	0.0008
	Tembotrione	4.61	47.56	<0.0001
GO	Carfentrazone-ethyl	0.22	0.19	0.8502
	Clomazone	79.52	103.34	0.0084
	Clopyralid	42.54	88.89	<0.0001
	Metribuzin	6.42	75.93	<0.0001
	Oxadiazon	5.65	39.69	<0.0001
	Oxyfluorfen	2.76	33.13	<0.0001
	Rimsulfuron	93.83	110.08	0.0134
	Tembotrione	10.76	75.31	<0.0001

[a] GY—Givat Yoav and GO—Ginegar; [b] All herbicides were applied at the labeled field rate as specified in Table 3; [c] Plants were sprayed at two different growth stages (4–5 cm and 8–9 cm); Shoot fresh weight was recorded 21 days after herbicide application.

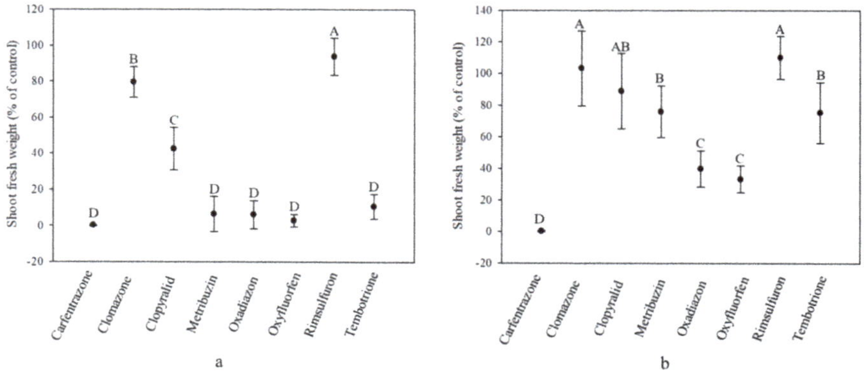

Figure 2. Shoot fresh weight (% of control) of *Solanum rostratum* plants from the Ginegar population treated with the recommended field rate of various herbicides at the 4–5 cm (**a**) and 8–9 cm (**b**) growth stages. Shoot fresh weight was recorded 21 days after herbicide application. Different uppercase letters indicate statistically significant differences among treatments, as determined by a Tukey–Kramer HSD test (α = 0.05).

2.2. Herbicide Response of Givat Yoav (GY) Plants Treated at Different Growth Stages

The same trend of herbicide response was recorded for plants of the GY population (Figure 3a,b). As in the case of GO plants, clomazone, clopyralid and rimsulfuron, showed low efficacy on GY plants treated at the 4–5 cm growth stage. However, in contrast to the GO responses, clomazone was more effective than clopyralid, as exhibited by the mean shoot fresh weight values of 36.5% vs. 60.1%, respectively, as compared to untreated controls (Table 1). Plants treated at the 8–9 cm growth stage showed low response to herbicides, with mean shoot fresh weights not dropping below 40% for most treatments (Figure 3b; Table 1). The only herbicides that induced shoot fresh weight reductions below 40% were the two PPO inhibitors, oxadiazon and carfentrazone ethyl (30.5% and 0.3%, respectively).

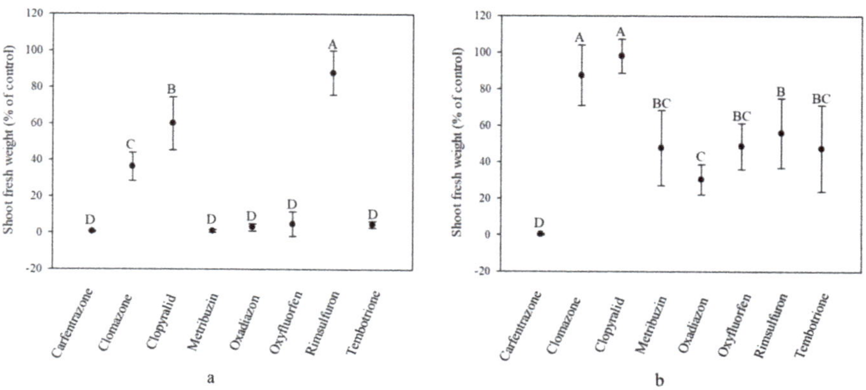

Figure 3. Shoot fresh weight (% of control) of *Solanum rostratum* plants from the Givat Yoav population treated with the recommended field rate of various herbicides at the 4–5 cm (**a**) and 8–9 cm (**b**) growth stages. Shoot fresh weight was recorded 21 days after herbicide application. Different uppercase letters indicate statistically significant differences among treatments, as determined by a Tukey–Kramer HSD test (α = 0.05).

2.3. Herbicide Effect on Plants Treated at Different Growth Stage

A negative correlation between plant growth stage and herbicide efficacy was recorded in both GY and GO populations for all herbicides, except for carfentrazone-ethyl (Table 1), with lower control rates observed when plants were sprayed at later growth stage (8–9 cm). However, the GY population response to rimsulfuron showed an opposite trend when plants were treated at the 8–9 cm growth stage; plants exhibited lower shoot fresh weight as compared to plants treated at the 4–5 cm growth stage

2.4. Variation in Herbicide Response among Populations

For most herbicides, no significant differences in plant response were recorded at the 4–5 cm growth stage (Table 1). Clomazone was more effective in controlling plants from the GY population, while clopyralid was more effective in controlling plants of the GO population. However, while differences between populations, overall, both herbicides showed poor performances in controlling GO and GY plants at the 4–5 cm growth stage (clomazone 79.5% vs. 36.4% and clopyralid 42.5% vs. 60.1%, respectively). For the 8–9 cm growth stage, differences among populations were more significant, with metribuzin, oxyfluorfen rimsulfuron and tembutrione all showing differences in control efficacy of GO vs. GY plants. Although differences were significant, herbicide efficacy rates were still low, as shown by the high mean shoot fresh weight (Table 1).

2.5. Synergism of the Surfactant with PPO Inhibitors

The addition of surfactant to the herbicide treatments resulted in an increased sensitivity of the large size (8–9 cm) *S. rostratum* plants to both oxadiazon and oxyfluorfen (Table 2). The application of 0.25% or 0.5% surfactant alone did not result in a significant reduction in shoot fresh weight compared with the control. Remarkably, the use of 1% surfactant alone resulted in an approximate 35% reduction in plant biomass, without affecting plant survival. The effective surfactant concentration that increased herbicide activity was different for the two herbicides. When applying oxadiazon with surfactant, shoot biomass declined by to 28 ± 8.0% of untreated controls when applied with 1% surfactant dose and to 9.9 ± 3.6% when applied with 0.5% surfactant (Table 2). When combined with oxyfluorfen, the 0.5% surfactant concentration resulted in a final shoot fresh weight of 6.6 ± 4.4% in comparison to the controls, while the 1% dose brought to a final shoot biomass that was 3.4 ± 3.4% of the untreated control (Table 2).

Table 2. Reduction of shoot fresh weight (mean ± SE) and survival rate of 8–9 cm tall *Solanum rostratum* Givat Yoav plants following oxadiazon and oxyfluorfen application, with or without surfactant.

Treatment [a]	Mean (SE) [b]	Lower 95%	Upper 95%	Survival (%)
0.5% surfactant	103.40 (7.30) a	84.65	122.16	100%
Control	102.14 (5.51) a	89.67	114.61	100%
0.25% surfactant	94.04 (9.04) ab	71.92	116.16	100%
1% surfactant	65.50 (4.89) b	53.52	77.48	100%
Oxadiazon + 1% surfactant	28.00 (8.01) c	9.87	46.13	70%
Oxadiazon + 0.25% surfactant	26.18 (6.41) c	11.69	40.68	80%
Oxifluorfen	25.54 (9.73) c	2.54	48.54	50%
Oxadiazon	19.38 (7.26) c	2.94	35.81	50%
Oxadiazon + 0.5% surfactant	9.84 (3.54) c	1.84	17.85	40%
Oxifluorfen + 0.5% surfactant	6.57 (4.42) c	−3.63	16.77	30%
Oxifluorfen + 0.25% surfactant	4.10 (4.08) c	−5.14	13.34	20%
Oxifluorfen + 1% surfactant	3.38 (3.36) c	−4.37	11.13	20%

[a] All herbicides were applied at the labeled field rate as specified in Table 3; [b] Different lowercase letters indicate statistically significant differences among treatments, as determined by a Tukey–Kramer HSD test (α = 0.05); Shoot fresh weight was recorded 21 days after herbicide application.

Table 3. List of herbicides used in this study and their labeled field rates.

Common Name	Trade Name	MOA [a]	Manufacturer	Rate (g ai ha^{-1})
Oxyfluorfen	Galigan®	PPO	ADAMA-Agan	352.5
Carfentrazone-ethyl	Spotlight®	PPO	FMC	0.9
Oxadiazon	Ronstar®	PPO	Bayer	875
Clopyralid	Lontrel®	Auxinic herbicide	Corteva	50
Clomazone	Comand®	Carotenoid biosynthesis	FMC	540
Tembotrione	Laudis®	HPPD	Bayer	99
Metribuzin	Sencor®	PSII	Bayer	175
Rimsulfuron	Titus®	ALS	Corteva	25 [b]

[a] Mode of action; [b] Addition of a nonionic surfactant (Shatah 90®, ADAMA-Makhteshim, 0.05%) as part of the manufacturer recommendation.

3. Discussion

In this work, plants of two different populations were treated with several herbicides at two growth stages. As evident from the presented data, growth stage had a significant effect on herbicide control of *S. rostratum*. The fact that growth stage has a significant effect on herbicide efficacy is known and was shown for both crop and weed plant species [12,13]. However, as *S. rostratum* may germinate in several germination flashes throughout the season, both young and mature plants can coexist in the same field, thereby reducing the overall efficacy of herbicide treatment. Thus, herbicides that effectively control a wider range of growth stages are more desirable for the farmer.

Surfactants increase herbicide activity as they reduce the droplet surface tension, thus increasing herbicide permeability and mobility through the leaf cuticular layer [14]. Surfactant interactions with both the herbicide and leaf cuticle may differ as a function of their electrical charge (nonionic, ionic and anionic), thereby generating various levels of synergism. Several herbicides, such as the acetolactate inhibitors trifloxysulfuron-methyl [15] and bispyribac-sodium [16], as well as the PPO inhibitor fomesafen [11], require the addition of surfactant to the spraying tank mix. In this study, the effect of surfactant addition on oxadiazon and oxyfluorfen efficacy in larger size (8–9 cm tall) *S. rostratum* plants, was evaluated. Overall, addition of the surfactant induced a synergistic effect, manifesting by higher herbicide activity (Table 2). However, different surfactant concentrations showed diverse synergism levels. For oxadiazon, the lower concentration provided better outcomes, while for oxyfluorfen, the higher surfactant concentration was more effective. Previous studies have demonstrated the synergistic effect of herbicide and surfactant combinations [17,18]. However, evaluation of surfactants as adjuvant should include phytotoxicity tests to ensure crop safety.

One of the most successful integrated weed management approaches is the use of different herbicides in combination with smart crop rotation. Crop rotation is a highly important component of good agricultural practice and can be used to increase productivity and optimize weed control. For instance, nonlegume crop yield can be improved by introducing a legume into the cropping sequence. In comparison to a continuous cotton or wheat-cotton sequence, growing cotton in rotation with vetch (*Vicia sativa* L.), led to a 20% increase in lint yield, even in the absence of N fertilizer [19]. In sugar beet (*Beta vulgaris* L.), yield was 5% higher after introduction of pea (*Pisum sativum* L.), as compared to maize, as a sequence crop [20]. Crop rotation may also be used to reduce weed infestation. A sequence of corn–soybean vs. soybean–soybean crop was shown to decrease *Conyza canadensis* (L.) Cronq. escapes [21]. However, crop rotation with no proper weed management approaches may not be very efficient in reducing weed density [22]. Thus, the chemical management tools that were successful in this study should be combined with adequate crop rotation, in order to reduce *S. rostratum* threat. For instance, using metribuzin in tomato, followed by clopyralid in onion and carfentrazone-ethyl in chickpea, may serve as a long-term crop-herbicide combination that can effectively control *S. rostratum* in these fields. Besides the crops mentioned above, as a summer irrigated crop, corn (*Zea mays* L.) is also an important part of crop rotation in the Israeli agriculture [23]. Although *S. rostratum* shows no marked effect on corn yield, it can be found in high densities at field margins (Figure 1d), where it may enrich the field seed bank for upcoming years. Moreover, this study included herbicides that are registered for use in corn, such as tembutrione. High effectiveness of this herbicide in controlling *S. rostratum*, mainly at the 4–5 cm growth stage, may serve as an advantageous tool to prevent further buildup of *S. rostratum* field populations.

In conclusion, the presented experiments identified several herbicides with varying levels of efficacy, dependent on the plant growth stage, tested on plants from two geographically distinct *S. rostratum* population. In addition, the combination of PPO inhibitors with surfactant increased the susceptibility of *S. rostratum* to herbicides, even at a later growth stage, increasing the overall efficacy of oxadiazon and oxyfluorfen. These findings should be further evaluated under field conditions in order to validate their efficacy. Other approaches, such as nonchemical weed management tools, should also be explored to prevent future damage of agricultural habitats by *S. rostratum*.

4. Materials and Methods

4.1. Plant Material

As *S. rostratum* seeds possess high dormancy, which may prevent their germination [24], the presented herbicide application tests used field-collected seedlings. Seedling were collected from two fields presenting high infestation of *S. rostratum*, located near Moshav Givat Yoav (GY) (32.8012266548, 35.6977272346) and Kibbutz Ginegar (GO) (32.6543271802, 35.2488696828) (Figure 1). To ensure appropriate representation of the field population, seedlings at the three-four leaf stage were randomly selected from each field population. Seedlings were then transplanted into 250 mL pots filled with commercial potting medium (Tuff, Marom Golan, Israel), including Osmocote® slow-release fertilizer, at the New-Yaar Research Center. Plants were maintained in a greenhouse under Israeli summer conditions and watered daily.

4.2. Herbicides Application

Experiments were carried out in the summer of 2020, in a greenhouse at the Newe-Yaar Research Center, under natural conditions. Plants from both GY and GO populations were sprayed when reaching heights of 4–5 cm and 8–9 cm. Experiments were spaced in time such that plants from the first group (4–5 cm) were sprayed two weeks prior to plants from the second group (8–9 cm). Plants were treated with eight different herbicides at the recommended field rate as specified in Table 3, using a chain-driven sprayer delivering 300 L ha^{-1}, with a flat-fan 8001E nozzle (TeeJet®, Spraying Systems Co., Wheaton, IL, USA). All experiments were arranged in a fully randomized-factorial design, with eight to ten replicates for each treatment. Shoot fresh weight of each plant was recorded 21 days after treatment (DAT).

4.3. Surfactant Effect

Both oxyfluorfen and oxadiazon, were applied in a tank mix with the nonionic surfactant alkyl phenol ethylene oxide (Shatah 90®, ADAMA-Makhteshim, Israel). The effect of increasing surfactant concentrations on *S. rostratum* control was evaluated. Plants from both populations, GY and GO, reacted similarly to oxadiazon at heights of 8–9 cm (Table 1). However, plants from the GY population were less responsive to oxyfluorfen at the later growth stage. Thus, advanced growth stage plants from the GY population were chosen for this experiment. Plants were treated with three different surfactant concentrations (0.25%, 0.5% and 1%), applied with or without oxyfluorfen and oxadiazon (at recommended field rate; Table 3). Experiments were conducted as specified above. Shoot fresh weight and survival rate were recorded for each plant 21 DAT.

4.4. Statistical Analyses

Shoot fresh weight data were analyzed using ANOVA in JMP (ver. 15) statistical package (SAS Institute Inc., Cary, NC, USA). Data was visualized using SigmaPlot (ver. 13) software (Systat Software Inc., San Jose, CA, USA). The assumptions of homoscedasticity and normality were met using Levene's tests. For both experiments, i.e., the response of GO and GY populations to herbicide treatments at different growth stages and the synergistic effect of surfactant application, interactions between experimental parameters were observed. Thus, data from each experiment were analyzed separately as experiment-by-treatment. For all experiments, shoot fresh weight was analyzed as percent of untreated control. All parameters were subjected to one-way ANOVA and means were separated using the Tukey-HSD test ($\alpha = 0.05$).

Author Contributions: J.A.-N. and M.M. both authors contributed equally to conceptualization, original draft preparation, review and editing. All authors have read and agreed to the published version of the manuscript.

Funding: This research received no external funding.

Acknowledgments: The authors would like to thank Ran Lati for his valuable comment and suggestions.

Conflicts of Interest: The authors declare no conflict of interest.

References

1. Frodin, D.G. History and concepts of big plant genera. *Taxon* **2004**, *53*, 753–776. [CrossRef]
2. Knapp, S.; Vorontsova, M.S.; Särkinen, T. Dichotomous keys to the species of *Solanum* L. (Solanaceae) in continental Africa, Madagascar (incl. the Indian Ocean islands), Macaronesia and the Cape Verde Islands. *PhytoKeys* **2019**, *127*, 39–76. [CrossRef]
3. Bryson, C.T.; Reddy, K.N.; Byrd, J.D. Growth, development, and morphological differences among native and nonnative prickly nightshades (*Solanum* spp.) of the southeastern United States. *Invasive Plant Sci. Manag.* **2012**, *5*, 341–352. [CrossRef]
4. Danin, A. Flora of Israel Online. Available online: http://www.flora.huji.ac.il (accessed on 14 January 2021).
5. Bassett, I.J.; Munro, D.B. The biology of Canadian weeds. 78. Solunum carolinense L. and Solanum rostratum Dunal. *Can. J. Plant Sci.* **1986**, *66*, 977–991. [CrossRef]
6. Zhao, J.; Solís-Montero, L.; Lou, A.; Vallejo-Marín, M. Population structure and genetic diversity of native and invasive populations of *Solanum rostratum* (Solanaceae). *PLoS ONE* **2013**, *8*, e79807. [CrossRef]
7. Celesti-Grapow, L.; Alessandrini, A.; Arrigoni, P.V.; Banfi, E.; Bernardo, L.; Bovio, M.; Brundu, G.; Cagiotti, M.R.; Camarda, I.; Carli, E.; et al. Inventory of the non-native flora of Italy. *Plant Biosyst.* **2009**, *143*, 386–430. [CrossRef]
8. Rushing, D.W.; Murray, D.S.; Verhalen, L.M. Verhalen. Weed interference with cotton (*Gossypium hirsutum*). I. Buffalobur (*Solanum rostratum*). *Weed Sci.* **1985**, *33*, 810–814. [CrossRef]
9. Lyon, D.J.; Kniss, A.; Miller, S.D. Carfentrazone improves broadleaf weed control in proso and foxtail millets. *Weed Technol.* **2007**, *21*, 84–87. [CrossRef]
10. Salas, R.A.; Burgos, N.R.; Tranel, P.J.; Singh, S.; Glasgow, L.; Scott, R.C.; Nichols, R.L. Resistance to PPO-inhibiting herbicide in *Palmer amaranth* from Arkansas. *Pest Manag. Sci.* **2016**, *72*, 864–869. [CrossRef]
11. Rangani, G.; Salas-Perez, R.A.; Aponte, R.A.; Knapp, M.; Craig, I.R.; Mietzner, T.; Langaro, A.C.; Noguera, M.M.; Porri, A.; Roma-Burgos, N. A novel single-site mutation in the catalytic domain of protoporphyrinogen oxidase ix (PPO) confers resistance to PPO-inhibiting herbicides. *Front. Plant Sci.* **2019**, *10*, 1–12. [CrossRef]
12. Jensen, P.K. Tolerance to foliage-applied herbicides in combining peas: Effect of growth stage, cultivar type and herbicide. *Crop Prot.* **1993**, *12*, 214–218. [CrossRef]
13. Falk, J.S.; Shoup, D.E.; Al-Khatib, K.; Peterson, D.E. Protox-resistant common waterhemp (*Amaranthus rudis*) response to herbicides applied at different growth stages. *Weed Sci.* **2006**, *54*, 793–799. [CrossRef]
14. Hess, F.D.; Foy, C.L. Interaction of surfactants with plant cuticles. *Weed Tech.* **2000**, *14*, 807–813. [CrossRef]
15. Matzrafi, M.; Herrmann, I.; Nansen, C.; Kliper, T.; Zait, Y.; Ignat, T.; Siso, D.; Rubin, B.; Karnieli, A.; Eizenberg, H. Hyperspectral technologies for assessing seed germination and trifloxysulfuron-methyl response in *Amaranthus palmeri* (palmer amaranth). *Front. Plant Sci.* **2017**, *8*, 474. [CrossRef]
16. Iwakami, S.; Uchino, A.; Kataoka, Y.; Shibaike, H.; Watanabe, H.; Inamura, T. Cytochrome P450 genes induced by bispyribac-sodium treatment in a multiple-herbicide-resistant biotype of *Echinochloa phyllopogon*. *Pest Manag. Sci.* **2014**, *70*, 549–558. [CrossRef]
17. Riechers, D.E.; Wax, L.M.; Liebl, R.A.; Bush, D.R. Surfactant-increased glyphosate uptake into plasma membrane vesicles isolated from common lambsquarters leaves. *Plant Physiol.* **1994**, *105*, 1419–1425. [CrossRef]

18. Kirkwood, R.C.; Hetherington, R.; Reynolds, T.L.; Marshall, G. Absorption, localisation, translocation and activity of glyphosate in barnyardgrass (*Echinochloa crus-galli* (L) Beauv): Influence of herbicide and surfactant concentration. *Pest Manag. Sci.* **2000**, *56*, 359–367. [CrossRef]
19. Peoples, M.B.; Brockwell, J.; Herridge, D.F.; Rochester, I.J.; Alves, B.J.R.; Urquiaga, S.; Boddey, R.M.; Dakora, F.D.; Bhattarai, S.; Maskey, S.L.; et al. The contributions of nitrogen-fixing crop legumes to the productivity of agricultural systems. *Symbiosis* **2009**, *48*, 1–17. [CrossRef]
20. Koch, H.J.; Trimpler, K.; Jacobs, A.; Stockfisch, N. Crop rotational effects on yield formation in current sugar beet production—Results from a farm survey and field trials. *Front. Plant Sci.* **2018**, *9*, 231. [CrossRef]
21. Davis, V.M.; Gibson, K.D.; Mock, V.A.; Johnson, W.G. In-field and soil-related factors that affect the presence and prediction of glyphosate-resistant horseweed (*Conyza canadensis*) populations collected from indiana soybean fields. *Weed Sci.* **2009**, *57*, 281–289. [CrossRef]
22. Doucet, C.; Weaver, S.E.; Hamill, A.S.; Zhang, J. Separating the effects of crop rotation from weed management on weed density and diversity. *Weed Sci.* **1999**, *47*, 729–735. [CrossRef]
23. Rubin, B.; Matzrafi, M. Weed management in israel-challenges and approaches. In *Weed Science in the Asian-Pacific Region*; Indian Society of Weed Science: Jabalpur, India, 2015; pp. 253–270.
24. Shalimu, D.; Qiu, J.; Tan, D.; Baskin, C.C.; Baskin, J.M. Seed biology of the invasive species buffalobur (*Solanum rostratum*) in Northwest China. *Weed Sci.* **2012**, *60*, 219–224. [CrossRef]

plants

MDPI

Article

Point Mutations and Cytochrome P450 Can Contribute to Resistance to ACCase-Inhibiting Herbicides in Three *Phalaris* Species

José G. Vázquez-García [1,*], Joel Torra [2], Candelario Palma-Bautista [1], Ricardo Alcántara-de la Cruz [3] and Rafael De Prado [1,*]

1 Department of Agricultural Chemistry, Edaphology and Microbiology, University of Córdoba, 14014 Córdoba, Spain; z82pabac@uco.es

2 Department d'Hortofruticultura, Botànica i Jardineria, Agrotecnio, Universitat de Lleida, 25198 Lleida, Spain; joel.torra@udl.cat

3 Centro de Ciências da Natureza, Campus Lagoa do Sino, Universidade Federal de São Carlos, Buri 18290-000, Brazil; ricardo.cruz@ufscar.br

* Correspondence: z82vagaj@uco.es (J.G.V.-G.); qe1pramr@uco.es (R.D.P.); Tel.: +34-95-721-8600 (R.D.P.)

Citation: Vázquez-García, J.G.; Torra, J.; Palma-Bautista, C.; Alcántara-de la Cruz, R.; Prado, R.D. Point Mutations and Cytochrome P450 Can Contribute to Resistance to ACCase-Inhibiting Herbicides in Three *Phalaris* Species. *Plants* **2021**, *10*, 1703. https://doi.org/10.3390/plants10081703

Academic Editors: Alessia Restuccia and Aurelio Scavo

Received: 19 July 2021
Accepted: 16 August 2021
Published: 19 August 2021

Abstract: Species of *Phalaris* have historically been controlled by acetyl-coenzyme A carboxylase (ACCase)-inhibiting herbicides; however, overreliance on herbicides with this mechanism of action has resulted in the selection of resistant biotypes. The resistance to ACCase-inhibiting herbicides was characterized in *Phalaris brachystachys*, *Phalaris minor*, and *Phalaris paradoxa* samples collected from winter wheat fields in northern Iran. Three resistant (R) biotypes, one of each *Phalaris* species, presented high cross-resistance levels to diclofop-methyl, cycloxydim, and pinoxaden, which belong to the chemical families of aryloxyphenoxypropionates (FOPs), cyclohexanediones (DIMs), and phenylpyrazolines (DENs), respectively. The metabolism of ^{14}C-diclofop-methyl contributed to the resistance of the *P. brachystachys* R biotype, while no evidence of herbicide metabolism was found in *P. minor* or *P. paradoxa*. ACCase in vitro assays showed that the target sites were very sensitive to FOP, DIM, and DEN herbicides in the S biotypes of the three species, while the R *Phalaris* spp. biotypes presented different levels of resistance to these herbicides. ACCase gene sequencing confirmed that cross-resistance in *Phalaris* species was conferred by specific point mutations. Resistance in the *P. brachystachys* R biotype was due to target site and non-target-site resistance mechanisms, while in *P. minor* and *P. paradoxa*, only an altered target site was found.

Keywords: herbicide resistance; resistance mechanisms; NTSR mechanisms; TSR mechanisms; metabolism

1. Introduction

The genus *Phalaris* L. has a complicated taxonomic history. This genus comprises 22 species of annual and perennial grasses found in open habitats of temperate regions around the world, affecting cereal, pasture fodder, and vegetable crops [1]. *Phalaris* spp. are among the most frequent annual winter weeds in Iran, and they are represented mainly by *Phalaris minor* Retz., *Phalaris paradoxa* L., and *Phalaris brachystachys* Link. [2]. These species are distributed in various regions of the country, invading mainly wheat fields and other arable crops [3,4]. In Iran, wheat is the most important crop, while weeds, mainly *Avena* spp., *Lolium* spp., and *Phalaris* spp. grasses, can reduce the annual yield by ~23% [5]. In addition, *Phalaris* spp. are highly competitive plants with high seed production [6–8]; therefore, managing these grasses is essential to avoid compromising crop yields.

Acetyl-coenzyme A carboxylase (ACCase)-inhibiting herbicides (WSSA/HRAC group 1/A) are graminicides widely used to control grass weeds, mainly in cereal fields [9]. Their post-emergence control of grass weeds in a wide variety of field crops accounts for their intensive use since their introduction [10]. These graminicides inhibit the plastid form of ACCase by blocking fatty acid biosynthesis, disrupting cell membrane integrity, and consequently causing metabolite leakage and rapid plant death [11]. Repeated applications of ACCase inhibitors, sometimes two or three times per crop season, have led to the selection of resistant plants of several grass weed species worldwide [12]. Herbicide resistance is an adaptive evolutionary process, and its dynamics and impacts depend on various factors, such as genetic diversity, weed biology, herbicidal and operational components, and other environmental factors [10]. The resistance of grass weeds to ACCase inhibitors is steadily increasing worldwide [4]. After the introduction of the chemical families of aryloxyphenoxypropionates (FOPs), cyclohexanediones (DIMs), and phenylpyrazolines (DENs), resistance to ACCase inhibitors was reported in *Lolium rigidum* and *Alopecurus myosuroides*, and within a few years resistance had spread to other grass weeds [4]. ACCase inhibitors have been the only postemergence herbicides available for selective control of grass weeds in Iran [13]. The most troublesome annual grass weeds found in Iranian wheat fields are *L. rigidum*, *Phalaris* spp. and *Avena sterilis* L. During the last decade, these species have evolved several ACCase-resistant populations showing differential patterns of cross-resistance [11,14,15].

Resistance to ACCase inhibitors in grass weeds may be due to two known mechanisms: (1) alterations of the gene encoding the herbicide target site (target-site-based resistance: TSR) [11,16]; (2) a reduction in the amount of active herbicide molecules reaching their target due to enhanced metabolism and reduced foliar penetration (non-target-site resistance—NTSR) [17,18]. Regarding TSR, different amino acid substitutions at key positions in the carboxyl transferase (CT) domain have been reported at amino acid positions 1781, 1999, 2027, 2041, 2078, 2088, and 2096 in different species [5,16,19,20]; however, there is only one documented case of ACCase overexpression [21].

Both TSR and NTSR mechanisms have been reported in *Phalaris* spp. populations with resistance to ACCase inhibitors, including in Iran and neighboring countries of the Middle East [22–24]. Because ACCase inhibitors have been applied for at least ten years in the northern regions of Iran, populations of *Phalaris* spp. may have developed increased resistance to ACCase-inhibiting herbicides. In this study, we describe and compare the biochemical and molecular aspects involved in the resistance to ACCase-inhibiting herbicides among three *Phalaris* species (*P. minor*, *P. paradoxa*, and *P. brachystachys*) collected in winter wheat fields in Iran in 2018.

2. Results

2.1. Dose Response Assay

Dry weight values gradually decreased in all *Phalaris* spp. populations as the doses of herbicides increased; however, the dry weight reductions differed between species depending on the herbicide. The S populations of the three *Phalaris* spp. were effectively controlled with lower doses than the field doses of diclofop-methyl (DM), cycloxydim, and pinoxaden (900, 250, and 40 g ai ha^{-1}, respectively). The GR_{50} values of the S populations ranged from 129.56 to 244.90 g ai ha^{-1} for DM, while for cycloxydim and pinoxaden, these values were below 10.5 g ai ha^{-1} (Table 1). The *P. brachystachys* R populations showed resistance to DM (RF = 10.31), although the RF values for cycloxydim and pinoxaden were 4.48 and 5.38, respectively. In addition, the GR_{50} values were much lower than the field doses of DIM and very similar to the field doses of DEN herbicides. R populations of *P. minor* and *P. paradoxa* showed cross-resistance to DM (RF = 7.48 and 11.87, respectively), cycloxydim (RF = 19.65 and 24.05, respectively), and pinoxaden (RF = 6.81 and 17.12, respectively).

Table 1. Estimated parameters of the log-logistic equation used to calculate the herbicide doses required for 50% reduction of the dry weights (GR_{50}) of R and S biotypes of *Phalaris* species.

Herbicides	Species	Biotype	d	b	GR_{50} (g ai ha^{-1})	RF [a]
Diclofop-methyl (FOP)	*P. brachystachys*	R	99.04 ± 2.09	3.66 ± 0.79	1336.45 ± 109.29	10.31 ± 0.68
		S	98.13 ± 2.10	1.21 ± 0.57	129.56 ± 7.52	-
	P. minor	R	95.21 ± 3.09	3.85 ± 0.77	1832.34 ± 103.98	7.48 ± 0.73
		S	96.18 ± 3.81	2.88 ± 0.75	244.90 ± 19.71	-
	P. paradoxa	R	95.69 ± 11.15	2.74 ± 0.93	2831.61 ± 129.18	11.87 ± 1.57
		S	92.17 ± 7.43	1.63 ± 0.45	238.62 ± 31.87	
Cyclocydim (DIM)	*P. brachystachys*	R	99.94 ± 2.14	1.73 ± 0.49	39.61 ± 2.12	4.48 ± 0.25
		S	95.48 ± 1.10	1.42 ± 0.57	8.84 ± 0.95	-
	P. minor	R	97.3 ± 5.55	1.77 ± 0.30	198.44 ± 25.13	19.65 ± 1.89
		S	98.98 ± 5.73	1.33 ± 0.19	10.10 ± 1.63	
	P. paradoxa	R	96.19 ± 4.44	2.09 ± 0.37	236.85 ± 22.67	24.05 ± 2.74
		S	99.44 ± 4.67	1.47 ± 0.18	9.85 ± 1.21	-
Pinoxaden (DEN)	*P. brachystachys*	R	98.50 ± 4.74	0.69 ± 0.11	41.02 ± 3.89	5.38 ± 0.74
		S	99.71 ± 2.57	1.23 ± 0.46	7.63 ± 0.41	-
	P. minor	R	97.21 ± 5.88	1.48 ± 0.25	56.87 ± 9.25	6.81 ± 1.58
		S	96.87 ± 9.25	1.34 ± 0.21	8.35 ± 1.38	-
	P. paradoxa	R	96.81 ± 4.56	1.72 ± 0.37	116.95 ± 17.28	17.12 ± 1.62
		S	97.37 ± 6.04	1.63 ± 0.28	6.83 ± 1.03	-

d is the coefficient corresponding to the upper limits of the asymptotes and *b* is the slope of the curve; [a] Resistance factor (RF = GR_{50} resistant biotype (R)/GR50 susceptible biotype (S)). FOP, aryloxyphenoxypropionates; DIM, cyclohexanediones; and DEN, phenylpyrazolines.

2.2. ^{14}C-DM Metabolism

When plants were not incubated with ABT, the ^{14}C-DM metabolism patterns were similar between the *P. minor* and *P. paradoxa* R and S populations but not in *P. brachystachys* populations. For the latter populations, the transformation rate of DM acid (toxic form) into polar conjugates of ^{14}C-DM (non-toxic metabolites) was ~3 times greater in the R biotype than in its S counterpart (Table 2). Pretreatment with ABT solution severely decreased this metabolic rate in the *P. brachystachys* R population, and the concentrations of DM acid and polar conjugates reached concentrations similar to those observed in the S population and in the other R and S populations of *P. minor* and *P. paradoxa*. In these populations, the percentages of DM, DM acid, and D-conjugates ranged from 29 to 35%, from 55 to 67%, and from 11 to 16%, respectively.

Table 2. ^{14}C-Diclofop-methyl (DM) percentage and metabolites retrieved from shoots after application to leaves in resistant (R) and susceptible (S) biotypes of *Phalaris* species 48 h after treatment (HAT).

Phalaris Species	% Extracted Radioactivity					
	DM		D-Acid		D-Conjugates	
	R	S	R	S	R	S
P. brachystachys †	12.53 ± 2.13 b	26.72 ± 2.48 a	22.43 ± 1.83 b	67.50 ± 3.17 a	65.12 ± 3.42 a	16.83 ± 1.31 a
P. brachystachys ‡	30.41 ± 4.24 a	28.71 ± 3.52 a	59.87 ± 1.05 a	58.65 ± 2.19 a	16.80 ± 0.81 b	15.61 ± 0.54 a
P. minor †	33.33 ± 1.45 b	35.18 ± 1.73 a	55.33 ± 4.33 a	56.66 ± 2.60 a	11.33 ± 2.90 b	12.37 ± 2.40 a
P. minor ‡	34.33 ± 2.02 b	33.33 ± 1.45 a	62.04 ± 2.30 a	62.13 ± 1.52 a	12.66 ± 1.45 b	11.33 ± 0.66 a
P. paradoxa †	29.11 ± 0.47 b	32.33 ± 2.22 a	57.33 ± 2.37 a	56.33 ± 1.18 a	16.43 ± 1.88 b	14.56 ± 1.08 a
P. paradoxa ‡	29.66 ± 1.18 b	30.66 ± 1.18 a	67.66 ± 1.65 a	66.66 ± 0.72 a	13.66 ± 0.54 b	12.39 ± 0.82 a

1-Aminobenzotriazole (ABT) was applied via root at 7.5 mg L^{-1} one week before DM application. † Without ABT and ‡ with ABT. Different letters per column refer to treatments that are significantly different based on the Tukey test at the 95% probability. Mean values ± standard errors of the mean are shown (n = 6).

2.3. ACCase Enzyme Activity Assay

R and S plants presented similar ACCase basal activity profiles in the absence of herbicides within *Phalaris* spp. (Figure 1). ACCase assays showed that the target site of the S populations was very sensitive to the three herbicides tested, and in all cases I_{50} was $\leq$1 µL in the R populations of *Phalaris* spp. (Table 3). ACCase insensitivity was variable, showing I_{50} values that ranged from 2.4 to 14.43 µL of herbicide. The RF values for R biotypes ranged from 12.99 to 20.78 for DM, from 5.16 to 10.91 for cycloxydim, and from 13.71 to 19.36 for pinoxaden.

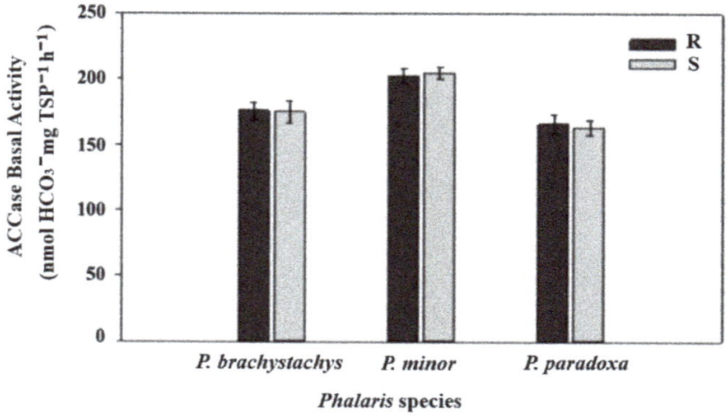

Figure 1. Basal ACCase activity (absence of herbicides) in R and S biotypes of *Phalaris* species expressed as nmol HCO_3 per mg of total soluble protein (TSP) per hour.

Table 3. Estimated parameters of the log-logistic equation used to calculate the enzyme activity levels (I_{50}) of R and S biotypes of *Phalaris* species.

Herbicide	Species	Biotype	d	b	I_{50} (µM)	RF [a]
Diclofop-methyl (FOP)	*P. brachystachys*	R	99.64 ± 0.83	0.67 ± 0.03	6.65 ± 0.78	20.78
		S	101.99 ± 0.23	0.72 ± 0.02	0.32 ± 0.05	-
	P. minor	R	94.90 ± 3.68	0.56 ± 0.09	11.04 ± 4.57	12.99
		S	96.49 ± 4.56	0.49 ± 0.07	0.85 ± 0.03	-
	P. paradoxa	R	93.98 ± 3.46	0.55 ± 0.08	9.73 ± 3.80	16.37
		S	97.91 ± 4.26	0.48 ± 0.06	0.59 ± 0.23	-
Cyclocydim (DIM)	*P. brachystachys*	R	103.11 ± 1.40	0.52 ± 0.03	2.40 ± 0.34	10.91
		S	100.53 ± 1.58	0.69 ± 0.04	0.22 ± 0.02	-
	P. minor	R	96.98 ± 2.90	1.76 ± 0.43	3.46 ± 1.87	5.16
		S	97.87 ± 2.81	0.98 ± 0.57	0.67 ± 0.91	-
	P. paradoxa	R	98.76 ± 0.76	0.87 ± 0.32	2.86 ± 0.98	6.21
		S	96.67 ± 0.67	1.11 ± 0.12	0.46 ± 0.06	-
Pinoxaden (DEN)	*P. brachystachys*	R	100.60 ± 1.40	0.90 ± 0.09	14.43 ± 1.90	18.50
		S	102.93 ± 2.22	0.62 ± 0.06	0.78 ± 0.01	
	P. minor	R	97.76 ± 2.76	0.78 ± 0.98	8.09 ± 2.80	13.71
		S	96.56 ± 1.98	0.87 ± 0.19	0.59 ± 0.07	-
	P. paradoxa	R	97.87 ± 2.98	0.76 ± 0.09	12.78 ± 2.87	19.36
		S	95.56 ± 3.98	1.09 ± 0.09	0.66 ± 0.09	-

d is the coefficient corresponding to the upper limits of the asymptotes and *b* is the slope of the curve; [a] Resistance factor (RF = GR_{50} resistant biotype (R)/GR50 susceptible biotype (S)). FOP, aryloxyphenoxypropionates; DIM, cyclohexanediones; and DEN, phenylpyrazolines.

2.4. Molecular Analysis

CT-domain sequencing of the *ACCase* gene revealed the occurrence of several amino acid substitutions in the R biotypes of *Phalaris* spp., all of them associated with resistance to ACCase-inhibiting herbicides at the target site level (Figure 2). The amino acid substitutions found in the R biotypes were Ile-1781-Leu (L) or Thr (T), Trp-2027-Cys (C), Ile-2041-Asn (N), Asp-2078-Gly (G), and Gly-2096-Ser (S). In *P. brachystachys*, only the substitution of Thr for Ile at position 1781 was found. *P. minor* showed Ile-1781-Leu, Trp-2027-Cys, and Asp-2078-Gly substitutions, while *P. paradoxa* showed the substitutions of Ile-2041-Asn, Asp-2078-Gly, and Gly-2096-Ser.

```
                                  1770       1780       1790       1800
                            ----|----| ----|----| ----|----| ----|----|
Consensus                   EIRWVIDSVV GKEDGLGVEN IHGSAAIASA YSRAYEETFT
A. myosuroides (AJ310767)   ********** ********** ********** **********
P. brachystachys R          ********** ********** T********* **********
P. brachystachys S          ********** ********** ********** **********
P. minor R                  ********** ********** L********* **********
P. minor S                  ********** ********** ********** **********
P. paradoxa R               ********** ********** ********** **********
P. paradoxa S               ********** ********** ********** **********

                                  2030       2040       2050       2060
                            ----|----| ----|----| ----|----| ----|----|
Consensus                   LFILANWRGF SGGQRDLFEG ILQAGSTIVE NLRTYNQPAF
A. myosuroides (AJ310767)   ********** ********** ********** **********
P. brachystachys R          ********** ********** ********** **********
P. brachystachys S          ********** ********** ********** **********
P. minor R                  ******C*** ********** ********** **********
P. minor S                  ********** ********** ********** **********
P. paradoxa R               ********** ********** N********* **********
P. paradoxa S               ********** ********** ********** **********

                                  2070       2080       2090       2100
                            ----|----| ----|----| ----|----| ----|----|
Consensus                   VYIPKAAELR GGAWVVIDSK INPDRIECYA ERTAKGNVLE
A. myosuroides (AJ310767)   ********** ********** ********** **********
P. brachystachys R          ********** ********** ********** **********
P. brachystachys S          ********** ********** ********** **********
P. minor R                  ********** *******G** ********** **********
P. minor S                  ********** ********** ********** **********
P. paradoxa R               ********** *******G** ********** *****S****
P. paradoxa S               ********** ********** ********** **********
```

Figure 2. Amino acid substitutions in the A and B regions of the carboxyl transferase (CT) domain of the homomeric chloroplast acetyl CoA carboxylase (ACCase) gene in susceptible (S) and resistant (R) *Phalaris* biotypes.

3. Discussion

Since the first study to detect weed biotypes resistant to herbicides in Iran in 1997, several species, including *Avena sterilis*, *L. rigidium*, *P. minor,* and *P. paradoxa,* have shown resistance to ACCase inhibitors in this country [25]. These reports of resistant weeds were found in winter cereals in Iran, mostly wheat (*Triticum aestivum* L.) and barley (*Hordeum vulgare L.*) [26]. Herbicide-resistant biotypes of *P. minor* are widely distributed around the world, while resistant biotypes of *P. paradoxa* and *P. brachystachys* are less frequent; however, these three species have been reported as having cross-resistance to all ACCase-inhibitor herbicide chemical families [4]. It is probable that the few grass herbicide options for wheat in Iran and the lack of crop rotation aided in the selection of target site mutations in *Phalaris* species and their consequent resistance to ACCase inhibitors [13].

In this study, dose–response experiments confirmed that the three *Phalaris* spp. assayed at the whole-plant level were resistant to DM, cycloxydim, and pinoxaden. These results are in agreement with those of other studies that reported similar resistance patterns to the three groups of ACCase-inhibiting herbicides, such as those in *P. minor* [16,23], *P. paradoxa* [22,26,27], and *P. brachystachys* [24].

Enhanced metabolism, regulated by detoxification enzymes Cyt-P450, glutathione-S-transferase (GST), and Glycosyl-transferase (GT), which are able to metabolize herbicides into non-toxic metabolites, may confer resistance to herbicides [28–32]; however, we found no evidence that enhanced metabolism conferred resistance in *P. minor* and *P. paradoxa*, in agreement with the results found by Cruz-Hipolito et al. [16,27], while for *P. brachystachys*, this NTSR mechanism, mediated by the enzyme complex of Cyt-P450 monooxygenases, was responsible for resistance to DM. This statement is based on the response observed in *P. brachystachys* R plants incubated with ABT, a potent inhibitor of Cyt-P450, which after incubation showed ^{14}C-DM metabolism patterns similar to those of S plants. In general, in susceptible plants, DM is de-esterified (activated) by an esterase enzyme into diclofop acid, a compound more toxic than DM [17,33]. On the other hand, in R plants, DM is metabolized into non-toxic compounds that are more polar, such as sugar ester conjugates of diclofop acid and sugar conjugates of hydroxyl-diclofop, by Cyt-P450 [24]. DM metabolization into sugar conjugates of hydroxyl-diclofop seems to be the main detoxification route of this herbicide in *P. brachystachys*. Enhanced herbicide metabolism is an NTSR mechanism documented in several grass weeds around the world, such as *L. rigidum* [17,34], *Alopecurus myosuroides* [35], and *Echinochloa phyllopogon* [36], and in recent years this mechanism has gained more attention because it is becoming an increasingly common resistance mechanism, although comprehension of this mechanism is still limited [31]. Cyt-P450-regulated herbicide metabolism can confer not only cross-resistance but also multiple resistance to modes of action of herbicides that are never used [37]. Although multiple resistance was not evaluated in the R biotype of *P. brachystachys*, its management may require more diversification than simply rotating the mode of action.

The ACCase enzyme activity results showed clear cross-resistance to the three ACCase-inhibiting herbicides in the *Phalaris* spp. R biotypes at the target site level. The resistance was high to pinoxaden and diclofop methyl and moderate to cycloxydim. According to these results, a less sensitive form of ACCase was responsible for cross-resistance to ACCase-inhibiting herbicides in R populations; however, each R population of *Phalaris* spp. showed a different susceptibility level for each herbicide. The level of insensitivity of the target enzyme to ACCase-inhibiting herbicides depends on amino acid substitutions at key positions that modify the binding of the herbicide [10,18,33]; therefore, the mutations governing herbicide resistance may differ among *Phalaris* spp.

Target site resistance is essentially caused by amino acid changes in the CT domain, which impact the effective binding of ACCase-inhibiting herbicides [10]. Substitutions in seven locations (Ile-1781, Trp-1999, Trp-2027, Ile-2041, Asp-2078, Cys-2088, and Gly-2096) in the ACCase gene have been described in grass weeds as conferring resistance to ACCase inhibitors [18]. In this study, six different substitutions were found at five key amino acid positions (Ile-1781-Leu/Thr, Trp-2027-Cys, Ile-2041-Asn, Asp-2078-Gly, and Gly-2096-Ser) in the ACCase genes of the R biotypes of the *Phalaris* spp. The substitutions found at key position 1781 occurred in *P. minor* (Ile by Leu) and *P. brachystachys* (Ile replaced by Thr). Mutation Ile-1781-Leu has been reported in R populations of *P. minor* and other grass species resistant to ACCase herbicides [16], conferring cross-resistance to the three chemical families (FOPs, DIMs, and DENs). The Ile-1781-Thr mutation has been found in *P. brachystachys* and *A. myosuroides* [18,24] and was suggested to confer resistance mainly to FOP and DEN herbicides. In the same form, Trp-2027-to-Cys substitution endows resistance to ACCase inhibitors [23], mainly to FOPs [18,19,38,39], although also confers resistance to DENs [20]. This mutation was found in *P. minor* but not *P. paradoxa* or

P. brachystachys. In addition to the mutation at the Ile-1781 position, the Asp-2078-Gly mutation was found to confer resistance to the three families of ACCase inhibitors [22,38]; therefore, this mutation explains the cross-resistance to DM, cycloxydim, and pinoxaden in our *P. minor* and *P. paradoxa* R biotypes. On the other hand, the Ile-2041-Asn and Gly-2096-Ser mutations were found only in *P. paradoxa*. These mutations have been previously reported in *P. paradoxa* from Israel [22] and Mexico [27]. The first mutation was related to resistance only to FOPs [20], while the second was associated with cross-resistance to FOPs, DIMs, and DENs [27].

4. Materials and Methods

4.1. Plant Materials

Seeds of resistant (R) populations of *Phalaris* spp. were collected from winter wheat fields from Golestan Province in Iran (Figure S1, Table S1, Supplementary Materials). The R *P. brachystachys* and *P. minor* plants survived field applications of diclofop-methyl (DM), cycloxydim, and pinoxaden herbicides, while *P. paradoxa* presented medium resistance levels to these herbicides, with low frequencies of R individuals; therefore, generation 0 (G0) plants from R *P. paradoxa* population were sown directly into trays ($40 \times 60 \times 15$ cm^3) containing a mixture of sand and peat (2:1, *v/v*) and placed in a greenhouse at 28/18 °C day/night under a 16 h photoperiod with 850 μmol m^{-2} s^{-1} photon flux density and 80% relative humidity. At the four-leaf stage, plants were treated with DM at 300 g ai ha^{-1}, one hour later with cycloxydim at 100 g ai ha^{-1}, and finally one hour later with pinoxaden at 30 g ai ha^{-1}, using a laboratory spray chamber equipped with a flat fan nozzle (TeeJet 8002 EVS) with a total output volume of 250 L ha^{-1} water at a pressure of 200 kPa. Four weeks after herbicide treatment, plant survival of the resistant accessions was estimated, and seeds produced from surviving plants were collected and stored in paper bags for subsequent recurrent selection trials. Five months later, G1 seeds of *P. paradoxa* were treated as above, although in this case with field doses of 900, 250, and 40 g ai ha^{-1} of DM, cycloxydim, and pinoxaden, respectively. A similar approach was followed to produce G2 and G3. In January 2020, dose–response assays were run to determine the multiple resistance levels in these three *Phalaris* species. Seeds of susceptible (S) populations were harvested within the same region in sites where herbicides had never been applied. Mature R and S seeds of *P. brachystachys* and *P. minor* and F3 *P. paradoxa* populations were germinated in Petri dishes (15 cm diameter) with filter paper moistened with distilled water. The Petri dishes were placed in a growth chamber at 28/18 °C (day/night) with a photoperiod of 16 h, 850 μmol m^{-2} s^{-1} photosynthetic photon flux, and 80% relative humidity. Seedlings from each population were transplanted individually into plastic pots (448 cm^3) containing sand/peat at a 1:2 (*v/v*) ratio, then they were then placed in a greenhouse at 28/18 °C (day/night) with the same photoperiod.

4.2. Dose–Response Assays

Plants of the R and S biotypes of *Phalaris* spp. were treated at the BBCH 13–14 stage [40] with different doses of DM, cycloxydim, and pinoxaden (Table 4). Ten plants of each biotype were treated per herbicide dose in a laboratory spray chamber equipped with a Tee Jet 8002E-VS flat-fan nozzle at a pressure of 200 kPa calibrated to deliver 250 L ha^{-1} of herbicide solution. Twenty-one days after treatment (DAT), the plants were cut and oven-dried to constant mass at 70 °C, then the dry weight of each plant was recorded. Dry weight data were expressed as percentages relative to the untreated controls. The herbicide rates required to reduce the dry weight by 50% (GR_{50}) were determined for each biotype and the resistance factors (RFs) were determined for each herbicide within each species as $GR_{50}(R)/GR_{50}(S)$. Dose–response assays were repeated twice.

Table 4. Herbicide treatments for dose–response assays in *Phalaris* species.

Herbicides	Rate (g a.i. ha^{-1})	
	Biotype S	Biotype R
Diclofop-methyl [a]	0, 45, 90, 180, 360, 720, 1080	0, 1000, 1500, 2000, 3000, 3500, 4000,
Cycloxydim [b]	0, 5, 10, 20, 40, 60, 100, 200	0, 50, 100, 200, 300, 400, 800,1200
Pinoxaden [c]	0, 4, 8, 16, 32, 64, 128, 256	0, 25, 50, 100, 200, 400, 600, 800

Trademark: [a] Iloxan (Bayer). [b] Focus Ultra 10% (BASF). [c] Axial (Syngenta)

4.3. ^{14}C-DM Metabolism

^{14}C-DM metabolism was studied following the methodologies described by Cruz-Hipolito et al. [27] and De Prado et al. [17]. A labeled herbicide emulsion was prepared by mixing commercially formulated DM plus ^{14}C-DM (specific activity, 95.5 kBq μmol^{-1} provided by Bayer CropScience, Leverkusen, Germany) (Table 2). The ^{14}C-DM emulsion had a specific activity of 5000 Bq μL^{-1} and was applied to the adaxial surface of the second leaves (10 droplets of 0.5 μL each) of *Phalaris* spp. plants (BBCH 13–14 stage) using a microapplicator (mod. PB600-1 Hamilton, Reno, NV, USA). Sampling of plants was carried out 48 h after treatment (HAT), starting by washing the non-absorbed ^{14}C-herbicide from the treated leaves with 1.5 mL of acetone. An aliquot of leaf wash solution was assayed for radioactivity and the remaining solution was stored (−20 °C) until analysis. Then, the shoots of each plant were ground in liquid nitrogen in a cold mortar. ^{14}C-DM and its possible metabolites were extracted from the fine powder with 4 mL of methanol (80% methanol, 4 °C). The homogenate was centrifuged (20,000× *g* for 20 min). The resulting pellet was subjected to two new extractions with methanol (80%) to recover as much ^{14}C as possible. The pellets were oven-dried and combusted in a biological oxidizer (Packard 307, Packard Instruments, Meriden, CT, USA). Supernatants were combined and evaporated to dryness (40 °C) under a stream of N_2 (10 kPa), then samples were redissolved in 500 μL of methanol (80%). DM and its metabolites in the supernatant were identified using thin-layer chromatography on 20 cm × 20 cm × 250 μm silica gel plates (Merck, Darmstadt, Germany; silica gel 60). A toluene–ethanol–acetic acid mixture (150/7/7; *v/v/v*) was used as the mobile phase. Radioactive zones were detected using a radio chromatogram scanner (Berthold LB 2821). Their chemical identity was identified by comparing RF values to those of standards (DM, 0.70; diclofop acid, 0.44; hydroxy-diclofop, 0.34; polar conjugates, 0.00). The experiment was performed twice with three replicates.

In a second experiment, three plants (BBCH 12–13 stage) per *Phalaris* spp. biotype were collected from the pots and the roots were washed with distilled water. Subsequently, plants were placed into 50 mL containers filled with a constantly aerated nutrient solution containing 7.5 mg L^{-1} 1-aminobenzotriazole (ABT), a potent inhibitor of the enzyme necessary for the metabolism of DM to non-toxic forms in plants, i.e., Cyt-P450 monooxygenases [27]. After 1 week of incubation with ABT, the second leaf of each plant was treated with ^{14}C-DM 48 HAT and the methodology described above was followed again.

4.4. ACCase Enzyme Activity Assay

Young and fully expanded leaves (6 g fresh weight) of *Phalaris* spp. R and S populations were harvested from plants at the BBCH 13–14 stage. Leaf tissue samples were ground in liquid N_2 in a mortar and added to extraction buffer (24 mL) composed of 0.1 M *N*-2-hydroxyethylpiperazine-*N'*-2-ethanesulfonic acid-KOH at pH 7.5, 0.5 M glycerol, 2 mM EDTA, and 0.32 mM PMSF. The homogenate was mixed for 3 min with a magnetic stirrer and filtered sequentially through four layers of cheesecloth. The crude extract was centrifuged (24,000× *g*, 30 min, 4 °C). The supernatant was fractionated with $(NH_4)_2SO_4$ and centrifuged (12,000× *g*, 10 min, 4 °C). Material precipitating between 35% and 45% $(NH_4)_2SO_4$ saturation was resuspended in 1 mL of S400 buffer (0.1 M Tricine–KOH at pH 8.3, 0.5 M glycerol, 0.05 M KCl, 2 mM EDTA, and 0.5 mM DTT). The clarified supernatant was applied to a desalting column previously equilibrated with S400 buffer. ACCase enzyme was eluted from the column in 2 mL of S400 buffer.

The enzyme activity was assayed by measuring the ATP-dependent incorporation of NaH [^{14}C]O_3 into an acid-stable [^{14}C]-product. The reaction product had previously been shown to be [^{14}C] malonyl-CoA [41]. Assays were conducted in (7 mL) scintillation vials containing 0.1 M tricine-KOH (pH 8.3), 0.5 M glycerol, 0.05 M KCl, 2 mM EDTA, 0.5 mM DTT, 1.5 mM ATP, 5 mM $MgCl_2$, 15 mM NaH [^{14}C]O_3 (1.22 MBq µmol^{-1}), 50 µL of the enzyme fraction, 5 mM acetyl-CoA, and different concentrations of DM acid for a final volume of 0.2 mL. Activity was assayed for 5 min at 34 °C and the reaction was stopped by adding HCl (30 µL at 4 N). A piece of filter paper was added to the reaction vial and the sample was dried (40 °C) under a stream of air. After the sample was dried, an ethanol–water mixture (1:1, *v/v*, 0.5 mL) was added to the vial. This was followed by the addition of a scintillation cocktail (5 mL). Radioactivity was determined by liquid scintillation spectrometry (LSS) on a Beckman LS 6000 TA. Background radioactivity values, measured as acid-stable counts (dpm) in the absence of acetyl-CoA, were subtracted from each treatment (16, 17, 24, 27). One unit of ACCase activity was defined as 1 µmol malonyl-CoA formed per min. Herbicide concentrations resulting in 50% inhibition of enzyme activity (I_{50} values) were estimated for each herbicide using the obtained concentration response curves. The experiment was repeated twice with three replicates. Data were pooled and a non-linear regression model (Equation (1)) was fitted to the data.

4.5. Molecular Analysis

Samples (~100 mg) of young foliar tissue were taken from 20 plants of the *Phalaris* spp. R and S biotypes. Then, the plants were treated with doses of DM, cycloxydim, and pinoxaden (900, 250, and 40 g ai ha^{-1}, respectively), as described in the dose–response assays. S plants died 21 DAT and surviving R-biotype plants were used for molecular analysis. DNA was extracted using the Speedtools Plant DNA Extraction Kit (Biotools B&M Labs S.A., Spain) following the manufacturer's instructions, and the DNA amount was quantified with a NanoDrop spectrophotometer (ThermoFisher, NanoDrop Products, Wilmington, DE, USA). The DNA sample was diluted to a final concentration of 10 ng/µL and was immediately used for polymerase chain reaction (PCR) or stored at −20 °C until use. Primers were designed to amplify regions in the CT domain known to be involved in the sensitivity to ACCase herbicides using Primer Premier 5.0 software (Premier Biosoft International, Palo Alto, CA, USA). Two sets of primers covering all seven known mutation sites in regions A (1781) and B (1999, 2027, 2041, 2078, 2088, and 2096) were designed based on the sequences of chloroplastic ACCase from *A. myosuroides* (AJ310767). Ten individual plants of each biotype were sequenced following the methods described by Golmohammadzadeh et al. [24].

4.6. Statistical Analysis

Dose–response and enzyme activity data were pooled and fitted to a non-linear regression analysis using a three-log-logistic model (Equation (1)), with the lower asymptote (c) fixed at 0 for GR_{50} and I_{50}.

$$Y = c + \frac{d - c}{1 + \exp(b(\log(x) - \log(e)))} \quad (1)$$

In the logistic model, *d* and *c* are the coefficients corresponding to the upper and lower (fixed at 0) asymptotes, respectively; *b* is the slope of the curve; *e* is the herbicide rate (or concentrations) at the point of inflection halfway between the upper and lower asymptotes; *x* (independent variable) is the herbicide dose (or concentration). Non-linear regression analyses were carried out in R software using the "drc" statistical package [42]. Data on DM metabolism and basal activity were subjected to analysis of variance (ANOVA). The percentage data were previously transformed (arcsine of the square root) to comply with the model assumptions of normally distributed errors and homogeneity of variances. The assumptions of the model were graphically inspected. Values of $p < 0.05$ were considered statistically significant and mean comparisons were made using Tukey's HSD test with a probability of 5%. ANOVA was conducted using Statistix software version 10.0 (Analytical Software, Tallahassee, FL, USA).

5. Conclusions

This study characterized cross-resistance patterns to ACCase-inhibiting herbicides in populations of *Phalaris* spp. collected in wheat fields in Iran. It should be noted that the metabolic resistance governed by Cyt-P450 found in *P. brachystachys* is a cause for concern because it may provide widespread resistance to other modes of action and compromise crop productivity; therefore, its management may require more diversification than simply rotating the herbicide's mode of action. On the other hand, choosing a suitable herbicide to control the biotypes of *Phalaris* spp. with cross-resistance patterns to ACCase inhibitors is risky due to the fact that a single amino acid substitution may lead to different levels of resistance. In addition, the continued use of ACCase inhibitors may facilitate the appearance of new Cyt-P450 genes in *P. minor* and *P. paradoxa*, as reported here for *P. brachystachys*, or new ACCase mutations able to confer higher levels of resistance to these herbicides. These results demonstrated the participation of TSR and NTSR mechanisms in the resistance to ACCase-inhibiting herbicides in the *P. brachystachys* R population, while only the TSR mechanism was involved in the resistant populations of *P. minor* and *P. paradoxa*.

Supplementary Materials: The following are available online at https://www.mdpi.com/article/10.3390/plants10081703/s1, Figure S1: Geographical position of Golestan Province in Iran and distribution map of *Phalaris* genus, Circles represent *P. minor*, squares represent *P. brachystachys*, triangles represent *P. paradoxa*, Table S1: Geographical location, collection site of *Phalaris* genus biotypes

Author Contributions: Conceptualization R.D.P.; methodology, J.G.V.-G., C.P.-B. and R.A.-d.l.C.; validation, J.G.V.-G., J.T., C.P.-B., R.A.-d.l.C. and R.D.P.; formal analysis, J.G.V.-G., J.T., C.P.-B. and R.A.-d.l.C.; investigation, J.G.V.-G., J.T., C.P.-B., R.A.-d.l.C., J.T. and R.D.P.; resources, J.T. and R.D.P.; data curation, writing—original draft preparation, writing—review and editing, and visualization, J.G.V.-G., J.T., C.P.-B., R.A.-d.l.C. and R.D.P.; supervision R.D.P.; project administration, R.D.P. and J.T.; funding acquisition, R.D.P. and J.T. All authors have read and agreed to the published version of the manuscript.

Funding: This research has also been supported by the Spanish Government, through project AGL2017-83325-C4-2-R (AEI/FEDER/UE), This research has been funded by the Asociación de Agroquímicos y Medioambiente (Spain).

Institutional Review Board Statement: Not applicable.

Informed Consent Statement: Not applicable.

Data Availability Statement: Not applicable.

Acknowledgments: Joel Torra acknowledges support from the Spanish Ministry of Science, Innovation, and Universities (grant Ramon y Cajal RYC2018-023866-I). The authors are grateful to Sajedeh Golmohammadzadeh (Department of Agronomy, Gorgan University of Agricultural Sciences and Natural Resources, Gorgan, Iran), and María J. Giménez (Departamento de Mejora Genetica Vegetal, Instituto de Agricultura Sostenible (IAS-CSIC), Cordoba, Spain) for her valuable help in reviewing this manuscript.

Conflicts of Interest: The authors declare no conflict of interest.

References

1. Alcantara, C.; Jimenez-Hidalgo, M.J.; Saavedra, M. Responses of *Phalaris minor* Rezt. and *Phalaris brachystachys* to different levels of soil water availability. *Span. J. Agric. Res.* **2010**, *8*, 1074–1083. [CrossRef]
2. Keshavarzi, M.; Khaksar, M.; Ghadam, P. Biosystematic study of *Phalaris* L. species (Poaceae) in Iran. *Taxon. Biosyst.* **2013**, *4*, 25–30.

3. Mirkamali, H. Control of *Phalaris brachystachys* and *Phalaris minor* in wheat grown in northern Iran. *Proc. Br. Crop Prot. Conf. Weeds* **1987**, *2*, 407–412.
4. Heap, I. International Survey of Herbicide Resistant Weeds. 2021. Available online: http://www.weedscience.org/ (accessed on 15 July 2021).
5. Gherekhloo, J.; Rashed Mohassel, M.H.; Mahalati, M.N.; Zand, E.; Ghanbari, A.; Osuna, M.D.; De Prado, R. Confirmed resistance to aryloxyphenoxypropionate herbicides in *Phalaris minor* populations in Iran. *Weed Biol. Manag.* **2011**, *11*, 29–37. [CrossRef]
6. Oad, F.C.; Siddiqui, M.H.; Buriro, U.A. Growth and yield losses in wheat due to different weed densities. *Asian J. Plant Sci.* **2007**, *6*, 173–176. [CrossRef]
7. Siddiqui, I.; Bajwa, R.; Huma, Z.E.; Javaid, A. Effect of six problematic weeds on growth and yield of wheat. *Pak. J. Bot.* **2010**, *42*, 2461–2471.
8. Gherekhloo, J.; Alcántara-de la Cruz, R.; Osuna, M.D.; Sohrabi, S.; De Prado, R. Assessing genetic variation and spread of *Phalaris minor* resistant to ACCase inhibiting herbicides in Iran. *Planta Daninha* **2020**, *38*. [CrossRef]
9. Délye, C.; Zhang, X.Q.; Michel, S.; Matéjicek, A.; Powles, S.B. Molecular bases for sensitivity to Acetyl-Coenzyme A Carboxylase inhibitors in black-grass. *Plant Physiol.* **2005**, *137*, 794–806. [CrossRef]
10. Powles, S.B.; Yu, Q. Evolution in Action: Plants Resistant to Herbicides. *Annu. Rev. Plant Biol.* **2010**, *61*, 317–347. [CrossRef]
11. Petit, C.; Duhieu, B.; Boucansaud, K.; Délye, C. Complex genetic control of non-target-site-based resistance to herbicides inhibiting Acetyl-Coenzyme A Carboxylase and Acetolactate-Synthase in *Alopecurus myosuroides* Huds. *Plant Sci.* **2010**, *178*, 501–509. [CrossRef]
12. Du, L.; Liu, W.; Yuan, G.; Guo, W.; Li, Q.; Wang, J. Cross-resistance patterns to ACCase-inhibitors in American sloughgrass (*Beckmannia syzigachne* Steud.) homozygous for specific ACCase mutations. *Pestic. Biochem. Physiol.* **2016**, *126*, 42–48. [CrossRef]
13. Gherekhloo, J.; Oveisi, M.; Zand, E.; De Prado, R. A review of herbicide resistance in Iran. *Weed Sci.* **2016**, *64*, 551–561. [CrossRef]
14. Sasanfar, H.; Zand, E.; Baghestani, M.A.; Mirhadi, M.J.; Mesgaran, M.B. Cross-resistance patterns of winter wild oat (*Avena ludoviciana*) populations to ACCase inhibitor herbicides. *Phytoparasitica* **2017**, *45*, 419–428. [CrossRef]
15. Zangeneh, H.S.; Chamanabad, H.R.M.; Zand, E.; Alcántara-de la Cruz, R.; Travlos, I.S.; De Prado, R.; Alebrahim, M.T. Clodinafop-propargyl resistance genes in *Lolium rigidum* Guad. populations are associated with fitness costs. *Agronomy* **2018**, *8*, 106. [CrossRef]
16. Cruz-Hipolito, H.; Fernandez, P.; Alcantara, R.; Gherekhloo, J.; Osuna, M.D.; de Prado, R. Ile-1781-Leu and Asp-2078-Gly mutations in ACCase gene, endow cross-resistance to APP, CHD, and PPZ in *Phalaris minor* from Mexico. *Int. J. Mol. Sci.* **2015**, *14*, 415–421. [CrossRef] [PubMed]
17. De Prado, J.L.; Osuna, M.D.; Heredia, A.; De Prado, R. *Lolium rigidum*, a pool of resistance mechanisms to ACCase inhibitor herbicides. *J. Agric. Food Chem.* **2005**, *53*, 2185–2191. [CrossRef] [PubMed]
18. Kaundun, S.S. Resistance to acetyl-CoA carboxylase-inhibiting herbicides. *Pest Manag. Sci.* **2014**, *70*, 1405–1417. [CrossRef]
19. Yu, Q.; Collavo, A.; Zheng, M.Q.; Owen, M.; Sattin, M.; Powles, S.B. Diversity of acetyl-coenzyme A carboxylase mutations in resistant *Lolium* populations: Evaluation using clethodim. *Plant Physiol.* **2007**, *145*, 547–558. [CrossRef]
20. Raghav, N.; Singh, R.; Chhokar, R.S.; Sharma, D.; Kumar, R. Mutations in the plastidic ACCase gene endowing resistance to ACCase-inhibiting herbicide in *Phalaris minor* populations from India. *3 Biotech* **2016**, *6*, 12. [CrossRef]
21. Laforest, M.; Soufiane, B.; Simard, M.J.; Obeid, K.; Page, E.; Nurse, R.E. Acetyl-CoA carboxylase overexpression in herbicide-resistant large crabgrass (*Digitaria sanguinalis*). *Pest Manag. Sci.* **2017**, *73*, 2227–2235. [CrossRef]
22. Hochberg, O.; Sibony, M.; Rubin, B. The response of ACCase-resistant *Phalaris paradoxa* populations involves two different target site mutations. *Weed Res.* **2009**, *49*, 37–46. [CrossRef]
23. Gherekhloo, J.; Osuna, M.D.; De Prado, R. Biochemical and molecular basis of resistance to ACCase-inhibiting herbicides in Iranian *Phalaris minor* populations. *Weed Res.* **2012**, *52*, 367–372. [CrossRef]
24. Golmohammadzadeh, S.; Rojano-Delgado, A.M.; Vázquez-García, J.G.; Romano, Y.; Osuna-Ruíz, M.D.; Gherekhloo, J.; De Prado, R. Cross-resistance mechanisms to ACCase-inhibiting herbicides in short-spiked canarygrass (*Phalaris brachystachys*). *Plant Phys. Biochem.* **2020**, *151*, 681–688. [CrossRef] [PubMed]
25. Zand, E.; Baghestani, M.A.; AghaAlikhani, M.; Soufizadeh, S.; Khayami, M.M.; PourAzar, R.; Sabeti, P.; Jamali, M.; Bagherani, N.; Forouzesh, S. Chemical control of weeds in wheat (*Triticum aestivum* L.) in Iran. *Crop Prot.* **2010**, *29*, 1223–1231. [CrossRef]
26. Collavo, A.; Panozzo, S.; Lucchesi, G.; Scarabel, L.; Sattin, M. Characterisation and management of *Phalaris paradoxa* resistant to ACCase-inhibitors. *Crop Prot.* **2011**, *30*, 293–299. [CrossRef]
27. Cruz-Hipolito, H.; Domínguez-Valenzuela, J.A.; Osuna, M.D.; De Prado, R. Resistance mechanism to acetyl coenzyme A carboxylase inhibiting herbicides in *Phalaris paradoxa* collected in Mexican wheat fields. *Plant Soil* **2012**, *355*, 121–130. [CrossRef]
28. Yuan, J.S.; Tranel, P.J.; Stewart, C.N. Non-target-site herbicide resistance: A family business. *Trends Plant Sci.* **2007**, *12*, 6–13. [CrossRef] [PubMed]
29. Owen, M.J.; Goggin, D.E.; Powles, S.B. Non-target-site-based resistance to ALS-inhibiting herbicides in six *Bromus rigidus* populations from Western Australian cropping fields. *Pest Manag. Sci.* **2012**, *68*, 1077–1082. [CrossRef]
30. Hatami, Z.M.; Gherekhloo, J.; Rojano-Delgado, A.M.; Osuna, M.D.; Alcántara, R.; Fernández, P.; Sadeghipour, H.R.; de Prado, R. Multiple mechanisms increase levels of resistance in *Rapistrum rugosum* to ALS herbicides. *Front. Plant Sci.* **2016**, *7*, 169. [CrossRef]
31. Ghanizadeh, H.; Harrington, K.C. Non-target site mechanisms of resistance to herbicides. *Crit. Rev. Plant Sci.* **2017**, *36*, 24–34. [CrossRef]

32. Nakka, S.; Jugulam, M.; Peterson, D.; Asif, M. Herbicide resistance: Development of wheat production systems and current status of resistant weeds in wheat cropping systems. *Crop J.* **2019**, *7*, 750–760. [CrossRef]
33. Fernández, P.; Alcántara de la Cruz, R.; Cruz-Hipólito, H.; Osuna, M.D.; De Prado, R. Underlying resistance mechanisms in the *Cynosurus echinatus* biotype to acetyl CoA carboxylase-inhibiting herbicides. *Front. Plant Sci.* **2016**, *7*, 449. [CrossRef]
34. Preston, C.; Tardif, F.J.; Christopher, J.T.; Powles, S.B. Multiple resistance to dissimilar herbicide chemistries in a biotype of *Lolium rigidum* due to enhanced activity of several herbicide degrading enzymes. *Pestic. Biochem. Physiol.* **1996**, *54*, 123–134. [CrossRef]
35. Letouzé, A.; Gasquez, J. Enhanced activity of several herbicide-degrading enzymes: A suggested mechanism responsible for multiple resistance in blackgrass (*Alopecurus myosuroides* Huds.). *Agronomie* **2003**, *23*, 601–608. [CrossRef]
36. Yun, M.S.; Yogo, Y.; Miura, R.; Yamasue, Y.; Fischer, A.J. Cytochrome P-450 monooxygenase activity in herbicide-resistant and -susceptible late watergrass (*Echinochloa phyllopogon*). *Pestic. Biochem. Physiol.* **2005**, *83*, 107–114. [CrossRef]
37. Jugulam, M.; Shyam, C. Non-target-site resistance to herbicides: Recent developments. *Plants* **2019**, *8*, 417. [CrossRef]
38. Liu, W.; Harrison, D.K.; Chalupska, D.; Gornicki, P.; O'Donnell, C.C.; Adkins, S.W.; Haselkorn, R.; Williams, R.R. Single-site mutations in the carboxyltransferase domain of plastid acetyl-CoA carboxylase confer resistance to grass-specific herbicides. *Proc. Natl. Acad. Sci. USA* **2007**, *104*, 3627–3632. [CrossRef]
39. Délye, C.; Matéjicek, A.; Michel, S. Cross-resistance patterns to ACCase-inhibiting herbicides conferred by mutant ACCase isoforms in *Alopecurus myosuroides* Huds. (black-grass), re-examined at the recommended herbicide field rate. *Pest Manag. Sci.* **2008**, *64*, 1179–1186. [CrossRef] [PubMed]
40. Zadoks, J.C.; Chang, T.T.; Konzak, C.F. A decimal code for the growth stages of cereals. *Weed Res.* **1974**, *14*, 415–421. [CrossRef]
41. Gronwald, J.W.; Eberlein, C.V.; Betts, K.J.; Baerg, R.J.; Ehlke, N.J.; Wyse, D.L. Mechanism of diclofop resistance in an Italian ryegrass (*Lolium multiflorum* Lam.). *Pestic. Biochem. Physiol.* **1992**, *44*, 126–139. [CrossRef]
42. Knezevic, S.Z.; Streibig, J.C.; Ritz, C. Utilizing R Software Package for Dose–response Studies: The Concept and Data Analysis. *Weed Technol.* **2007**, *21*, 840–848. [CrossRef]

plants MDPI

Article

Mechanistic Aspects and Effects of Selected Tank-Mix Partners on Herbicidal Activity of a Novel Fatty Acid Ester

Javier Campos [1,2,*], Luciana Bodelon [1], Mercedes Verdeguer [2] and Peter Baur [3]

[1] Global Innovation & Technology, Clariant, 65926 Frankfurt am Main, Germany; Luciana.Bodelon@clariant.com
[2] Instituto Agroforestal Mediterráneo, Universitat Politècnica de València, 46022 Valencia, Spain; merversa@eaf.upv.es
[3] CropPromotion, 86938 Schondorf am Ammersee, Germany; profpeterbaur@icloud.com
* Correspondence: Javier.camposcuevas@clariant.com; Tel.: +49-15735528819

Abstract: Only a limited number of contact herbicides exist in agricultural production. While systemic herbicides are more efficient also at suboptimum spray coverage with long-lasting weed control, contact herbicides provide several advantages. There is no translocation to fruits or roots of plantation and other crop, low risk for resistance development, and minor risk for spray-drift damage. Besides, synthetic products that often have toxicological or residues issues, natural fatty acids, particularly pelargonic acid (PA), have contact activity and are safer for home and garden use. We recently described a methyl capped polyethylene glycol ester of pelargonic acid (PA-MPEG) that acts independent of acid formation. Both, PA-MPEG and PA are applied at high rates per hectare to achieve excellent weed control. Here, we report about potential additives to increase PA-MPEG efficacy. The herbicidal active, 1-decanol, and the non-phytotoxic alkylated seed oil-based adjuvant, Hasten™, improved performance and outperformed a commercial PA herbicide. Both, PA-MPEG and PA appear to mainly act by the disintegration of bio-membranes besides having effects on transpiration. The main suggested effect is desiccation due to cutting the water continuum at the site of evaporation in the intercellular spaces. The synergistic action of the adjuvant Hasten™ and its practical uses are also discussed.

Keywords: contact herbicide; pelargonic acid; esterified seed oil; foliar penetration; adjuvant; tank-mix partner

Citation: Campos, J.; Bodelon, L.; Verdeguer, M.; Baur, P. Mechanistic Aspects and Effects of Selected Tank-Mix Partners on Herbicidal Activity of a Novel Fatty Acid Ester. *Plants* **2022**, *11*, 279. https://doi.org/10.3390/plants11030279

Academic Editors: Alessia Restuccia and Aurelio Scavo

Received: 30 November 2021
Accepted: 18 January 2022
Published: 21 January 2022

Publisher's Note: MDPI stays neutral with regard to jurisdictional claims in published maps and institutional affiliations.

1. Introduction

Most herbicides normally used for agricultural weed control are based on synthetic active ingredients (AIs) and possess systemic properties [1,2]. The majority of foliar AIs are low to moderately water-soluble non-electrolytes, with an octanol/water partition coefficient (log *P*) below 4, that allows acropetal movement in the xylem. Other active substances are weak organic acids or form such acids from pre-drug esters that move both basipetally and acropetally through the plant (mobility in the phloem) [2]. In many cases, systemic soil-applied herbicides are only xylem mobile after root or hypocotyl uptake, and/or when sufficiently volatile to also distribute in the gas phase of soils [2,3]. Selective weed control by such systemic herbicides is based on several complex and sophisticated plant-herbicide interactions, such as herbicide-tolerant transgenic crops, combinations with safeners, timing of applications or tolerance by a higher biomass, or the quick growth compensation of damaged assimilation areas [1,2,4–6].

In contrast to more effective and systemic AIs, only a few available herbicides are not translocated in plants. They are commonly known as contact herbicides and act only on treated organs [1,6]. The most important ones are quaternary ammonium compounds of bipyridines, and 1,1′-dimethyl-4,4′-bipyridylium dichloride (paraquat) is the most widely used [7]. The advantages of such compounds are the rare occurrence of herbicide-resistant weeds (paraquat resistance is documented in 30 plant species in 72 situations vs. 53 species in 339 scenarios for glyphosate use) and the low risk that sensitive weeds become resistant over generation [8]. Other benefits are that they are harmless to non-target plants by spray drift and their lack of translocation, e.g., to fruits in orchards [9]. However, given the bipyridines' mode of action (MoA)—formation of reactive oxygen species after accepting electrons from photosystem I, causing the inhibition of photosynthesis—there are concerns regarding safety. Nonselective, nonspecific damage and continuous action occur due to the active ingredient regeneration causing oxidation of cell components, including membranes. The toxicity and side effects of these herbicides on human and non-target organisms are substantial [7]. Therefore, paraquat and related products are increasingly banned in various regions [7,10].

Contact herbicides also include natural alternatives with practically zero toxicity such as nonanoic or pelargonic acid (PA) and related octanoic (caprylic) and decanoic (capric) acids [11–13]. Several commercial formulations of short chain fatty acids (FA) and their salts are available for weed control [14–16]. With the advantage of having extremely rapid action and being rainfast, they do not pose residual problems, and no resistant weed biotypes have been reported [17–19]. However, FA herbicides are volatile, have an unpleasant odor and are difficult to formulate [20,21]. For good and long-lasting effects on weed control, FA should be applied at extremely high rates, and repeated applications must be performed within short time intervals, which makes them very expensive for users [22–24]. Given their fast herbicidal activity, the combination with other synthetic or natural AIs is a challenge. It is often impossible to achieve a synergistic or an additive effect on weed control, particularly with systemic AIs [25,26].

We have recently shown that novel short chain FA derivatives, particularly the methyl polyethylene glycol esters (MPEG) of C_8–C_{10} FA, are as effective as the free acid, and do not merely act as pre-drug [23]. Pelargonic acid ester of methyl polyethylene glycol (PA-MPEG) is the preferred candidate [23]. PA-MPEG is liquid at the relevant temperature range, not volatile and can be used as a straight product without further formulation efforts [23]. It is also combinable in-can or in tank-mix with other herbicides and acts as a wetting agent on its own [27]. With its very low animal and human toxicity as known to date, the use is very encouraging in environmentally friendly and organic farming. The use rate and water volumes of PA-MPEG are lower than those of various current PA formulations, but are still higher than conventional herbicides [17,23]. Therefore, further reductions in the use rate and spray volume, and increasing PA-MPEG performance, are essential for it to become an alternative to the traditional contact herbicides.

Adjuvants and natural additives are often added to the spray tank of the herbicides to enhance final performance. They can modify the characteristics of the spray mixture or improve herbicidal activity [28,29]. Thus, adjuvants also have the potential to enhance PA-MPEG activity by affecting spray deposition, bioavailability and/or the effect in the transport across cuticles [28–31]. For example, a strong selected wetting agent can offer better coverage, which is fundamental for contact herbicides, or an alcohol ethoxylated can increase the mobility of a solute in cuticles [29,30].

In this study, we present the results of sustainable adjuvants and natural additives as potential enhancers of PA-MPEG weed control efficacy. New insights into the likely mode of action are also discussed.

2. Materials and Methods

2.1. Plant Species and Biological Test Conditions

Seeds of velvetleaf (*Abutilon theophrasti* M.), large crabgrass (*Digitaria sanguinalis* L.) and black nightshade (*Solanum nigrum* L.) were acquired from Herbiseed (Reading, UK). Tomato (*Solanum lycopersicum* L.), bell peppers (*Capsicum annuum* L.), soybean (*Glycine max* L.) and maize (*Zea mays* L.) seeds were kindly provided by a local farmer in Bad Soden am Taunus (Germany). Plant species seeds were sown separately in plastic pots (9 cm × 9 cm × 10 cm) containing an artificial substrate named Typ B Hawita Fruhstorfer from Hawita Gruppe

GmbH (Vechta, Germany). One week after emergence, plants were carefully thinned to one plant per pot. Weeds and crops were grown in the Clariant phytotron (Frankfurt am Main, Germany) under natural light and augmented with supplemental sodium vapor lights that produced a photosynthetic photon flux density (PPFD) of 200 mE m^{-2} s^{-1}. The photoperiod was 16/8 h light/dark. Daytime temperature was 23 ± 1 °C, and night-time temperature was kept at 18 ± 1 °C. Relative humidity (R.H.) fell within 55 ± 5% range. Enough moisture was maintained in soil until the end of trials to avoid water stress and keep plants in the optimum stage. Crop plants were irrigated with a standard fertilizer solution once a week to prevent nutrient deficiencies.

2.2. Experimental Design of the Phytotron Trials

Trials were conducted as a randomized complete block (RCB) design with four replicates per weed species. An untreated control was always included for comparison purposes. Spray preparations were applied to *D. sanguinalis* in phenological stage 22 (with two tillers) and *S. nigrum* in stage 16 (true six leaves) according to the *Biologische Bundesanstalt, Bundessortenamt und Chemische Industrie* (BBCH) scale. They were approximately 18–20 cm tall. Applications were carried out with a custom-built spray chamber (Ing-Büro CheckTec, Braunschweig, Germany) equipped with two off-center flat nozzles and a mobile carrier of the spray tank. Spray volumes of 200 and 400 L ha^{-1} were set by adapting the application carrier speed, using OC2 nozzles from Lechler GmbH (Metzingen, Germany) mounted 50 cm above the weed canopy. Spray pressure was 300 kPa.

Herbicide efficacy was visually assessed at 1, 2 and 7 days after application (DAA) on a percentage scale, where the value "0%" represents no weed control (weeds alive) and one of "100%" denotes complete weed control (weeds killed) [23].

2.2.1. Herbicidal Compound and Tank-Mix Partners Tested

The experimental herbicide was the pelargonic acid ester of methylated polyethylene glycol (PA-MPEG) which was synthesized by Clariant (Gendorf, Germany). This active is liquid and was diluted directly in tap water. For comparison purposes, PA-MPEG content is 340 g of PA acid equivalent (a.e.) per liter. Based on previous knowledge, PA-MPEG was used at 7.5% *v/v*, alone or with selected tank-mix partners, at a spray volume of 200 L ha^{-1} [17]. Phosphoric acid, D-glucose, potassium carbonate and *1*-decanol were selected as non-synthetic amendments. They were purchased from Sigma-Aldrich Chemie GmbH (Merk KGaA, Darmstadt, Germany). The tested commercial adjuvants were Synergen® TS 7, Polyglykol 400, Genapol® C 050 from Clariant ((Muttenz, Switzerland) and Hasten™ from Victorian Chemicals (Victoria, Australia) [30,31]. Table 1 is a more detailed description of the tested compounds and the applied rates.

Table 1. Test compounds, applied concentrations used and the pH of spray solutions with PA-MPEG.

Test Compounds	Description	Use Rate (% *v/v*) [1]	pH Spray Mixture [2]
Phosphoric acid	Solution–85 wt. % in H_2O.	0.60	1.9
D-glucose	97.5% purity.	1.00	5.8
Potassium carbonate	99.0% purity.	1.00	10.3
1-decanol	99.0% purity.	1.00	6.1
Synergen® TS 7 [3]	Blend of docusate sodium and ethoxylated fatty alcohol (sum 100%).	0.15	5.8
Polyglykol 400 [3]	Polyethylene glycol (PEG) with a molar weight of 400.	1.50	5.9
Genapol® C 050 [3]	Coconut fatty alcohol polyglycol ether with 5 EO.	1.00	5.8
Hasten™ [4]	Emulsifiable concentrate of esterified vegetable oil and non-ionic surfactants.	2.50	6.1

[1] Rate based on label recommendation and previous trials. [2] Pelargonic acid ester of methyl polyethylene glycol (PA-MPEG) at 7.5% *v/v*. PA-MPEG pH: 5.8. [3] Clariant, (Muttenz, Switzerland). [4] Victorian Chemicals (Coolaroo, Australia).

2.2.2. Interaction of the Hasten Concentration and the PA-MPEG Rate and Spray Volume

Two experiments were carried out to check the optimum conditions for the PA-MPEG and Hasten. The first trial was conducted with a factorial arrangement of three Hasten use concentrations (0, 1, 2, and 2.5% *v/v*), two spray volumes (200 and 400 L ha^{-1}) and a single PA-MPEG concentration of 7.5%. A commercial emulsifiable concentrate (EC) formulation of PA (Beloukha, 680 g AI L^{-1}, Belchim Crop Protection, Londerzeel, Belgium) according to the label recommendation (10.9 kg a.i ha^{-1}), was used as a standard reference [14]. In the second experiment, different PA-MPEG concentrations alone or with 2.5% Hasten were applied at 200 L ha^{-1}. Based on earlier studies, the following herbicide concentrations were employed: 5, 6, 7, 8, 9, and 10% [17]. No commercial reference was used because we explored Hasten enhancement at different PA-MPEG concentrations and PA-MPEG and the commercial PA herbicide gave closer weed control values at the selected 200 L ha^{-1} in previous trials [17].

2.2.3. Phytotoxicity of Spray Tank Partners after Spraying and Single Droplet Application

The species used for this experiment were *D. sanguinalis* and *S. nigrum*, as described in 2.1. The tested tank-mix partners were mixed in tap-water at the aforementioned concentrations (Table 1). No herbicide (PA-MPEG) was employed in the spray solutions at this time. Test preparations were applied by spraying the weeds and also using 10 µL droplets. In the first experiment, spray applications were performed in the customized spray chamber with the parameters described in Section 2.2. (OC2; 300 kPA; 200 L ha^{-1}) Treatments were replicated four times per weed species. The second trial evaluated the phytotoxicity of a single droplet application on the adaxial leaf surface of weeds at room temperature (25 °C and 56% RH). In addition, 10 µL droplets were also applied on the adaxial leaf of *A. theophrasti* (*BBCH 14*), tomato (BBCH 16), soybean (BBCH 16), and maize (BBCH 14), whose characteristics and wettability differ. An adjustable volume pipette (Eppendorf, Hamburg, Germany) was used for droplet applications. Two leaves were treated per plant, and two plants were treated for each plant species (four leaves in tall for each plant species). After droplet evaporation, plants were placed in the phytotron Phytotoxicity was visually evaluated 1 day after treatment and was then assessed as described in Table 2.

Table 2. Phytotoxicity assessment.

Rate	Description
1	No damage
2	Slight symptoms (discoloration of tissue)
3	Slight necrotic spots
4	Strong symptoms (Complete necrosis)

2.3. *Laboratory Experiments*

2.3.1. Cuticular Penetration

The penetration tests of PA-MPEG and free PA (99%, Matrica, Porto Torres, Italy), with and without additive, were studied with enzymatically isolated cuticular membranes as described in detail in the literature [32,33]. Mature leaves of apple trees (*Malus domestica* B.) cv. Gala, from plantations in Hofheim am Taunus (Germany), were taken in June, and after a quick transfer to the laboratory, 2-cm diameter discs were punched with cork borers. Leaf discs were vacuum-infiltrated in a pectinase-cellulase solution. After incubation in the enzymatic solution for about 2 weeks, cuticles were separated, cleaned with deionized water, and dried on Teflon plates.

Adaxial cuticles (stomata-free) were mounted on stainless steel chambers with original outer surfaces exposed to air, and the inner cuticle surface came into contact with the aqueous-acceptor solution from the chamber's interior [32]. Under controlled conditions (25 °C and 56% R.H.), 10 µL droplets of the spray solution were applied to the external cuticle surface of the cuticles and dried in room ambient with air circulation (approx. 25 min.). The aliquots of the acceptor solution that were drawn after different time points were analyzed by a 1290 Infinity HPLC (Agilent, Santa Clara, CA, USA).

A geometric mean of the penetration values per treatment was obtained from 10 repetitions and three measurements (6, 24, 48 h after application). The kinetics indicated the mean of active ingredient penetration across the cuticle at different times.

2.3.2. Characterization of Spray Deposits on Glass Slides

Spray deposits of PA-MPEG and PA with and without inert ingredients were characterized on silanized glass slides on parallel to the cuticular penetration test. The physical appearance of the 10 µL droplet was analyzed with a research light microscope (DM4000M, Leica, Wetzlar, Germany) in the polarized light modus connected to a high-resolution color digital camera (DFC450, Leica).

2.3.3. Scanning Electron Microscope (SEM)

The adaxial leaf samples of *D. sanguinalis* and *S. nigrum* were observed by a scanning electron microscope JSM-5600 LV from JEOL (Tokyo, Japan). Test preparations (0.3 µL droplets) were applied to leaves. After allowing for water evaporation under room conditions (25 °C and 56% R.H.), for approx. 30 min., samples were prepared as described in detail by Pathan et al. 2010 [34], frozen at −170 °C and sputtered with gold. Then samples were analyzed at different magnifications. The resulting image of the adaxial leaf surface showed minimal distortion, which allowed the product deposit characteristics on leaves to be examined.

2.3.4. Cuticular Transpiration

The effect of PA and PA-MPEG on cuticular transpiration was measured with the enzymatically isolated cuticles of mature ivy (*Hereda helix* L.) leaves. The method first determined transpiration in the steady state before treatment, and then after applying and drying the test compounds. In this experiment, 10 repetitions (individual cuticles) were performed, where each cuticle is control (without treatment) and later treated, allowing paired observations. This method is described in detail in the literature (e.g., Geyer and Schönherr [35].

2.3.5. Stomatal Conductance

The impact of PA-MPEG on leaf transpiration was investigated on bell pepper leaves because they have stomata on both leaf sides, with lower adaxial density, which are similar to *S. nigrum*, but they do not have trichomes (*S. nigrum* leaves have them) that can interfere with porometer measurements. Stomatal conductance was measured adaxially and abaxially with an SC-1 Leaf Porometer (Meter, Pullman, US) at room temperature (25 °C and 56% R.H.). PA-MPEG was only adaxially applied as 10 µL droplet, which was spread over an area of about 1 cm^2. Porometry measurement was carried out after droplet evaporation on the treated surface (adaxial) and on the abaxial side for the first four hours after application.

2.4. Statistical Analysis

The results of the efficacy trials were subjected to an analysis of variance (ANOVA) using the ARM software (Gylling Data Management Inc., Brookings, OR, USA). The individual treatment means were compared by the Student-Newman-Keuls least significant difference (LSD) test at the 5% level of significance ($p < 0.05$). Prior to the analysis, data normality and homoscedasticity were verified using the software's functionalities. Data were automatically transformed by the software whenever necessary. Data transformations are indicated in the Tables as footnotes.

3. Results and Discussion

3.1. PA-MPEG Herbicidal Activity Affected by the Test Compounds Added to the Spray Tank

We have previously reported that PA-MPEG is not just a pre-drug of PA, in contrast to the esters of auxins [23]. For example, the *iso*-octyl ester form of 2,4-D (2,4-dichlorophenoxyacetic acid) is rapidly hydrolyzed to free acid, which is the active [36]. Other PA ester derivatives have not shown any herbicidal activity [23]. A comparable extremely rapid action with symptoms of wilt and necrosis on treated organs only a few hours after application, is observed with both PA and PA-MPEG [23].

Various adjuvants have been tested with PA in other studies [37,38]. The impact of salts vs. free acid was also tested [39]. As far as we know, no significant economically reasonable PA efficacy enhancement is known by means of formulation or using tank-mix adjuvants. Since PA-MPEG is different from PA, being potentially both, pre-drug and drug, nonelectrolyte, surface-active, and having a molecular weight 2.5-fold higher [23], we also explored some potential enhancers of its herbicidal activity. Previous works have shown that PA-MPEG efficacy is best at 10% concentration with a 500 L ha^{-1} spray volume on tested weeds [17,23]. We evaluated the weed control of the test compound at the 7.5% concentration with 200 and 400 L ha^{-1} volumes on medium-sized plants (approximately 18–20 cm tall). Table 3 and Table S1 report the effect of the test compounds on weed control 2 days after PA-MPEG application. We also determined weed control after 1 week, but values were not significantly different and gave no further information on the performance of the test compounds.

Table 3. Impact of the test compounds on weed control (*Digitaria sanguinalis* and *Solanum nigrum*) 2 days after applications with 7.5% pelargonic acid ester of methylated polyethylene glycol (PA-MPEG) at 200 L ha^{-1} spray volume.

Test Compound	Concentration (%) [1]	Weed Control (%)	
		D. sanguinalis	*S. nigrum*
None		29 d *	50 bc *
1-Decanol	1.00	43 a	74 a
Phosphoric acid	0.63	33 cd	59 b
D-Glucose	1.00	29 d	47 c
Potassium Carbonate	1.00	30 cd	48 c
Genapol C 050	1.00	31 cd	52 bc
Polyglycol 400	1.50	31 cd	53 bc
Synergen TS 7	0.15	36 bc	58 b
Hasten	2.50	39 ab	68 a

[1] Concentration based on label recommendation and previous trials. * Means followed by common letters in a column are not significantly different by the Student–Newman–Keuls test at the 5% level of significance.

Except for *1*-decanol and Hasten, the proved products did not enhance the PA-MPEG activity at the selected and tested concentrations. These concentrations were chosen based on economic and potential maximum activity considerations. The extreme spray pH values of pH 2 with phosphoric acid and pH 10 with potassium carbonate (Table 1) did not affect PA-MPEG activity. Previous stability trials suggested that there is no hydrolysis of PA-MPEG in the spray liquid until two days at pH below 10 [37]. Acid hydrolysis did even not occur on the time scale of weeks to months. Spray droplet evaporation was extremely quick and bulk droplet evaporation took only minutes [40]. However, the resulting spray deposit was hydrated because PA-MPEG is a liquid. Alkaline hydrolysis was particularly considered a possible process in the more concentrated spray deposit, resulting in free PA with differentially response. Nevertheless, the obtained results (Table 1) suggested that this was not the case. Thus, potential pH changes in the apoplast by phosphoric acid or potassium carbonate did not affect PA-MPEG stability or efficacy, respectively.

Glucose, as both an osmotic agent and a potential energy source for epiphytic organisms by enhancing the decay of damaged leaves, and PEG 400 as a hygroscopic liquid were neither effective. Since coverage is the key for the herbicidal activity of contact herbicides, another promising candidate was Synergen TS 7, which is a very strong wetting and spreading agent. However, it did not sufficiently improve PA-MPEG efficacy, but obtained better results on *D. sanguinalis,* with a slight increase in weed control than in *S. nigrum,* for which no additional control was observed. No superior spreading probably took place in the presence of the high already PA-MPEG concentration, which is also a wetting agent on its own [23,27]. The lack of any significant influence on PA-MPEG efficacy by strong penetration enhancer, Genapol C 050, was at first surprising [41,42]. However, in the presence of 7.5% PA-MPEG in the spray, and assuming a cuticle/water partition coefficient of Genapol C 050 close to 1, the explanation of these results was that Genapol C 050 probably did not simply enter through leaf cuticles [43,44]. An 7.5% PA-MPEG application at a spray volume of 300 L ha^{-1} by assuming 10% coverage on leaf area, resulted in an AI load of approximately 3 mg cm^{-2}. This was 10- to 100-fold more than the cuticle specific mass and was, therefore, more than 10-fold in favor of the spray deposit [45]. Hence, the amount of Genapol C 050 sorbed in cuticles was only 1% of the cuticle mass, or lower. This was too low to increase AI mobility in cuticles, where 5% is needed to obtain significant effects [33,44]. Hasten and *1*-decanol are much more lipophilic, with a log *P* values of 4.5 for *1*-decanol and one above 8 for Hasten. Therefore, both products have a better potential to be quickly sorbed in cuticles after spray applications.

3.2. Phytotoxicity of 1-Decanol and Hasten

These two test compounds were also examined for their own phytotoxicity after spraying on complete plants and also applying individual 10 µL droplets on the leaves of the selected plants. The droplet volume was large, and thus doses per area were much higher than with real sprays. Coarse spray droplets have mean droplet diameters close to 500 µm, and most droplets are typically below the 0.5 µL volume [46,47]. While *1*-decanol is usually applied as a plant growth regulator for the sucker control of tobacco (*Nicotiana tabacum* L.) plants and also can act as a contact herbicide, Hasten is an adjuvant that boosts the efficacy of many AIs without own activity [31,48,49]. This was also reflected by the phytotoxicity results of both these products on the selected weeds and crops, respectively.

On all tested weeds and crops, *1*-decanol at 1% caused phytotoxicity symptoms when applied as an individual droplet (Figure 1A), while spray application did not lead any damage. The typical *1*-decanol use concentration, e.g., for sucker control in tobacco is above 3% [48]. In this experiment, necrotic tissues were already observed at the 1% use concentration due to the 2- to 3-fold higher dose rate per area with 10 µL droplets. *1*-decanol volatility is very high with a vapor pressure of 1387 mPa, while PA-MPEG is non-volatile [23,50]. The log *P* of PA-MPEG is around 2.5, while, *1*-decanol has 100-fold higher lipophilicity and has therefore, completely different bioavailability characteristics. Alcohols with chain lengths of C_8-C_{12} also increase mobility in the cuticles of other solutes such as PA-MPEG [44,51]. Adding of *1*-decanol to PA-MPEG enhanced its herbicidal efficacy probably by causing additional penetration besides the desiccation effect provoked by *1*-decanol itself.

The situation was completely different with the adjuvant Hasten. This product is very safe according to safety data sheet information and also possesses no herbicidal activity at typical use concentration up to 1% [31,49,52,53]. When sprayed on plants, no phytotoxicity symptoms were observed up to the highest tested concentration (5%), which also indicates no own herbicidal activity. No phytotoxicity symptoms were observed on five of the six selected weeds and crops tested after droplet application of 2.5% Hasten (Figure 1B). Only *S. nigrum* showed necrotic spots after applying 10 µL droplets, while spraying Hasten at even 5% exhibited no symptoms such as leaf curling, yellowing, or necrosis. The *S. nigrum* results were relevant because it is considered a strong allelopathic plant with herbicidal active secondary metabolites [54]. The glandular trichomes (Figure 1C) that exist on both leaf sides contain products such as flavonoids and alkaloids that could be released and enter leaf tissue to cause phytotoxicity even on *S. nigrum* itself [54]. Therefore, it is likely that the presence of Hasten caused the release of the allelopathic herbicidal compounds from *S. nigrum* trichomes, and it enabled to enter the mesophyll tissue of the leaf. When trichomes were damaged by a razor blade, no symptoms such as necrosis were developed in the absence of Hasten, but symptoms appeared when 2.5% Hasten was later applied. Apparently, Hasten also acted as a penetration enhancer for substances in glandular trichomes, but it did not cause phytotoxicity on its own [53].

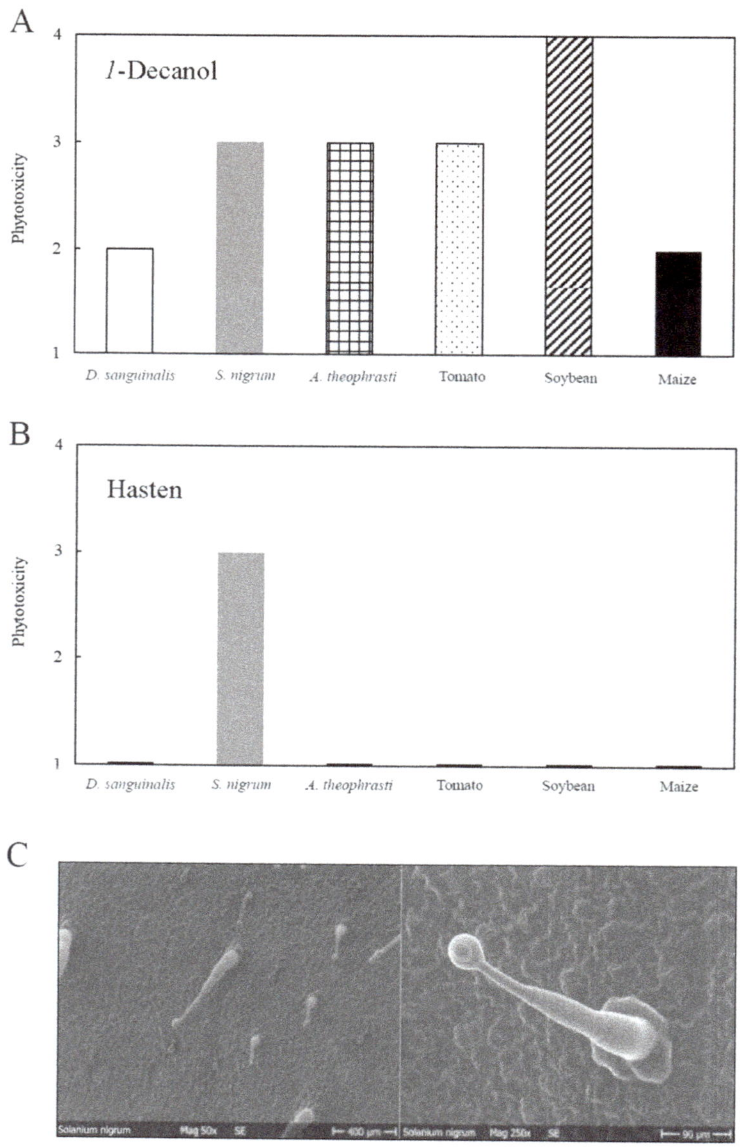

Figure 1. Phytotoxicity of the 10 µL droplet application on the adaxial side of mature leaves after 24 h. (**A**) 1-Decanol at 1%. (**B**) Hasten at 2.5%. (**C**). SEM micrographs showing the glandular trichomes on the adaxial leaf of *Solanum nigrum*.

3.3. Concentration Dependence of the Adjuvant Effect on PA-MPEG Herbicidal Activity

After considering the positive effect of Hasten on the herbicidal activity of PA-MPEG, this adjuvant was used in further experiments. As previously mentioned, PA-MPEG was tested at 7.5% instead of at the previously reported optimum 10% PA-MPEG to better differentiate its herbicidal activity [23]. The employed benchmark, Beloukha, was a high load PA formulation (680 g/L) applied at the recommended use rate and water volumes (Figure 2 and Table S2). With a 400 L ha^{-1} spray volume, Beloukha (10.9 kg a.i ha^{-1}) showed an inferior efficacy than PA-MPEG at 7.5% (10.2 kg a.e ha^{-1}). However, at 200 L ha^{-1} both PA-MPEG 7.5% (5.1 kg a.e ha^{-1}) and Beloukha achieved very low level weed control. The addition of Hasten at 1.0–2.5% to the spray tank positively affected the PA-MPEG efficacy in a concentration-dependent way. Weed control increased for both the water volumes tested up to 20%, with slightly stronger effects on weed control percentage at 400 L ha^{-1}. With Hasten, PA-MPEG performance was clearly boosted and superior to the commercial PA formulation (Figure 2 and Table S2). Obviously, the effect of water volumes dominated the differences in PA vs. PA-MPEG, and the adjuvant's impact on PA-MPEG efficacy [17]. On the other hand, PA-MPEG efficacy was increased by the adjuvant even at 400 L ha^{-1}, reaching a higher weed control level after 2 days, and being approximately 30% better than the benchmark.

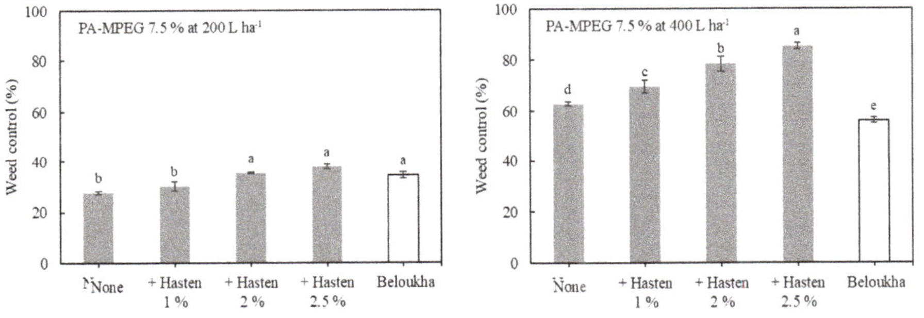

Figure 2. Effect of the Hasten rate and spray volume on the weed control of *Digitaria sanquinalis* with pelargonic acid ester of methylated polyethylene glycol (PA-MPEG) at 7.5%. Visual assessment at 2 days after application. Common letters above bars indicate that the means are not significantly different by the Student-Newman-Keuls test at the 5% level. Bars represent standard errors. Beloukha's rate was 10.9 kg a.i. ha^{-1}.

The beneficial impact of the larger water volume was not related to coverage *per se*, i.e., the absolute area of the treated weed plant surfaces. This was practically complete at 200 L ha^{-1} for *D. sanguinalis* after treatment with both products, PA-MPEG and the benchmark. Both, spray liquid adhesion and capillary wetting of monocots with surfactant solutions below critical surface tension (35 mM m^{-1}) ensure full treated leaf area coverage [55]. Spraying with fluorescent tracers displayed full coverage [17]. At a higher load liquid (400 L ha^{-1}) there was more run-off to the leaf angles of the vertical grass leaves [39]. So, the better performance of 7.5% PA-MPEG at the 400 L ha^{-1} spray volume could be caused by the higher dose per area of PA-MPEG and its basipetal run-off of spray liquid with uneven distribution [17,55].

At the 2.5% adjuvant concentration and the 200 L ha^{-1} spray volume, we examined the optimum use concentrations of PA-MPEG on *D. sanguinalis* and *S. nigrum* (Figure 3 and Table S3). Previous results have not shown either herbicidal activity or phytotoxicity at 3% PA-MPEG [17]. While the maximum control with 10% PA-MPEG was not exceeded much by adding Hasten at 2.5%, there was a consistent increase at lower use PA-MPEG concentration. The results suggest that 7.5% PA-MPEG plus the adjuvant was comparable to 10% PA-MPEG. The enhancing effect of Hasten was given at all PA-MPEG concentrations for both weeds, but there was no hint for a particular ratio for optimum increases.

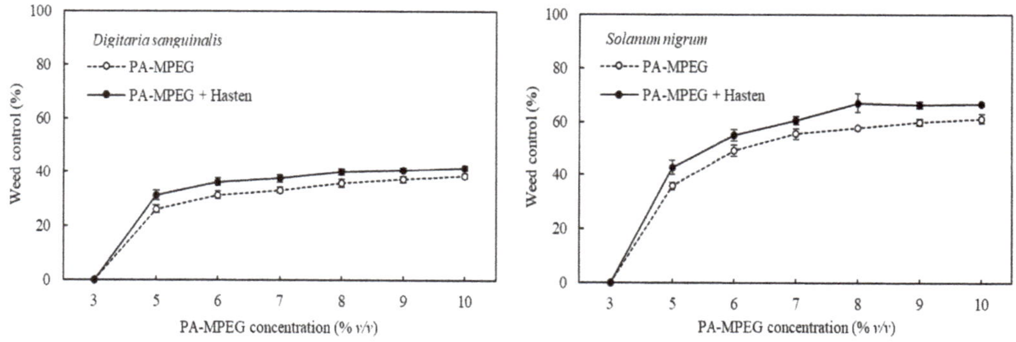

Figure 3. Effect of 2.5% Hasten on the weed control of *Digitaria sanguinalis* and *Solanum nigrum* at different concentrations of pelargonic acid ester of methylated polyethylene glycol (PA-MPEG), 2 days after application. Spray volume of 200 L ha^{-1}. Bars represent standard errors.

3.4. Pelargonic Acid and PA-MPEG Cuticular Penetration

Previous studies have demonstrated that Hasten does not significantly enhance PA activity [37]. We have also found that it is neutral and sometimes antagonistic for salts such as ammonium PA and C_8–C_{10} FA at equal amounts of active substance per hectare [39]. In contrast, PA-MPEG performance was significantly improved by Hasten (Figures 2 and 3 and Table S3). As wetting agent related effects and others such as drift or volatility can be excluded, we checked the potential effects on PA-MPEG penetration compared to free PA. Figure 4 illustrates how Hasten acts as penetration enhancer of PA-MPEG but conversely decreased PA penetration, which was faster penetrating than PA, with about 30 % more absorbed a few hours after application. The penetration level, at the very high doses per area, corresponding to the 25 g a.e. L^{-1} solute concentration, was also generally high. The difference in penetration correlated well with the observed shifts in herbicidal activity in the presence of the adjuvant. Hasten belongs to the class of alkylated or methylated seed oil (MSO) type adjuvants that are swelling agents for cuticles [32]. This increases the mobility of the AI and a several-fold faster penetration through more liquid-like cuticles [44].

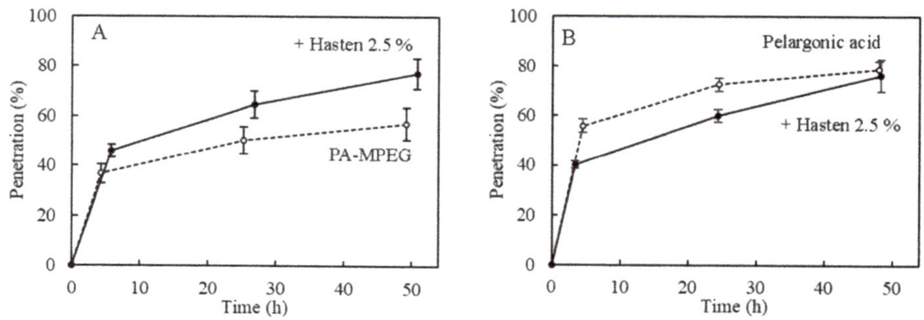

Figure 4. The impact of Hasten at 2.5% on the cuticular penetration of (**A**) pelargonic acid ester of methylated polyethylene glycol (PA-MPEG) at 25 g a.e. L^{-1} and (**B**) straight pelargonic acid (PA) at 25 g a.i. L^{-1}. Each curve is the mean of seven to nine repetitions. (Temperature was 25 °C and relative humidity was 56%). Bars represent standard errors.

In the presence of Hasten, PA-MPEG reached the PA penetration level after 2 days, while it was still slightly below PA at shorter times. The penetration of both PA and PA-MPEG was very fast and similar to the quickly penetrating alcohol ethoxylates, such as the previously mentioned Genapol C 050, where a fraction of 60–80% of the applied amount penetrates within one day the cuticle of different species [51]. Free PA is a very small molecule with 110 cm^3 mol^{-1} [20]. The PA mobility is so high that adjuvants such as Hasten cannot increase mobility [44]. The negative effect of Hasten on PA penetration was similar to the one observed effect with Genapol C 050 on PA-MPEG, and it is probably related to a change of partitioning coefficient [32,44]. The mixtures of Hasten with large amounts of PA reduced the sorption in cuticles. In contrast to the free PA, PA-MPEG is a bulkier molecule with a molecular weight over 400 g mol^{-1} which results in a 3-fold bigger molar volume than PA. Although linear molecules were found to have higher mobilities, this does not apply to PA-MPEG, which has a central ester group and, thus, sp2 hybridization. It was found that ethoxylates of fatty acids (esters) having the same degree of ethoxylation penetrate much slower than alcohol ethoxylates (ethers). This structure and molecular weight caused a 10-fold lower mobility in cuticles. Clearly, the swelling effect of Hasten increased mobility in such a way that enhanced cuticular permeability resulted in better weed control [44].

3.5. Characterization of Spray Deposits on Glass Slides

Spray deposits showed a homogeneous and amorphous PA-MPEG and Hasten mixture, which indicates good bioavailability (Figure 5). The light microscopic evaluation of deposits on glass slides suggested that PA always forms some crystalline particles with counterions from water, which were also visible in the presence of Hasten (Figure 5).

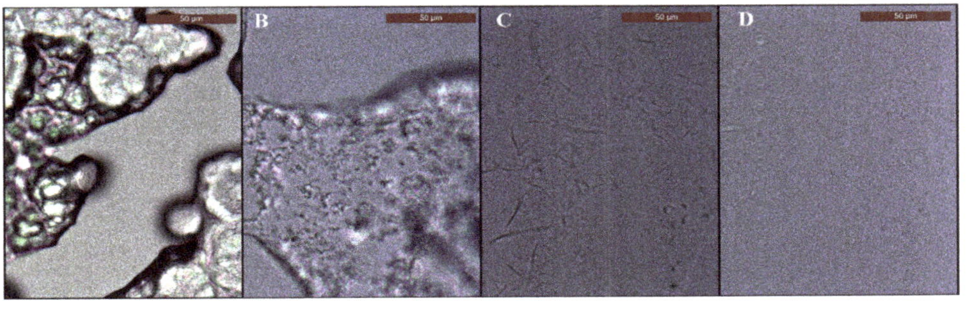

Figure 5. Optical microscope images of spray deposits on glass slides. (**A**) Straight pelargonic acid (PA) at 25 g a.i. L^{-1}, (**B**) PA plus Hasten 2.5%, (**C**) pelargonic acid ester of methyl polyethylene glycol (PA-MPEG) at 25 g a.e. L^{-1}, and (**D**) PA-MPEG plus Hasten 2.5%

3.6. Scanning Electron Microscope (SEM)

Figure 6 shows for single droplet application that Hasten closed the gaps not covered with PA-MPEG in the spray droplet deposits on *S. nigrum* leaves, which could result in the recovery of the leaf tissue below. On *D. sanguinalis*, the deposit area also appeared more homogeneous than with the straight PA-MPEG application.

3.7. The Mechanistic Aspects and High Use Rates of PA-MPEG and PA

The high use rates of both PA-MPEG and PA are still a limiting factor for their use in conventional crop production, even though the products are comparable in costs per hectare to for example, glufosinate, and are more environmentally friendly. PA is quite volatile, and loss of product could be one reason for the necessity of high PA rates. However, also for non-volatile PA-MPEG, high rates are needed for good weed control [17]. So, this property does not appear to be very relevant. The phytotoxic symptoms with both PA and PA-MPEG typically start with wilting several hours after application and desiccation of the treated plant parts and, if sufficiently extensive, weed death. A second application is sometimes needed to exhaust weeds. The generally suggested MoA for PA and PA-MPEG are changes in the leaf epidermal structure, such as erosion of surface waxes, a related higher leaf transpiration, the disintegration of bio-membranes, and likely as a consequence, decreased photosynthesis [13,37].

Figure 6. SEM micrographs of the *Solanum nigrum* (upper row) and *Digitaria sanguinalis* (lower row) leaves, 2 h after the 0.3 μL droplet application of pelargonic acid ester of methyl polyethylene glycol (PA-MPEG) at 25 g a.e L^{-1} with (**C**,**F**) and without (**B**,**E**) Hasten 2.5%. Untreated leaves (**A**,**D**).

Unexpectedly, the application of individual droplets at the very high PA and PA-MPEG concentrations of did not lead to striking changes in the epidermal fine structure (Figure 6). Later observable epidermal changes were the consequences of the destruction of the mesophyll structure, and thus, the quick initial increase in transpiration was not causing lethal desiccant effects. This was also suggested by the fact that the still high used PA-MPEG concentration of about 30 g L^{-1} increased transpiration, but did not cause any phytotoxicity, even though this concentration resulted (see the calculation above) in a 5-fold higher dose per area than the cuticle mass (0.03–0.3 mg cm^{-2} for different species). Neither the increased efficacy with the used concentration nor the Hasten effect suggested a key role of transpiration. Although the transpiration effects of PA and PA-MPEG were measurable, they did not give a clear picture. For example, about 2 h after the adaxial application of PA-MPEG to amphistomatic pepper plants (experimentally preferred to *S. nigrum* due to lack of trichomes), adaxial transpiration rose from 20 (SD 4.6) mmol m^{-2} s^{-1} for untreated plants to 35 (SD 6.8) mmol m^{-2} s^{-1}. In contrast, on the abaxial side with a higher stomatal number, transpiration rates decreased on average from about 66 mmol m^{-2} s^{-1} for the untreated plants to 44 mmol m^{-2} s^{-1}, and the daily maximum transpiration rates were generally much higher with values of around 200 mmol m^{-2} s^{-1}.

Cuticular transpiration was also measured with enzymatically isolated cuticles [56]. The addition of PA-MPEG to very dense common ivy cuticles increased transpiration by more than 10-fold. Cuticular transpiration is only minor contributing to total leaf water loss (a few percent with open stomata) of mature leaves. Not even significantly increased cuticular transpiration alone can explain phytotoxicity. However, for young expanding leaves or growing weeds, cuticular transpiration is the main source of water loss, and at least juvenile plant organs might be completely damaged.

Thus, we conclude that even when cuticular and/or stomatal transpiration increased, the observed wilting and desiccation symptoms were not caused by them. Instead, we suggest a combination of three factors that makes PA and PA-MPEG contact herbicides with desiccant action. First, high amounts of both herbicides are needed because the main targets are thylakoid membranes or chloroplast lipids. Plants can flexibly react to temperature or other stress factors and permanently repair membranes, having distinct membrane lipids and lipid metabolism with galactolipids and sulfolipids that directly come from photosynthetic products [57,58]. Disturbing this key plant function and large lipid compartment and the permanently running repair mechanism, requires high amounts, and a 30% load of the lipid. To cause such damage, PA and PA-MPEG need to reach that target. To do so, not only cuticular penetration, but also migration in the apoplast of cell walls and the xylem are required. Some alcohol ethoxylates (non-ionic surfactants) have been reported to increase transpiration at 0.5%, which cause phytotoxicity as necrotic tissues [43]. However, not even very high use concentrations (of these surfactants) such as that typical for PA or PA-MPEG, bring about a comparable desiccant effect such as that with PA and PA-MPEG. The putative reason is that such alcohol ethoxylates are not mobile in the mesophyll, and do not even allow locosystemic movement in treated leaves. In contrast, PA and PA-MPEG are probably more mobile, given their lower affinity to bio-membranes, but high amounts are still needed to disturb the thylakoid assembly. PA is a small anionic solute that is particularly mobile. We still do not know whether PA-MPEG is hydrolyzed after entering the epidermis or the mesophyll to form free PA, but with an octanol/water partition coefficient of a log *P* value of 2.5, it is already as nonmetabolized PA-MPEG a very mobile solute once it has penetrated the cuticle [59]. Yet still, high amounts of PA and PA-MPEG continue to be needed to disturb the thylakoid assembly.

Another aspect that we consider to be highly relevant, and even causal for desiccant action, is that as large amounts of PA or PA-MPEG enter the plant tissue, the bio-membranes in the chloroplast disintegrate and cause the release of lipids galactolipids and sulfolipids as well as FA from membrane lipids and/or PA to the cell walls and xylem and thus the total apoplast [11,57]. The surface tension of PA-MPEG is below 30 mN m^{-1}, and both phospholipids and soaps have surface tensions of 35–40 mN m^{-1}. This breaks the cohesion of water such that water supply for transpiration is reduced and causes wilting rather than increased transpiration rates at the cuticle or epidermal level. This could be the real cause of the observed desiccation effect. Further research is already underway to confirm these findings.

This should not be mixed up with recent observations showing that surfactants such as phospholipids in the xylem can contribute to stabilize water flow at negative pressure [60]. Schenk et al. [60] suggest that xylem surfactants have a high affinity to lipophilic areas in vessels, and while surfactants increase the probability and number of air bubbles, they can reduce embolism by their action to limit bubble size and ease re-bubble dissolution in xylem sap. In contrast, we suggest that the large amounts of surfactants resulting from either PA or PA-MPEG applications together with those released from membrane disintegration, potentially destroy this very stable mechanism of water uptake via the cohesion-tension. The site of actions is in the cell wall where the capillaries of cellulose fibers supply water that evaporates at the interface to the intercellular. Negative effects of surfactants on water transport by increased embolism have been shown to occur in the xylem, but not in the context of the MoA for killing weeds by affecting the capillary system in mesophyll cell [61]. With PA, water can no longer follow the steep gradient to the more negative water potential of unsaturated air. This could be a new target for other novel contact herbicide principles by interfering with the water cohesion-tension in the apparent free space of the cell wall.

4. Conclusions

Contact herbicides based on free PA have several disadvantages such as high use rates and water volumes, bad smell, and irritant factors, and they also require complex formulation. A novel ester, PA-MPEG, can reach and outperform the best PA benchmark, and is a ready-to-apply liquid with exceptional use properties. In contrast to free PA, the PA-MPEG is a significantly larger molecule that benefits from penetration enhancing adjuvants. An alkylated seed oil product (Hasten) increased cuticular permeability, directly giving better weed control. Even though performance was boosted with this adjuvant, the high application rates remained almost unchanged. Therefore, the product is preferred for precision application to specific sites, such as furrow application or with weed detection application systems, and drone application appears particularly interesting.

Supplementary Materials: The following supporting information can be downloaded at: https://www.mdpi.com/article/10.3390/plants11030279/s1, Table S1. Weed control percentage (visual control ratings) of *Digitaria sanguinalis* and *Solanum nigrum*, 2 days after application of pelargonic acid ester methyl polyethylene glycol (PA-MPEG) at 7.5% alone and with tested compounds. Spray Volume 200 L ha^{-1}; Table S2. Weed control percentage (visual control ratings) of *Digitaria sanguinalis* 2 days after application of pelargonic acid ester methyl polyethylene glycol (PA-MPEG) at 7.5% with the addition into the spray tank of Hasten at different concentrations; Table S3. Weed control percentage (visual control ratings) of *Digitaria sanguinalis* and *Solanum nigrum*, 2 days after application of pelargonic acid ester methyl polyethylene glycol (PA-MPEG) at 7.5% alone and with tested compounds. Spray Volume 400 L ha^{-1}.

Author Contributions: Conceptualization, J.C., L.B. and P.B.; methodology, J.C.; formal analysis, J.C. L.B., P.B; investigation, J.C. and P.B.; data curation, J.C., and P.B.; writing—original draft preparation J.C., M.V. and P.B.; writing—review and editing, J.C., L.B., M.V. and P.B.; supervision, P.B. All authors have read and agreed to the published version of the manuscript.

Funding: This work was supported by Clariant.

Institutional Review Board Statement: Not applicable.

Informed Consent Statement: Not applicable.

Data Availability Statement: Not applicable.

Conflicts of Interest: The authors declare no conflict of interest.

References

1. Hatcher, P.E.; Froud-Williams, R.J. *Weed Research: Expanding Horizons*, 1st ed.; John Wiley & Sons Ltd.: West Sussex, UK, 2017; p. 456.
2. Jeschke, P.; Witschel, M.; Kräamer, W.; Schirmer, U. *Modern Crop Protection Compounds*, 3rd ed.; Wiley-VCH Verlag GmbH: Weinheim, Germany, 2019; p. 1784.
3. Baur, P.; Aponte, J. Co-penetration of actives and adjuvants and its significance for the matched pair liaison. In *Retention, Uptake and Translocation of Agrochemicals in Plants*; Myung, K., Satchivi, N.M., Kingston, C., Eds.; American Chemical Society: Washington, DC, USA, 2014; pp. 23–39.
4. Duke, S.O.; Powles, S.B. Glyphosate: A once-in-a-century herbicide. *Pest. Manag. Sci.* **2008**, *64*, 319–325. [CrossRef]
5. Gatzweiler, E.; Krähmer, H.; Hacker, E.; Hills, M.; Trabold, K.; Bonfig-Picard, G. *Weed Spectrum and Selectivity of Tembotrione under Varying Environmental Conditions. 25. Deutsche Arbeitsbesprechung über Fragen der Unkrautbiologie und -bekämpfung, Braunschweig, Germany*; Nordmeyer, H., Ulber, L., Eds.; Julius-Kühn-Archiv: Braunschweig, Germany, 2012; pp. 385–391.
6. Zimdahl, R. *Fundamentals of Weed Science*, 5th ed.; Academic Press: London, UK, 2018; p. 758.
7. Eddleston, M. Paraquat and Diquat. In *Critical Care Toxicology*, 2nd ed.; Brent, J., Burkhart, K., Dargan, P., Hatten, B., Bruno Megarbane, B., Palmer, R., White, J., Eds.; Springer: Cham, Switzerland, 2017; pp. 1855–1874.
8. The International Survey of Herbicide Resistant Weeds. Available online: https://weedscience.org (accessed on 1 October 2021).
9. World Health Organization (WHO). *Environmental Health Criteria 39, Paraquat and Diquat*; WHO: Geneva, Switzerland, 1984.
10. Dinham, B. Why Paraquat should be banned. *Outlooks Pest Manag.* **2004**, *15*, 268–271. [CrossRef]
11. Fukuda, M.; Fujimori, T.; Tsujino, Y.; Wakabayashi, K.; Böger, P. Phytotoxic activity of middle-chain fatty acids I: Effects on cell constituents. *Pesti. Biochem. Physiol.* **2004**, *80*, 143–150. [CrossRef]
12. Lederer, B.; Fujimori, T.; Tsujino, Y.; Wakabayashi, K.; Böger, P. Phytotoxic activity of middle-chain fatty acids II: Peroxidation and membrane effects. *Pesti. Biochem. Physiol.* **2004**, *80*, 151–156. [CrossRef]
13. Coleman, R.; Penner, D. Desiccant Activity of Short Chain Fatty Acids. *Weed Technol.* **2006**, *20*, 410–415. [CrossRef]
14. Belchim Crop Protection. Beloukha®. Available online: https://belchim-agro.de (accessed on 21 October 2018).
15. Westbridge Agricultural Products. SUPPRESS®HERBICIDE EC. Available online: https://westbridge.com (accessed on 20 June 2020).
16. Certis Europe. Finalsan. Available online: https://www.certiseurope.com (accessed on 20 June 2020).
17. Campos, J.; Mansour, P.; Verdeguer, M.; Baur, P. Rate, Spray Volume, and Application Parameters of a Novel Contact Herbicide. 2021; submitted.
18. Ciriminna, R.; Fidalgo, A.; Ilharco, L.; Pagliaro, M. Herbicides based on pelargonic acid: Herbicides of the bioeconomy. *Biofuels Bioprod. Biorefin.* **2019**, *13*, 1476–1482. [CrossRef]
19. Chiotti, D.; Leonard, R.; Schlenk, D.; Wilen, C.; Schiff, K. *Alternatives to Glyphosate for Vegetation Management in Los Angeles County*; Technical Committee Report. Southern California Coastal Water Research Project; Southern California Coastal Water Research: Costa Mesa, CA, USA, 2020.
20. BPDB: Bio-Pesticides DataBase—Pelargonic Acid. Available online: http://sitem.herts.ac.uk (accessed on 1 October 2021).

21. Baur, P.; Bauer, M.; Bodelon, L.; Campos-Cuevas, J.; Giessler, S.; Hövelmann, F. Fatty Acid Derivatives for Use as Herbicides. International Patent WO 2019/162484 A1, 25 February 2019.
22. Merritt, J.; Webber, C.; Shrefler, J. The Economic Value of Pelargonic Acid as a Natural Herbicide in Sweet Bell Peppers. In Proceedings of the Southern Agricultural Economics Association's 2016 Annual Meeting, San Antonio, TX, USA, 6–9 February 2016.
23. Campos, J.; Verdeguer, M.; Baur, P. Capped polyethylene glycol esters of fatty acids as novel active principles for weed control. *Pest Manag. Sci.* **2021**, *77*, 4648–4657. [CrossRef] [PubMed]
24. Crmaric, I.; Keller, M.; Krauss, J.; Delabays, N. *Efficacy of Natural Fatty Acid Based Herbicides on Mixed Weed Stands. 28. Deutsche Arbeitsbesprechung über Fragen der Unkrautbiologie und -bekämpfung, Braunschweig, Germany*; Nordmeyer, H., Ulber, L., Eds.; Julius-Kühn-Archiv: Braunschweig, Germany, 2018; pp. 328–333.
25. Wehtje, G.; Altland, J.; Gilliam, C. Interaction of glyphosate and pelargonic acid in ready-to-use weed control products. *Weed Technol.* **2009**, *23*, 544–549. [CrossRef]
26. Little, K.; Nadel, R. Testing pelargonic acid and pyraflufen-ethyl with glyphosate as alternatives to paraquat dichloride for the preparation of fire-break tracer lines at Underberg, South Africa. *South. For. J. For. Sci.* **2014**, *74*, 67–73. [CrossRef]
27. Baur, P.; Bauer, M.; Bodelon, L.; Campos-Cuevas, J.; Giessler, S.; Hövelmann, F. Fatty Acid Derivatives for Improving the Effect of Agrochemical Actives. International Patent WO 2020/173717 A1, 14 February 2020.
28. Hazen, J.L. Adjuvants—Terminology, Classification and Chemistry. *Weed Technol.* **2000**, *14*, 773–784. [CrossRef]
29. Pacanoski, Z. Herbicides and Adjuvants. In *Herbicides, Physiology of Action, and Safety*; Price, A., Kelton, J., Sarunaite, L., Eds.; IntechOpen: London, UK, 2015; pp. 125–147.
30. Clariant. Brochure Crop Solutions. Available online: https://clariant.com (accessed on 13 October 2021).
31. Victoria Chemical Company. HASTEN Spray Adjuvant. Available online: https://vicchem.com/ (accessed on 20 June 2020).
32. Baur, P.; Bodelon, L.; Lowe, B. Novel Adjuvant Blend for In-Can and Tank-Mix Application Potential Combining Sustainability and Performance as Versatile Penetration Enhancer. In *Pesticide Formulation and Delivery Systems: 33rd Volume, "Sustainability: Contributions from Formulation Technology"*; Sesa, C., Ed.; ASTM International: West Conshohocken, PA, USA, 2014; pp. 87–106.
33. Cronfeld, P.; Lader, K.; Baur, P. Classification of Adjuvants and Adjuvant Blends by Effects on Cuticular Penetration. In *Pesticide Formulations and Application Systems*; Viets, A.K., Tann, R.S., Mueninghoff, J.C., Eds.; ASTM International: West Conshohocken, PA, USA, 2001; Volume 20, pp. 80–94.
34. Pathan, A.; Bond, J.; Gaskin, R. Sample preparation for SEM of plant surfaces. *Mater. Today* **2010**, *12*, 32–43. [CrossRef]
35. Geyer, U.; Schönherr, J. In Vitro Test for Effects of Surfactants and Formulations on Permeability of Plant Cuticles. In *Pesticides Formulation. Innovations and Developments*; Comstock, J., Ed.; American Chemical Society: Washington, DC, USA, 1988; pp. 22–33. [CrossRef]
36. Zhang, S.; Ji, M.; Gu, Z.; Yang, Q. The hydrolytic and photolytic properties of 2,4-D isooctyl ester. *Chin. J. Pestic. Sci.* **2019**, *21*, 125–130. [CrossRef]
37. Coleman, R.; Penner, D. Organic Acid Enhancement of Pelargonic Acid. *Weed Technol.* **2008**, *22*, 38–41. [CrossRef]
38. Webber, C.; Shrefler, J.W.; Taylor, M.J. Adjuvants Affect Duckweed (*Lemna minor*) Control with Pelargonic Acid. *J. Agric. Sci.* **2014**, *12*, 1–8. [CrossRef]
39. Campos, J.; (Global Innovation & Technology, Clariant, Frankfurt am Main, Germany). Unpublished Work. 2019.
40. Yu, Y.; Zhu, H.; Frantz, J.M.; Redubf, M.E.; Chan, K.C.; Ozkan, H.E. Evaporation and coverage area of pesticide droplets on hairy and waxy leaves. *Biosyst. Eng.* **2009**, *104*, 324–334. [CrossRef]
41. Baur, P.; Grayson, B.T.; Schönherr, J. Polydisperse ethoxylated fatty alcohol surfactants as accelerators of cuticular penetration. 1. Effects of ethoxy chain length and the size of the penetrants. *Pestic. Sci.* **1997**, *51*, 131–152. [CrossRef]
42. Baur, P.; Schönherr, J.; Grayson, B.T. Polydisperse ethoxylated fatty alcohol surfactants as accelerators of cuticular penetration. 2: Separation of effects on driving force and mobility and reversibility of surfactant action. *Pestic. Sci.* **1999**, *55*, 831–842. [CrossRef]
43. Baur, P. Surfactant effects on Cuticular penetration of neutral polar compounds: Dependence on humidity and temperature. *J. Agric. Food Chem.* **1999**, *47*, 753–761. [CrossRef] [PubMed]
44. Baur, P. Mechanistic aspects of foliar penetration of agrochemicals and the effect of adjuvants. *Recent Res. Develop. Agric. Food Chem.* **1998**, *2*, 809–837.
45. Baur, P.; Buchholz, A.; Schönherr, J. Diffusion in plant cuticles as affected by temperature and size of organic solutes: Similarity and diversity among species. *Plant Cell Environ.* **1997**, *20*, 982–994. [CrossRef]
46. American Society of Agricultural and Biological Engineers (ASABE). Spray Nozzle Classification by Droplet Spectra, ANSI/ASAE S572.2. Available online: https://www.asabe.org/ (accessed on 10 October 2021).
47. Smith, D.; Askew, S.; Morris, W.; Shaw, D.; Boyette, M. Droplet size and leaf morphology effects on pesticide spray deposition. *Trans. ASAE* **2000**, *43*, 255–259. [CrossRef]
48. Janardhan, K.V.; Janakiraman, N.; Nataraju, S.P.; Setty, M.V.N. Decanol—An effective suckericide for flue-cured tobacco production in Karnataka. *Tob. Res.* **1990**, *16*, 57–60.
49. Grains Research & Development Corporation. Adjuvants—Oils, Surfactants and Other Additives for Farm Chemicals—Revised 2012 Edition. Available online: https://grdc.com.au/ (accessed on 20 June 2021).
50. Lewis, K.A.; Tzilivakis, J.; Warner, D.; Green, A. An international database for pesticide risk assessments and management. *Hum. Ecol. Risk Assess. Int. J.* **2016**, *22*, 1050–1064. [CrossRef]

51. Baur, P.; Schönherr, J. Penetration of an ethoxylated fattyalcohol surfactant across lead cuticles as affected by concentration additives, and humidity. *J. Plant Dis. Prot.* **1997**, *104*, 380–393.
52. Dubovik, V.; Dalinova, A.; Berestetskiy, A. Effect of Adjuvants on Herbicidal Activity and Selectivity of Three Phytotoxins Produced by the Fungus, Stagonospora cirsii. *Plants* **2020**, *9*, 1621. [CrossRef] [PubMed]
53. Wernecke, A.; Eckert, J.H.; Forster, R.; Kurlemann, N.; Odemer, R. Inert agricultural spray adjuvants may increase the adverse effects of selected insecticides on honeybees (*Apis mellifera* L.) under laboratory conditions. *J. Plant Dis. Prot.* **2021**. [CrossRef]
54. Henriques, J.; Lima, M.; Rosa, S.; Días, A.; Días, L. Allelopathic Plants. XVIII. *Solanum nigrum* L. *Allelopath. J.* **2006**, *17*, 1–16.
55. Baur, P.; Pontzen, R. Basic features of plant surface wettability and deposit formation and the impact of adjuvants. In *Proceedings of the 8th International Symposium on Adjuvants for Agrochemicals (ISAA)*; International Society for Agrochecmicals Adjuvants Columbus, OH, USA, 2007.
56. Schreiber, L.; Schönherr, J. *Water and Solute Permeability of Plant Cuticles: Measurement and Data Analysis*; Springer Berlin/Heidelberg, Germany, 2009; p. 299.
57. Cook, R.; Lupette, J.; Benning, C. The Role of Chloroplast Membrane Lipid Metabolism in Plant Environmental Responses. *Cells* **2021**, *10*, 706. [CrossRef] [PubMed]
58. Hernandez, M.; Cejudo, F.J. Chloroplast Lipids Metabolism and Function. A Redox Perspective. *Front. Plant Sci.* **2021**, *12*, 1636 [CrossRef] [PubMed]
59. Briggs, G.; Bromilow, R. Influence of Physicochemical Properties on Uptake and Loss of Pesticides and Adjuvants from the Leaf Surface. In *Interactions between Adjuvants, Agrochemicals and Target Organisms*; Holloway, P.J., Rees, R.T., Stock, D., Eds.; Springer Berlin/Heidelberg, Germany, 1994; Volume 12, pp. 1–26. [CrossRef]
60. Schenk, J.; Espino, S.; Romo, D.; Nima, N.; Do, A.; Michaud, J.; Papahadjopoulos-Sternberg, B.; Yang, J.; Zuo, Y.; Steppe, K.; et al Xylem Surfactants Introduce a New Element to the Cohesion-Tension Theory. *Plant Physiol.* **2017**, *173*, 1177–1196. [CrossRef] [PubMed]
61. Sperry, J.; Tyree, M. Mechanism of Water Stress-Induced Xylem Embolism. *Plant Physiol.* **1988**, *88*, 581–587. [CrossRef] [PubMed]

plants

MDPI

Article

Controlling Woody Weed Chinese Elm (*Celtis sinensis* Pers.) with Stem-Implanted Herbicide Capsules

Ciara Jade O'Brien, Vincent Mellor and Victor Joseph Galea *

School of Agriculture and Food Sciences, Gatton Campus, The University of Queensland, Gatton, QLD 4343, Australia; ciara.obrien@uqconnect.edu.au (C.J.O.); vam103@maths.uq.edu.au (V.M.)
* Correspondence: v.galea@uq.edu.au; Tel.: +61-(07)-5460-1282

Abstract: Chinese elm [*Celtis sinensis* Pers.] is an emerging environmental weed naturalised throughout the coastal and riparian (creek-banks, river margins, and streams) regions of eastern Australia. Throughout this introduced range, its management is limited to the application of synthetic herbicides and mechanical clearing operations (terrain and soil type permitting). The current mechanisms of chemical control (basal bark spraying, stem-injection, and cut-stump applications) often result in collateral damage to non-target native species (such as *Eucalyptus* spp. and *Casuarina cunninghamiana* Miq.) through herbicidal drift, runoff or leaching into adjacent habitats. This has raised concerns regarding the suitability of synthetic herbicides in ecologically sensitive (e.g., riparian zones, rainforest margins, and woodlands) or low-value habitats, thereby promoting significant developments in the fields of integrated weed management. This study investigated the effectiveness of a novel stem-implantation system for controlling woody weed species in the context of a conserved habitat. A replicated trial (n = 315) was established among a naturally occurring population of *C. sinensis*. This trial involved the mapping, measurement, and treatment of this invasive species with five encapsulated synthetic herbicides, as well as an untreated control and benchmark treatment (diesel + Access^{TM}). A significant effect ($p < 0.05$) on plant vigour and functional canopy was discerned for each assessment period following trial establishment. The highest incidence of mortality was observed among the individuals treated with glyphosate (245 mg/capsule), aminopyralid and metsulfuron-methyl (58.1 and 37.5 mg/capsule) and picloram (10 mg/capsule), achieving a similar response to the basal bark application of diesel and Access^{TM} (240 g/L triclopyr, 120 g/L picloram, and 389 g/L liquid hydrocarbon). This was also evidenced by a rapid reduction in functional canopy (i.e., no or little living leaf tissue) from three weeks after treatment. Unlike their industry counterparts, these encapsulated herbicides are immediately sealed into the vascular system of the target species by a plug. This significantly minimises the possibility of environmental or operator exposure to synthetic compounds by providing a targeted, readily calibrated herbicide application.

Keywords: Chinese elm; woody weed; weed management; chemical control; stem implantation

Citation: O'Brien, C.J.; Mellor, V.; Galea, V.J. Controlling Woody Weed Chinese Elm (*Celtis sinensis* Pers.) with Stem-Implanted Herbicide Capsules. *Plants* **2022**, *11*, 444. https://doi.org/10.3390/plants11030444

Academic Editors: Alessia Restuccia and Aurelio Scavo

Received: 27 December 2021
Accepted: 31 January 2022
Published: 6 February 2022

1. Introduction

Chinese elm (*Celtis sinensis* Pers.) is a deciduous or semi-deciduous tree native to the slopes of eastern Asia, most notably China, Korea, Taiwan and Japan [1,2]. However, this species has spread from its endemic habitat to the coastal and sub-coastal regions of Australia, New Zealand and South Africa through its deliberate introduction as an ornamental plant [2]. In Australia, its naturalisation throughout the riparian zones (creek-banks, river margins, and streams) of south-eastern Queensland (Brisbane, Nambour, Toowoomba, Dalby) and north-eastern New South Wales (Lismore, Kyogle, Tweed Harbour, Coffs Harbour) has caused the displacement of existing native vegetation, thereby threatening the biodiversity, resilience, and integrity of natural ecosystems [1–4]. This adversely affects populations of resident fauna (e.g., Koala *Phascolarctos cinereus*, Common Brushtail Possum *Trichosurus vulpecula*, Greater Glider *Petauroides volans*, Rufous

Rat-Kangaroo *Aepyprymnus rufescens*, and Black Wallaby *Wallabia bicolor*) by altering habitat conditions or resource availability (foliage, seeds, nectar, or sap) [1,4–6]. It has also formed dense infestations in disturbed sites such as urbanised bushlands, parklands and roadsides [1,3]. Given this evidence of invasiveness, *C. sinensis* is ranked in the ten highest invasive species in south-east Queensland alongside other notable woody weeds such as Lantana (*Lantana camara* var. *camara*), Camphor Laurel (*Cinnamomum camphora*) and Broad-Leaf Pepper Tree (*Schinus terebinthifolius*) [7].

The management of *C. sinensis* is currently limited to the application of synthetic herbicides and mechanical clearing operations (terrain and soil type permitting). The manual removal of individual plants may be practical for the initial clearing of higher-density (>150 plants/ha) or isolated infestations [4]. This can be achieved through the hand pulling of small seedlings (height < 30 cm), bulldozing or controlled grazing [4,5,8] However, these manual attempts at control are largely ineffective due to the vigorous resprouting capacity of severed plants [4,5].

Although the herbicides registered for its management are limited, the minor use of agricultural and veterinary (AGVET) chemical products is authorised under a permit (APVMA Permit PER11463) for the control of environmental weeds in non-agricultural areas [5,9]. In particular, the cut-stump or basal application of synthetic auxin chemicals (i.e., fluroxypyr, triclopyr or picloram) is recommended with compliance to label directions and permit conditions [5,8,9]. The latter is performed at the base of the target species (plants with <20 cm basal diameter) with a mixture of oil-soluble herbicide and diesel distillate to assist penetration through the bark [9,10]. This has been proven effective for the management of scattered, lower-density infestations of parkinsonia (*Parkinsonia aculeata* L.) [11], mimosa (*Mimosa pigra* L.) [12], mesquite (*Prosopis* L. species) [13], bellyache bush (*Jatropha gossypiifolia* L.) [14], calotrope (*Calotropis procera*) [15], yellow oleander (*Cascabela thevetia* (L.) Lippold) [9] and white weeping broom (*Ratama raetam* (Forssk.) Webb) [16] in the Australian landscape. For larger woody weeds, the cut-stump method involves the painting or spraying of herbicide to the exposed surface of a felled stump [10]. (Whilst their efficacy is undisputed, there are concerns regarding the suitability of these application methods in ecologically sensitive (e.g., riparian zones, rainforest margins, and woodlands) or low-value habitats [10,17].) The imprecise or excessive application of herbicides may result in collateral damage to non-target native species (such as *Eucalyptus* spp. and *Casuarina cunninghamiana* Miq.) through herbicidal drift, runoff or leaching into adjacent habitats [10,17]. The movement of herbicide between the anastomosed roots of neighbouring plants has also been documented [10,18] with conventional stem-injection methods such as 'frill and fill' [19] or 'drill and fill' [9]. This greater appreciation for environmental stewardship has promoted significant developments in the field of woody weed management by reducing dosage or improving application methods [9,10].

This study investigated the effectiveness of BioHerbicides Australia's (www.bioherbicides.com.au) proprietary stem-implantation system (InJecta 800®) and Di-Bak® range of synthetic herbicides for controlling *C. sinensis* in the context of a conserved habitat. This lightweight, handheld device was initially developed for the encapsulated delivery of three endophytic fungal species (*Lasiodiplodia pseudotheobromae*, *Macrophomina phaseolina*, *Neoscytalidium novaehollandiae*) for the management of parkinsonia (*P. aculeata*) on Australian rangelands [20,21]. This novel technology has since been expanded for the application for other endophytic organisms, as well as synthetic compounds (herbicides, fungicides, and insecticides) available in dry formulations [10,20,21]. More recently, the synthetic herbicide formulations have been trialed against a range of woody weed species such as parkinsonia (*P. aculeata*), prickly acacia (*Vachellia nilotica*), leucaena (*Leucaena leucocephala*), camphor laurel (*Cinnamomum camphora*) and privet (*Ligustrum lucidum*) [10,21,22]. These studies demonstrated that encapsulated glyphosate (~350 mg per capsule) was highly efficacious against all species except for parkinsonia (*P. aculeata*) [10]. Other formulations under evaluation include metsulfuron-methyl, picloram and imazapyr [10]. Unlike its industry counterparts, these encapsulated synthetic herbicides are immediately sealed into the target species, thereby minimising the possibility of unintentional chemical exposure to the neighbouring native vegetation or human operator [10].

2. Results

2.1. Weather Data

A record of monthly rainfall (mm) from January 2018 to March 2020 was retrieved from Old Hidden Vale Station, Grandchester [23] (Figure 1). A significant rainfall event (180 mm) was recorded during the establishment of the trial in mid-March 2019. Although this corresponds with the wet season (November to April) in Queensland, the intensity of rainfall was greater than the previous (2018) and subsequent year (2020). The following nine months (April 2019 to December 2019) were unusually dry with a total rainfall of 80.5 mm relative to the 312 mm of rainfall recorded in the previous year (2018). However, the rainfall returned to expected levels in the latter months of the wet season (January, February, and March of 2020).

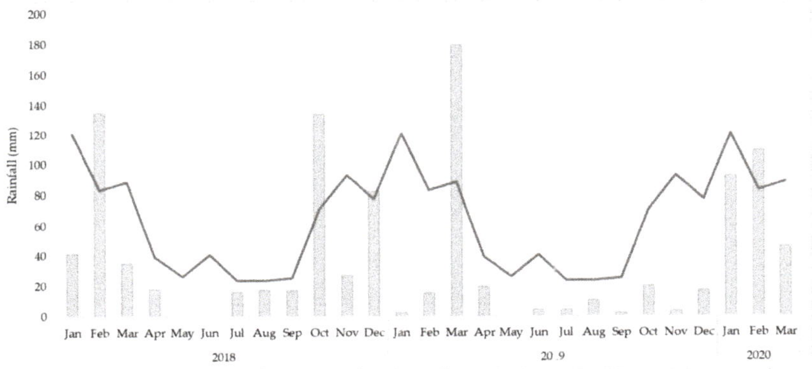

Figure 1. The monthly rainfall (mm) records at Old Hidden Vale Station, Grandchester, Queensland from January 2018 to March 2020. The line is indicative of the long-term (2000–2019) monthly rainfall means (μ).

2.2. Encapsulated Synthetic Herbicide Trial

A significant effect ($p < 0.05$) on plant vigour was discerned for each assessment period following trial establishment (week 3, 8, 15, 20, 25, 35, and 52) (Table 1). The benchmark treatment (diesel + Access^{TM}) and glyphosate had the most immediate effect on plant vigour, whereby all treated individuals were deemed 'dead' (stress score of three) or 'dying' within fifteen weeks (Table 1, Table 2). A similar trend in plant mortality was observed with aminopyralid and metsulfuron-methyl and picloram, as evidenced by their steadily increasing stress scores (Table 1). However, there was no significant difference ($p > 0.05$) between these four treatments at the conclusion of the trial (i.e., week 52) (Table 1, Table 2). There was also a high incidence of mortality (82.22%) among the individuals treated with metsulfuron-methyl (Table 1). However, the transition from being 'distressed' (stress score of two) to 'dead' (stress score of three) was slower relative to the preeminent treatments (Table 1). Whilst the effect of imazapyr plateaued from week eight to week twenty-five, achieving the lowest degree of mortality (46.67% at week fifty-two) among the encapsulated treatments (Table 1, Table 2). The health of the untreated plants (i.e., control treatment) was unaffected throughout the trial period (Table 2). Hence, a change in the condition of the target species was attributed to the explanatory variable (i.e., synthetic herbicide) rather than an unaccounted-for factor, such as drought stress, nutrient deficiency or plant disease.

Similarly, a significant effect $p < 0.05$ was discerned for functional canopy at each assessment period weeks 3, 8, 15, 20, 25, 35, and 52 Table 3. This value is referring to the aboveground portion of the plant with photosynthetic capacity i.e., healthy, living foliage. The benchmark treatment diesel + Access^{TM} and glyphosate caused a rapid reduction in functional canopy, whereby no living tissue 0% was remaining at fifteen weeks after treatment Table 3, Figure 2. A similar downward trend in functional canopy was also observed with aminopyralid and metsulfuron-methyl and picloram, as shown in Table 3 However, there was no significant difference $p > 0.05$ between these four treatments from week twenty onwards in terms of foliage loss or functional canopy Table 3, Figure 2. The individuals treated with metsulfuron-methyl and imazapyr experienced a more progressive steady reduction in functional canopy Table 3. Despite there being no living tissue 0% remaining at week twenty, their canopies recovered slightly at the conclusion of the trial Table 3, Figure 2. This may be epicormic growth in response to herbicidal injury or distress rather than a flashback attempt. The untreated plants i.e., control were also affected between week twenty and week thirty-five Table 3, Figure 2. This is characteristic behaviour in the autumn March, April, and May and winter June, July, and August months i.e., prolonged dry conditions given the deciduous and semi-deciduous nature of this tree species [24]. The condition of the untreated plants was restored following consistent rainfall in the summer months January, February, and March Table 3.

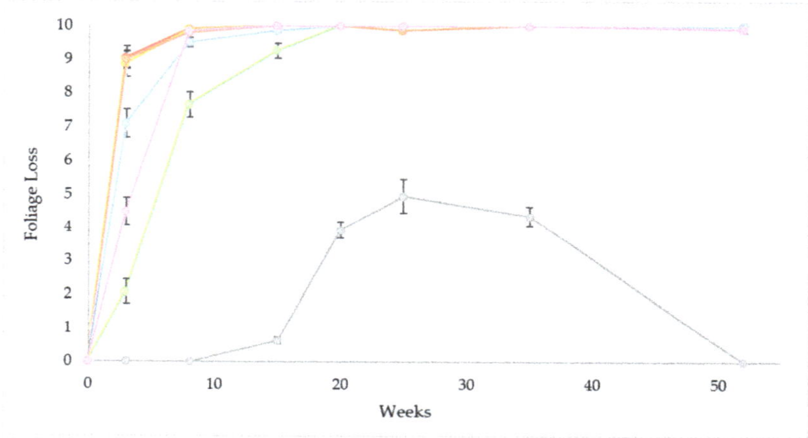

Figure 2. The mean (μ) foliage loss (0 = 0%, 1 = 1–10%, 2 = 11–20%, 3 = 21–30%, 4 = 31–40%, 5 = 41–50%, 6 = 51–60%, 7 = 61–70%, 8 = 71–80%, 9 = 81–90%, 10 = 91–100%) of the six chemical treatments (—●— = diesel + Access®; —●— = glyphosate; —○— = picloram; —○— = imazapyr; —○— = aminopyralid + metsulfuron-methyl; —○— = metsulfuron-methyl) and the control treatment (—○—) for each assessment period (week 0, 3, 8, 15, 20, 25, 35 and 52) under field conditions. The error bars represent the standard error (SE).

Table 1. One-way analysis of variance, estimated marginal means (EMM), and standard error (SE) of stress score for each assessment period (week 0, 3, 8, 15, 20, 25, 35, and 52). The superscript letters (i.e., compact letter displays) denote all pairwise comparisons among treatment means (μ).

	Week															
	0		3		8		15		20		25		35		52	
***p*-Value**	0.468		4.70×10^{-5} ***		1.12×10^{-7} ***		8.11×10^{-9} ***		3.02×10^{-9} ***		2.82×10^{-10} ***		1.12×10^{-9} ***		1.35×10^{-12} ***	
	EMM	**SE**	**EMM**	**SE**	**EMM**	**SE**	**EMM**	**SE**	**EMM**	**SE**	**EMM**	**SE**	**EMM**	**SE**	**EMM**	**SE**
Control	1.00^{a}	0	1.00^{b}	0	1.00^{d}	0	1.02^{c}	0.022	1.02^{c}	0.022	1.00^{c}	0	1.00^{c}	0	1.00^{c}	0
Diesel + Access TM	1.00^{a}	0	1.96^{a}	0.031	2.87^{a}	0.051	2.98^{a}	0.022	2.98^{a}	0.022	2.96^{a}	0.031	2.98^{a}	0.022	3.00^{a}	0
Glyphosate	1.00^{a}	0	1.96^{a}	0.031	2.84^{a}	0.055	2.87^{a}	0.051	2.91^{a}	0.043	2.93^{a}	0.038	2.98^{a}	0.022	3.00^{a}	0
Picloram	1.00^{a}	0	2.00^{a}	0	2.56^{ab}	0.075	2.78^{a}	0.063	2.80^{a}	0.060	2.78^{a}	0.063	2.96^{a}	0.031	3.00^{a}	0
Aminopyralid + Metsulfuron-Methyl	1.00^{a}	0	1.93^{a}	0.038	2.44^{abc}	0.075	2.64^{a}	0.072	2.69^{a}	0.070	2.71^{a}	0.068	2.82^{a}	0.058	2.98^{a}	0.022
Metsulfuron-Methyl	1.00^{a}	0	1.98^{a}	0.022	2.07^{c}	0.038	2.22^{b}	0.063	2.20^{b}	0.060	2.18^{b}	0.066	2.36^{b}	0.072	2.82^{a}	0.058
Imazapyr	1.00^{a}	0	1.42^{b}	0.074	2.18^{bc}	0.058	2.11^{b}	0.047	2.09^{b}	0.043	2.13^{b}	0.051	2.27^{b}	0.067	2.47^{b}	0.075

Significance value *** = 0.

Table 2. Percentage (%) mortality of each encapsulated treatment at the final assessment (week 52).

Treatment	Mortality %
Control	0
Diesel + AccessTM	100
Glyphosate	100
Picloram	100
Aminopyralid + Metsulfuron-Methyl	97.78
Metsulfuron Methyl	82.22
Imazapyr	46.67

Table 3. One-way analysis of variance, estimated marginal means (EMM), and standard error (SE) of functional canopy for each assessment period week 0, 3, 8, 15, 20, 25, 35, and 52. The superscript letters i.e., compact letter displays denote all pairwise comparisons among treatment means μ. Significance value *** = 0.

	Week															
	0		3		8		15		20		25		35		52	
***p*-Value**	-		1.66×10^{-9} ***		3.58×10^{-13} ***		1.99×10^{-9} ***		$<2 \times 10^{-16}$ ***		7.17×10^{-12} ***		2.79×10^{-8} ***		1.42×10^{-15} ***	
	EMM	**SE**	**EMM**	**SE**	**EMM**	**SE**	**EMM**	**SE**	**EMM**	**SE**	**EMM**	**SE**	**EMM**	**SE**	**EMM**	**SE**
Control	-	0	1.00 ^{a}	0	0.99 ^{a}	0.003	0.96 ^{a}	0.01	0.64 ^{a}	0.01	0.53 ^{a}	0.05	0.59 ^{a}	0.03	1.00 ^{a}	0
Diesel + Access ™	-	0	0.06 ^{de}	0.029	0.003 ^{d}	0.005	0.0002 ^{bc}	0	0.0 ^{b}	0	0.0 ^{b}	0	0.0 ^{b}	0	0.0 ^{c}	0
Glyphosate	-	0	0.02 ^{e}	0.009	0.004 ^{cd}	0.004	0.0 ^{c}	0	0.0 ^{b}	0	0.0 ^{b}	0	0.0 ^{b}	0.001	0.0 ^{c}	0
Picloram	-	0	0.04 ^{e}	0.016	0.02 ^{cd}	0.005	0.0002 ^{bc}	0	0.0 ^{b}	0	0.0 ^{b}	0.001	0.0 ^{b}	0	0.0 ^{c}	0
Aminopyralid + Metsulfuron-Methyl	-	0	0.20 ^{d}	0.038	0.03 ^{c}	0.009	0.005 ^{bc}	0.005	0.0 ^{b}	0	0.0 ^{b}	0	0.003 ^{b}	0.002	0.0 ^{c}	0.001
Metsulfuron-Methyl	-	0	0.45 ^{c}	0.042	0.02 ^{cd}	0.004	0.003 ^{bc}	0.001	0.0 ^{b}	0	0.001 ^{b}	0	0.005 ^{b}	0.003	0.013 ^{b}	0.01
Imazapyr	-	0	0.72 ^{b}	0.040	0.22 ^{b}	0.034	0.07 ^{b}	0.022	0.0 ^{b}	0	0.0 ^{b}	0	0.006 ^{b}	0.01	0.023 ^{b}	0.004

3. Discussion

The result of this study suggests that the successful management of *C. sinensis* in conserved habitats (e.g., riparian zones, woodlands, and rainforest margins) is possible through the implantation of encapsulated synthetic herbicides. The highest incidence of mortality was observed among the individuals treated with glyphosate (245 mg/capsule), aminopyralid plus metsulfuron-methyl (58.1 and 37.5 mg/capsule) and picloram (10 mg/capsule), achieving a similar response to the industry accepted standard (i.e., basal bark application of diesel + Access^{TM}). This is evidenced by their rapidly increasing stress scores that translated to entirely (100%) 'dead' *C. sinensis* plants by the conclusion of the trial. Other symptoms of herbicidal injury were also apparent such as the puckering, longitudinal cracking and bleaching of the outer bark tissue. Despite causing considerable distress, the least effective synthetic treatments were metsulfuron-methyl (198 mg/capsule) and imazapyr (262.5 mg/capsule) on a comparative basis. The health of the untreated plants was unaffected (0% mortality) throughout the trial period. However, a slight reduction in functional canopy was recorded from week twenty to week thirty-five. This is characteristic behaviour in the autumn and winter months (i.e., prolonged dry conditions) given the deciduous and semi-deciduous nature of this species [24]. As expected, the condition of the untreated plants was restored following consistent rainfall (247.5 mm cumulative) in the summer months (January, February, and March).

There are a limited number of herbicides registered by the Australian Pesticides and Veterinary Medicines Authority (APVMA) for the management of *C. sinensis*. However, the use of agricultural and veterinary (AGVET) chemical products is authorised in non-agricultural areas under an off-label use permit (APVMA Permit PER11463) [5]. In particular, the basal bark application of fluroxypyr (200 g/L, 333 g/L or 400 g/L) or triclopyr (240 g/L) plus picloram (120 g/L) with diesel distillate is recommended for the treatment of saplings or regrowth with a basal diameter of $\leq$20 cm [3,5,25]. The latter (triclopyr + picloram) is also registered for the management of other woody weed species such as parkinsonia (*Parkinsonia aculeata* L.), prickly acacia (*Vachellia nilotica*), lantana (*Lantana camara*), leucaena (*Leucaena leucocephala*), mimosa bush (*Mimosa pigra* L.), camphor laurel (*Cinnamomum camphora*) and brigalow (*Acacia harpophylla*) [22,25]. This supported its selection as the benchmark treatment for this study. Unsurprisingly, the benchmark treatment caused a rapid deterioration in plant health (100% mortality at fifty-two weeks). However, there was no significant difference ($p > 0.05$) between the most efficacious encapsulated synthetic herbicides (glyphosate, aminopyralid + metsulfuron-methyl, picloram) and the benchmark treatment at the conclusion of the trial. This infers that the stem-implantation technology meets the current industry standard.

The delivery of encapsulated synthetic herbicides has been proven successful for the management of other woody weed species such as prickly acacia (*V. nilotica*), leucaena (*L. leucocephala*), mimosa bush (*M. pigra*), camphor laurel (*C. camphora*) and privet (*Ligustrum lucidum*) [10,21,22]. These trials followed the guidelines of the Australian Pesticides and Veterinary Medicines Authority (APVMA) for research on pesticide efficacy [21]. Typically, this efficacy criterion requires ($\geq$) fifteen plants per treatment group with a minimum of three replications in a randomized complete block design (RCBD) [21]. These studies demonstrated that encapsulated glyphosate (~350 mg/capsule) was highly efficacious against all species except for parkinsonia (*P. aculeata*) [10,21]. The effectiveness of stem-injected (i.e., 'drill and fill' and 'frill and fill') glyphosate has also been documented in mimosa bush (*M. pigra*), yellow oleander (*Cascabela thevetia*) and tree of heaven (*Ailanthus altissima*) [9,22,26]. The results of this study support these previous findings that encapsulated glyphosate (~245 mg/capsule) is a promising candidate for woody weed management.

The stem-implantation technology (InJecta®) has many benefits by providing a targeted, readily calibrated herbicide application. This methodology delivers a minimum recommended lethal dose of chemical directly into the vascular system of the target species, thereby fully capturing (100%) the active agent internally [10,21]. The possibility of environmental or operator exposure (through dermal absorption or respiratory inhalation) to synthetic compounds is greatly reduced compared to its industry counterparts, such as canopy application or stem spraying [21]. In our study, there was no visual indication of herbicidal injury or distress (e.g., distorted growth, foliage loss, interveinal chlorosis and necrosis) [27] among the untreated *C. sinensis* plants or adjacent non-target vegetation. This indicates that little or no translocation occurred between treated and untreated plants. Hence, this method is deemed broadly appropriate for the management of woody weed species in ecologically sensitive habitats (e.g., riparian zones, rainforest margins, national parks, woodlands, and wetlands).

Despite being highly efficacious, these encapsulated chemical formulations also contain less (20% to 30%) herbicidal active ingredients relative to other control options (e.g., basal bark spraying, cut stump, and 'drill and fill') [10]. For example, a single dose (1 mL) of undiluted RoundUp® has 360 mg of active ingredient (glyphosate) for 'axe-cut' applications. Whilst a single glyphosate capsule (size 0 hypromellose capsule) contains 245 mg of active ingredient. The occurrence of 'flashback' will be reduced under these lowered dosages [10], as well as the development of herbicide resistance or tolerance in targeted species [28].

The primary intent of this study was to investigate the effectiveness of BioHerbicides Australia's (BHA Pty Ltd., Brisbane, QLD, Australia) proprietary stem-implantation system and Di-Bak® range of synthetic herbicides for controlling *C. sinensis* in the context of a conserved habitat. It was found that the most effective encapsulated herbicides were glyphosate (245 mg/capsule), aminopyralid and metsulfuron-methyl (58.1 and 37.5 mg/capsules) and picloram (10 mg/capsule), achieving similar degree of plant mortality relative to the benchmark treatment (i.e., basal bark application of diesel + Access™). Unlike its industry counterparts, this novel technology (InJecta®) delivers concentrated dry formulations directly into the vascular system of the target species where all (100%) of the active ingredient is captured internally [10,21]. This has the potential for (i) reducing the amount of active agent required, (ii) preventing environmental exposure to plant protection chemicals and (iii) improving operator safety [10,21]. Hence, this methodology could be a replacement for stem-injection or cut-stump applications in ecologically sensitive habitats (riparian zones, rainforest margins, national parks, woodlands, wetlands) [10], as well as for the management of root and stem disorders in plantation crops (e.g., rubber *Hevea brasiliensis* and oil palm *Elaeis guineensis*) [21]. Further research to optimise the dosage level and placement of the most effective treatments (glyphosate, aminopyralid and metsulfuron-methyl) is currently underway.

4. Materials and Methods

4.1. Experimental Site, Design and Treatments

A replicated trial (n = 315) was established among a naturally occurring population of *C. sinensis* located on the banks of Franklin Vale Creek (near Grandchester, southeast Queensland: 27°44′46″ S, 152°27′17″ E). This trial involved the mapping, measurement and treatment of individual plants with five encapsulated synthetic herbicides sourced from BioHerbicides Australia's (Bioherbicides Australia Pty Ltd., Brisbane, QLD, Australia) Di-Bak® range of registered and developmental products (Table 4). A control (untreated plants) and benchmark treatment were also included for performance comparison, this being the basal bark application of diesel + Access™ herbicide (240 g/L triclopyr, 120 g/L picloram, and 389 g/L liquid hydrocarbon).

Table 4. Treatment name, active ingredient(s) concentration and dosage (mg/capsule) of the five encapsulated herbicides. All capsules were sourced from BioHerbicides Australia (BHA Pty Ltd.).

Treatment	Active Ingredient(s)	Active Ingredient Concentration (g/kg)	Dose (mg/Capsule)
Di-Bak AM®	Aminopyralid + Metsulfuron Methyl	375 & 300	58.1 & 37.5
Di-Bak M®	Metsulfuron Methyl	600	198
Di-Bak G®	Glyphosate	700	245
Di-Bak I®	Imazapyr	750	262.5
Di-Bak P®	Picloram	100	10

The trial was established in mid-March 2019 using a randomised complete block design (RCBD) with three blocks. Within each block, the seven treatments were randomly assigned to a total of fifteen plants (of similar age) complying with the recommendations of the Australian Pesticides and Veterinary Medicines Authority (APVMA) for efficacy evaluation on woody weed species. The plants (stem circumference range of $\geq$15 cm to 90 cm) within each treatment plot were clustered into small groupings or rows along the creekbank. Treatment plots were differentiated from one another by coloured flagging tape, clearly labelled and their respective GPS waypoints determined using a handheld Garmin® 62s GPS device (Garmin Australasia Pty Ltd., Eastern Creek, NSW, Australia).

4.2. Treatment Application

The encapsulated synthetic herbicides were administered via the InJecta® handheld device (Bioherbicides Australia Pty Ltd., Brisbane, QLD, Australia) Figure 3a]. This applicator is powered by a cordless drill using an 8 mm drill bit creating a hole (25 mm depth) into the plant stem at an approximate height of one metre (Figure 3b) [29]. The withdrawal of the drill backwards is followed by the rotation of the magazine, thereby priming a single capsule (21.6 mm $\times$ 7.6 mm) containing the dry herbicide formulation and plug for delivery (which are in-tandem within each of the thirty chambers of the magazine) (Figure 3c) [22,29]. The capsule and plug are then simultaneously inserted into the drilled hole through the forward movement of the non-rotating drill [29]. The synthetic herbicide is immediately sealed into the target species by a polypropylene plastic plug (Figure 3d) [22,29]. This exclusion of an oxidizing atmosphere to the wound tissues facilitates the absorption of xylem and phloem fluids by the capsule (i.e., dissolving the herbicide) [29].

The applied dosage was determined by the stem circumference of the plant or each branch (multiple-stemmed plant) at chest height. A single capsule was administered for every 15 cm incremental increase in stem circumference. In the case of multiple doses, the capsules were spaced evenly around the plant stem.

Figure 3. (a) Implanting a synthetic herbicide capsule into the stem of a *C. sinensis* plant using the InJecta® handheld device; (b) rotating drill bit (8 mm) creating a hole into the plant stem; (c) loading the magazine with synthetic herbicide capsules and polypropylene plugs; (d) polypropylene plug partially protruding from the implantation site of a treated *C. sinensis* plant.

The basal application of Access™ (Corteva Agriscience Pty Ltd., Sydney, NSW, Australia) herbicide (240 g/L triclopyr, 120 g/L picloram, and 389 g/L liquid hydrocarbon) with diesel (dilution rate of 1 L/60 L) was achieved with a manual pressure sprayer (Nylex 8 L Heavy Duty Shoulder Sprayer; Ames Australia, Doncaster, Victoria, Australia). The entirety of the stem and root collar area was treated liberally from ground level to an approximate height of 60 cm (as per manufacturer's instructions) for sufficient penetration through the bark. The appropriate safety equipment (Bossweld elbow-length gloves (Dynaweld, Prestons, NSW, Australia), valved activated carbon respirator (3M Australia Pty Ltd., North Rhyde, NSW, Australia, and covered clothing) was worn during the preparation and application of the solution.

4.3. Trial Assessment

The trial was rated at approximate monthly intervals by recording the percentage of foliage loss, the colour composition (percentage green, yellow and brown) of the remaining canopy and the overall vigour of each individual plant. Based upon visual observation, a rating of 0 to 10 (0 = 0%, 1 = 1–10%, 2 = 11–20%, 3 = 21–30%, 4 = 31–40%, 5 = 41–50%, 6 = 51–60%, 7 = 61–70%, 8 = 71–80%, 9 = 81–90%, 10 = 91–99%, 11 = 100%) was given for each plant indicating the percentage (%) of total foliage loss since the establishment of the trial. The overall vigour of each plant was also recorded and expressed as a stress score (1 = healthy, 2 = distressed, 3 = dead). This was discerned by removing the outermost layer of the bark with a rasp to reveal the colour of the tissue beneath. Additionally, an auditory assessment of the degree of hydration was conducted by tapping the stem with a hammer. Other observable symptoms of stress were recorded such as the splitting or discolouration of the bark, sap seepage from the implantation site and insect damage.

The 'functional canopy' of each plant was also calculated from the percentage (%) of foliage loss and the colour composition of the remaining canopy.

$$Functional\ Canopy = \left(\frac{Percentage\ Existing\ Canopy}{100}\right) \times \left(\frac{Percentage\ Green\ Canopy}{100}\right)$$

This rating refers to the aboveground portion of the tree canopy that is functional, living tissue. This is expressed on a scale of zero to one, whereby a value of one is indicative of a highly functional or healthy canopy (i.e., full, green canopy).

4.4. Data Analysis

The treatment effects on stress score and functional canopy were analysed using RStudio® (RStudio Inc., Boston, MA, USA). Although stress score is an ordinal scale (i.e., quantitative data), a one-way analysis of variance (ANOVA) was performed by taking the mean (μ) value from each replicate. The functional canopy was also analysed using the same approach (i.e., one-way ANOVA). All pairwise comparisons among treatment means (μ) were estimated with the emmeans (estimated marginal means) package.

Author Contributions: Conceptualization, C.J.O. and V.J.G.; methodology, C.J.O. and V.J.G.; formal analysis, V.M. and C.J.O.; resources: C.J.O. and V.J.G.; writing—original draft preparation, C.J.O.; writing—review and editing, V.J.G.; visualization, C.J.O.; supervision, V.J.G.; project administration, V.J.G.; funding acquisition, V.J.G. All authors have read and agreed to the published version of the manuscript.

Funding: This research was funded by The University of Queensland with technical advice provided by BioHerbicides Australia.

Data Availability Statement: Not Applicable.

Acknowledgments: The authors would like to thank Megan Brady for her assistance with site logistics, as well as the team at Hidden Vale Wildlife Centre for allowing the research to be undertaken at Franklin Vale Creek. We also acknowledge Allan Lisle for his initial recommendations with the statistical analysis and Lachlan Fowler for his technical support.

Conflicts of Interest: The authors declare no conflict of interest. The funders had no role in the design of the study; in the collection, analyses, or interpretation of data; in the writing of the manuscript, or in the decision to publish the results.

References

1. Csurhes, S.; Edwards, R. *Potential Environmental Weeds in Australia: Candidate Species for Preventative Control*; Queensland Deparment of Natural Resources: Brisbane, QLD, Australia, 1998; p. 145.
2. Panetta, F.D. Seedling Emergence and Seed Longevity of the Tree Weeds *Celtis sinensis* and *Cinnamomum camphora*. *Weed Res.* **2001**, *1*, 83–95. [CrossRef]
3. Brisbane City Council. Chinese Celtis (*Celtis sinensis*). Available online: https://weeds.brisbane.qld.gov.au/weeds/chinese-celtis (accessed on 27 March 2019).
4. Ensbey, R. Chinese Celtis (*Celtis sinensis*). Available online: https://weeds.dpi.nsw.gov.au/Weeds/ChineseCeltis (accessed on 8 March 2019).
5. Queensland Government Department of Agriculture and Fisheries. *Chinese Celtis (Celtis sinensis)*; Department of Agriculture and Fisheries: Brisbane, QLD, Australia, 2020; pp. 1–3.
6. Adair, R.J.; Groves, R.H. *Impact of Environmental Weeds on Biodiversity: A Review and Development of a Methodology*; National Parks and Wildlife: Canberra, ACT, Australia, 1998; pp. 1–5.
7. Batianoff, G.N.; Butler, D.W. Assessment of Invasive Naturalized Plants in South-East Queensland. *Plant Prot. Q.* **2002**, *17*, 27–34.
8. Armstrong, T.R.; Keegan, S.L. *Celtis sinensis* and Its Control. In Proceedings of the 11th Australian Weeds Conference, Melbourne, VIC, Australia, 30 September–3 October 1996; Shepherd, R.C.H., Ed.; Weed Science Society of Victoria Inc.: Bundoora, VIC, Australia, 1996.

9. McKenzie, J.; Brazier, D.; Owen, A.; Vitelli, J.; Mayer, B. Stem-Injection: A Control Technique Often Overlooked for Exotic Woody Weeds. In Proceedings of the 17th Australasian Weeds Conference, Christchurch, New Zealand, 26 September–30 September 2010; New Zealand Plant Protection Society: Christchurch, New Zealand, 2010.
10. Goulter, K.C.; Galea, V.J.; Riikonen, P. Encapsulated Dry Herbicides: A Novel Approach for Control of Trees. In Proceedings of the 21st Australasian Weeds Conference, Sydney, NSW, Australia, 9–12 September 2018.
11. Van Klinken, R.D.; Campbell, S.D.; Heard, T.A.; McKenzie, J.; March, N. The Biology of Australian Weeds 54. *Parkinsonia aculeata* L. *Plant Prot. Q.* **2009**, *24*, 100–117.
12. Wingrave, S. Herbicides and Their Application for the Control of Mimosa in the Northern Territory. In *Research and Management of Mimosa Pigra*; Julien, M., Flanagan, G., Heard, T., Hennecke, B., Paynter, Q., Wilson, C., Eds.; CSIRO Entomology: Clayton South, VIC, Australia, 2004; pp. 96–101.
13. Van Klinken, R.D.; Campbell, S.D. The Biology of Australian Weeds 37. Prosopis L. Species. *Plant Prot. Q.* **2001**, *16*, 2–20.
14. Bebawi, F.F.; Vitelli, J.S.; Campbell, S.D.; Vogler, W.D.; Lockett, C.J.; Grace, B.S.; Lukitsch, B.; Heard, T.A. The Biology of Australian Weeds 47. *Jatropha gossypiifolia* L. *Plant Prot. Q.* **2007**, *22*, 42–58.
15. Vitelli, J.; Madigan, B.; Wilkinson, P.; van Haaren, P. Calotrope (*Calotropis procera*) Control. *Rangel. J.* **2008**, *30*, 339–348. [CrossRef]
16. Bettink, K.A.; Brown, K.L. Determining Best Control Methods for the National Environmental Alert List Species, '*Retama raetam*' (Forssk.) Webb (White Weeping Broom) in Western Australia. *Plant Prot. Q.* **2011**, *26*, 36–38.
17. Marrs, R.H.; Williams, C.T.; Frost, A.J.; Plant, R.A. Assessment of the Effects of Herbicide Spray Drift on a Range of Plant Species of Conservation Interest. *Environ. Pollut.* **1989**, *59*, 71–89. [CrossRef]
18. Ashton, M.S.; Kelty, M.J. *The Practice of Silviculture: Applied Forest Ecology*, 10th ed.; John Wiley and Sons: London, UK, 2018.
19. DiTomaso, J.M.; Kyser, G.B.; Oneto, S.R.; Wilson, R.G.; Orloff, S.B.; Anderson, L.W.; Wright, S.D.; Roncoroni, J.A.; Miller, T.L.; Prather, T.S.; et al. *Weed Control in Natural Areas in the Western United States*; Weed Research and Information Centre, University of California: Los Angeles, CA, USA, 2013.
20. Morin, L. Progress in Biological Control of Weeds with Plant Pathogens. *Annu. Rev. Phytopathol.* **2020**, *58*, 201–223. [CrossRef] [PubMed]
21. Galea, V.J. Stem Implanted Capsules—A Target, Efficient and Sustainable Method for Delivery of Synthetic and Biological Agents to Plantation Crops. In Proceedings of the 2nd International Plantation Conference, Universiti Teknologi MARA, Selangor, Malaysia, 9–10 March 2021.
22. Limbongan, A.A.; Campbell, S.D.; Galea, V.J. Novel Encapsulated Herbicide Delivery Mechanism: Its Efficacy in Mimosa Bush (*Vachellia farnesiana*) Control. *Plants* **2021**, *10*, 2505. [CrossRef] [PubMed]
23. Oakes, M.; (Old Hidden Vale Station, Grandchester, QLD, Australia). Personal Communication, 2020.
24. Nathan, R.; Katul, G.G. Foliage Shedding in Deciduous Forests Lifts Up Long-Distance Seed Dispersal By Wind. *Proc. Natl. Acad. Sci. USA* **2005**, *102*, 8251–8256. [CrossRef] [PubMed]
25. Corteva Agriscience. Access®Herbicide. Available online: https://www.corteva.com.au/products-and-solutions/crop-protection/access.html (accessed on 26 December 2021).
26. Badalamenti, E.; La Mantia, T. Stem-Injection of Herbicide for Control of *Ailanthus altissima* (Mill.) Swingle: A Practical Source of Power for Drilling Holes in Stem. *Iforest* **2013**, *6*, 123–126. [CrossRef]
27. Feucht, J.R. Herbicide Injuries to Trees—Symptoms and Solutions. *J. Arboric.* **1988**, *14*, 215–219.
28. Roy, B.A. Rounding Up the Costs and Benefits of Herbicide Use. *Proc. Natl. Acad. Sci. USA* **2004**, *101*, 13974–13975. [CrossRef] [PubMed]
29. Galea, V.J. Use of Stem Implanted Bioherbicide Capsules to Manage an Infestation of *Parkinsonia aculeata* in Northern Australia. *Plants* **2021**, *10*, 9109. [CrossRef] [PubMed]

plants

MDPI

Review

Bioherbicidal Potential of the Essential Oils from Mediterranean Lamiaceae for Weed Control in Organic Farming

Giuseppe De Mastro [1,*], Jihane El Mahdi [2] and Claudia Ruta [1]

[1] Department of Agricultural and Environmental Science, Università degli Studi di Bari Aldo Moro, Via Amendola 165/a, 70125 Bari, Italy; claudia.ruta@uniba.it
[2] CIHEAM Centre International des Hautes Études en Agriculture Méditerranéenne, IAM Mediterranean Agronomic Institute of Bari, Via Ceglie 9, 70010 Valenzano, Italy; elmahdijihane@gmail.com
* Correspondence: giuseppe.demastro@uniba.it; Tel.: +39-33-5876-2824

Abstract: In all farming systems, weeds are the most expensive pest to manage, accounting for 30% of potential losses. In organic farming, the problem may be further amplified by restrictions on herbicides, thus making weeds the main problem faced by organic farmers in the field. In this sense, much research is focusing on the allelopathic potential of plants as an ecological weed control tool. Many plant species can release allelopathic compounds with high phytotoxicity that can be used in weed control. Species belonging to the Lamiaceae family have been studied widely for this purpose, and their essential oils (EOs) appear to be promising bioherbicides. However, there are still many challenges for their development. Considering these aspects, a review of the bioherbicidal effect of EOs from Mediterranean Lamiaceae could help identify the most effective ones and the challenges for their actual development.

Keywords: terpenes; mechanism of action; germination inhibitors; crops

Citation: De Mastro, G.; El Mahdi, J.; Ruta, C. Bioherbicidal Potential of the Essential Oils from Mediterranean Lamiaceae for Weed Control in Organic Farming. *Plants* **2021**, *10*, 818. https://doi.org/10.3390/plants10040818

Received: 10 March 2021
Accepted: 9 April 2021
Published: 20 April 2021

1. Introduction

The emerging worldwide need to find alternatives to synthetic herbicides for sustainable weed control has prompted considerable interest in exploiting the natural herbicidal potential in plants [1]. Bioherbicide sources are sought out by both conventional and organic farming systems: the former wish to identify new sites of action to cope with weed resistance, the latter seek potent alternatives to synthetic herbicides that can be integrated in an overall management approach [2]. In this context, weed control research has recently focused on extracts from allelopathic species. These are species that can release secondary metabolites able to interfere with the growth and functions of surrounding plants [3].

A well-established group of allelopathic plants is that of the Lamiaceae family. They are known to contain high concentrations of volatile allelochemicals, which are responsible for their aroma, and are reported to give the species a competitive advantage in their natural habitats [4]. In this context, extracts of different Lamiaceae species were studied extensively and found to inhibit the germination and growth of many weed species [5–12]. Essential oils (EOs) from species such as oregano, thyme, rosemary, sage and mint are reported to be particularly strong bioherbicide candidates.

The phytotoxic effect of these species extracts, notably EOs, has mainly been linked to the presence of volatile bio-active compounds such as α-pinene, limonene, 1,8-cineole, carvacrol, camphor and thymol, which have been shown to have varying individual phytotoxicity levels [4,11–15]. Some phenolic compounds present in the EOs were also reported to be involved in allelopathic interactions and were even used to develop commercial bioherbicides. The mechanisms by which these allelochemicals can affect weeds was not discussed in detail. Only a few individual compounds were studied [16–18], in addition to the mechanism behind some naturally occurring allelopathic interactions [16,19–21].

Although there are numerous studies reporting on the successful use of EOs in weed control, to date there are still many constraints limiting their practical application in commercial bioherbicides. For instance, the role of the EO composition is still not clearly described. The mechanisms of action and the observed selectivity are also very poorly understood, limiting their rational implementation. Moreover, studies concerning the possible side effects of these EOs on beneficial soil microorganisms are still lacking.

This review will address all the above-mentioned issues pertaining to the use of EOs from Lamiaceae species in weed control in order to further highlight their potential uses and perspectives for future studies. It will also review the literature on certain species most frequently studied.

2. Weed Management in Organic Agriculture

2.1. Objectives and Methods of Weed Management in Organic Agriculture

To understand the aims of weed control in organic farming one must understand the overall objective of this production system. How to maximize yields and economic gain are major concerns for organic farmers, like others. However, in this system the emphasis is on the long-term outcome and overall health of the soil, plants, animals, and humans rather than just immediate maximum profitability [22]. In this context, many operational techniques have been defined to meet what could be regarded as the main goals of organic farming. Kirschenmann [23] presented four techniques related to the different aspects of management: nutrients, insects, plant disease and weeds. As for the latter, the overall goal was "to achieve weed control using crop rotation systems to deprive weeds of favorable growing conditions". Liebman et al. [24] defined more detailed, equally important objectives. These can be summarized as follows:

- reducing weed density to a tolerable level, instead of targeting 100% control or total suppression;
- reducing the damage that a given density of weeds can cause, by increasing the competitive ability of crops and minimizing that of weeds through different preventive and cultural tools (competitive varieties, fertilization, irrigation and false seed beds);
- shifting the composition of weed communities to less aggressive, more easily managed species.

These goals may be achieved through a knowledge-intensive process. A good understanding of weed ecology, of the site and of the crop-weed interactions is required.

Kirschenmann [23] claims in this context that the organic system seeks to farm like nature, which implies knowing and understanding the natural processes and incorporating those principles on the farm.

Managing weeds in an organic system is more complicated than in a conventional system, mainly because of restrictions on the use of herbicides [25]. The latter are easy to apply and aggressively marketed, although in recent years there has been a tendency to restrict the use of chemicals in agriculture to preserve human health and the environment. [24]. Therefore, a combination of tools and practices that take into account the natural system's cycles and interactions are increasingly being adopted to manage weeds. The management system is consequently an integrated approach, one that adopts different preventive, cultural, and direct control methods to achieve the goals detailed in the paragraph above [25,26].

- Preventive methods

Prevention aims at reducing the density of the actual weed vegetation by exhausting the potential weed vegetation (e.g., weed seedbank in the soil). This means reducing in-crop weed emergence and weed seed dispersal. Operational techniques include crop rotation, tillage systems, the false seedbed technique, cover crops, mulching and soil solarization.

- Cultural methods

Cultural methods are commonly used to reduce the need for direct weed control (e.g., herbicides) and increase its effectiveness. This is achieved by choosing cultural techniques that favor the competitive ability of crops against weeds. Cultural weed management techniques include crop genotype choice, planting pattern, polyculture production systems, fertilization, and irrigation strategies.

- Direct methods

Direct methods aim at intervening directly during the crop cycle to eliminate the weeds, mainly using physical or chemical tools.

Physical tools include mechanical weeding or cultivation, which is based on a variety of equipment. Recent technical innovations focusing on intra-row weed control in arable and vegetable crops have proven to be effective [26]. Robotic control is another technological innovation increasingly adopted by organic farmers. Post-emergence flame weeding, which can be used after planting or crop emergence, is another physical technique. Flaming can be used to eliminate weeds within the row where cultivation is difficult or can considerably damage the crop. Lastly, manual weeding is also widely used in organic management, notably when other measures are not feasible, such as within rows or when the crop is susceptible to damage by cultivation.

Bioherbicides are the main chemical tools. The latter are compounds and secondary metabolites derived from microbes, phytotoxic plant extracts or single compounds [2].

2.2. Challenges of the Current Weed Management Methods in Organic Agriculture

Effective weed management in the organic system necessarily involves integration of the highest number of available tools and approaches [27,28]. Preventive methods based on ecological principles and building biodiversity such as rotation or cover cropping are of particular interest to this system. However, direct methods are still contributing the most to weed control in many organic farms [27,29,30]. Therefore, serious weed competition problems may arise when few direct control methods are available or applicable. Moreover, those currently available present serious limitations (Table 1).

By studying existing limitations in direct weed control methods, it is possible to define research needs and opportunities. This is particularly true of bioherbicides, especially considering the limitations and the numerous potential sources of active compounds in nature. This research field is increasingly important due to increasing consumer awareness and environmental problems related to synthetic herbicides (residues, weed resistance). Usually, the EOs have various modes of action, and therefore it is more complicated for weeds to develop easily resistance against them [10,31]. This aspect increased the attention to their bioherbicidal potential, widely investigated in the hope of finding effective, viable products that can meet registration requirements.

Table 1. Limitations to current direct weed control methods in organic farming.

Direct Method	Limitations	Reference
Mechanical weeding (Tillage/Cultivation)	Weather and soil moisture conditions Excessive soil disturbance Difficult to control perennial weeds Damages to crops root system No/reduced tillage systems Difficult to control within rows Stimulation of latent weed seed germination	[27,28,30]
Flame weeding	Most energy-consuming task (Fossil fuel) Possible damage to the crop Effectiveness depends on weed tolerance to heat and weather conditions High machine cost initially	[27,30]
Manual weeding	Large surfaces High cost	[27]
Bioherbicides	Limited products available (only 13 registered products for organic farming; only one is based on plant extract) Nonselective products Too expensive considering the necessary rates Marginal efficacy	[2,27,32]

3. The Use of Plant-Based Bioherbicides

3.1. Bioherbicidal Potential in Plants

The interest in exploiting the natural herbicidal potential of plants stems from a worldwide need to find new sustainable weed control strategies [1].

As plants are the richest source of active organic compounds on Earth, the bioherbicidal potential of a long list of plant species has been explored [32]. A Scopus literature search using the keywords « Bioherbicides AND plant extracts » and « Bioherbicides AND Essential oils » found 130 articles (excluding review articles) on the bioherbicidal potential of species from 38 different families (Figure 1).

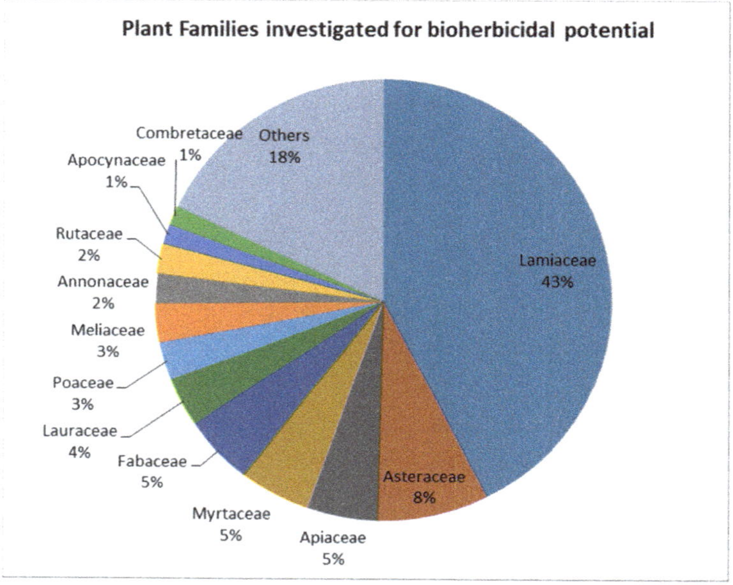

Figure 1. Plant families studied for their bioherbicidal potential. (Source: elaborated from a search on Scopus, 2019).

Plant species can be considered for investigations due to their known composition in terms of biologically active compounds, or an observed allelopathic effect in their natural environment. Allelopathy is a characteristic of many plant species, and can be defined as a form of interaction between plants through chemical inhibitors released from living or decaying tissues [3]. Evidence that some plants are able to inhibit the growth of other plants in their surroundings has long been known and reported, and studies have linked these interactions to the presence of compounds named "allelochemicals" [4,21,33]. This can justify the high interest in the Lamiaceae family species—accounting for 43% of the total studied species from the papers covered by the search—which are known to possess high concentrations of volatile allelochemicals. In this context, much effort has been made to extract allelochemicals from plants and test their bioherbicidal activity in bioassays; many were successful (Table 2). Other studies evaluated the activity of single compounds isolated from plants, such as flavonoids, alkaloids and terpenoids. These different classes of secondary metabolites have different importance in term of bioherbicidal activity. In extracts from the same species, they can occur in different proportions, depending on the type of extract and the extraction method [34].

Table 2. Examples of frequently tested families for bioherbicidal activity.

Species	Family	Bio Herbicidal Effect	Reference
Xanthium strumarium L.	Asteraceae	Significant inhibition of germination and growth of the noxious weed *Bidens pilosa* L.	[35]
Thymus fontanesii Boiss. et Reut. *Satureja calamintha* subsp. *nepeta* Briq.	Lamiaceae	Wide herbicidal effect on seed germination and 3–4 leaf stage of *Sinapis arvensis* L, *Avena fatua* L., *Sonchus oleraceus* L., *Xanthium strumarium* L., *Cyperus rotundus* L.	[36]
Ulex europaeus L. *Cytisus scoparius* L.	Fabaceae	Pure volatile organic compounds extracted caused irreversible phytotoxicity for *Digitaria sanguinalis* L.	[37]
Trachyspermum copticum L.	Apiaceae	Germination and shoot/root length of *Zea mays* L. and *Lepidium sativum* L. significantly reduced by all concentrations of EO and methanol extract.	[38]
Eucalyptus citriodora Hook	Myrtaceae	*Parthenium hysterophorus* L.: Germination completely inhibited. Chlorophyll content and respiratory activity decreased for 4-week-old plants.	[39]

3.2. Types of Active Compounds and Plant Extracts Tested as Bioherbicides

3.2.1. Active Compounds with Bioherbicidal Potential

The term active compounds usually refers to secondary metabolites occurring in plants, known for having diverse biological activities. These are the compounds with no relevance to vital functions (like respiration, photosynthesis and reproduction), but involved in interactions between plants and their surrounding environment, notably as part of their mechanism of defense against stress [40,41]. Secondary metabolites in plants have been classified differently by different authors; a recent review by Yasri et al., [42] defined four main groups: terpenoids, phenolics, sulphur-containing secondary metabolites and nitrogen-containing secondary metabolites. Not all of these groups of secondary metabolites were found to be implicated in allelopathic interactions or showed a bioherbicidal potential. Although some authors included amino acids and proteins among the phytotoxic compounds, terpenoids and phenolics were the ones most frequently studied [4]. Only these two groups will therefore be considered in detail in this paragraph.

(a)—Terpenoids

The terpenoid group is present in the majority of secondary metabolite classifications and is reported to be very important in allelopathic interactions [4,21,43]. The compounds of this group, sometimes referred to as volatile allelochemicals, can be divided into monoterpenes, sesquiterpenes, diterpene, triterpenes and polyterpenes [42]. Monoterpenes are the major constituents of essential oils and have been shown to inhibit seed germination and seedling growth [14]. They are the most frequently described secondary metabolites for bioherbicidal activity [4,14,42–46]. Some monoterpene-based commercial herbicides have been developed such as cinmethylin, which is a derivative of 1,4-cineole [14]. Compounds having different chemical functions (Table 3) belonging to this sub-group were found to have varied inhibition effects.

Most authors considered that ketone-containing compounds such as camphor and pulegone are the most toxic, followed by alcohol compounds such as cineol and citronellol, and by ether, diene and monoene compounds such as α-pinene, which are the least toxic [47]. This was confirmed by many other authors [14,47]. Considering that plant species and chemotypes have different monoterpene composition, the phytotoxicity of extracts can vary between plant materials with different percentages of effective compounds (e.g camphor). However, there is no clear evidence reported in the literature as to how the active compounds of a plant extract define its activity level. In other words, it is not clear whether the observed toxic effect of plant extracts is due to the potent phytotoxicity of a single compound or to the synergic action of many constituents.

Table 3. Examples of monoterpenes and their phytotoxic effect.

Monoterpene	Representative Structure	Chemical Function	Containing Plant Species	Germination Inhibition	Reference
α-pinene	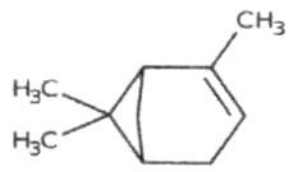	Monoene	*Eucalyptus grandis* W. Hill ex Maiden *Rosmarinus officinalis* L.	*Amaranthus hybridus* L. * *Portulaca oleracea* L. * *Pisum sativum* L. *Cicer arietinum* L.	[47] [48]
Limonene	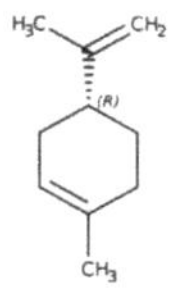	Monoene	*Citrus limon* (L.) Burm. f. *Apium graveolens* L.	*Amaranthus viridis* L. *	[49]
1,8-cineole	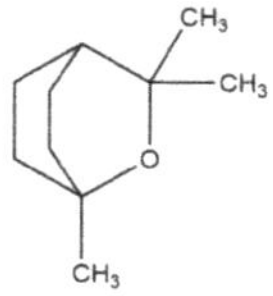	Ether	*Eucalyptus* spp. *R. officinalis* L.	*Ageratum conyzoides* L. *	[50]
α-phellandrene	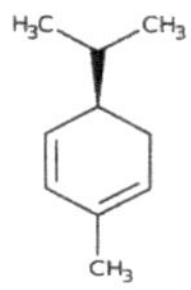	Diene	*Ligusticum marginatum* C.B. Clarke	*Raphanus sativus* L. *	[44]
Linalool	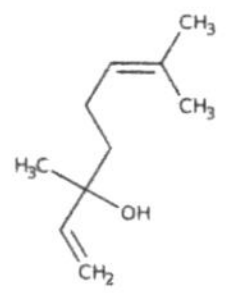	Alcohol	*Mentha* spp. *Lavandula hybrida* L.	*Echinochloa crus-galli* L. ** at highest concentration	[51]
Camphor	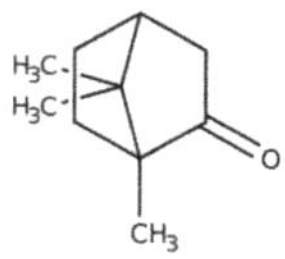	Ketone	*Lavandula abrialis* L. *R. officinalis* L.	*Amaranthus retroflexus* L. ** *L. multiflorum* L.** at low concentation	[14]
Pulegone	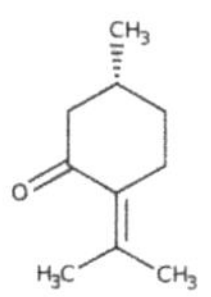	Ketone	*Mentha piperita* L. *Calamintha arkansana* (Nutt.) Shinners	*R. sativus* L. ** at low concentation	[44]
Menthol	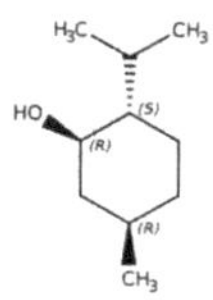	Alcohol	*Helianthus annuus* L. *Mentha* spp.	*A. retroflexus* L. ** *Lolium multiflorum* L. ** *Lactuca sativa* L.*	[14] [49]
Citronellol	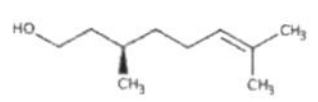	Alcohol	*Rosa* spp. *Eucalyptus* spp.	*Chenopodium album* L. * *A. retroflexus* L. * *E. crus-galli* L. ** at highest concentration	[18] [52]
Borneol	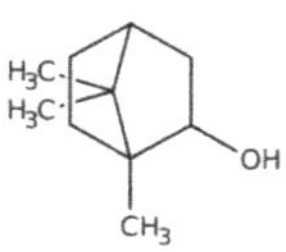	Alcohol	*Salvia officinalis* L. *R. officinalis* L.	*Lepidium sativum* L. *R. sativus* L. *	[53]

Table 3. *Cont.*

Monoterpene	Representative Structure	Chemical Function	Containing Plant Species	Germination Inhibition	Reference
Carvacrol	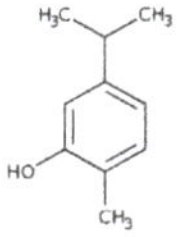	Alcohol	*Origanum vulgare* L. *Thymus capitatus* L.	*L. perenne* L. ** *A. retroflexus* L. **	[31]

* Significant effect; ** Total inhibition; *N*: no significant effect.

(b)—Phenolics

Plant phenolics include phenolic acids, flavonoids and tannins. They are synthetized by plants as a response to ecological and physiological conditions, mainly when they are under biotic or abiotic stress [54]. Like the terpenoids, an important focus exists on the identification of phenols with bioherbicidal activity. This was attributed to the easiness of their extraction and their water solubility [47]. They are usually the main components in aqueous and organic solvents extracts, and their polarity determines the type and amount of phenols extracted. An example of a well-studied phenolic for this effect is juglone (Figure 2) produced by walnuts [55].

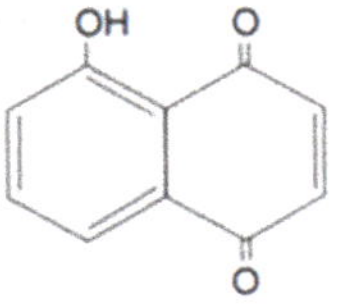

Figure 2. Representative structure of juglone.

3.2.2. Types of Plant Extracts Tested for their Bioherbicidal Activity

The extraction method is a determining factor in the recovery of active compounds from plants, considering that secondary metabolites of different groups have varying chemical properties (volatility, polarity etc.). For instance, anthocyanins, tannins, saponins and terpenoids can be recovered using water, whereas polyphenols, flavonoids, flavones and alkaloids require organic solvents [34]. Like the terpenoids, much research focuses on identifying phenols with bioherbicidal potential. This is because they are easily extracted and are soluble in water [43]. They are usually the main components in aqueous and organic solvent extracts, the polarity of which determines the type and amount of phenols extracted. An example of a well-studied phenolic with this effect is juglone (Figure 2), produced by walnuts [55]. These different extracts often show different levels of toxicity. In a study conducted to test *Calamintha nepeta* L. (Savi) as a source of phytotoxic compounds, solvents of varying polarity (*n*-hexane, chloroform, ethyl acetate and n-butanol) were used to fractionate the leaves' methanol extract. The study defined the following hierarchical phytotoxicity: ethyl acetate > *n*-hexane > chloroform > *n*-butanol [56].

In general, three main groups of extracts can be found in the literature: essential oils (EOs), aqueous extract and organic solvent extracts.

- Essential oils: Sometimes called volatile oils, these are natural substances that can be extracted from aromatic plants by distillation or by appropriate mechanical process without heating. EOs mainly contain compounds that can be volatilized during extraction, making this an effective means of extracting plant terpenoids in the purest form [18,57]. These, the most frequently tested extracts from aromatic plants, can cause higher phytoxicity compared to aqueous or organic solvent extracts [38,56,58].
- Aqueous extracts: These are obtained simply by soaking in water ground dry material from plants, from which water-soluble compounds are extracted. Several phenols

are water soluble and can successfully be extracted using this method. Aqueous extracts have been used to investigate the bioherbicidal potential of many plants and have been found to produce significant effects mainly at the highest tested concentrations [1,59,60].

- Organic solvent extracts. This group consists mainly of phenols. As the type of solvent (mainly differing in polarity) affects the amount and type of phenols extracted, authors have used various solvents. Methanol, ethanol, acetone, ethyl acetate, n-hexane and chloroform are among those used most frequently [54]. The choice of solvent depends on the types of phenol present in the tested plant, and many authors have tested different ones simultaneously in order to compare the composition and phytotoxicity of the resulting extract [56,61].

3.3. *Modes of Action of Plant Allelochemicals*

After investigating the type of active compounds in the plants' extracts, research has also explored the mechanisms of toxicity to weeds. The most frequently described effects are from single allelochemicals rather than whole plant extracts; the modes of action of terpenoids and phenolic acids, which are reported to be the most relevant secondary metabolites in allelopathic interactions, have been studied by many authors. However, studies on this topic are still lacking and the mechanism of only a few phytotoxic compounds has been described. This paragraph therefore focuses mainly on toxicity mechanisms reported for single allelochemicals, as well as allelopathic mechanisms observed in nature. Note that to determine the stage in which plants are most sensitive to allelochemicals, the latter were often tested in two different periods (pre-germination and post-emergence), and different features and mechanisms were analyzed accordingly.

3.3.1. Effect on Cells Division, Elongation and Structure

The size and weight of weed seedlings are the features most often measured to assess their reaction to the application of allelopathic compounds. The application of plant extracts usually results in a significant decrease of these parameters compared to the control. The substances undoubtedly affect the responsible physiological processes: cell division and elongation [21]. In this sense some studies have reported that some allelochemicals affect mitosis: the process was either slowed down [19], interrupted in the anaphases or hindered altogether [16,19,20,62,63]. All the cited studies measured the number of cells and their ultrastructure at specific times as indicators. Muller [16] also reported that volatile terpenes extracted from *Salvia leucophylla* Greene (mainly cineol and camphor), prevented the elongation of root and hypocotyl cells. Cineole is in fact the most widely described of all monoterpenes [64]. It is generally reported to strongly inhibit all stages of mitosis. The suggested mechanism can therefore result in considerable damage to weeds by reducing their growth or retarding it, which can give the crop a competitive advantage.

3.3.2. Effect on the Cells Membrane Integrity and Permeability

Cell membrane integrity is critical for cell functions and survival. Any alteration may compromise its role as a barrier, affecting permeability to nutrients or toxins or inducing the leakage of solutes [65,66]. A number of allelochemicals seem to alter plant cell membranes. Due to lipophilic nature of the cell membranes, monoterpenes can cause their destruction by increasing permeability or inhibiting enzymes [18]. Moreover, some monoterpenes are reported to induce oxidative stress; α-pinene, for example, caused lipid peroxidation when applied to young seedlings of *Cassia occidentalis* L., resulting in an increase in solute leakage [48]. Furthermore, some compounds produced changes to the permeability of membranes; Varona et al. [67] found that linalool caused an increase in permeability, whereas Muller et al. [16] found that permeability decreased after applying cineole and dipentene from *S. leucophylla*. This suggests that allelochemicals can result in important damage to weeds by acting at the membrane level.

3.3.3. Effect on Photosynthesis

There is evidence of a relationship between the visible effects on weeds and photosynthetic functions. Early studies found a correlation between the reduction in growth caused by a plant-extracted phenolic substance, "scopoletin", and net photosynthesis in *Amarantus retroflexus* L. [68]. Many other studies found that a number of phenolic acids affect photosynthesis, and this was linked to changes to stomatal conductance or to plant chlorophyll contents [68,69]. Furthermore, many monoterpenes were also found to inhibit photosynthesis and chlorophyll synthesis [70]. Citronellol and 1,8-cineole, for example, showed a similar effect on the invasive weed species *Ageratum conyzoides* L.: its chlorophyll content decreased by 60% and 66%, respectively [18,50]. Eugenol, another monoterpene, has a similar effect: it induced photosynthetic inhibition by reducing chlorophyll content in *C. occidentalis* and *Bidens pilosa* L. [71]. These examples suggest that photosynthesis-related processes could be behind the observed damage. However, only a few of the allelochemicals were tested, and the actual cause-effect between the described processes is not yet well understood.

3.3.4. Effect on Nutrients Availability and Uptake

Because of the observed effects on the root appearance, some research has focused on whether allelochemicals inhibit nutrient uptake [21]. The uptake of phosphorous, potassium, calcium and zinc, for example, was affected either by the direct application of some phenolic acids or by growing plants in association with allelopathic species [72–77]. Moreover, some early studies found that toxic excretions from plants reduce the availability of nutrients by affecting nutrient cycling mechanisms; mineralization, for example, was suppressed by the root excretion of some natural forest vegetation due to its toxicity to the nitrification process [78]. This suggests that phytotoxic compounds from plants may affect soil microbial activity, which plays an essential role in making important nutrients like nitrogen available to plants.

All the presented modes of action suggest that allelochemicals have a strong potential as weed control tools. However, they also highlight the many challenges to their practical application. For instance, no clear selectivity can be concluded from the reported mechanisms, which means that crops may also be susceptible. Moreover, the impact on crop and soil health is also of concern if the allelochemicals have a detrimental effect on beneficial soil microbes. This, in addition to other possible challenges, will be detailed in the next paragraph.

3.4. Challenges and Perspectives to the Use of Plant Extracts as Bioherbicides

3.4.1. Challenges to the Use of Plant Extracts as Bioherbicides

(a)—Unclear selectivity

Although allelochemicals may affect specific functions like photosynthesis or respiration, they lack site specificity, which excludes their use as selective bioherbicides. This also means they could be phytotoxic to crops and must be managed carefully when applied. However, many studies that tested plant extracts on different weed species revealed varying degrees of sensitivity. In most cases monocots were more resistant than dicots [7,8,79]. Moreover, it was frequently reported that many crops were less affected than weeds; for instance, when applying the EO of *Satureja hortensis* L. and *Laurus nobilis* L. at low concentrations, *A. retroflexus* germination decreased significantly whereas tomatoes were unaffected. However, at the highest tested concentration, tomato germination also decreased, albeit at a lower rate than *A. retroflexus* [8]. Similar results were obtained when applying *Origanum onites* L. and *Rosmarinus officinalis* L. on *Avena sterilis* L., *Sinapis arvensis* L. and a number of wheat cultivars, where the latter were less affected [7]. This suggests that careful dosage may resolve phytotoxicity to crops. Nevertheless, studies were not able to explain this variation in sensitivity, which makes it difficult to predict and exploit. Further research is required to better understand the mechanism of action of different allelochemicals and the synergies by which they operate in plant extracts.

(b)—Toxicity to soil microorganisms

Organic farming relies on soil health and natural soil processes to satisfy crop nutrient needs and ensure long term fertility. In fact, one of the serious drawbacks of synthetic chemicals is their impact on soil biodiversity and their harmful effect on beneficial organisms Plant extracts with similar effects cannot be recommended regardless of their possible effectiveness on weeds. Only a few studies have addressed this important aspect. As mentioned in the "effect on nutrients uptake and availability" paragraph, some allelochemicals may be detrimental to nitrification bacteria [78]. Moreover, many plant extracts, notably those from the Lamiaceae family, have been shown to possess antimicrobial properties [80–82] Doubts may thus arise about their possible harmful effects on soil microbes. However, other studies have reported a positive impact in this respect; volatile substances from alfalfa (*Medicago sativa* L.), for example, induced a rapid increase in microbial respiration and fungi mycelium growth when added to the soil. Results thus suggest a possible beneficial effect on the initial colonization stage of plant residue decomposers [83]. The different findings may be ascribed to variations in the concentration of compounds in contact with microorganisms.

In summary, the soil microbial community seems to be affected by allelochemicals (either negatively or positively). Hence, when assessing the use of plant extracts as agrochemicals, care should also be taken to detect any possible negative repercussions on soil life, a crucial component of any sustainable management strategy.

(c)—Degradation of plant extracts in the environment

While the incorporation of allelopathic plant species biomass into the soils is constrained by the difficulty in accumulating active concentrations [21], the direct use of concentrated extracts is mainly limited by susceptibility to environmental elements. Once released in the environment, the extracts are subject to decomposition either by microorganisms or by chemical reactions. Blum [84] reported in the book chapter «Fate of phenolic allelochemicals in soils – the role of soil and rhizosphere microorganisms» that because microorganisms use phenolic acids as a source of carbon or energy, they are thus more subject to microbial transformation and utilization than to other processes (ionization, oxidation, sorption onto soil particles, fixation into the recalcitrant organic matter (e.g., polymerization)). In this chapter the author reports results from many studies suggesting that this degradation is very likely and that phenolic acids are unlikely to produce any phytotoxic effects. Moreover, Marmulla and Harder [85] report that monoterpenes such as d-limonene, α-pinene, γ-terpinene and terpinolene are readily biodegradable. They also found that different monoterpenes show different susceptibility to degradation. In addition, many allelochemicals are highly susceptible to spontaneous decomposition; abiotic photochemical processes in the atmosphere can result in lifetimes of minutes to hours, as cited by the same authors [21]. However, very little is known about their abiotic degradation in soil [81]. These aspects suggest that allelochemicals may lack the necessary persistence to be effective bioherbicides. This may be remedied by selecting critical stages of weed growth. Even a brief period of phytotoxicity could affect the competitive ability of weeds with respect to crops [81]. Another approach recently under study is the use of innovative formulations that could regulate the rate of release without compromising the desired concentration levels. For instance, experiments with rosemary EO encapsulated in a starch matrix were successful [86].

3.4.2. Perspectives for the Use of Plant Extracts as Bioherbicides

Despite the many constraints, the use of plant extracts for weed control is still considered a field with great potential. However, to address limitations, research should focus on better understanding the phenomena in terms of:

- Linking the observed effects of extracts to the action of specific compounds and their synergies;

- Defining the mechanisms behind the phytotoxicity to enhance it and understand the selectivity;
- Defining the most sensitive stages of weed development to increase effectiveness and tackle the problem of the limited duration of the effect;
- Defining innovative formulations that take into consideration the interactions of the extracts with field conditions (soil texture, microorganisms and abiotic factors such as light and temperature);
- Defining innovative techniques for the cultivation and extraction of essential oils to guarantee the commercial feasibility of a mass production large quantity of EOs;
- Defining formulations that allow for containing the concentrations of EOs within technical limits for an easy application on an agricultural scale.

4. Examples of Lamiaceae Species with Bioherbicidal Potential

4.1. Oregano

In the literature, oregano is used to refer to a number of species in different genera and families, the leaves and flowers of which have a common characteristic odor and flavor [87,88]. The major oregano species belonging to the Lamiaceae family, *Poliomintha longiflora* L., *Origanum vulgare* L. and *O. onites*, are mainly found in the Mediterranean basin [87,88]. *O. vulgare*, and *O. vulgare* subsp. *hirtum* (Link) Iestwart plants in particular are extremely rich in essential oils (up to 8% dry weight) [87]. The famous odor and flavor of these species are mainly linked to their carvacrol content, which in addition to thymol, p-cymene and γ-terpinene, is known to be the major component of oregano essential oil [31,89,90]. As for the bioherbicidal effect, pure carvacrol and the EO from some species (mainly *O. vulgare* and *O. onites*) have been widely investigated, with interesting findings. In a study by De Mastro et al. [31] on *A. retroflexus* and *Lolium perenne* L., carvacrol at the concentration of 0.3 μL/mL completely inhibited the germination of both species. The same study assessed the application of dry biomass of an oregano hybrid (*O. vulgare* ssp. *virilidum* x *O. vulgare* L. ssp. *hirtum*) in a pot trial, and promising results were obtained using 20 g per kg of soil. In another investigation by Atak et al. [7], *O. onites* EO was tested against *A. sterilis* and *S. arvensis* L., and severe inhibition was observed on both species starting from 0.2 μL/mL. In this experiment the EO was also tested on a number of wheat cultivars, which were found to be less sensitive: this led the authors to suggest a possible dosed application of the EO as a bioherbicide in wheat fields. Ibáñez and Blázquez [91] also tested an EO dominated by carvacrol (60.42%) against *Portulaca oleracea* L., *Lolium multiflorum* L. and *Echinochloa crus-galli* L. They found that germination was completely inhibited in all the species starting from the lowest tested dose: 0.125 μL/mL. Hanana et al. [92] found the same high effectiveness at low doses on some important weed species (*S. arvensis* L., *Phalaris paradoxa* L. and *Lolium rigidum* Gaud) using a carvacrol- and δ-terpinene- rich EO from *O. vulgare*. Its high yield and strong anti-germination and phytotoxic effect make oregano EO a promising bioherbicide candidate.

4.2. Rosemary

Rosemary (*R. officinalis*), an evergreen shrub that can grow up to 2 m high, has aromatic leaves and flowers rich in essential oils [87,93]. Native to the Mediterranean region, it is characterized by high tolerance to heat, drought, and poor, dry, sandy and rocky soil types [87]. It is grown in different parts of the world, such as Europe, Africa and Asia, and is used in various culinary, pharmaceutical and cosmetics industries [94]. There are three species in the *Rosmarinus* genus, *R. officinalis*, *R. eryocalix* and *R. tomentosus*, but *R. officinalis* is the most widely distributed and important for its valuable EO, which can be extracted in amounts ranging from 0.9 to 2.5%, depending on many factors [93,95,96]. Rosemary EO is appreciated in cosmetics for its strong camphorous aroma, and in medicine for its content in biological compounds of high value [87,97].

The compounds generally found to be dominant in *R. officinalis* EO are 1,8-cineol, camphor, α-pinene, borneol, p-cymene and verbenone, as reported in Hernández et al. [96]. The proportions of these, however, vary considerably among chemotypes [98,99]. *R. officinalis* is one of the Lamiaceae species that has received considerable attention in plant-based bioherbicide research. Many recent studies have investigated the phytotoxicity of its extracts to weeds, and in this respect significant results were obtained on a number of important weed species, such as *A. retroflexus* L., *Bromus tectorum* L., *Cynodon dactylon* L., *Digitaria sanguinalis* L. and *L. perenne* [5,8,99]. All the studies reported a concentration-dependent effect some even found that at very low concentrations, such as 100 or 200 μL/L, the extracts had a stimulatory effect rather than a phytotoxic one [8]. Nevertheless, for some species significant decreases in germination were found at concentrations as low as 400 μL/L [8,99]. More resistant species (mostly monocots), however, were only sensitive to higher tested doses [5,7,8,99]. A range of concentrations were therefore always tested. As EOs are very susceptible to environmental conditions, recent studies are now investigating innovative formulations to support their practical implementation as bioherbicides. For instance nanoformulation and encapsulation in starch were successfully tested as germination and early growth inhibitors [100].

4.3. Thyme

The genus *Thymus* L. from the Lamiaceae family consists of over 200 species of herbaceous perennials and small shrubs [101,102]. Many of these species are widespread in the world, but the center of the genus is considered to be in the Mediterranean region [101,103] *Thymus* is one of the most studied genera for bioactive activity due to its wide use in folk medicine [102]. This bioactive activity can be linked to its high phenolic monoterpene content (e.g., carvacrol and thymol are the major compounds in the species EO), in addition to α-terpinene, α-cymene and borneol [6,92,101]. One of the emerging bioactivity investigations of *Thymus* spp. EO is in weed control research. Similar to oregano, low doses of EO from thyme were reported to have potent phytotoxic effects on a number of problematic weed species. For instance, Hanana et al. [92] tested the EO from *Thymus capitatus* against *S. arvensis*, *P. paradoxa* and *L. rigidum*, and germination inhibition was significant at concentrations as low as 0.25 μL/mL. A recent study by Sarić-Krsmanović et al. [104] also tested the bioherbicidal effect of *T. vulgaris* EO, which is dominated by carvacrol (17.0%), thymol (11.6%) and p-cymene (11.6%), against Abutilon theophrasti Medik. Complete germination inhibition was obtained at a concentration of 1% EO. After applying *Thymus* spp. EO, other weed species such as *A. retroflexus*, *Avena fatua* L., *Datura stramonium* L., *Lepidium sativum* L and *Agrostemma githago* L. were also significantly affected, either at germination or at the seedling growth stage [6,105]. These findings indicate that different species of *Thymus* (*T. fallax* Fisch. et Mey, *T. vulgaris* L., *T. capitatus* (L.) Hoffmanns. & Link, *T. daenensis* Celak.), even if somewhat varied in chemical composition, have great potential as bioherbicides Kashkooli and Saharkhiz [6] tested different ecotypes of the same species (*T. daenensis* Celak.), and despite the important chemical variations in the EOs, no significant differences were observed in their effect on weeds.

4.4. Mint

Mentha L. is another important genus of the Lamiaceae family that is widely distributed and cultivated in most parts of the world thanks to its adaptation to diverse environments. About 42 species and 15 hybrids fall under this genus: they are commonly characterized by odorous secondary metabolites that make its EO famous [106,107]. In fact, most *Mentha* species are industrially cultivated for EO production. Members of the genus *Mentha* show a great variability in chemical composition, both intra- and inter-species, resulting in different chemotypes. Nevertheless, most of the species are either C3-oxygenated p-menthane types (e.g., pulegone, menthone, menthol) or C6-oxygenated p-menthane (e.g., carvone) types [107,108]. The EOs of species from both types figure in bioherbicide research. For instance, the EOs of *M. dumetorum* Schult, *Mentha* × *piperita* L. cv. Mitcham (Peppermint), *M. pulegium* L. and *M. spicata* L. were successfully tested on different weed species as germination and growth inhibitors. Different studies tested them on a variety of weed species with varying results. Onaran et al. [105] found that *M. dumetorum* EO suppressed the germination of *A. theophrasti* better than other tested EOs such as *O. vulgare* and *T. fallax*. Another test used the EO from a specific Peppermint cultivar (*Mentha* × *piperita* L. cv. Mitcham) with 35% menthol on tomato (*Lycopersicon esculentum* Mill.) and radish (*Raphanus sativus* L.), in addition to three weeds: *Convolvulus arvensis* L., *P. oleracea* and *Echinochloa colonum* L. The EO caused varying degrees of inhibition, depending on the species; tomato was particularly sensitive, and germination was completely suppressed at 900μL/L, whereas field bindweed and purslane were still able to germinate even at 1500μL/L [109]. Similarly, a study by Argyropoulos et al. [110] assessed the effects of a *Mentha* species EO (*M. spicata*) on two horticultural crops (cotton and tomato), besides weeds (*A. retroflexus*, *E. crus-galli*, *Oryza sativa* L., *P. oleracea* and *Setaria verticillata*). The EO containing 82% trans piperitone oxide severely inhibited all the tested species, but with a greater effect on cotton. The EOs of *Mentha*, like that of the other species discussed above, mainly have dose- and species-dependent effects. The reason for this selectivity and the mechanisms involved are still unclear.

4.5. Other Species

With fewer occurrences in the literature than those detailed above, other species in the Lamiaceae family were used to produce bioherbicidal extracts, mainly EOs. Among these number the *Salvia*, *Satureja*, *Nepeta* and *Lavandula* species [8,79,111,112]. Findings reveal differences among tested species and in resulting toxicity levels, but all conclude that the tested monoterpene-rich EOs are promising.

5. Conclusions

In summary, many Lamiaceae species are valuable for the bioactive compounds they contain. They are widely distributed throughout the Mediterranean area and are currently cultivated in most parts of the world. Many yield significant amounts of EOs (up to 8%) rich in terpenes, which are considered important in allelopathic interactions, and may thus have potential as bioherbicides. To this end, a considerable amount of research is still needed on technical aspects, such as exploring the mechanism of action, understanding selectivity, investigating side effects on beneficial plants, and exploring innovative formulations for effective application. Furthermore, studies are required to assess cost-benefits and define the target value for the class crops, as well as the environmental impact of production.

Author Contributions: Conceptualization, G.D.M.; methodology, J.E.M.; investigation, J.E.M.; writing—original draft preparation, J.E.M.; review and editing, J.E.M., C.R., G.D.M.; project administration, G.D.M. All authors have read and agreed to the published version of the manuscript.

Funding: This research received no external funding.

Conflicts of Interest: The authors declare no conflict of interest.

References

1. Puig, C.G.; Reigosa, M.J.; Valentão, P.; Andrade, P.B.; Pedrol, N. Unravelling the bioherbicide potential of Eucalyptus globulus Labill: Biochemistry and effects of its aqueous extract. *PLoS ONE* **2018**, *13*, e0192872. [CrossRef]
2. Cordeau, S.; Triolet, M.; Wayman, S.; Steinberg, C.; Guillemin, J.-P. Bioherbicides: Dead in the water? A review of the existing products for integrated weed management. *Crop. Prot.* **2016**, *87*, 44–49. [CrossRef]
3. Zimdahl, R.L. Allelopathy. In *Fundamentals of Weed Science*, 5th ed.; Zimdahl, R.L., Ed.; Academic Press: Fort Collins, CO, USA, 2018; pp. 253–270.
4. Ramezani, S.; Saharkhiz, M.J.; Ramezani, F.; Fotokian, M.H. Use of Essential Oils as Bioherbicides. *J. Essent. Oil Bear. Plants* **2008**, *11*, 319–327. [CrossRef]
5. Chen, F.; Peng, S.; Chen, B.-M.; Ni, G.; Liao, H. Allelopathic potential and volatile compounds of *Rosmarinus officinalis* L. against weeds. *Allelopath. J.* **2013**, *32*, 57–66.
6. Kashkooli, A.B.; Saharkhiz, M.J. Essential Oil Compositions and Natural Herbicide Activity of Four Denaei Thyme (Thymus daenensis Celak.) Ecotypes. *J. Essent. Oil Bear. Plants* **2014**, *17*, 859–874. [CrossRef]
7. Atak, M.; Mavi, K.; Uremis, I. Bioherbicidal effects of oregano and rosemary essential oils on germination and seedling growth of bread wheat cultivars and weeds. *Rom. Biotechnol. Lett.* **2016**, *21*, 11149–11158.

8. Hazrati, H.; Saharkhiz, M.J.; Moein, M.; Khoshghalb, H. Phytotoxic effects of several essential oils on two weed species and Tomato. *Biocatal. Agric. Biotechnol.* **2018**, *13*, 204–212. [CrossRef]
9. Matković, A.; Marković, T.; Vrbničanin, S.; Sarić-Krsmanović, M.; Božić, D. Chemical composition and in vitro herbicidal activity of five essential oils on Johnson grass (*Sorghum halepense* [L.] Pers.). *Lek. Sirovine* **2018**, *38*, 44–50. [CrossRef]
10. Jouini, A.; Verdeguer, M.; Pinton, S.; Araniti, F.; Palazzolo, E.; Badalucco, L.; Laudicina, V. Potential Effects of Essential Oils Extracted from Mediterranean Aromatic Plants on Target Weeds and Soil Microorganisms. *Plants* **2020**, *9*, 1289. [CrossRef]
11. Abd-Elgawad, A.M.; El Gendy, A.E.-N.G.; Assaeed, A.M.; Al-Rowaily, S.L.; Alharthi, A.S.; Mohamed, T.A.; Nassar, M.I.; Dewir Y.H.; ElShamy, A.I. Phytotoxic Effects of Plant Essential Oils: A Systematic Review and Structure-Activity Relationship Based on Chemometric Analyses. *Plants* **2020**, *10*, 36. [CrossRef]
12. Maccioni, A.; Santo, A.; Falconieri, D.; Piras, A.; Farris, E.; Maxia, A.; Bacchetta, G. Phytotoxic effects of Salvia rosmarinus essential oil on Acacia saligna seedling growth. *Flora-Morphol. Distrib. Funct. Ecol. Plants* **2020**, *269*, 151639. [CrossRef]
13. Muller, W.H.; Lorber, P.; Haley, B.; Johnson, K. Volatile Growth Inhibitors Produced by Salvia leucophylla: Effect on Oxygen Uptake by Mitochondrial Suspensions. *Bull. Torrey Bot. Club* **1969**, *96*, 89–96. [CrossRef]
14. Vaughn, S.F.; Spencer, G.F. Volatile Monoterpenes as Potential Parent Structures for New Herbicides. *Weed Sci.* **1993**, *41*, 114–119 [CrossRef]
15. Alipour, M.; Saharkhiz, M.J. Phytotoxic activity and variation in essential oil content and composition of Rosemary (*Rosmarinus officinalis* L.) during different phenological growth stages. *Biocatal. Agric. Biotechnol.* **2016**, *7*, 271–278. [CrossRef]
16. Muller, W.H. Volatile Materials Produced by Salvia leucophylla: Effects on Seedling Growth and Soil Bacteria. *Int. J. Plant Sci.* **1965**, *126*, 195–200. [CrossRef]
17. Macías, F.A.; Marín, D.; Oliveros-Bastidas, A.; Varela, R.M.; Simonet, A.M.; Carrera, C.; Molinillo, J.M. Allelopathy as a new strategy for sustainable ecosystems development. *Biol. Sci. Space* **2003**, *17*, 18–23. [CrossRef]
18. Grana, E.; Sanchez-Moreiras, A.M.; Reigosa, M.J. Mode of Action of Monoterpenes in Plant-Plant Interactions. *Curr. Bioact. Compd.* **2012**, *8*, 80–89. [CrossRef]
19. Cornman, I. The responses of onion and lily mitosis to coumarin and parasorbic acid. *J. Exp. Biol.* **1947**, *23*, 292–297. [PubMed]
20. Jensen, T.E.; Welbourne, F. The Cytological Effects of Growth Inhibitors on Excised Roots of Vicia Faba and Pisum Sativum. 1962. Available online: https://sdaos.org/wp-content/uploads/pdfs/Vol%2041%201962/62p131.pdf (accessed on 22 July 2019).
21. Rice, E.L. *Allelopathy*, 2nd ed.; Rice, E.L., Ed.; Academic Press: San Diego, CA, USA, 1984; pp. 8–344.
22. Drinkwater, L. Ecological knowledge: Foundation for sustainable organic agriculture. In *Organic Farming: The Ecological System*; Francis, C., Ed.; American society of agronomy: New York, NY, USA, 2009; pp. 19–47.
23. Kirschenmann, F.L. *Cultivating an Ecological Conscience: Essays from a Farmer Philosopher*; Falk, C., Ed.; University Press of Kentucky: Lexington, KY, USA, 2010; pp. 48–61. [CrossRef]
24. Liebman, M.; Mohler, C.L.; Staver, C.P. Weed management: A need for ecological. In *Ecological Management of Agricultural Weeds*; Liebman, M., Mohler, C.L., Staver, C.P., Eds.; Cambridge University Press: Cambridge, UK, 2001; pp. 1–20.
25. Bastiaans, L.; Paolini, R.; Baumann, D.T. Focus on ecological weed management: What is hindering adoption? *Weed Res.* **2008**, *48*, 481–491. [CrossRef]
26. Barberi, P. Weed management in organic agriculture: Are we addressing the right issues? *Weed Res.* **2002**, *42*, 177–193. [CrossRef]
27. Leslie, A.; Chen, G.; Hooks, C.R. Organic Integrated Weed Management in Vegetables. 2016. Available online: https://extension.umd.edu/sites/extension.umd.edu/files/_images/programs/mdvegetables/OrganicWeedMgntVeg21May2016(1).pdf (accessed on 25 July 2019).
28. Bàrberi, P. Integrated weed management in organic cropping systems. In *Improving Organic Crop Cultivation*; Köpke, U., Ed.; Burleigh Dodds: Bonn, Germany, 2018; pp. 323–341.
29. Smith, R.; Lanini, W.T.; Gaskell, M.; Mitchell, J.; Koike, S.T.; Fouche, C. Weed Management for Organic Crops. 2000, Volume 7250, p. 2. Available online: https://anrcatalog.ucanr.edu/pdf/7250.pdf (accessed on 3 September 2019).
30. Bond, W.; Grundy, A.C. Non-chemical weed management in organic farming systems. *Weed Res.* **2001**, *41*, 383–405. [CrossRef]
31. Da Mastro, G.; Fracchiolla, M.; Verdini, L.; Montemurro, P. Oregano and its Potential Use as Bioherbicide. *Acta Hortic.* **2006**, *723*, 335–346. [CrossRef]
32. Flamini, G. Natural Herbicides as a Safer and More Environmentally Friendly Approach to Weed Control: A Review of the Literature Since 2000. In *Studies in Natural Products Chemistry*; Elsevier BV: Amsterdam, The Netherlands, 2012; Volume 37, pp. 353–396.
33. Molisch, H. Der Einfluss einer Pflanze auf die Andere, Allelopathie. *Nat. Cell Biol.* **1938**, *141*, 493. [CrossRef]
34. Azmir, J.; Zaidul, I.; Rahman, M.; Sharif, K.; Mohamed, A.; Sahena, F.; Jahurul, M.; Ghafoor, K.; Norulaini, N.; Omar, A. Techniques for extraction of bioactive compounds from plant materials: A review. *J. Food Eng.* **2013**, *117*, 426–436. [CrossRef]
35. El-Gawad, A.A.; ElShamy, A.; El Gendy, A.E.-N.; Gaara, A.; Assaeed, A. Volatiles Profiling, Allelopathic Activity, and Antioxidant Potentiality of Xanthium Strumarium Leaves Essential Oil from Egypt: Evidence from Chemometrics Analysis. *Molecules* **2019**, *24*, 584. [CrossRef]
36. Benchaa, S.; Hazzit, M.; Zermane, N.; Abdelkrim, H. Chemical composition and herbicidal activity of essential oils from two Labiatae species from Algeria. *J. Essent. Oil Res.* **2019**, *31*, 335–346. [CrossRef]
37. Pardo-Muras, M.; Puig, C.G.; López-Nogueira, A.; Cavaleiro, C.; Pedrol, N. On the bioherbicide potential of Ulex europaeus and Cytisus scoparius: Profiles of volatile organic compounds and their phytotoxic effects. *PLoS ONE* **2018**, *13*, e0205997. [CrossRef]

38. Kayanifard, M.; Mohsenzadeh, S. Allelopathic Analysis of Four Ecotypes of Ajowan. *Iran. J. Sci. Technol. Trans. A Sci.* **2017**, *41*, 971–978. [CrossRef]
39. Singh, H.P.; Batish, D.R.; Setia, N.; Kohli, R.K. Herbicidal activity of volatile oils from Eucalyptus citriodora against Parthenium hysterophorus. *Ann. Appl. Biol.* **2005**, *146*, 89–94. [CrossRef]
40. Bernhoft, A. A Brief Review on Bioactive Compounds in Plants. In *Bioactive Compounds in Plants—Benefits and Risks for Man and Animals*; Bernhoft, A., Ed.; The Norwegian Academy of Science and Letters: Oslo, Norway, 2010; pp. 11–17.
41. Yasri, A.; Naboulsi, I.; Aboulmouhajir, A.; Kouisni, L.; Bekkaoui, F. Plants extracts and secondary metabolites, their extraction methods and use in agriculture for controlling crop stresses and improving productivity: A review. *Acad. J. Med. Plants* **2018**, *6*, 223–240.
42. Kumar, A.; Memo, M.; Mastinu, A. Plant behaviour: An evolutionary response to the environment? *Plant Biol.* **2020**, *22*, 961–970. [CrossRef]
43. Echeng, F.; Echeng, Z. Research Progress on the use of Plant Allelopathy in Agriculture and the Physiological and Ecological Mechanisms of Allelopathy. *Front. Plant Sci.* **2015**, *6*, 1020. [CrossRef]
44. Asplund, R. Monoterpenes: Relationship between structure and inhibition of germination. *Phytochemistry* **1968**, *7*, 1995–1997. [CrossRef]
45. Reynolds, T. Comparative Effects of Alicyclic Compounds and Quinones on Inhibition of Lettuce Fruit Germination. *Ann. Bot.* **1987**, *60*, 215–223. [CrossRef]
46. Dudai, N.; Ben-Ami, M.; Chaimovich, R.; Chaimovitsh, D. Essential oils as allelopathic agents: Bioconversion of monoterpenes by germinating wheat seeds. *Acta Hortic.* **2004**, *629*, 505–508. [CrossRef]
47. Elakovich, S.D. Terpenoids as Models for New Agrochemicals. In *Biologically Active Natural Products*; Cutler, H.G., Ed.; ACS Symposium Series; American Chemical Society: Hattiesburg, MS, USA, 1988; Volume 380, pp. 250–261.
48. Singh, H.P.; Batish, D.R.; Kaur, S.; Arora, K.; Kohli, R.K. α-Pinene Inhibits Growth and Induces Oxidative Stress in Roots. *Ann. Bot.* **2006**, *98*, 1261–1269. [CrossRef]
49. Vaid, S.; Batish, D.R.; Singh, H.P.; Kohli, R.K. Phytotoxicity of limonene against *Amaranthus viridis* L. *Bioscan* **2011**, *6*, 163–165.
50. Singh, H.P.; Batish, D.R.; Kohli, R.K. Allelopathic effect of two volatile monoterpenes against bill goat weed (*Ageratum conyzoides* L.). *Crop. Prot.* **2002**, *21*, 347–350. [CrossRef]
51. Singh, D.H.; Batish, D.R.; Kaur, S.; Vaid, S.; Kohli, R. Weed suppressing ability of some monoterpenes. *J. Plant Dis. Prot.* **2004**, *19*, 821–828.
52. Vokou, D.; Douvli, P.; Blionis, G.J.; Halley, J.M. Effects of monoterpenoids, acting alone or in pairs, on seed germination and subsequent seedling growth. *J. Chem. Ecol.* **2003**, *29*, 2281–2301. [CrossRef] [PubMed]
53. De Martino, L.; Mancini, E.; De Almeida, L.F.R.; De Feo, V. The Antigerminative Activity of Twenty-Seven Monoterpenes. *Molecules* **2010**, *15*, 6630–6637. [CrossRef]
54. Dai, J.; Mumper, R.J. Plant Phenolics: Extraction, Analysis and Their Antioxidant and Anticancer Properties. *Molecules* **2010**, *15*, 7313–7352. [CrossRef]
55. Stewart, A.J.; Stewart, R.F. Phenols. In *Encyclopedia of Ecology*; Jørgensen, S.E., Fath, B.D., Eds.; Academic Press: Oxford, UK, 2008; pp. 2682–2689.
56. Araniti, F.; Lupini, A.; Mercati, F.; Statti, G.A.; Abenavoli, M.R. Calamintha nepeta L. (Savi) as source of phytotoxic compounds: Bio-guided fractionation in identifying biological active molecules. *Acta Physiol. Plant* **2013**, *35*, 1979–1988. [CrossRef]
57. Eslahi, H.; Fahimi, N.; Sardarian, A. Chemical composition of essential oils: Chemistry, safety and applications. In *Essential Oils in Food Processing: Chemistry, Safety and Applications*; Hashemi, S.M.B., Khaneghah, A.M., Sant'Ana, A.S., Eds.; Wiley: Hoboken, NJ, USA, 2017; pp. 119–171.
58. Kordali, S.; Tazegul, A.; Cakir, A. Phytotoxic effects of Nepeta meyeri Benth. Extracts and essential oil on seed germinations and seedling growths of four weed species. *Rec. Nat. Prod.* **2015**, *9*, 404–418.
59. Islam, A.M.; Hasan, M.; Musha, M.H.; Uddin, K.; Juraimi, A.S.; Anwar, P. Exploring 55 tropical medicinal plant species available in Bangladesh for their possible allelopathic potentiality. *Ann. Agric. Sci.* **2018**, *63*, 99–107. [CrossRef]
60. Scavo, A.; Restuccia, A.; Pandino, G.; Onofri, A.; Mauromicale, G. Allelopathic effects of Cynara cardunculus L. leaf aqueous extracts on seed germination of some Mediterranean weed species. *Ital. J. Agron.* **2018**, *11*, 119–125. [CrossRef]
61. Tigre, R.; Silva, N.; Santos, M.; Honda, N.; Falcão, E.; Pereira, E. Allelopathic and bioherbicical potential of Cladonia verticillaris on the germination and growth of Lactuca sativa. *Ecotoxicol. Environ. Saf.* **2012**, *84*, 125–132. [CrossRef] [PubMed]
62. Issa, M.; Chandel, S.; Singh, H.P.; Batish, D.R.; Kohli, R.K.; Yadav, S.S.; Kumari, A. Appraisal of phytotoxic, cytotoxic and genotoxic potential of essential oil of a medicinal plant Vitex negundo. *Ind. Crops Prod.* **2020**, *145*, 112083. [CrossRef]
63. Dutra, Q.P.; Christ, J.A.; Carrijo, T.T.; Alves, T.D.A.; Alves, T.D.A.; Mendes, L.A.; Praça-Fontes, M.M. Phytocytotoxicity of volatile constituents of essential oils from Sparattanthelium Mart. species (Hernandiaceae). *Sci. Rep.* **2020**, *10*, 1–11. [CrossRef] [PubMed]
64. Duke, S.O.; Oliva, A. Mode of Action of Phytotoxic Terpenoids. In *Allelopathy: Chemistry and Mode of Action of Allelochemicals*; Macias, F.A., Galindo, J.C.G., Molinillo, J.M.G., Cutler, H.G., Eds.; CRC Press: Boca Raton, FL, USA, 2003; pp. 201–216. ISBN 0-8493-1964-1.
65. McNeil, P.L.; Steinhardt, R.A. Loss, Restoration, and Maintenance of Plasma Membrane Integrity. *J. Cell Biol.* **1997**, *137*, 1–4. [CrossRef] [PubMed]

66. Pieracci, J.P.; Armando, J.W.; Westoby, M.; Thommes, J. Industry review of cell separation and product harvesting methods. In *Biopharmaceutical Processing*; Jagschies, G., Lindskog, E., Łącki, K., Galliher, P., Eds.; Elsevier: Cambridge, CA, USA, 2018; pp. 165–206.
67. Varona, S.; Martín, Á.; Cocero, M.J. Formulation of a natural biocide based on lavandin essential oil by emulsification using modified starches. *Chem. Eng. Process. Process. Intensif.* **2009**, *48*, 1121–1128. [CrossRef]
68. Einhellig, F.A.; Rice, E.L.; Risser, P.G.; Wender, S.H. Effects of Scopoletin on Growth, CO_2 Exchange Rates, and Concentration of Scopoletin, Scopolin, and Chlorogenic Acids in Tobacco, Sunflower, and Pigweed. *Bull. Torrey Bot. Club* **1970**, *97*, 22–23. [CrossRef]
69. Einhellig, F.A.; Rasmussen, J.A. Effects of three phenolic acids on chlorophyll content and growth of soybean and grain sorghum seedlings. *J. Chem. Ecol.* **1979**, *5*, 815–824. [CrossRef]
70. Pouresmaeil, M.; Nojadeh, M.S.; Movafeghi, A.; Maggi, F. Exploring the bio-control efficacy of Artemisia fragrans essential oil on the perennial weed Convolvulus arvensis: Inhibitory effects on the photosynthetic machinery and induction of oxidative stress. *Ind. Crops Prod.* **2020**, *155*, 112785. [CrossRef]
71. Vaid, S.; Batish, D.; Singh, D.H.; Kohli, R. Phytotoxic effect of eugenol towards two weedy species. *Bioscan* **2010**, *5*, 339–341.
72. Chambers, E.E.; Holm, L.G. Phosphorus Uptake as Influenced by Associated Plants. *Weeds* **1965**, *13*, 312–314. [CrossRef]
73. Iii, C.E.O.; Rice, E.L. Relative Effects of Known Plant Inhibitors on Species from First Two Stages of Old-Field Succession. *Southwest. Nat.* **1970**, *15*, 165. [CrossRef]
74. Glass, A.D.M. Influence of Phenolic Acids on Ion Uptake: I. Inhibition of phosphate uptake. *Plant Physiol.* **1973**, *51*, 1037–1041. [CrossRef]
75. Glass, A.D.M. Influence of Phenolic Acids upon Ion Uptake: III. Inhibition of potassium absorption. *J. Exp. Bot.* **1974**, *25*, 1104–1113. [CrossRef]
76. Harper, J.R.; Balke, N.E. Characterization of the Inhibition of K+ Absorption in Oat Roots by Salicylic Acid. *Plant Physiol.* **1981**, *68*, 1349–1353. [CrossRef]
77. Glass, A.D.M. The allelopathic potential of phenolic acids associated with the rhizosphere of Pteridium aquilinum. *Can. J. Bot.* **1976**, *54*, 2440–2444. [CrossRef]
78. Greenland, D.J. Nitrate fluctuations in tropical soils. *J. Agric. Sci.* **1958**, *50*, 82–92. [CrossRef]
79. Cavalieri, A.; Caporali, F. Effects of essential oils of cinnamon, lavender and peppermint on germination of Mediterranean weeds. *Allelopath. J.* **2010**, *25*, 1–5.
80. Al-Mariri, A.; Safi, M. In Vitro Antibacterial Activity of Several Plant Extracts and Oils against Some Gram-Negative Bacteria. *Iran. J. Med Sci.* **2014**, *39*, 36–43.
81. Elisha, I.L.; Botha, F.S.; McGaw, L.J.; Eloff, J.N. The antibacterial activity of extracts of nine plant species with good activity against Escherichia coli against five other bacteria and cytotoxicity of extracts. *BMC Complement. Altern. Med.* **2017**, *17*, 1–10. [CrossRef]
82. Mostafa, A.A.; Al-Askar, A.A.; Almaary, K.S.; Dawoud, T.M.; Sholkamy, E.N.; Bakri, M.M. Antimicrobial activity of some plant extracts against bacterial strains causing food poisoning diseases. *Saudi, J. Biol. Sci.* **2018**, *25*, 361–366. [CrossRef]
83. Menzies, J.D.; Gilbert, R.G. Responses of the Soil Microflora to Volatile Components in Plant Residues. *Soil Sci. Soc. Am. J.* **1967**, *31*, 495–496. [CrossRef]
84. Blum, U. Fate of phenolic allelochemicals in soils—The role of soil and rhizosphere microorganisms. In *Allelopathy: Chemistry and Mode of Action of Allelochemicals*; Macías, F.A., Galindo, J.C.G., Molinillo, J.M.G., Cutler, H.G., Eds.; CRC Press: Boca Raton, FL, USA, 2004; pp. 57–76.
85. Emarmulla, R.; Eharder, J. Microbial monoterpene transformations—A review. *Front. Microbiol.* **2014**, *5*, 346. [CrossRef]
86. Alipour, M.; Saharkhiz, M.J.; Niakousari, M.; Damyeh, M.S. Phytotoxicity of encapsulated essential oil of rosemary on germination and morphophysiological features of amaranth and radish seedlings. *Sci. Hortic.* **2019**, *243*, 131–139. [CrossRef]
87. Kokkini, S.; Karousou, R.; Hanlidou, E. HERBS | Herbs of the Labiatae. In *Encyclopedia of Food Sciences and Nutrition*, 2nd ed.; Caballero, B., Ed.; Academic Press: Oxford, UK, 2003; pp. 3082–3090.
88. Rodriguez-Garcia, I.; Silva-Espinoza, B.; Ortega-Ramirez, L.; Leyva, J.; Siddiqui, M.; Cruz-Valenzuela, M.; Gonzalez-Aguilar, G.; Ayala-Zavala, J.F. Oregano Essential Oil as an Antimicrobial and Antioxidant Additive in Food Products. *Crit. Rev. Food Sci. Nutr.* **2016**, *56*, 1717–1727. [CrossRef]
89. Figiel, A.; Szumny, A.; Gutiérrez-Ortíz, A.; Carbonell-Barrachina, Á.A. Composition of oregano essential oil (*Origanum vulgare*) as affected by drying method. *J. Food Eng.* **2010**, *98*, 240–247. [CrossRef]
90. Teixeira, B.; Marques, A.; Ramos, C.; Serrano, C.; Matos, O.; Neng, N.R.; Nogueira, J.M.F.; Saraiva, J.A.; Nunes, M.L. Chemical composition and bioactivity of different oregano (Origanum vulgare) extracts and essential oil. *J. Sci. Food Agric.* **2013**, *93*, 2707–2714. [CrossRef]
91. Ibáñez, M.D.; Blázquez, M.A. Herbicidal value of essential oils from oregano-like flavour species. *Food Agric. Immunol.* **2017**, *28*, 1168–1180. [CrossRef]
92. Hanana, M.; Mansour, M.B.; Algabr, M.; Amri, I.; Gargouri, S.; Romane, A.; Jamoussi, B.; Hamrouni, L. Potential use of es-sential oils from four Tunisian species of Lamiaceae: Biological alternative for fungal and weed control. *Rec. Nat. Prod.* **2017**, *11*, 258–269.
93. Sasikumar, B. Rosemary. In *Handbook of Herbs and Spices*; Peter, K.V., Ed.; CRC Press: Boca Raton, FL, USA, 2004; pp. 243–255.
94. Ribeiro-Santos, R.; Carvalho-Costa, D.; Cavaleiro, C.; Costa, H.S.; Albuquerque, T.G.; Castilho, M.C.; Ramos, F.; Melo, N.R.; Sanches-Silva, A. A novel insight on an ancient aromatic plant: The rosemary (*Rosmarinus officinalis* L.). *Trends Food Sci. Technol.* **2015**, *45*, 355–368. [CrossRef]

95. Ghanmi, M.; Satrani, B.; Aberchane, M.; Ismaili, M.R.; Aafi, A.; Abid, A.E. Nature valorisation et caractérisation des PAM des strates arbustives et herbacées. In *Plantes Aromatiques et Médicinales au Maroc: Les Milles et une Vertus*; Bayed, G.E., Ed.; Centre de Recherche Forestière: Rabat, Morocco, 2011; p. 28.
96. Hernández, M.D.; Sotomayor, J.A.; Hernandez, A.S.M.; Jordán, M.J. Rosemary (*Rosmarinus officinalis* L.) Oils. In *Essential Oils in Food Preservation, Flavor and Safety*; Elsevier BV: Amsterdam, The Netherlands, 2016; pp. 677–688.
97. Andrade, J.M.; Faustino, C.; Garcia, C.; Ladeiras, D.; Reis, C.P.; Rijo, P. Rosmarinus officinalis L.: An update review of its phytochemistry and biological activity. *Futur. Sci. OA* **2018**, *4*, FSO283. [CrossRef]
98. Li, G.; Cervelli, C.; Ruffoni, B.; Shachter, A.; Dudai, N. Volatile diversity in wild populations of rosemary (Rosmarinus officinali s L.) from the Tyrrhenian Sea vicinity cultivated under homogeneous environmental conditions. *Ind. Crops Prod.* **2016**, *84*, 381–390. [CrossRef]
99. El Mahdi, J.; Tarraf, W.; Ruta, C.; Piscitelli, L.; Aly, A.; De Mastro, G. Bio-Herbicidal Potential of The Essential Oils from Different *Rosmarinus officinalis* L. Chemotypes in Laboratory Assays. *Agronomy* **2020**, *10*, 775. [CrossRef]
100. de Barros Fernandes, R.V.; Borges, S.V.; Botrel, D.A. Gum arabic/starch/maltodextrin/inulin as wall materials on the microencapsulation of rosemary essential oil. *Carbohydr. Polym.* **2014**, *101*, 524–532. [CrossRef]
101. Ghasemi Pirbalouti, A.; Bistghani, Z.; Malekpoor, F. An overview on genus Thymus. *J. Herb. Drugs.* **2015**, *6*, 93–100.
102. Leal, F.; Taghouti, M.; Nunes, F.; Silva, A.C.C.A.; Matos, M. Thymus Plants: A Review—Micropropagation, Molecular and Antifungal Activity. In *Active Ingredients from Aromatic and Medicinal Plants*; El-Shemy, H., Ed.; IntechOpen: London, UK, 2017; pp. 107–126.
103. Morales, R. The history, botany and taxonomy of the genus Thymus. In *Thyme, the Genus Thymus*; Stahl-Biskup, E., Sáez, F., Eds.; Taylor & Francis: London, UK, 2002; pp. 1–43. ISBN 0415284880.
104. Sarić-Krsmanović, M.; Gajić Umiljendić, J.; Radivojević, L.; Šantrić, L.; Potočnik, I.; Đurović-Pejčev, R. Bioherbicidal effects of five essential oils on germination and early seedling growth of velvetleaf (*Abutilon theophrasti* Medik.). *J. Environ. Sci. Health B* **2019**, *54*, 247–251. [CrossRef]
105. Onaran, A.; Yilar, M.; Belguzar, S.; Bayan, Y.; Aksit, H. Antifungal and Bioherbicidal Properties of Essential Oils of Thymus fallax Fish & Mey., Origanum vulgare L. and Mentha dumetorum Schult. *Asian J. Chem.* **2014**, *26*, 5159–5164. [CrossRef]
106. Fatih, B.; Madani, K.; Chibane, M.; Duez, P.; Brahmi, F.; Khodir, M.; Mohamed, C.; Pierre, D. Chemical Composition and Biological Activities of Mentha Species. In *Aromatic and Medicinal Plants—Back to Nature*; El-Shemy, H.A., Ed.; IntechOpen: London, UK, 2017; Chapter 3. [CrossRef]
107. Salehi, B.; Armstrong, L.; Rescigno, A.; Yeskaliyeva, B.; Seitimova, G.; Beyatli, A.; Sharmeen, J.; Mahomoodally, M.F.; Sharopov, F.; Durazzo, A.; et al. Lamium Plants-A Comprehensive Review on Health Benefits and Biological Activities. *Molecules* **2019**, *24*, 1913. [CrossRef]
108. Croteau, R. Metabolism of Monoterpenes in Mint (Mentha) Species. *Planta Med.* **1991**, *57*, S10–S14. [CrossRef] [PubMed]
109. Mahdavikia, F.; Saharkhiz, M.J. Phytotoxic activity of essential oil and water extract of peppermint (Mentha × piperita L. CV. Mitcham). *J. Appl. Res. Med. Aromat. Plants* **2015**, *2*, 146–153. [CrossRef]
110. Argyropoulos, E.; Eleftherohorinos, I.; Vokou, D. In vitro evaluation of essential oils from Mediterranean aromatic plants of the Lamiaceae for weed control in tomato and cotton crops. *Allelopath. J.* **2008**, *22*, 69–78.
111. Mutlu, S.; Atici, O.; Esim, N. Bioherbicidal effects of essential oils of Nepeta meyeri Benth. on weed spp. *Allelopath. J.* **2010**, *26*, 291–300.
112. Taban, A.; Saharkhiz, M.J.; Hadian, J. Allelopathic potential of essential oils from fourSaturejaspp. *Biol. Agric. Hortic.* **2013**, *29*, 244–257. [CrossRef]

plants

MDPI

Article

Phytochemical Constituents and Allelopathic Potential of *Parthenium hysterophorus* L. in Comparison to Commercial Herbicides to Control Weeds

Mst. Motmainna [1], Abdul Shukor Juraimi [1,*], Md. Kamal Uddin [2], Norhayu Binti Asib [3], A. K. M. Mominul Islam [4], Muhammad Saiful Ahmad-Hamdani [1] and Mahmudul Hasan [1]

[1] Department of Crop Science, Faculty of Agriculture, Universiti Putra Malaysia, Serdang 43400, Selangor, Malaysia; gs51794@student.upm.edu.my (M.M.); s_ahmad@upm.edu.my (M.S.A.-H.); gs53801@student.upm.edu.my (M.H.)

[2] Department of Land Management, Faculty of Agriculture, Universiti Putra Malaysia, Serdang 43400, Selangor, Malaysia; mkuddin@upm.edu.my

[3] Department of Plant Protection, Faculty of Agriculture, Universiti Putra Malaysia, Serdang 43400, Selangor, Malaysia; norhayuasib@upm.edu.my

[4] Department of Agronomy, Faculty of Agriculture, Bangladesh Agricultural University, Mymensingh 2202, Bangladesh; akmmominulislam@bau.edu.bd

* Correspondence: ashukur@upm.edu.my

Abstract: The allelopathic effect of various concentrations (0, 6.25, 12.5, 50 and 100 g L^{-1}) of *Parthenium hysterophorus* methanol extract on *Cyperus iria* was investigated under laboratory and glasshouse conditions. No seed germination was recorded in the laboratory when *P. hysterophorus* extract was applied at 50 g L^{-1}. In the glasshouse, *C. iria* was mostly injured by *P. hysterophorus* extract at 100 g L^{-1}. The phytochemical constituents of the methanol extract of *P. hysterophorus* were analyzed by LC-ESI-QTOF-MS=MS. The results indicated the presence of phenolic compounds, terpenoids, alkaloids, amino acids, fatty acids, piperazines, benzofuran, indole, amines, azoles, sulfonic acid and other unknown compounds in *P. hysterophorus* methanol extract. A comparative study was also conducted between *P. hysterophorus* extract (20, 40 and 80 g L^{-1}) with a synthetic herbicide (glyphosate and glufosinate ammonium at 2 L ha^{-1}) as a positive control and no treatment (negative control) on *Ageratum conyzoides*, *Oryza sativa* and *C. iria*. The growth and biomass of test weeds were remarkably inhibited by *P. hysterophorus* extract. Nevertheless, no significant difference was obtained when *P. hysterophorus* extract (80 g L^{-1}) and synthetic herbicides (glyphosate and glufosinate ammonium) were applied on *A. conyzoides*.

Keywords: allelopathy; phytochemicals; *P. hysterophorus*; germination; growth

Citation: Motmainna, M.; Juraimi, A.S.; Uddin, M.K.; Asib, N.B.; Islam, A.K.M.M.; Ahmad-Hamdani, M.S.; Hasan, M. Phytochemical Constituents and Allelopathic Potential of *Parthenium hysterophorus* L. in Comparison to Commercial Herbicides to Control Weeds. *Plants* **2021**, *10*, 1445. https://doi.org/10.3390/plants10071445

Academic Editors: Alessia Restuccia and Aurelio Scavo

Received: 17 May 2021
Accepted: 11 June 2021
Published: 15 July 2021

1. Introduction

Cyperus iria L. (family: Cyperaceae) is a smooth, tufted sedge weed of lowland rice worldwide and is also a common weed in upland fields of 22 countries [1]. This weed is also reported to appear in dry, direct-seeded rice fields in 21 countries and wet-seeded rice in 11 countries [2]. The roots of *C. iria* are numerous, yellowish-red, short and fibrous. The leaves are usually shorter than culm, 1–8 mm wide and the inflorescence is simple or compound. A prolific nature (5000 seeds from a single plant) and a very short life cycle of *C. iria* help it to establish a second generation in the same growing season [3,4]. It is estimated that approximately 64% of rice yield reduction occurs due to this weed [5].

Weed management in the crop field is a challenging task in agriculture. Chemical herbicides are mainly preferred by the farmers to control weeds due to their higher efficacy, affordable cost and more rapid out return. The migration of labor away from agriculture to industries or other countries for employment is also a major concern for dependence in some countries [6]. However, the excessive use of synthetic herbicides can lead to an increase in the number of herbicide-resistant biotypes [7], low agricultural production environmental pollution and health hazards [8,9]. On the other hand, the introduction of allelopathic plants or bio-herbicide develop from allelochemicals can play an important role as a substitute for the chemical dependence on synthetic chemical herbicides to control weeds in sustainable agriculture [10].

Invasive weed species have the potential to release allelopathic substances to the surrounding environments to suppress their neighboring competing plants [11–15]. *Parthenium hysterophorus* L. has taken the shape of a noxious weed and is becoming a threat to crop production, animal husbandry and human health due to its strong allelopathic effects [16–19]. The isolation and identification of the allelopathic substances from *P. hysterophorus* could be used as a tool for the development of a natural-product-based herbicide for weed control.

Bioassays are generally designed to test the allelopathic properties of a plant species. However, a plant that shows strong phytotoxicity on the target plant species in laboratory conditions might not be so strong in the field condition due to the influence of several environmental factors [20,21]. In this context, two experiments were conducted in both laboratory and glasshouse conditions to evaluate the allelopathic properties of *P. hysterophorus* with a view to developing natural-product-based bioherbicides. The identification of its phytochemical constituents was analyzed by using LC-ESI-QTOF-MS=MS.

2. Results

2.1. Laboratory Experiment

Effect of Methanol Extracts on Germination and Initial Growth of *C. iria*

The results showed that *P. hysterophorus* extracts significantly ($p \leq 0.05$) reduced the germination percentage as well as coleoptile and radicle length of *C. iria* (Table 1). The inhibitory activity was concentration-dependent. By the application of methanol extracts, the seed germination was significantly ($p \leq 0.05$) reduced. No seed germination was recorded when *P. hysterophorus* extract was applied at 50 g L^{-1}.

Parthenium hysterophorus extract decreased the coleoptile and radicle elongation of *C. iria*. The magnitude of inhibition increased with an increase in extract concentration. At a concentration of 50 g L^{-1} or above, *P. hysterophorus* extract reduced the coleoptile and radicle length of *C. iria* by 100%.

Table 1. Effects of *P. hysterophorus* on germination, coleoptile and radicle length of *C. iria*.

Dose (g L^{-1})	Germination (%)	Coleoptile Length (cm)	Radicle Length (cm)
0.00	100.00a (0)	1.51a (0)	1.66a (0)
6.25	80.00b (20)	1.20b (20.72)	1.10b (33.68)
12.5	47.00c (53)	0.86c (43.14)	0.60c (64.02)
25	19.00d (81)	0.36d (76.24)	0.24d (85.65)
50	0.00e (100)	0.00e (100)	0.00e (100)
100	0.00e (100)	0.00e (100)	0.00e (100)

Data are expressed as means. Means with same letters in the column for concentrations are not significantly different at $p > 0.05$. Values inside the parenthesis are inhibition percentages relative to the control.

2.2. Glasshouse Experiment

2.2.1. Effect of Methanol Extract on Plant Height, Leaf Area and Dry Weight of *C. iria*

Table 2 showed the effect of *P. hysterophorus* methanol extract on the plant height, leaf area and dry weight of tested weeds. Dose-dependent inhibitory activity was also observed here. *Parthenium hysterophorus* showed significant inhibition on plant height at the highest concentration (100 g L^{-1}). At the concentration of 100 g L^{-1}, *P. hysterophorus* extract 44.40% inhibition was observed on the plant height of *C. iria*. A decline in leaf area of the tested weed was also observed with an increase in *P. hysterophorus* methanol extract concentration. The leaf area inhibition of *C. iria* ranged from 7.63 to 52.03% from 6.25 g L^{-1} to 100 g L^{-1}

concentrations of *P. hysterophorus* extract. The control obtained the highest dry weight. The extract reduced 60.81% of the dry weight of *C. iria* at 100 g L^{-1} compared to the control.

Table 2. Effect of *P. hysterophorus* methanol extracts on the plant height (cm), leaf area (cm^2) and dry weight (g pot^{-1}) of *C. iria*.

Dose (g L^{-1})	Plant Height	Leaf Area	Dry Weight
0	64.75a (0)	151.05a (0)	5.12a (0)
6.25	63.37ab (2.13)	139.52b (7.63)	4.89ab (4.46)
12.5	62.02ab (4.20)	132.24c (12.44)	4.53b (11.44)
25	57.42b (11.29)	115.22d (23.70)	3.86c (24.55)
50	50.31c (22.31)	91.15e (39.63)	3.00d (41.20)
100	36.00d (44.40)	72.45f (52.03)	2.00e (60.81)

Data are expressed as means. Means with same letters in the column for each extract concentrations are not significantly different at $p > 0.05$. Values inside the parenthesis are inhibition percentages relative to the control.

2.2.2. Effect of Methanol Extract on Fv/Fm, Photosynthesis Rate, Stomatal Conductance and Transpiration Rate of *C. iria*

No significant difference was observed when *C. iria* was treated with 6.25 and 12.5 g L^{-1} of *P. hysterophorus* extract (Table 3). The extract reduced the Fv/Fm value by 46.32% at 100 g L^{-1}. The significant effect of extracts concentrations was observed on the photosynthesis, stomatal conductance and transpiration rate of *C. iria*. The photosynthesis rate of *C. iria* was inhibited by 44.41% when treated with the highest concentrations (100 g L^{-1}) of *P. hysterophorus* extract. The lowest stomatal conductance (0.25 mol m^{-2} s^{-1}) was recorded at 100 g L^{-1}, and the inhibition value was 39.63% (Table 4). The lowest transpiration rate was observed at the highest concentration (100 g L^{-1}), and the inhibition value was 40.98%.

Table 3. Effects of *P. hysterophorus* methanol extract on Fv/Fm, photosynthesis rate (μmol m^{-2} s^{-1}), stomatal conductance (mol m^{-2} s^{-1}) and transpiration rate (mmol m^{-2} s^{-1}) of *C. iria*.

Dose (g L^{-1})	Fv/Fm	Photosynthesis Rate	Stomatal Conductance	Transpiration Rate
0	1.47a (0)	45.14a (0)	0.42a (0)	11.50a (0)
6.25	1.41a (3.90)	43.50ab (3.64)	0.41ab (3.43)	10.83b (5.82)
12.5	1.34a (8.56)	42.50ab (5.86)	0.40ab (6.04)	10.41c (9.52)
25	1.20ab (17.84)	40.00b (11.37)	0.38b (10.07)	9.35d (18.69)
50	1.08ab (26.19)	35.29c (21.86)	0.34c (20.31)	8.20e (28.67)
100	0.79b (46.32)	25.13d (44.41)	0.25d (39.63)	6.79f (40.98)

Data are expressed as means. Means with same letters in the column for each extract concentrations are not significantly different at $p > 0.05$. Values inside the parenthesis are inhibition percentages relative to the control.

Table 4. LC-MS profile of methanol extract of *P. hysterophorus*

Sl. No	RT (min)	Proposed Compound	Molecular Formula	Mass Fragment (m/z)	Polarity
1	1.436	Valine	$C_5H_{11}NO_2$	117.0802	Positive
2	1.418	Glyceryl sulfoquinovoside	$C_9H_{18}O_{10}S$	318.063	Negative
3	1.575	Lotaustralin	$C_{11}H_{19}NO_6$	261.1215	Positive
4	3.162	Trazolopride	$C_{20}H_{23}N_5O_2$	365.1851	Positive
5	3.571	Pirenzepine	$C_{19}H_{21}N_5O_2$	351.1694	Positive
6	3.92	1-Cyclopropyl-3-[[1-(4-hydroxybutyl)benzimidazol-2-yl]methyl]imidazo [4,5-c]pyridin-2-one	$C_{21}H_{23}N_5O_2$	377.1848	Positive
7	4.239	Umbelliferone	$C_9H_6O_3$	162.0317	Positive
8	4.244	Quinic Acid	$C_7H_{12}O_6$	192.0638	Negative
9	4.941	Atevirdine	$C_{21}H_{25}N_5O_2$	379.2002	Positive

Table 4. *Cont.*

Sl. No	RT (min)	Proposed Compound	Molecular Formula	Mass Fragment (m/z)	Polarity
10	5.253	Dihydrophaseic acid 4-O-beta-D-glucoside	$C_{21}H_{32}O_{10}$	444.1998	Negative
11	5.536	2-(2-Ethoxyethoxy)ethanol;4-methylbenzenesulfonic acid	$C_{13}H_{22}O_6S$	306.1136	Negative
12	5.475	4-Azidobenzyl benzyl 1,4-butanediylbiscarbamate	$C_{20}H_{23}N_5O_4$	397.175	Positive
13	5.823	4-(N-hydroxyamino)-2r-isobutyl-2s-(2-Thienylthiomethyl)succinyl-L-Phenylalanine-N-Methylamide	$C_{20}H_{31}NO_3S_2$	397.176	Positive
14	6.08	Branaplam	$C_{22}H_{27}N_5O_2$	393.2162	Positive
15	6.257	Pulchellamine G	$C_{21}H_{31}NO_6$	393.2151	Positive
16	6.503	Hymenoxynin	$C_{21}H_{34}O_9$	430.2208	Negative
17	6.939	Chlorogenic acid	$C_{16}H_{18}O_9$	354.0957	Negative
18	7.006	Parthenin	$C_{15}H_{18}O_4$	262.1202	Positive
19	7.006	Gaillardilin	$C_{17}H_{22}O_6$	322.1415	Positive
20	7.006	Dehydroleucodine	$C_{15}H_{16}O_3$	244.1095	Positive
21	7.264	N-Propyl-3-(1,3-thiazol-2-yl)thian-3-amine	$C_{11}H_{18}N_2S_2$	242.0928	Positive
22	7.266	Oleacein	$C_{17}H_{20}O_6$	320.1252	Positive
23	7.49	Bendazac lysine	$C_{22}H_{28}N_4O_5$	428.2053	Negative
24	7.641	Lajollamide A	$C_{30}H_{55}N_5O_5$	565.4206	Positive
25	7.673	Isochlorogenic acid A	$C_{25}H_{24}O_{12}$	516.127	Negative
26	7.673	Chlorogenic acid	$C_{16}H_{18}O_9$	354.0958	Negative
27	7.897	4-[(6-Chloro-2-naphthalenyl)sulfonyl]-1-[[1-(4-pyridinyl)-4- piperidinyl]methyl]-2 piperazinecarboxylic acid	$C_{27}H_{41}ClN_4O_6$	552.2699	Positive
28	7.905	N-Chloro-9-(diaminomethylideneamino)-3-hydroxynonanamide	$C_{10}H_{21}ClN_4O_2$	264.1358	Positive
29	7.908	1-(N-6-Amino-n-hexyl)carbamoylimidazole	$C_{10}H_{19}ClN_4O$	246.1253	Positive
30	8.042	2,4-Toluene Diisocyanate Dimer	$C_{18}H_{12}N_4O_4$	348.0862	Positive
31	8.044	Alaptide	$C_9H_{14}N_2O_2$	182.1063	Positive
32	8.05	Carbocyclic-3′-amino-ara-adenosine	$C_{11}H_{16}N_6O_2$	264.1339	Positive
33	8.054	Tris(pyrazolyl)ethane	$C_{11}H_{12}N_6$	228.1118	Positive
34	8.055	Descyclopropyl Abacavir	$C_{11}H_{14}N_6O$	246.1225	Positive
35	8.058	1-Boc-3-oxopiperazine	$C_9H_{16}N_2O_3$	200.1162	Positive
36	8.13	Teroxalene hydrochloride	$C_{28}H_{42}Cl_2N_2OS$	524.2364	Positive
37	8.132	Ethane;(3-oxo-6′-sulfanylcarbonyloxyspiro [2 -benzofuran-1,9′-xanthene]-3′-yl)oxymethanethioicS-acid;propane	$C_{31}H_{38}O_7S_2$	586.206	Positive
38	8.133	(2-Aminoethylamino) 2,2-diaminooxyacetate	$C_4H_{12}N_4O_4$	180.0845	Positive
39	8.134	N-[(S)-2-Benzo[1,3]dioxol-5-yl-4-(4-phenyl-piperidin-1-yl)-butyl]-N-methyl-benzenesulfonamide	$C_{29}H_{34}N_2O_4S$	506.2237	Positive
40	8.135	3-Diazo-1-hexylsulfanyl-1-methylurea	$C_8H_{16}N_4OS$	216.1055	Positive
41	8.135	Ethylene oxide-b-maleic hydrazide	$C_6H_{12}N_8O_3$	244.103	Positive
42	8.136	N-[3-(1H-Imidazol-4-yl)propyl]-N′-methylthiourea	$C_8H_{14}N_4S$	198.0952	Positive
43	8.136	1-Methylpiperazine-1,4-Diium Bis	$C_5H_{14}N_4O_6$	226.0914	Positive
44	8.136	3-(2-Methylpropylthio)-1H-1,2,4-triazol-5-amine	$C_6H_{12}N_4S$	172.0801	Positive
45	8.136	Benzylamidinoisothiourea	$C_9H_{12}N_4S$	208.0792	Positive
46	8.136	1-Amino-3-(propylamino)thiourea	$C_4H_{12}N_4S$	148.0798	Positive
47	8.136	9-hydroxyellipticine	$C_{17}H_{14}N_2O$	262.1122	Positive
48	8.136	4-Phenylamino-3-quinolinecarbonitrile deriv. 28	$C_{27}H_{30}Cl_2N_4O_4$	544.16	Positive
49	8.136	1-(3-ethyl-1,2,4-thiadiazol-5-yl)azetidin-3-amine	$C_7H_{12}N_4S$	184.0793	Positive
50	8.413	1,8,15,22,29,36-Hexaazacyclodotetracontane-2,7,16,21,30,35-hexone	$C_{36}H_{66}N_6O_6$	678.504	Positive
51	8.415	2,4,6-tris(3-methylbutoxy)-1,3,5-triazine	$C_{18}H_{33}N_3O_3$	339.2522	Positive
52	8.435	Arginyl-tyrosyl-aspartic acid	$C_{19}H_{28}N_6O_7$	452.2022	Positive
53	8.636	8-(2,4,6-Trimethoxyphenyl)-9H-purine-2,6-diamine	$C_{14}H_{16}N_6O_3$	316.1282	Positive
54	8.818	Dimethyl 2-(heptane-1-sulfonyl)butanedioate	$C_{13}H_{24}O_6S$	308.1298	Negative
55	8.721	AC-Ala-gln-ala-pna	$C_{19}H_{26}N_6O_7$	450.1864	Positive
56	9.065	Laciniatin	$C_{17}H_{14}O_8$	346.0693	Positive
57	9.067	2-[(3,5-Dinitrobenzoyl)amino]benzoic acid	$C_{14}H_9N_3O_7$	331.0461	Negative

Table 4. *Cont.*

Sl. No	RT (min)	Proposed Compound	Molecular Formula	Mass Fragment (m/z)	Polarity
58	9.243	3-Ethyl-1-propyl-8-(1H-pyrazol-4-yl)-1H-purine-2,6(3H,7H)-dione	$C_{13}H_{16}N_6O_2$	288.134	Positive
59	11.645	Apnea	$C_{18}H_{22}N_6O_4$	386.1696	Positive
60	11.844	Thyroliberin N-ethylamide	$C_{18}H_{26}N_6O_4$	390.2011	Positive
61	11.996	Hexadecasphinganine	$C_{16}H_{35}NO_2$	273.2672	Positive
62	12.034	Phytosphingosine	$C_{18}H_{39}NO_3$	317.2935	Positive
63	12.176	Dihydroxyethyllauramine oxide	$C_{16}H_{35}NO_3$	289.262	Positive
64	12.193	Lauramine oxide	$C_{14}H_{31}NO$	229.2405	Positive
65	12.308	Rishitin	$C_{14}H_{22}O_2$	222.161	Negative
66	12.316	Dioctylnitrosamine	$C_{16}H_{34}N_2O$	270.2673	Positive
67	12.343	Dodecylacrylamide	$C_{15}H_{29}NO$	239.2251	Positive
68	12.349	Tetrabutylurea	$C_{17}H_{36}N_2O$	284.2832	Positive
69	12.703	Aminopregnane	$C_{21}H_{37}N$	303.2934	Positive
70	12.778	Tridecylglycerol	$C_{16}H_{34}O_3$	274.2512	Positive
71	13.164	2,3,3-Tris(1,2-diaminoethyl)-2-ethylhexanoic acid	$C_{14}H_{34}N_6O_2$	318.2769	Positive
72	13.633	4-dodecylbenzenesulfonic acid	$C_{18}H_{30}O_3S$	326.1916	Negative
73	14.691	Angoletin	$C_{18}H_{20}O_4$	300.1357	Positive
74	14.694	Phthalic anhydride	$C_8H_4O_3$	148.069	Positive
75	15.406	Eicosasphinganine	$C_{20}H_{43}NO_2$	329.3298	Positive
76	16.483	Lauryl sulfate	$C_{12}H_{26}O_4S$	266.1551	Negative
77	16.957	Dodecandial-disemicarbazon	$C_{14}H_{28}N_6O_2$	312.2282	Positive
78	18.267	Benzenesulfonic acid, tridecyl-	$C_{19}H_{32}O_3S$	340.2072	Negative
79	19.135	3-[5-(3-Dimethylamino-1,2,4-thiadiazol)-yl] quinuclidine	$C_{11}H_{18}N_4S$	238.125	Positive
80	19.496	Benzenesulfonic acid, undecyl-	$C_{17}H_{28}O_3S$	312.176	Negative
81	19.918	N,N-bis(2-hydroxyethyl)stearylamine	$C_{22}H_{47}NO_2$	357.3609	Positive
82	20.245	Benzoyl benzenecarboperoxoate;dodecane-1-thiol;toluene	$C_{33}H_{44}O_4S$	536.2965	Positive

2.3. *Identification of Phytotoxic Components from Methanol Extract of P. hysterophorus*

LC-MS analyses of *P. hysterophorus* methanol extract revealed the presence of 82 known compounds that appeared between 1 and 20 mins. The list of proposed compounds with their retention times, molecular formula, polarity and mass fragment (m/z) is shown in Table 4. For most of the constituents, $[M-H]^+$ and $[M-H]^-$ ions were observed. The total ion current chromatography in positive and negative ESI mode is shown in Figures 1 and 2. Eight amino acids (Valine, Lajollamide A, Alaptide, Arginyl-tyrosyl-aspartic acid, Thyroliberin N-ethylamide, Hexadecasphinganine, Phytosphingosine and Eicosasphinganine) were identified, which usually provides $[M-H]^+$ ions as the best peak positive ESI mode. The amino acids were identified at 1.436, 7.641, 8.004, 8.435, 11.844, 11.996, 12.034, 15.406 min, with 117.0802, 565.4206, 182.1063, 452.2022, 390.2011, 273.2672, 317.2935, 329.3298 m/z, respectively, in the positive ionization mode. A total of seven phenolic compounds (Umbelliferone, Quinic Acid, Chlorogenic acid, Oleacein, Isochlorogenic acid A, Laciniatinand Phthalic anhydride) and three terpenoids (Parthenin, Dehydroleucodine and Rishitin) were also identified. Among the phenolic compounds, chlorogenic acid ($C_{16}H_{18}O_9$) was detected with its $[M-H]^-$ ion at 6.939 min with 354.0957 m/z. In positive ionization mode, parthenin ($C_{15}H_{18}O_4$) was detected at 7.006 min with 262.1202 m/z. A fragment ion at 262.1122 m/z was displayed for 9-hydroxyellipticine (alkaloid) in positive ionization mode at 8.136 min. A number of other organic compounds were also detected in *P. hysterophorus* (Table 4). Descyclopropyl Abacavir ($C_{11}H_{14}N_6O$) is a carbohydrate and was detected from the extract at 8.055 min 246.1225 m/z. At 229.24 m/z, Lauramine oxide ($C_{14}H_{31}NO$) was identified as a detergent at 12.193 min. Glycolipid (Glyceryl sulfoquinovoside, $C_9H_{18}O_{10}S$) and glycoside (Dihydrophaseic acid 4-O-beta-D-glucoside, $C_{21}H_{32}O_{10}$) were identified at 1.418 and 5.253 min with 318.063 and 444.1998 m/z, respectively in the negative ionization mode. One ketone (Angoletin, $C_{18}H_{20}O_4$) was also identified in the positive ionization mode at 14.691 with 300.1357 m/z. Two sulfonic acids, namely, 4-dodecylbenzenesulfonic acid ($C_{18}H_{30}O_3S$) and Benzenesulfonic acid, tridecyl- ($C_{19}H_{32}O_3S$) at 13.633 and 18.267 min with 326.1916 and 312.2282 m/z in negative and positive ionization modes, respectively.

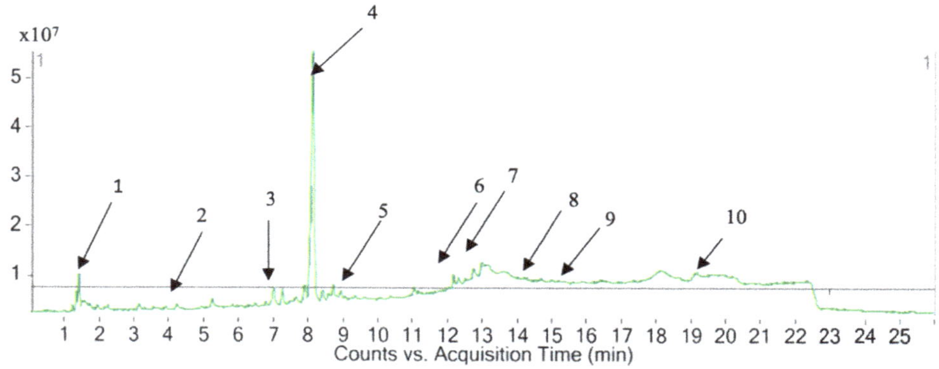

Figure 1. LC-MS chromatograms chemical compounds of *P. hysterophorus* in the positive ion mode (1. Valine, 2. umbelliferone, 3. parthenin, 4. 9-hydroxyellipticine, 5. laciniatin, 6. phytosphingosine, 7. tridecylglycerol, 8. phthalic anhydride, 9. eicosasphinganine, 10. N,N-bis (2-hydroxyethyl) stearylamine).

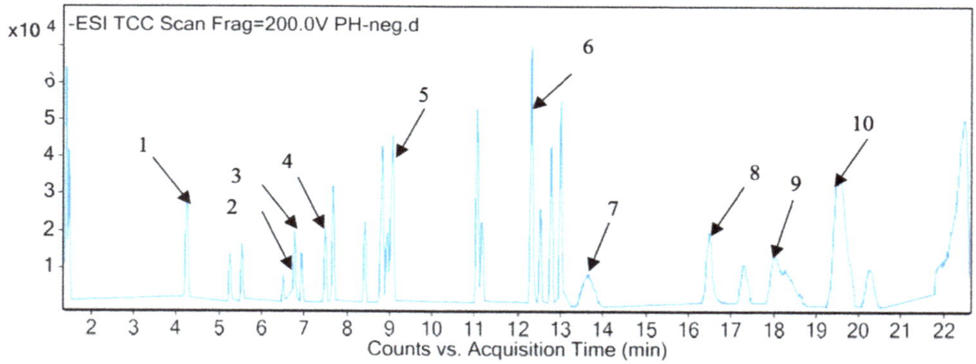

Figure 2. LC-MS chromatograms chemical compounds of *P. hysterophorus* in the negative ion mode (1. Quinic acid, 2. hymonoxynin, 3. chlorogenic acid, 4. isochlorogenic acid, 5. laciniatin, 6. Rishitin, 7. 4-dodecylbenzenesulfonic acid, 8. lauryl sulfate, 9. tridecyl-benzenesulfonic acid, 10. 4-undecyl benzene sulfonic acid).

2.4. Efficacy of P. hysterophorus Extract in Comparison with Commercial Herbicides

All treatments had significant effects ($p \leq 0.05$) on plant height and fresh and dry weight (Table 5). The phytotoxicity effects of *P. hysterophorus* and synthetic herbicide on *A. conyzoides*, *C. iria* and *O. sativa* were evaluated based on visual observation at 21 days after spray (Table 5). The visual injury of *A. conyzoides* was higher compared to *C. iria* and *O. sativa* at the applied concentrations of *P. hysterophorus* methanol extract. At the highest concentration (80 g L^{-1}), *A. conyzoides*, *C. iria* and *O. sativa* were injured severely with an injury rating scale of 9.00, 5.25 and 4.50, respectively. *Cyperus iria* and *O. sativa* were alive and showed either green foliage or minor chlorosis or minor leaf curling at the lowest concentration (20 g L^{-1}). All tested weeds died after treated with synthetic herbicide (glyphosate and glufosinate ammonium). However, only *A. conyzoides* died when *P. hysterophorus* was sprayed at 80 g L^{-1} (Figures 3 and 4).

Table 5. Effect of *P. hysterophorus* on the visual injury, plant height, fresh weight and dry weight of *A. conyzoides*, *C. iria* and *O. sativa*.

	Tested Weeds	*P. hysterophorus*				Synthetic Herbicides	
		0 g L^{-1}	20 g L^{-1}	40 g L^{-1}	80 g L^{-1}	Glyphosate	Glufosinate-Ammonium
Visual injury (Scale)	*A. conyzoides*	1.00d	2.75c	5.50b	9.00a	9.00a	9.00a
	C. iria	1.00e	2.50d	4.00c	5.25b	9.00a	9.00a
	O. sativa	1.00e	2.25d	3.00c	4.50b	9.00a	9.00a
Plant height (cm)	*A. conyzoides*	32.00a (0)	24.62b (23.02)	14.62c (54.32)	0.00d (100)	0.00d (100)	0.00d (100)
	C. iria	64.75a (0)	55.75b (13.58)	44.25c (37.71)	37.00d (42.97)	0.00e (100)	0.00e (100)
	O. sativa	67.00a (0)	58.50b (12.68)	49.50c (26.08)	39.53d (41.02)	0.00e (100)	0.00e (100)
Fresh weight (g pot^{-1})	*A. conyzoides*	26.45a (0)	18.34b (30.66)	3.14c (88.10)	0.45d (98.28)	0.22d (99.17)	0.27d (98.96)
	C. iria	25.95a (0)	20.21b (22.10)	15.70c (39.45)	12.80d (50.60)	0.30e (98.86)	0.50e (98.08)
	O. sativa	12.70a (0)	8.89b (29.97)	6.99c (44.93)	5.44d (57.13)	0.14e (98.92)	0.19e (98.48)
Dry weight (g pot^{-1})	*A. conyzoides*	5.13a (0)	3.04b (40.78)	0.50c (90.36)	0.07c (98.63)	0.03c (99.42)	0.05c (99.08)
	C. iria	6.29a (0)	4.95b (21.12)	3.98c (36.53)	2.28d (63.80)	0.06e (98.97)	0.10e (98.43)
	O. sativa	3.36a (0)	2.25b (32.27)	1.75bc (47.49)	1.24c (62.76)	0.03d (99.05)	0.04d (98.77)

Data are expressed as means. Means with same letters in the row are not significantly different at $p < 0.05$. Values inside the parenthesis are inhibition percentages relative to the control.

Figure 3. Effect of *P. hysterophorus* extract on *A. conyzoides* at 24 h after spray.

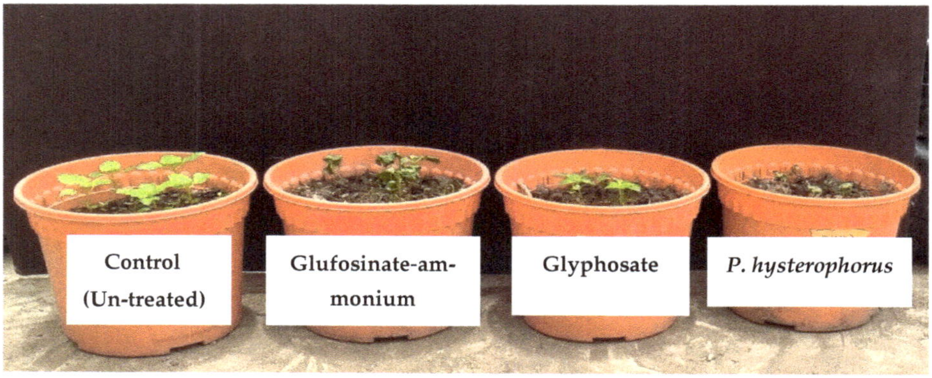

Figure 4. Effect of *P. hysterophorus* extract at 80 g L^{-1} concentration on *A. conyzoides* at 24 h after spray compared with glufosinate-ammonium and glyphosate herbicides.

The plant height of *A. conyzoides*, *C. iria* and *O. sativa* was inhibited by 54.32%, 37.71% and 26.08%, respectively, when treated with *P. hysterophorus* extract at 40 g L^{-1}. The complete inhibition of plant height of *A. conyzoides* was observed on those pots where 80 g L^{-1} of *P. hysterophorus* extract was sprayed, whereas 42.97% and 41.02% plant height inhibitions were observed for *C. iria* and *O. sativa*, respectively, at the same concentration In general, there was a reduction in the fresh and dry weights of treated weeds in pots receiving *P. hysterophorus* extract. The differences in inhibitory activity among the three doses, *viz.* 20, 40 and 80 g L^{-1} of *P. hysterophorus*, on the fresh and dry weight of weeds, were significant. The dry weights of *A. conyzoides*, *C. iria* and *O. sativa* were inhibited by 98.63%, 63.80% and 62.76%, respectively, when *P. hysterophorus* extract was sprayed at 80 g L^{-1}. This result exhibited that there is no significant difference between the foliar spray of *P. hysterophorus* at 80 g L^{-1} and positive control when applied on *A. conyzoides*, whereas *C. iria* and *O. sativa* were less sensitive to *P. hysterophorus* extract compared to the positive control.

3. Discussion

The allelopathic potential of *P. hysterophorus* on *C. iria* was studied in this study. The methanol extract of *P. hysterophorus* influenced *C. iria* seedling growth and germination percentages. The extracts had a dose-dependent effect on the germination percentage, coleoptile and radicle growth of the tested weed. Plant extracts are hypothesized to impede the germination process due to the osmotic effects on the fate of imbibition, which in turn reduce the commencement of germination and, in particular, cell elongation [22]. *C. iria* seed germination and seedling growth were completely suppressed by 50 g L^{-1} of *P. hysterophorus* extract. Batish et al. [23], Singh et al. [24] and Mersie and Singh [25] all observed that *P. hysterophorus* extract or its residues inhibited the growth and development of several field crops. Furthermore, when compared to germination percentage and the coleoptile length, the radicle length of the test species was more sensitive to extracts. As radicles are the first organ to be exposed to phytochemicals and have more permeable tissue than other organs [21,26,27], and/or low mitotic division in the root apical meristem [28], radicle growth is more sensitive to allelopathic plant extract. Furthermore, phytochemicals can inhibit the development of radicle tissues and endoderm by affecting genes involved in cellular characterization [29].

The glasshouse experiment gave more support for the high allelopathic potential of *P. hysterophorus* extract seen in the lab. The results revealed that extracts of *P. hysterophorus* at 50 and 100 g L^{-1} greatly showed the growth of 21-day-old *C. iria*. At the mature stage of *C. iria*, the maximum concentration (100 g L^{-1}) of *P. hysterophorus* extract resulted in the greatest decrease. Many researchers from all around the world have demonstrated dose-dependent inhibitory activity [21,27,30,31]. Only untreated *C. iria* continued flowering 21 days after spray, indicating that allelochemicals stress may have suppressed the other treated plants. Aslam et al. [32] investigated the phytotoxic effect of *Calatropis procera, Peganum harmala* and *Tamarix aphylla* on mustard and wheat shoot and root length, finding that wheat was susceptible to all three extracts at all dosages.

As the concentration of *P. hysterophorus* extract was raised, reduced dry weights and leaf area were reduced. The reduction in plant height and leaf area was discovered to be associated with a reduction in total dry weight. Several studies show that different extracts reduce the leaf area of plant species [33,34]. The dry weight of soybeans was greatly changed by the castor beans leaf aqueous extract, according to Da Silva et al. [35].

Foliar spray of *P. hysterophorus* extract reduced the Fv/Fm, photosynthesis rate, stomatal conductance and transpiration of *C. iria*. The value of Fv/Fm was significantly decreased by the foliar spray of *P. hysterophorus* extract. Thylakoid membrane damage and inhibition of energy transfer from antenna molecules to reaction centers can lead to photoinhibition damage and lower Fv/Fm [36]. Allelochemicals can significantly affect the performance of thylakoid electron transport during light reactions, stomatal control of carbon dioxide and the carbon cycle in dark reactions [37].

The reduction in leaf photosynthesis was attributed to a decrease in photosynthetic metabolites, carboxylation efficiency, impairment of chloroplast activity, increase in enzyme activities [38] and production of ROS caused impediment of photosynthetic mechanism [39]. Stomatal control is a vital property through which the plants limit water loss and gas exchange. These features are influenced by several determinants, including stress [40], and indicate the lower photosynthetic efficiency of plants. The carboxylation and water-use efficiency was also reduced in the plants subjected to *P. hysterophorus* extract.

The reduction in the transpiration rate is certainly associated with stomatal conductance. This study reveals that *P. hysterophorus* extract played a notable role in decreasing the transpiration rate for test plants at different exposure times. The concentration of phenolic acids resulted in a decline in overall water utilization and transpiration of cucumber seedlings in a linear manner [41]. The solution of cinnamic acid and benzoic acids decreased the stomatal conductance and transpiration of cucumber seedlings [42].

It was also observed in the present study that the application of plant extracts in laboratory conditions caused more inhibition compared to glasshouse as a foliar spray. Al-Humaid and El-Mergawi [43] also reported the same. The inhibition by foliar spray may occur through various mechanisms, such as a decreased rate of ion absorption, hormone and enzyme activity, cell membrane permeability and certain physiological processes, e.g., photosynthesis, respiration and protein formation [44]. Thus, the seedling and mature stage of target plants may vary in their sensitivities to plant extracts.

In this research, the methanol extract of *P. hysterophorus* was also investigated for the identification of active phytochemical constituents using LC-MS QTOF and also for their allelopathic potentiality on *C. iria*. Methanol was reported to be an efficient extraction solvent of lower molecular weight polyphenols [45] and a highly efficient solvent for extracting phenolic compounds compared to ethanol [46]. The results indicated the presence of phenolic compounds (flavonoids, phenols, coumarins, carboxylic acids, benzoic acids), terpenoids, alkaloids, amino acids, fatty acids, piperazines, benzofuran, indole, amines, azoles, sulfonic acid and other unknown compounds in *P. hysterophorus*. Among the proposed compounds, some of them have been reported as toxins in different studies. The hydroxyl group of phenolic compounds is directly attached to an aromatic ring. Phenolic allelochemicals are major allelochemicals that inhibited photosynthesis in plants [42] and modified the permeability of root cell membranes, decreased energy metabolism and inhibited cell division and root branching [47]. Research studies revealed that phenolic compounds from *Chenopodium murale* L. affect the growth and macromolecule content in chickpeas and peas [48].

Umbelliferone, a coumarin derivative, was found in *P. hysterophorus*, and, as Pan et al. [49] reported, it shows strong inhibition on lettuce and two field weeds, *Setaria viridis* and *Amaranthus retroflexus*. Phthalic anhydride, another compound of *P. hysterophorus*, formed Phthalic acid in the presence of water, which inhibited the fruit germination of *Lactuca sativa* L. [50]. Three terpenoids (Parthenin, Dehydroleucodine, Rishitin) and one alkaloid (9-hydroxyellipticine) were also found in *P. hysterophorus* extract. Many past and recent research reports revealed that terpenoids and alkaloids are also known for their allelopathic effect. Parthenin reduced the germination and growth of *Avena fatua* L. and *Bidens pilosa* L. and a dose–response relationship was observed by Batish et al. [51]. Valine is an amino acid found in *P. hysterophorus*, which significantly inhibited peach seedling growth [52]. Some fatty acids, amines and sulfonic acids were also observed in the LC-MS analysis of *P. hysterophorus*.

The efficacy of *P. hysterophorus* extract was increased with an increasing application rate. Similarly, the extract phytotoxicity level of *Zingiber officinale* increased with increasing concentration [53]. At 80 g L^{-1}, *P. hysterophorus* extract produced similar efficacy to glyphosate and glufosinate on *A. conyzoides*. Many researchers found the efficacy of bioherbicide for weed control. For instance, *Aglaia odorata* leaf extract has bioherbicide properties that can hinder the growth and development of weeds [54].

Furthermore, the results also indicated that the inhibition magnitude of applied methanol extract of *P. hysterophorus* was species-dependent. The selectivity of an herbicide depends on application rate, the growth stage and morphological characteristics of the target plants and other environmental factors, which might affect the absorption, translocation and metabolism of the herbicide [55].

4. Materials and Methods

Graphical scheme of experimental design was presented in Figure 5.

4.1. Test Plants

Cyperus iria L. (Rice flatsedge) *(voucher specimen#UPMWS019)*, *Ageratum conyzoides* L. (Billygoat-weed) *(voucher specimen#UPMWS001)*, Oryza sativa f. spontanea Roshev (Weedy rice) *(voucher specimen#UPMWS025)* were collected from the rice field of Sekinchan, Kuala Selangor, Selangor, Malaysia.

4.2. Extraction Procedure

The extraction was carried out conducted at Universiti Putra Malaysia's Weed Science Laboratory, which is a part of the Department of Crop Science. Methanol extracts were prepared using the method reported by Aslani et al. [56]. *Parthenium hysterophorus* (*voucher specimen#UPMWS0031*) was obtained at its matured stage in Ladang Infoternak, Sungai Siput, Perak, Malaysia. The plants were properly washed under running tap water to remove dust particles and other debris, and then air-dried for 3 weeks in open trays under shaded conditions at room temperature (25 ± 1 °C). In a Willey mill, the plants were then chopped and crashed. An amount of 100 g powder of *P. hysterophorus* was soaked in a conical flask with 1000 mL methanol: distilled water (80:20, *v*/*v*%) and the flask was wrapped in paraffin. An Orbital shaker was used to shake the flask for 48 h at room temperature (25 ± 1 °C). The solution was filtered through four layers of cheesecloth before being centrifuged at 3000 rpm for 1 hour. Then, a 0.2 mm Nalgene filter was used (Becton Dickinson Labware, Lincoln Park, NJ) to re-filter the solution. A rotary evaporator was used at 40 °C to evaporate the methanol from the extract. The mean extraction yield was 18.56 g from 100 g powdered sample of *P. hysterophorus*.

$$\text{Extraction percentage} = [\text{Extract weight (g)}/\text{powder weight (g)}] \times 100 \quad (1)$$

The crude sample (20 mg) was diluted into 100% HPLC GRADE methanol (20 mL) and filtered with 0.2-µm, 15-mm syringe filters (Phenex, Non-sterile, Luer/Slip, LT Resources, Malaysia) for LC-QTOF-MS/MS analysis.

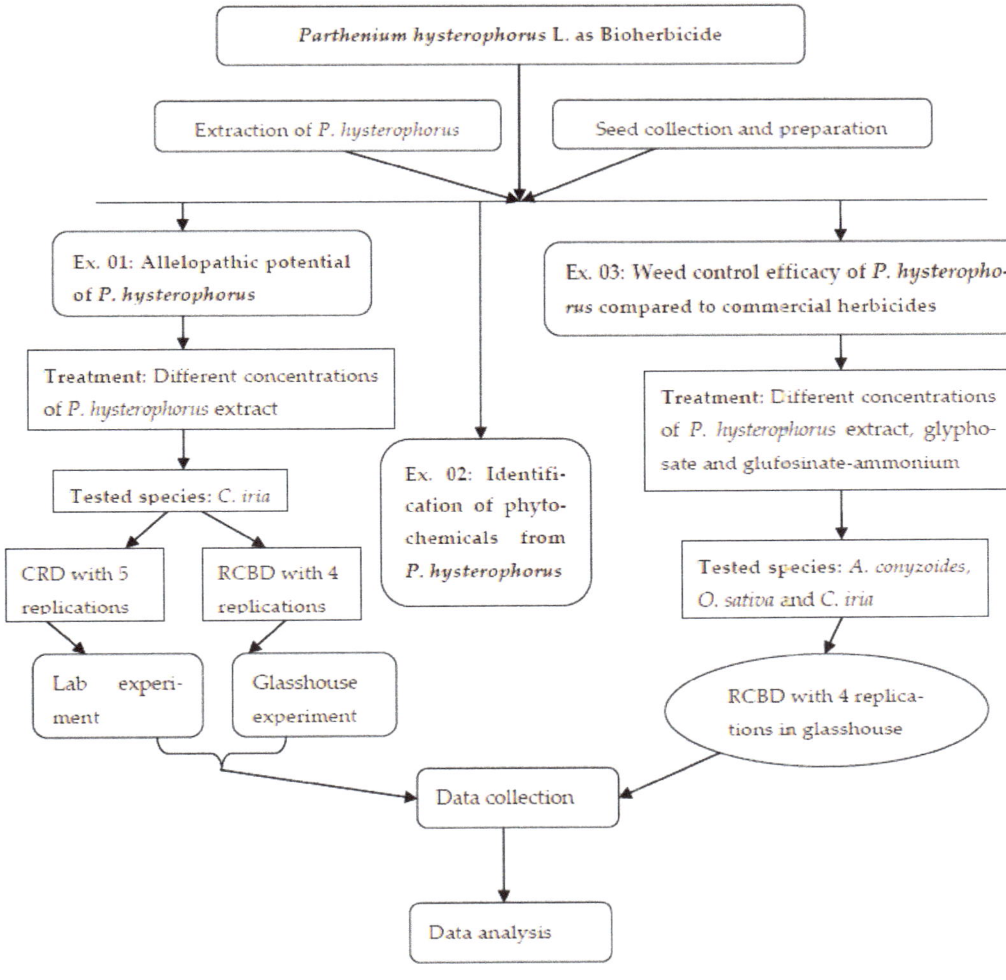

Figure 5. Graphical scheme of study design.

4.3. Laboratory Bioassay

From January to March 2019, the experiment was carried out in a growth chamber at the Seed Technology Laboratory, Department of Crop Science, Universiti Putra Malaysia (3°02′ N, 101°42′ E, 31 m elevation). Seeds were gathered that were healthy and uniform, then soaked for 24 h in 0.2 percent potassium nitrate (KNO_3), rinsed with distilled water and incubated at room (24–26 °C) temperature until the radicle emerged for about 1 mm. Twenty uniform pre-germinated *C. iria* seeds were inserted in disposable plastic Petri dishes with a 9.0-cm-diameter and two sheets of Whatman No. 1 filter paper. After that, the filter paper on the Petri dishes was wetted and soaked with 10 mL of *P. hysterophorus* methanol extracts at six different concentrations: 0 (distilled water only), 6.25, 12.5, 25, 50 and 100 g L^{-1}. The treatment was replicated 5 times in a completely randomized design. The Petri dishes were then incubated under fluorescent light (8500 lux) in a growth chamber at 30/20 °C (day/night) with a 12 h/12 h (day/night cycle). The relative humidity ranged from 30% to 50%. To facilitate gas exchange and avoid anaerobic conditions, the lids of the Petri dishes were not sealed.

All seedlings germination %, coleoptile and radicle length were assessed after 7 days. Image J software [57] was used to measure the length of the coleoptile and radicle, and the inhibitory effect was calculated using the equation below [56]:

$$I = 100\,(C - A)/C \quad (2)$$

where "I" represents the percent inhibition, "C" represents the mean length of coleoptile and radicle of the control and "A" is the mean length of coleoptile and radicle of the methanol extracts treated seeds.

4.4. Glasshouse Experiment

The glasshouse experiment took place at Universiti Putra Malaysia's Faculty of Agriculture in Ladang 15 from April to June 2020. The effects of foliar application of *P. hysterophorus* methanol extracts on the growth and development of *C. iria* were investigated. Pre-germinated seeds were placed in each pot (15 cm diameter × 12 cm height) and covered with 1 cm soil, then moistened with water. Only five healthy seedlings of equal size were maintained in each pot after germination. With four replications, the pots were arranged in a randomized complete block design. Methanol extracts of *P. hysterophorus* were sprayed on examined plants (2–3 leaf stage) at doses of 6.25, 12.5, 25, 50 and 100 g L^{-1} concentrations on tested plants (2–3 leaf stage) using a 1 L multipurpose sprayer (Deluxe pressure sprayer). Water was used to make spray volume (100 mL m^{-2}) [22]. At two-day intervals or when the soil became dry, plants in the control treatment were sprayed with 200 mL water without extract. Three weeks after spray, plant height, leaf area, dry weight, Fv/Fm, photosynthesis rate, transpiration and stomatal conductance were determined. Plant height was measured using 1 m ruler from the ground level in the pot. The leaf area was determined using leaf area meter (LI-3000, Li-COR, USA) and expressed as cm^2 $plant^{-1}$. Samples were dried in an oven at 60 °C for 72 h; then, dry weights were determined using a digital balance. The efficiency of photosystem II in each leaf was measured with a Multi-Function Plant Efficiency Analyser (Hansatech Instruments, King's Lynn, United Kingdom). The Fv parameter (variable fluorescence) was calculated as the difference between the Fm (maximum fluorescence) and Fo (minimum fluorescence). The rate of photosynthesis, transpiration and stomatal conductance were measured from randomly selected four leaves from each test weed species using LICOR (LI-6400XT) portable photosynthesis system, (LI-COR-Inc Lincoln, Nebraska, USA) between 9:00 am to 11:00 am under bright daylight. The measurements were taken on the abaxial surface at CO_2 flow rate of 400 µmol m^{-2} s^{-1} and the saturating photosynthetic photon flux density (PPFD) was 1000 mmol m^{-2} s^{-1} [58].

Another experiment was conducted to compare the phytotoxicity level of *P. hysterophorus* with synthetic herbicides. Therefore, the seeds of *A. conyzoides*, *C. iria* and *O. sativa* were seeded in the pots (15 cm diameter) and moistened with tap water. After germination, five equal-sized healthy seedlings were kept in each pot. The pots were arranged in a randomized complete block design with four replications. Methanol extracts of *P. hysterophorus* were sprayed with 20, 40 and 80 g L^{-1} concentration on tested plants (4–6 leaf stage for broadleaf and 2–3 for grasses and sedges). Plants in the negative control treatment were sprayed with 200 mL water without extract at 2 day intervals or when the soil became dry. Plants in the positive control treatment were sprayed with glyphosate 41% a.i. (Roundup®) and glufosinate-ammonium 13.5% a.i. (Basta®) without extract (2 L ha^{-1}/4.4 mL L^{-1}) at the same time when *P. hysterophorus* was sprayed.

Injury symptoms, plant height (cm) and fresh and dry weights (g pot^{-1}) were measured 3 weeks after spray. Injury symptoms were visually evaluated on test weeds using the European Weed Control and Crop Injury Evaluation scale (Table 6).

Table 6. Injury rating scale [59].

Scale	Injury (%)	Effects on Weeds
1	0	No effect (all foliage green and alive)
2	1–10	Very light symptoms
3	11–30	Light symptoms
4	31–49	Symptoms not reflected in yield
5	50	Medium
6	51–70	Fairly heavy damage
7	71–90	Heavy damage
8	91–99	Very heavy damage
9	100	Complete kill (dead)

4.5. LC-QTOF-MS/MS Analysis

Agilent 1290 Infinity LC system coupled to Agilent 6520 Accurate-Mass Q-TOF mass spectrometer with dual ESI source was used for analyzing chemical constituents from the methanol extract of *P. hysterophorus*. The types of the column, solvent systems and MS parameters were optimized for better analysis of the chemical profiling. ACQUITY UPLC BEH C18 column (150 mm × 2.1 mm × 3.5 μm) was selected and held at 50 °C with a constant flow rate of 0.4 mL min^{-1} for providing fast and efficient separations at lower column pressures [60] and total LC run time was 26 min. Sample elution was performed in a gradient manner using a mobile phase comprised of water (LC-MS Grade) containing 0.1% Formic acid (solvent A) and acetonitrile (LC-MS Grade) containing 0.1% Formic acid (solvent B). Nebulizer pressure was 40 psi, drying gas flow and temperature was set at 10 L min^{-1} and 325 °C, respectively, to perform the MS/MS experiments. In order to obtain the most sensitive ionization effect for analytes, positive and negative ion modes were investigated at different collision energy (CE) to optimize the signals and obtain maximal structure information from the ions for the mass range of 100–3200 m/z. Data processing was performed by Mass Hunter Qualitative Analysis software and peak identification was carried out based on comparison with literature values and online database [61].

4.6. Statistical Analysis

For all trials, a one-way analysis of variance (ANOVA) was used to see if there were any significant differences between the treatments and the control. The Tukey test with a 0.05 probability level was used to pool the differences between the treatment means. The analysis was carried out using SAS (Statistical Analysis System) software (version 9.4).

5. Conclusions

The current study reveals that the *P. hysterophorus* extract was capable of inhibiting the germination and growth of weeds and also confirmed the herbicidal potential compared with synthetic herbicides. The presence of 82 known compounds was also confirmed in the extract of *P. hysterophorus* and some of them have been reported as toxins in different studies. The great efficacy and selectivity of this weed could be characterized as a natural product to control weeds. The use of plant-based bioherbicide for weed management can increase crop yields as well as provide an alternative method of sustainable weed management. The most phytotoxic compounds from *P. hysterophorus* can be synthesize to develop new natural herbicides with novel modes of action. Metabolomics identification and the isolation of the major potential allelopathins, coupled with formulation techniques via multiple surfactants/nano-formulation, are also required to enhance the penetration and absorption of active compounds.

Author Contributions: Conceptualization, A.S.J. and N.B.A.; methodology, A.S.J., N.B.A. and A.K.M.M.I.; validation, A.S.J., M.K.U. and M.M.; formal analysis, M.M., M.H. and M.S.A.-H.; investigation, M.M. and M.H.; resources, A.S.J.; data curation, A.K.M.M.I., M.M. and M.H.; writing—original draft preparation, M.M. and M.H.; writing—review and editing, A.K.M.M.I., A.S.J. and M.S.A.-H. visualization, M.K.U., M.M. and M.H.; supervision, A.S.J., M.K.U., N.B.A., A.K.M.M.I. and M.S.A.-H. project administration, A.S.J.; funding acquisition, A.S.J. All authors have read and agreed to the published version of the manuscript.

Funding: This research received funding from the Fundamental Research Grant Scheme (FRGS), Malaysia (FRGS/1/2013/STWN03/UPM/01/1 and FRGS/1/2017/WAB01/UPM/01/2), Putra Grant UPM (GP-IPB/2017/9523400).

Institutional Review Board Statement: Not applicable.

Informed Consent Statement: Not applicable.

Data Availability Statement: Not applicable.

Acknowledgments: The authors are grateful to the Fundamental Research Grant Scheme (FRGS), Malaysia (FRGS/1/2013/STWN03/UPM/01/1 and FRGS/1/2017/WAB01/UPM/01/2), Putra Grant UPM (GP-IPB/2017/9523400) and also sincerely acknowledge the University Putra Malaysia for providing facilities.

Conflicts of Interest: The authors declare no conflict of interest.

References

1. Holm, L.G.; Plucknett, D.L.; Pancho, J.V.; Herberger, J.P. *The World's Worst Weeds: Distribution and Biology*; The University Press of Hawaii: Malabar, FL, USA, 1991; p. 609.
2. Rao, A.; Johnson, D.; Sivaprasad, B.; Ladha, J.; Mortimer, A. Weed Management in Direct-Seeded Rice. *Adv. Agron.* **2007**, *93*, 153–255. [CrossRef]
3. Chauhan, B.S.; Johnson, D. Ecological studies on *Cyperus difformis*, *Cyperus iria* and *Fimbristylis miliacea*: Three troublesome annual sedge weeds of rice. *Ann. Appl. Biol.* **2009**, *155*, 103–112. [CrossRef]
4. Chauhan, B.S.; Johnson, D.E. Responses of Rice Flatsedge (*Cyperus iria*) and Barnyardgrass (*Echinochloa crus-galli*) to Rice Interference. *Weed Sci.* **2010**, *58*, 204–208. [CrossRef]
5. Dhammu, H.S.; Sandhu, K.S. Critical period of *Cyperus iria* L. competition in transplanted rice. In Proceedings of the 13th Australian Weeds Con-ference: Weeds "Threats now and forever?", Perth, Western Australia, 8–13 September 2002.
6. Shrestha, A. Weed science as a new discipline and its status in some South Asian universities and colleges: Examples from Bangladesh, Bhutan, Nepal and Pakistan. *CAB Rev.* **2021**, *16*, 1–14. [CrossRef]
7. Heap, I. Herbicide Resistant Weeds. In *Integrated Pest Management*; Springer Science and Business Media LLC: Berlin/Heidelberg, Germany, 2014; pp. 281–301.
8. Hussain, M.I.; Reigosa, M.J. Higher peroxidase activity, leaf nutrient contents and carbon isotope composition changes in *Arabidopsis thaliana* are related to rutin stress. *J. Plant Physiol.* **2014**, *171*, 1325–1333. [CrossRef] [PubMed]
9. Islam, A.K.M.M.; Yeasmin, S.; Qasem, J.R.S.; Juraimi, A.S.; Anwar, P. Allelopathy of Medicinal Plants: Current Status and Future Prospects in Weed Management. *Agric. Sci.* **2018**, *9*, 1569–1588. [CrossRef]
10. Motmainna, M.; Juraimi, A.S.; Uddin, M.K.; Asib, N.B.; Islam, A.K.M.M.; Hasan, M. Assessment of allelopathic compounds to develop new natural herbicides: A review. *Allelopathy J.* **2021**, *52*, 19–37. [CrossRef]
11. Ridenour, W.M.; Callaway, R.M. The relative importance of allelopathy in interference: The effects of an invasive weed on a native bunchgrass. *Oecologia* **2001**, *126*, 444–450. [CrossRef] [PubMed]
12. Kato-Noguchi, H. Involvement of Allelopathy in the Invasive Potential of *Tithonia diversifolia*. *Plants* **2020**, *9*, 766. [CrossRef]
13. Kato-Noguchi, H.; Kobayashi, A.; Ohno, O.; Kimura, F.; Fujii, Y.; Suenaga, K. Phytotoxic substances with allelopathic activity may be central to the strong invasive potential of *Brachiaria brizantha*. *J. Plant Physiol.* **2014**, *171*, 525–530. [CrossRef]
14. Motmainna, M.; Juraimi, A.S.B.; Uddin, K.; Asib, N.B.; Islam, A.M.; Hasan, M. Allelopathic potential of Malaysian invasive weed species to control weedy rice (Oryza sativa f. spontanea Roshev). *Allelopath. J.* **2021**, *53*, 53–68. [CrossRef]
15. Qu, T.; Du, X.; Peng, Y.; Guo, W.; Zhao, C.; Losapio, G. Invasive species allelopathy decreases plant growth and soil microbial activity. *PLoS ONE* **2021**, *16*, e0246685. [CrossRef] [PubMed]
16. Gnanavel, I.; Natarajan, S.K. *Parthenium hysterophorus* L.: A major threat to natural and agro eco-systems in India. *Int. J. Agric. Environ. Biotech.* **2013**, *6*, 261–269.
17. Masum, S.M.; Hasanuzzaman, M.; Ali, M.H. Threats of *Parthenium hysterophorus* on agro-ecosystems and its management: A review. *Int. J. Agric. Crop Sci.* **2013**, *6*, 684.
18. Mekonnen, G. Threats and Management Options of Parthenium (*Parthenium hysterophorus* L.) in Ethiopia. *Agric. Res. Technol. Open Access J.* **2017**, *10*, 1–7. [CrossRef]

19. Motmainna, M.; Juraimi, A.S.; Uddin, M.K.; Asib, N.B.; Islam, A.K.M.M.; Hasan, M. Bioherbicidal Properties of *Parthenium hysterophorus*, *Cleome rutidosperma* and *Borreria alata* Extracts on Selected Crop and Weed Species. *Agronomy* **2021**, *11*, 643. [CrossRef]
20. Qasem, J.R. Differences in the allelopathy results from field observations to laboratory and glasshouse experiments. *Allelopath. J.* **2010**, *26*, 45–58.
21. Islam, A.H.; Yeasmin, S.; Kader, R.A. Bioassay screening of sawdust obtained from selected tropical tree species for allelopathic properties and their field performance against paddy weeds. *Fundam. Appl. Agric.* **2019**, *4*, 906–915. [CrossRef]
22. El-Mergawi, R.A.; Al-Humaid, A.I. Searching for natural herbicides in methanol extracts of eight plant species. *Bull. Natl. Res. Cent.* **2019**, *43*, 22. [CrossRef]
23. Batish, D. Allelopathic effects of parthenin against two weedy species, *Avena fatua* and *Bidens pilosa*. *Environ. Exp. Bot.* **2002**, *47*, 149–155. [CrossRef]
24. Singh, H.P.; Batish, D.R.; Pandher, J.K.; Kohli, R. Phytotoxic effects of *Parthenium hysterophorus* residues on three Brassica species. *Weed Biol. Manag.* **2005**, *5*, 105–109. [CrossRef]
25. Mersie, W.; Singh, M. Allelopathic effect of parthenium (*Parthenium hysterophorus* L.) extract and residue on some agronomic crops and weeds. *J. Chem. Ecol.* **1987**, *13*, 1739–1747. [CrossRef] [PubMed]
26. Nishida, N.; Tamotsu, S.; Nagata, N.; Saito, C.; Sakai, A. Allelopathic Effects of Volatile Monoterpenoids Produced by *Salvia leucophylla*: Inhibition of Cell Proliferation and DNA Synthesis in the Root Apical Meristem of *Brassica campestris* Seedlings. *J. Chem. Ecol.* **2005**, *31*, 1187–1203. [CrossRef] [PubMed]
27. Islam, A.M.; Hasan, M.; Musha, M.H.; Uddin, K.; Juraimi, A.S.; Anwar, P. Exploring 55 tropical medicinal plant species available in Bangladesh for their possible allelopathic potentiality. *Ann. Agric. Sci.* **2018**, *63*, 99–107. [CrossRef]
28. Levizou, E.; Karageorgou, P.; Psaras, G.K.; Manetas, Y. Inhibitory effects of water soluble leaf leachates from *Dittrichia viscosa* on lettuce root growth, statocyte development and graviperception. *Flora-Morphol. Distrib. Funct. Ecol. Plants* **2002**, *197*, 152–157. [CrossRef]
29. Franco, D.M.; Silva, E.M.; Saldanha, L.L.; Adachi, S.A.; Schley, T.R.; Rodrigues, T.M.; Dokkedal, A.L.; Nogueira, F.T.S.; De Almeida, L.F.R. Flavonoids modify root growth and modulate expression of SHORT-ROOT and HD-ZIP III. *J. Plant Physiol.* **2015**, *188*, 89–95. [CrossRef]
30. Mousavi, S.S.; Karami, A.; Haghighi, T.M.; Alizadeh, S.; Maggi, F. Phytotoxic Potential and Phenolic Profile of Extracts from *Scrophularia striata*. *Plants* **2021**, *10*, 135. [CrossRef] [PubMed]
31. Abd-Elgawad, A.M.; El Gendy, A.E.-N.G.; Assaeed, A.M.; Al-Rowaily, S.L.; Alharthi, A.S.; Mohamed, T.A.; Nassar, M.I.; Dewir, Y.H.; ElShamy, A.I. Phytotoxic Effects of Plant Essential Oils: A Systematic Review and Structure-Activity Relationship Based on Chemometric Analyses. *Plants* **2020**, *10*, 36. [CrossRef]
32. Alam, A.; Juraimi, A.S.; Rafii, M.Y.; Hamid, A.A.; Aslani, F.; Hakim, M.A. Salinity-induced changes in the morphology and major mineral nutrient composition of purslane (*Portulaca oleracea* L.) accessions. *Biol. Res.* **2016**, *49*, 1–19. [CrossRef]
33. Jabran, K.; Mahajan, G.; Sardana, V.; Chauhan, B.S. Allelopathy for weed control in agricultural systems. *Crop. Prot.* **2015**, *72*, 57–65. [CrossRef]
34. Algandaby, M.M.; El-Darier, S.M. Management of the noxious weed; *Medicago polymorpha* L. via allelopathy of some medicinal plants from Taif region, Saudi Arabia. *Saudi J. Biol. Sci.* **2018**, *25*, 1339–1347. [CrossRef] [PubMed]
35. Renathielly, F.D.S.; Rodrigo, T.B.; Bruno, M.Z.; Maurício, A.P.; Samuel, N.M.D.S.; Reginaldo, F.S.; Da Silva, R.F.; Bressan, R.T.; Zilli, B.M.; Pilatti, M.A.; et al. Allelopathic effect of aqueous extract of fresh leaf castor beans (*Ricinus communis* L.) applied to the beginning stage of soy (*Glycine max* L.) and safflower (*Carthamus tinctorius* L.). *Afr. J. Biotechnol.* **2016**, *15*, 2787–2793. [CrossRef]
36. Dalal, V.K.; Tripathy, B.C. Water-stress induced downsizing of light-harvesting antenna complex protects developing rice seedlings from photo-oxidative damage. *Sci. Rep.* **2018**, *8*, 1–16. [CrossRef]
37. Zhou, Y.; Yu, J. Allelochemicals and photosynthesis. In *Allelopathy*; Springer Science and Business Media LLC: Berlin, Germany, 2006; pp. 127–139.
38. Lawlor, D.W. Limitation to Photosynthesis in Water-stressed Leaves: Stomata vs. Metabolism and the Role of ATP. *Ann. Bot.* **2002**, *89*, 871–885. [CrossRef] [PubMed]
39. Foyer, C.H.; Shigeoka, S. Understanding Oxidative Stress and Antioxidant Functions to Enhance Photosynthesis. *Plant Physiol.* **2011**, *155*, 93–100. [CrossRef]
40. Casson, S.; Gray, J.E. Influence of environmental factors on stomatal development. *New Phytol.* **2008**, *178*, 9–23. [CrossRef] [PubMed]
41. Blum, U.; Gerig, T.M. Relationships between Phenolic Acid Concentrations, Transpiration, Water Utilization, Leaf Area Expansion, and Uptake of Phenolic Acids: Nutrient Culture Studies. *J. Chem. Ecol.* **2005**, *31*, 1907–1932. [CrossRef]
42. Li, Z.-H.; Wang, Q.; Ruan, X.; Pan, C.-D.; Jiang, D.-A. Phenolics and Plant Allelopathy. *Molecules* **2010**, *15*, 8933–8952. [CrossRef]
43. Al-Humaid, A.; El-Mergawi, R.A. Herbicidal activities of seven native plants on the germination and growth of *Phalaris minor*, *Echinochloa crus-galli*, *Portulaca oleracea* and *Lactuca sativa*. *J. Agric. Sci. Technol.* **2014**, *4*, 843–852.
44. Scognamiglio, M.; D'Abrosca, B.; Esposito, A.; Pacifico, S.; Monaco, P.; Fiorentino, A. Plant growth inhibitors: Allelopathic role or phytotoxic effects? Focus on Mediterranean biomes. *Phytochem. Rev.* **2013**, *12*, 803–830. [CrossRef]

45. Do, Q.-D.; Angkawijaya, A.E.; Tran-Nguyen, P.L.; Huynh, L.H.; Soetaredjo, F.E.; Ismadji, S.; Ju, Y.-H. Effect of extraction solven on total phenol content, total flavonoid content, and antioxidant activity of *Limnophila aromatica*. *J. Food Drug Anal.* **2014**, *22* 296–302. [CrossRef] [PubMed]
46. Pérez, M.B.; Calderón, N.L.; Croci, C.A. Radiation-induced enhancement of antioxidant activity in extracts of rosemary (*Rosmari nus officinalis* L.). *Food Chem.* **2007**, *104*, 585–592. [CrossRef]
47. Marchiosi, R.; Dos Santos, W.D.; Constantin, R.P.; De Lima, R.B.; Soares, A.R.; Finger-Teixeira, A.; Mota, T.R.; De Oliveira, D.M. Foletto-Felipe, M.D.P.; Abrahão, J.; et al. Biosynthesis and metabolic actions of simple phenolic acids in plants. *Phytochem. Rev* **2020**, *19*, 865–906. [CrossRef]
48. Batish, D.R.; Lavanya, K.; Singh, H.P.; Kohli, R.K. Phenolic allelochemicals released by Chenopodium murale affect the growth nodulation and macromolecule content in chickpea and pea. *Plant Growth Regul.* **2007**, *51*, 119–128. [CrossRef]
49. Pan, L.; Li, X.-Z.; Yan, Z.-Q.; Guo, H.-R.; Qin, B. Phytotoxicity of umbelliferone and its analogs: Structure–activity relationships and action mechanisms. *Plant Physiol. Biochem.* **2015**, *97*, 272–277. [CrossRef]
50. Ma, T.; Zhou, W.; Chen, L.; Wu, L.; Christie, P.; Liu, W. Toxicity of phthalate esters to lettuce (*Lactuca sativa*) and the soil microbial community under different soil conditions. *PLoS ONE* **2018**, *13*, e0208111. [CrossRef] [PubMed]
51. Batish, D.R.; Singh, H.P.; Pandher, J.K.; Arora, V.; Kohli, R. Phytotoxic effect of Parthenium residues on the selected soil properties and growth of chickpea and radish. *Weed Biol. Manag.* **2002**, *2*, 73–78. [CrossRef]
52. Li, S.; Peng, F.; Xiao, Y.; Gong, Q.; Bao, Z.; Li, Y.; Wu, X. Mechanisms of High Concentration Valine-Mediated Inhibition of Peach Tree Shoot Growth. *Front. Plant Sci.* **2020**, *11*, 603067. [CrossRef]
53. Han, C.-M.; Pan, K.-W.; Wu, N.; Wang, J.-C.; Li, W. Allelopathic effect of ginger on seed germination and seedling growth of soybean and chive. *Sci. Hortic.* **2008**, *116*, 330–336. [CrossRef]
54. Kato-Noguchi, H.; Suzuki, M.; Noguchi, K.; Ohno, O.; Suenaga, K.; Laosinwattana, C. A Potent Phytotoxic Substance in *Aglaia odorata* Lour. *Chem. Biodivers.* **2016**, *13*, 549–554. [CrossRef]
55. Blanco, F.M.G.; Ramos, Y.G.; Scarso, M.F.; Jorge, M.F.S.A.L.A.D.C. *Determining the Selectivity of Herbicides and Assessing Their Effect on Plant Roots-A Case Study with Indaziflam and Glyphosate Herbicides*; IntechOpen: London, UK, 2015; pp. 275–297.
56. Aslani, F.; Juraimi, A.S.; Ahmad-Hamdani, M.S.; Omar, D.; Alam, A.; Hashemi, F.S.G.; Hakim, A.; Uddin, M.K. Allelopathic effect of methanol extracts from *Tinospora tuberculata* on selected crops and rice weeds. *Acta Agric. Scand. Sect. B-Plant Soil Sci.* **2014**, *64* 165–177. [CrossRef]
57. Mirmostafaee, S.; Azizi, M.; Fujii, Y. Study of Allelopathic Interaction of Essential Oils from Medicinal and Aromatic Plants on Seed Germination and Seedling Growth of Lettuce. *Agronomy* **2020**, *10*, 163. [CrossRef]
58. Wani, A.; Hayat, S.; Ahmad, A.; Tahir, I. Efficacy of brassinosteroid analogues in the mitigation of toxic effects of salt stress in *Brassica juncea* plants. *J. Environ. Biol.* **2017**, *38*, 27–36. [CrossRef]
59. Burrill, L.C.; Cárdenas, J.; Locatelli, E. *Field Manual for Weed Control Research*; International Plant Protetion Center, Oregon State University: Corvallis, OR, USA, 1976.
60. Guiochon, G.; Gritti, F. Shell particles, trials, tribulations and triumphs. *J. Chromatogr. A* **2011**, *1218*, 1915–1938. [CrossRef]
61. Abu Bakar, F.I.; Abu Bakar, M.F.; Abdullah, N.; Endrini, S.; Fatmawati, S. Optimization of Extraction Conditions of Phytochemical Compounds and Anti-Gout Activity of *Euphorbia hirta* L. (Ara Tanah) Using Response Surface Methodology and Liquid Chromatography-Mass Spectrometry (LC-MS) Analysis. *Evid. Based Complementary Altern. Med.* **2020**, *2020*. [CrossRef] [PubMed]

plants

Article

Possibilities of Using Seed Meals in Control of Herbicide-Susceptible and -Resistant Biotypes of Rye Brome (*Bromus secalinus* L.) in Winter Wheat

Elżbieta Pytlarz [1,*] and Dorota Gala-Czekaj [2]

[1] Institute of Agroecology and Plant Production, Wrocław University of Environmental and Life Sciences, Grunwaldzki Square 24A, 50-363 Wrocław, Poland
[2] Department of Agroecology and Crop Production, University of Agriculture in Kraków, Mickiewicza 21 Ave, 31-120 Kraków, Poland; dorota.gala@urk.edu.pl
* Correspondence: elzbieta.pytlarz@upwr.edu.pl

Abstract: Rye brome is a rare and nuisance weed in winter wheat canopies. In recent years, farmers have complained about the inadequate chemical control of this species. This study aimed to assess the effectiveness of seed meals obtained from allelopathic crops as an environmentally-friendly alternative for the control of herbicide-susceptible (S) and -resistant (R) rye brome biotypes in winter wheat. The pot experiment was conducted in a greenhouse at the Swojczyce Research and Training Station in Wrocław (Poland) to determine the impact of seed meals from: *Fagopyrum esculentum*, *Sinapis alba*, *Phacelia tanacetifolia*, *Lupinus luteus*, *Raphanus sativus* var. *oleiformis* and *Ornithopus sativus*, at 1 and 3% doses. Wheat emergence (>90%) and early growth were not affected by the presence in the soil of seed meals (only at 1% concentration) from *P. tanacetifolia* and *R. sativus*. The efficacy of these meals (reduction of aboveground biomass) at rye brome control was the same as the herbicide or higher. Seed meals from *P. tanacetifolia* reduced the emergence of the S and R biotypes by approximately 70 percentage points (p.p.) and 30 p.p., respectively, and limited the initial growth of both biotypes. Addition to soil meals from *F. esculentum* and *R. sativus* generally reduced only initial weed growth.

Keywords: non-chemical weed management; rare weeds; herbicide resistance; weed control; allelopathy; winter wheat

Citation: Pytlarz, E.; Gala-Czekaj, D. Possibilities of Using Seed Meals in Control of Herbicide-Susceptible and -Resistant Biotypes of Rye Brome (*Bromus secalinus* L.) in Winter Wheat. *Plants* **2022**, *11*, 331. https://doi.org/10.3390/plants11030331

Academic Editors: Alessia Restuccia and Aurelio Scavo

Received: 16 December 2021
Accepted: 25 January 2022
Published: 26 January 2022

1. Introduction

Bromus L. is a genus belonging to the *Poaceae* family [1–3] and comprises about 150 species [4]. The most frequently occurring species of brome-grasses worldwide include: downy brome (*Bromus tectorum* L.), great brome (syn. ripgut brome; *B. diandrus* Roth, syn. *Anisantha diandra* (Roth) Tsvelev), meadow brome (*B. commutatus* Schrad.), rye brome (syn. cheat; *B. secalinus* L.), soft brome (*B. hordeaceus* L.), smooth brome (*B. racemosus* L.), sterile brome (*B. sterilis* L. syn. *Anisantha sterilis* (L.) Nevski) and rescue brome (*B. willdenowii* Kunth). All of these species pose a threat to arable crops as competitive weeds [1,5–7].

One of the most common and harmful weed species among the *Bromus* genus is rye brome. The infestation of numerous crops with *B. secalinus* can be observed on almost all continents. Rye brome is widespread in European countries such as the United Kingdom [1,7], France [2], Germany [8], Romania [9] and Poland [3,10–15]. It is a significant problem in North America [6], especially in the USA and Canada in winter wheat production areas of the Great Plains [16,17]. In the last decade, the occurrence of rye brome has been confirmed in Asia—including in Iran [18] and Taiwan [5].

Rye brome is an annual, speirochoric plant [12], which may grow in a spring or winter form; however, winter forms are more common [15]. In Poland, in the past, *B. secalinus* was a troublesome weed in winter crops [11]. Between the 1970s and the end of 20th century, it vanished from arable fields almost completely [19,20]. The regression of rye brome caused it to be classified as a rare species threatened with extinction [21]. In recent years, the weed has unexpectedly re-emerged on arable lands and poses a serious threat to crops. Nowadays, *B. secalinus* grows mostly in winter wheat and rye. It can be found sporadically in pastures and meadows, as well as in ruderal habitats such as field borders, wastelands or roadsides [9,11,19,20]. An increased occurrence of the species was observed in the southern [10,11], northern-eastern [20] and central [22] regions of Poland.

The reasons for this re-expansion of rye brome are to be found not only in the simplification of crop rotation, climate change, and the use of more selective herbicides, but also in the rapid co-evolution of the weed with winter crops and other segetal species [12,17,20]. Davies et al. [7] have pointed to the increased use of minimum tillage and the evolution of herbicide resistance as possible causes of the more frequent occurrence of rye brome in arable fields.

The increase of rye brome's occurrence in cereals is not reflected in the registration of herbicides for its control. Currently, in Poland, in the Recommendations for the Protection of Agricultural Plants of Institute of Plant Protection [23] for the control of *B. secalinus* in cereals, only four active ingredients are registered (mesosulfuron-methyl, propoxycarbazone-sodium, pyroxsulam, sulfosulfuron). All of them have one mode of action (MoA); they are ALS inhibitors and, according to the Herbicide Resistance Action Committee (HRAC), belong to Herbicide MoA Group 2. Moss et al. [24] classify active ingredients from the group of ALS inhibitors as substances with a high risk of developing resistance. In Herbicide MoA Group 2, the number of resistant biotypes is increasing most quickly, and, of the 524 resistant biotypes, as many as 32% are biotypes resistant to active ingredients belonging to this group. Wheat (in which brome grasses occur most frequently) is the crop with the highest confirmed number of biotypes of herbicide-resistant weeds (73 unique cases) [25]. Both worldwide and in Poland, it is a staple grain. In 2019, it was cultivated over an area of 239.6 and 2.5 M ha, respectively [26].

There are currently 24 confirmed herbicide-resistant biotypes of the *Bromus* genus worldwide. Half of these have been recorded in the last 10 years. In total, 13 resistant biotypes were found in the cultivation of wheat, of which 11 showed resistance to ALS inhibitors. There are currently two confirmed unique cases of herbicide-resistant biotypes of *B. secalinus*. They are characterized by resistance to active ingredients from Herbicide MoA Group 2 (imazamox, propoxycarbazone-sodium, pyroxsulam and sulfosulfuron) [25].

A relatively small variation in the mode of action (approx. 350 chemical compounds against weeds represent 26 different MoA) results in the selection of biotypes resistant to herbicide [27]. For many years, agricultural development has been focused on maximizing productivity, but currently the need to ensure the sustainable development of agroecosystems has become the dominant concept in agriculture. One of the goals of sustainable development and the Green Deal policy is to maintain biodiversity in the agroecosystem. The withdrawal of herbicides and the limitation of their use are coherent with the pro-ecological policy that is being promoted, as is the implementation of integrated weed management, which is considered to be the most desirable concept of weed control. There are thus several aspects to the management of weeds, such as their mechanical, cultural, ecological, biological and chemical aspects and allelopathy [28,29]. The use of the allelopathic potential of crops in the form of catch crops, living mulches, or seed meals, for example, may have a positive effect on the agroecosystem, including by providing an opportunity for inhibiting weed germination and development, and stimulating the growth of crops [30,31]. In agriculture and gardening, there is a growing interest in the production and addition into the soil of seed meals from plants belonging to various botanical families, as a non-chemical method of weed control [32–34]. In relation to the increase in the number of rye brome biotypes resistant to herbicides and the implementation of the EU Green Deal policy, the use of seed meals to control *B. secalinus* could be an interesting alternative.

The aim of this research was (1) to evaluate the effect of seed meals on the emergence and initial development of winter wheat and herbicide-susceptible or -resistant biotypes of rye brome; (2) to assess the possibility of using seed meals to reduce weed infestation with herbicide-susceptible or -resistant biotypes of rye brome in winter wheat; (3) to compare the effectiveness of rye brome control by seed meals with herbicide spraying.

The research hypothesis assumes that the presence of seed meals in the soil will limit the emergence and initial development of herbicide-susceptible and -resistant biotypes of rye brome, and will not affect the initial development of wheat.

2. Results

2.1. Influence of Seed Meals on Winter Wheat

The origin and dose of the seed meal already had a clearly differentiating effect on the development of winter wheat at the emergence stage (BBCH 09) (Figure 1).

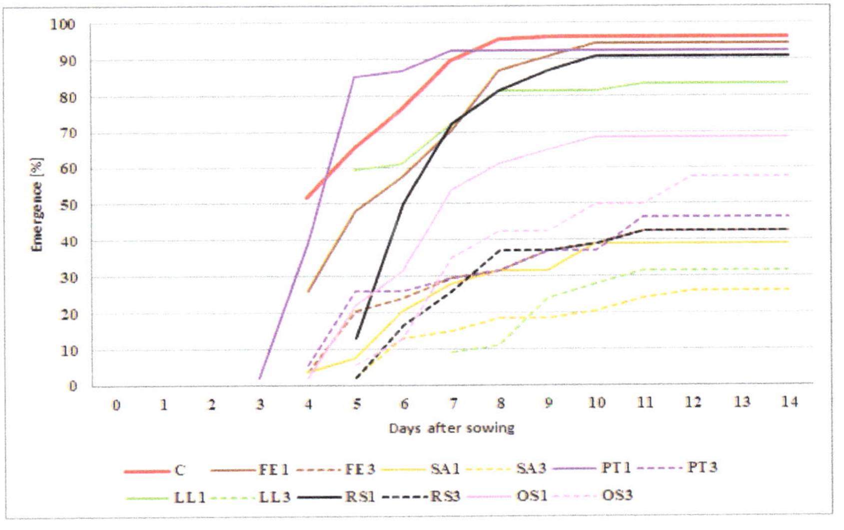

Figure 1. The mean emergence of winter wheat depending on type and concentration of seed meals applied. The symbols mean: C—control and seed meals from FE—*Fagophyrum esculentum*, SA—*Sinapis alba*, PT—*Phacelia tanacetifolia*, LL—*Lupinus luteus*, RS—*Raphanus sativus*, OS—*Ornithopus sativus*; 1—1% concentration of seed meals, 3—3% concentration of seed meals.

There was differentiation not only in the dynamics of emergence, but also in the number of wheat seedlings per pot. The wheat started to emerge earliest (day 3 after sowing) when treated with meal PT1; and latest (day 7 after sowing) when treated with meal LL3. The fastest rate of emergence (4 days) was found in wheat growing on the soil with meal LL1. In pots where a higher concentration of yellow lupine meal (LL3) was applied, a lengthening in the emergence of cereal shoots and a reduction in their number were observed. The fastest rate of emergence (8 days) was observed for wheat growing in the soil with the addition of FE3, SA3 and OS3 meal. The highest percentage of wheat seedlings was found in the pots without meal, i.e., in the control treatment (C)—96%. A similar percentage of emerging plants (>90%) was recorded for wheat mixed with FE1, PT1 and RS1 meals. In the case of meals of a lower concentration, the lowest percentage of grain seedlings (39%) was observed in the treatment with meal SA1. An increase in the concentration of the SA meal in the soil to 3% also resulted in the lowest percentage of seedlings among the meals tested (26%).

The addition of seed meals to the soil also led to a modification in the aboveground biomass per one plant of winter wheat (Figure 2).

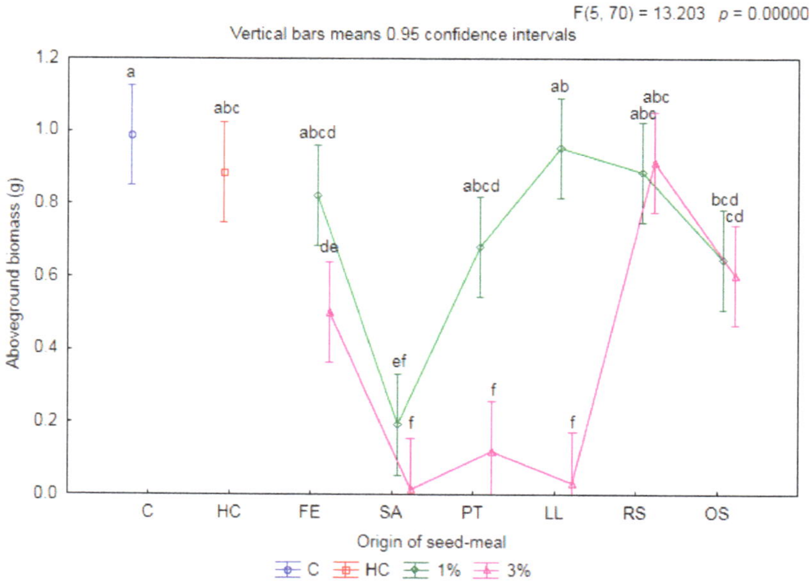

Figure 2. The mean aboveground biomass per one plant of winter wheat depending on origin of seed meals and their concentration. Means with various letters are significantly different, according to Tukey test ($p \leq 0.05$). The symbols mean: C—control, HC—herbicide control, and seed meals from FE—*Fagophyrum esculentum*, SA—*Sinapis alba*, PT—*Phacelia tanacetifolia*, LL—*Lupinus luteus*, RS—*Raphanus sativus*, OS—*Ornithopus sativus*; green line—1% concentration of seed meals, rose line—3% concentration of seed meals.

After the application of seed meals at a concentration of 1%, overall a non-significant reduction in the mass of aboveground parts of wheat was found, compared to the C (0.989 g) and HC (0.887 g) treatments. Only the SA1 meal significantly inhibited the growth of wheat biomass, by multiples of 5.1 and 4.6 respectively, compared to C and HC. In turn, the OS1 meal resulted in a significant limitation of the aboveground biomass of wheat by 34.7%, only in comparison to C. The addition to the soil of meals at a higher concentration inhibited growth of fresh mass of aboveground parts of wheat—compared both to C and HC. The exception was wheat growing on the substrate soil with the addition of RS3 meal. In this case, the aboveground biomass of wheat was on the same level as in the C and HC treatments. Moreover, the application of the OS3 meal resulted in a limitation in the aboveground biomass of wheat by 38.9%, only in comparison to C.

The type of meal and its concentration in the soil did not result in any differences in the belowground biomass per one plant of wheat compared to C and HC (Figure 3). Root biomass was found to be significantly lower only after application of the LL3 meal compared to FE1 and PT1—by multiples of 7.4 and 8.5, respectively.

The addition to the soil of seed meals from various crop species in differing concentrations had an impact on the average length of aboveground parts of the winter wheat plants (Figure 4).

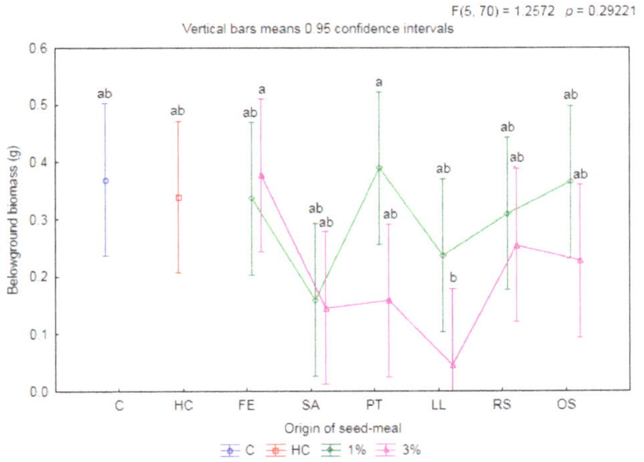

Figure 3. The mean belowground biomass per one plant of winter wheat depending on origin of seed meals and their concentration. Means with various letters are significantly different, according to Tukey test ($p \leq 0.05$). The symbols mean: C—control, HC—herbicide control, and seed meals from FE—*Fagophyrum esculentum*, SA—*Sinapis alba*, PT—*Phacelia tanacetifolia*, LL—*Lupinus luteus*, RS—*Raphanus sativus*, OS—*Ornithopus sativus*; green line—1% concentration of seed meals, rose line—3% concentration of seed meals.

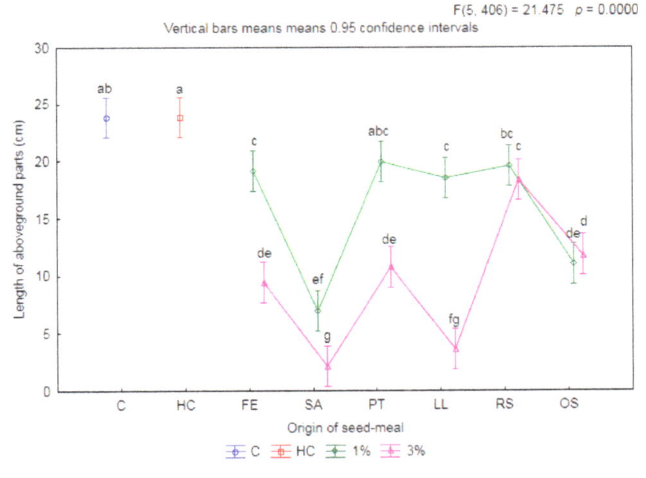

Figure 4. The mean length of aboveground parts of winter wheat depending on origin of seed meals and their concentration. Means with various letters are significantly different, according to Tukey test ($p \leq 0.05$). The symbols mean: C—control, HC—herbicide control, and seed meals from FE—*Fagophyrum esculentum*, SA—*Sinapis alba*, PT—*Phacelia tanacetifolia*, LL—*Lupinus luteus*, RS—*Raphanus sativus*, OS—*Ornithopus sativus*; green line—1% concentration of seed meals, rose line—3% concentration of seed meals.

In the majority of cases, the addition of meals led to a significant limitation of up to several centimeters in the length of the aboveground parts of the wheat. After the application of meals at a lower concentration (1%), less of an inhibitory impact on the tested parameter was found overall. On average, compared to treatment C, the inhibition of growth was 41%. Only the PT1 meal enabled the length of the aboveground parts of wheat to be maintained at the same level as in the C and HC treatments, while RS1 enabled it to be maintained at the level of C. It is worth emphasizing that in treatments with the same meals (PT1 and RS1), an emergence of wheat at the level of >90% (cf. Figure 1) was observed, as well as a non-significant limitation of the fresh mass of aboveground parts of the tested crop (cf. Figure 2). The shortest (7.0 cm) parts were found in wheat growing on the soil mixed with meal SA1. An increase in the concentration of the meals applied led to a significant limitation in the length of wheat overall. SA3 was found to have the most inhibitory effect on the increase in the length of aboveground parts of wheat (length 2.1 cm). After the application of RS and OS meals, the assessed parameter remained at the same level at both concentrations.

Meals from tested species of donor plants at a concentration of 1% and the application of herbicide (HC) significantly limited (by 11 cm^2 on average) development in the aboveground area of wheat compared to treatment C (21.2 cm^2) (Figure 5). In the case of meals FE1, LL1 and RS1, no decrease in the aboveground area of wheat was found compared to the treatment HC. Interestingly, these same meals (FE1, LL1, RS1) also did not cause any significant decrease in the mass of aboveground parts (cf. Figure 2) or in the mass of belowground parts (cf. Figure 3) of the wheat tested, compared both to HC and to C. The reduction in emergence was 13 percentage points (p.p.) at most for LL1 (cf. Figure 1). After application at a higher concentration, a further decrease in the area of the aboveground parts of the wheat (by 4 cm^2 on average) was found overall, compared to the treatments C and HC. Only the addition to the soil of the RS3 meal did not limit the aboveground area of wheat, compared to HC. The application to the soil of this meal also did not result in a reduction in the aboveground biomass of wheat (cf. Figure 3). This may be evidence of the neutral impact of this meal on the tested crop variety.

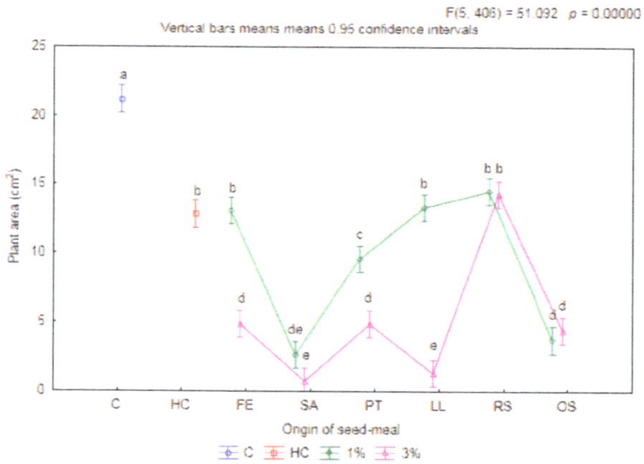

Figure 5. The mean plant area of winter wheat depending on origin of seed meals and their concentration. Means with various letters are significantly different, according to Tukey test ($p \leq 0.05$). The symbols mean: C—control, HC—herbicide control, and seed meals from FE—*Fagophyrum esculentum*, SA—*Sinapis alba*, PT—*Phacelia tanacetifolia*, LL—*Lupinus luteus*, RS—*Raphanus sativus*, OS—*Ornithopus sativus*; green line—1% concentration of seed meals, rose line—3% concentration of seed meals.

2.2. *Effectiveness of Seed Meals in Reduction of Rye Brome Growth*

The type and dose of meal added to the soil led to differences in the number of seedlings of rye brome of both the susceptible (Figure 6a) and the resistant (Figure 6b) biotype. The emergence of seedlings of the herbicide-susceptible biotype was inhibited most weakly on the soil with the addition of FE1 and RS1 and RS3 meals. The percentages of seedling emergence were 85%, 87% and 76% respectively. Independently of the biotype, the lowest percentage of seedling emergence was recorded after the application of the SA meal (2% or lack of seedling emergence). It should, however, be noted that these meals also significantly limited the development of the wheat (cf. Figures 1–5). For this reason, the use of SA meals in the cultivation of wheat to limit the development of rye brome may be of limited significance.

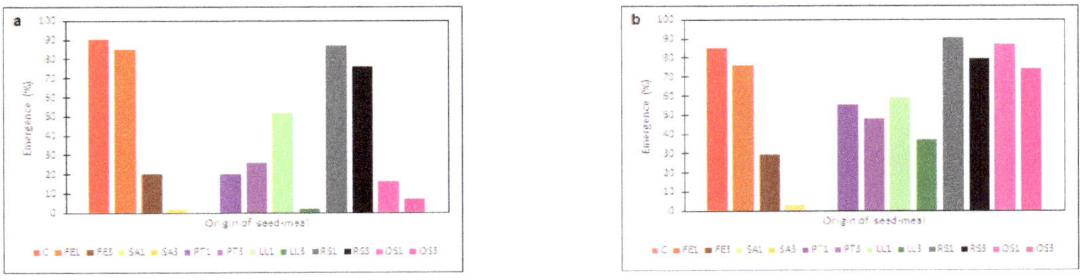

Figure 6. The mean emergence of herbicide-susceptible (**a**) and -resistant (**b**) biotypes of rye brome depending on origin of seed meals and their concentration (14 days after sowing). The symbols mean: C—control and seed meals from: FE—*Fagophyrum esculentum*, SA—*Sinapis alba*, PT—*Phacelia tanacetifolia*, LL—*Lupinus luteus*, RS—*Raphanus sativus*, OS—*Ornithopus sativus*; 1—1% concentration of seed meals, 3—3% concentration of seed meals.

Moreover, in the resistant biotype, a weaker reaction of emerging seedlings to the applied meals was observed in comparison to the susceptible biotype. The FE3 meal was fairly effective at limiting emergence (along with the SA meal). After the application of this meal, the percentage of seedling emergence for the resistant biotype of rye brome was 30%. With an increase in the concentration of the meals, an increase in the limitation of seedling emergence was observed, although the difference between the concentrations was not as big as in the case of the herbicide-susceptible biotype.

Application of herbicide resulted in a limitation in the length of aboveground parts of the herbicide-susceptible biotype of rye brome by 23% compared to C (15.9 cm) (Figure 7a).

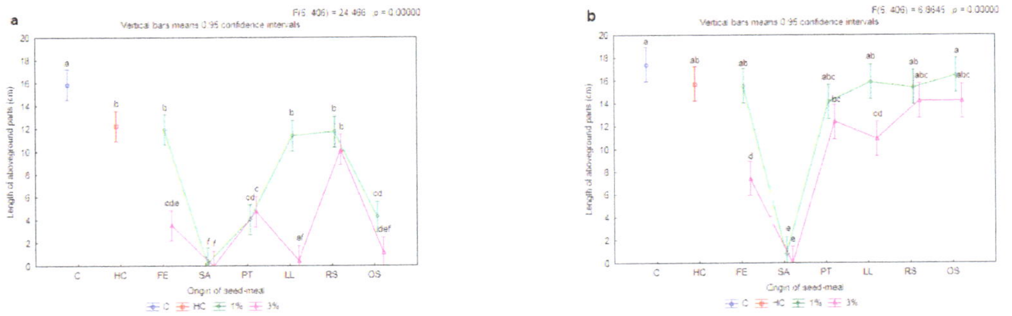

Figure 7. The mean length of aboveground parts of herbicide-susceptible (**a**) and -resistant (**b**) biotypes of rye brome depending on origin of seed meals and their concentration. Means with various letters are significantly different, according to Tukey test ($p \leq 0.05$). The symbols means: C—control, HC—herbicide control, and seed meals from: FE—*Fagophyrum esculentum*, SA—*Sinapis alba*, PT—*Phacelia tanacetifolia*, LL—*Lupinus luteus*, RS—*Raphanus sativus*, OS—*Ornithopus sativus*; green line—1% concentration of seed meals, rose line—3% concentration of seed meals.

In each case, the assessed parameter was found to have decreased in length after addition of the meal. The application of meals FE1, LL1 and RS1 allowed the weed plants to be shortened to the same level as with the spraying of herbicide (HC), i.e., by 26% on average, compared to C. This effect increased after the addition of the meals OS1, PT1 and SA1. It is worth underlining that the addition of the meals PT1 and RS1 did not have an impact on the length of the aboveground parts of the wheat (cf. Figure 4). For the FE and LL meals only, an increase in concentration from 1% to 3% caused a significant increase in the inhibition of the development of the length of the aboveground parts of the susceptible biotype of rye brome—by 70% and 96%, respectively.

There were no differences in the length of aboveground parts of the resistant biotype of rye brome as a result of the application of herbicide and the majority of meals at a concentration of 1% compared to C (17.4 cm) (Figure 7b). The exception was the meal SA1. It limited the tested parameter by 95% compared to C. Similarly, as in the case of the susceptible biotype of rye brome, it was only after an increase in the concentration of FE and LL meals from 1% to 3% that there was found to be a further decrease in the length of the aboveground parts of the weed; this decrease was by 52% and 31%, respectively.

The efficacy of the tested meals in the reduction of the aboveground biomass of the herbicide-susceptible biotype of rye brome was on the same level as the efficacy with the herbicide treatment (HC) (Figure 8a). On average, it was 80%. The application of the meals LL3, OS3, SA1 and SA3 resulted in a limitation on the biomass of the aboveground parts of over 95%. There was found to be a significant increase, by 29 p.p., in efficacy together with an increase in the concentration of meal only for the OS meal.

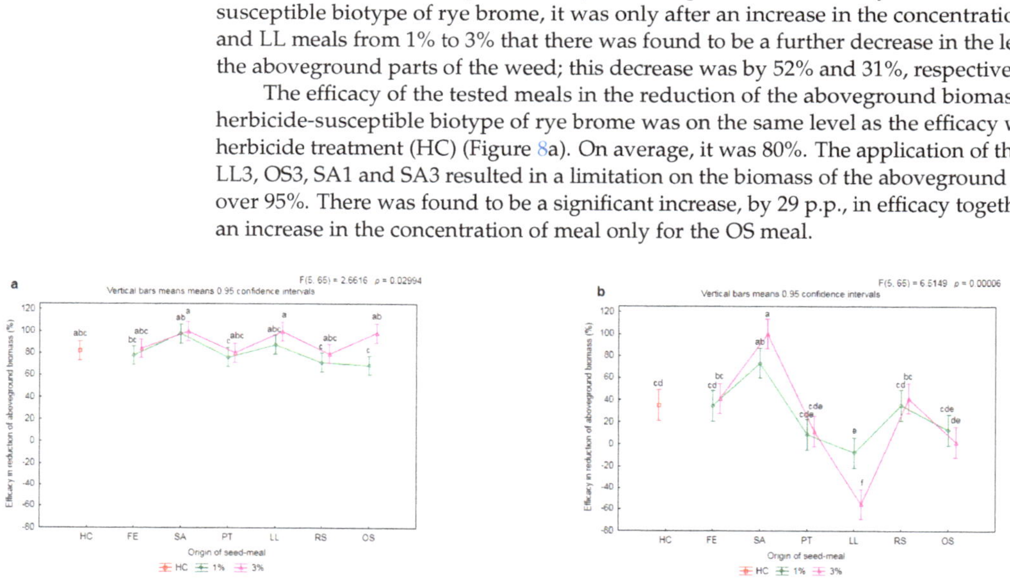

Figure 8. The mean efficacy in reduction of aboveground biomass of herbicide-susceptible (**a**) and -resistance (**b**) biotypes of rye brome depending on origin of seed meals and their concentration. Means with various letters are significantly different, according to Tukey test ($p \leq 0.05$). The symbols mean: C—control, HC—herbicide control, and seed meals from: FE—*Fagophyrum esculentum* SA—*Sinapis alba*, PT—*Phacelia tanacetifolia*, LL—*Lupinus luteus*, RS—*Raphanus sativus*, OS—*Ornithopus sativus*; green line—1% concentration of seed meals, rose line—3% concentration of seed meals.

There were differences in the effectiveness of limitation of the biomass of aboveground parts of the herbicide-resistant biotype of rye brome after the application of seed meals (Figure 8b). In the weeds growing on the soil with the addition of the FE1, PT1, RS1 and OS1 meals, the efficacy was found to be on the same level as with the chemical control, i.e., approximately 35%. Moreover, the application of the aforementioned meals did not cause any significant decrease in the biomass of the aboveground parts of the wheat in relation to HC (cf. Figure 2). The SA1 meal caused an increase in efficacy by 38 p.p. with reference to HC. In turn, the LL1 meal caused an increase in the biomass of aboveground parts of the weed by 8 p.p. in relation to C. Together with an increase the concentration of the LL meal to 3%, a further decrease in the efficacy of reduction of aboveground biomass by 47 p.p was observed.

The application to the soil of selected meals—namely SA and PT at a concentration of 1% caused a significant increase in efficacy in reduction of belowground biomass of the susceptible biotype of rye brome in relation to the application of herbicide (HC; 29.5%) (Figure 9a).

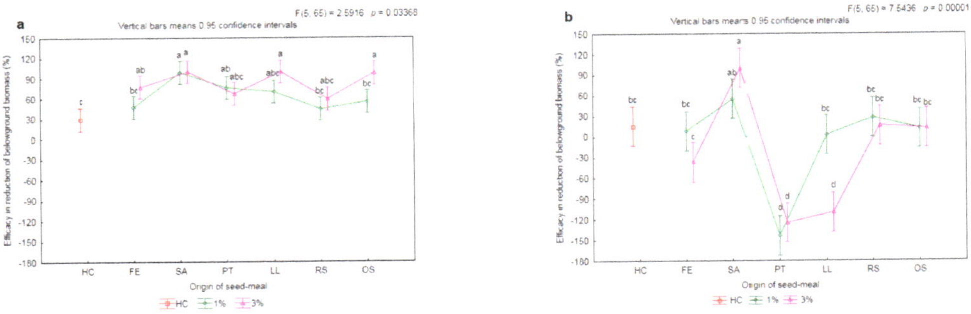

Figure 9. The mean efficacy in reduction of belowground biomass of herbicide-susceptible (**a**) and -resistance (**b**) biotypes of rye brome depending on origin of seed meals and their concentration. Means with various letters are significantly different, according to Tukey test ($p \leq 0.05$). The symbols mean: C—control, HC—herbicide control, and seed meals from: FE—*Fagophyrum esculentum*, SA—*Sinapis alba*, PT—*Phacelia tanacetifolia*, LL—*Lupinus luteus*, RS—*Raphanus sativus*, OS—*Ornithopus sativus*; green line—1% concentration of seed meals, rose line—3% concentration of seed meals.

The increase in the efficacy of the aforementioned meals compared to the HC treatment amounted to 70 and 47 p.p. respectively. A significant increase in the efficacy of the meal (by 42 p.p.) was found, together with an increase in its concentration in the soil for the OS meal only. Moreover, the meals FE3, SA3, LL3 and OS3 significantly limited the development of the belowground biomass of the weed with reference to HC. The increase in the efficacy of the reduction in the growth in mass amounted to: 48, 71, 70 and 68 p.p. respectively.

The application of meals from tested donor plants had a varying impact on the development of biomass of belowground parts of the herbicide-resistant biotype of rye brome (Figure 9b). After the application of nearly all the meals at a concentration of 1% (FE, LL, RS, OS), a reduction in the fresh mass of the aboveground parts of a level comparable to that seen after spraying with herbicide (HC; 12%) was observed. In the case of the SA1 meal, the efficacy was found to be over three times higher (55%) than after the application of herbicide. It should be emphasized that the SA1 meal also resulted in a significant limitation in the growth of the mass of the aboveground parts of the herbicide-susceptible biotype of rye brome compared to HC (cf. Figure 9a). In the case of the application of the SA meal at a higher concentration (3%), a significant increase, in relation to HC, was observed in the efficacy of the reduction of belowground biomass (by 85 p.p.). A significant increase in the efficacy of the meal was found, together with an increase in its concentration in the soil, only for the LL meal. It is worth underlining that, in the case of wheat (cf. Figure 3), the application of meals did not lead to any differences in its belowground biomass.

3. Discussion

On cereal fields, it is especially difficult to control monocotyledonous weed species, including those from the *Bromus* genus, which are also highly competitive with crop plants [35,36]. The chemical control of brome grasses has been the focus of much research worldwide. However, an effective and dependable solution is still to be found [6,16,17,37]. A significant problem in the management of *Bromus* spp. is the occurrence and evolution of herbicide resistance [7,38,39], as a consequence of the limited rotation of herbicides, as well as the application of simplifications to crop rotations and monocultures [40–42]. Additionally, herbicides pose toxicological and ecological threats, especially toward non-target organisms [43].

There are numerous works in the international body of research concerning non-chemical methods of control of brome-grasses [44–46]. Their authors show that rhizobacteria can be used in the biocontrol of *Bromus* spp., including—*B. secalinus* [47]—one of the most widespread and damaging weeds of the *Bromus* genus on a global scale [1,2,6–8,15]. Some authors point out that using crop rotation reduces the density of rye brome panicles per unit area, but does not eliminate it entirely [17]. Stone at al. [17], show that the rotation out of winter wheat for one growing season in comparison to continuous cropping winter wheat reduced up to 87% rye brome panicles. In non-chemical weed control, biological methods play a significant role alongside cultural methods, whereas suppressing weeds by using the allelopathic phenomenon is considered to be one of the most innovative methods of weed control [48,49]. Different agronomic methods enable the practical utilization of allelopathic plants in the form of seed meals [32–34,50].

Based on our own experiments, meal from white mustard (*Sinapis alba*; SA) proved to be the most effective at inhibiting initial growth of herbicide-susceptible and herbicide-resistant biotypes of rye brome. After its application, the percentage of germinating seeds of *B. secalinus* was 3.7% at most, and the average length of the aboveground parts of plants that emerged was less than 1 cm. Our results are consistent with the work of many authors [32,34,51], who also underline the strong inhibitory action of meal from *S. alba* in relation to weeds. These authors show that seed meals from white mustard reduced the growing seedlings of the following weeds: monocotyledonous (*Poa annua* L. *Digitaria ischaemum* (Schreb.), *Panicum dichotomiflorum* Michx.), dicotyledonous (*Stellaria media* L., *Oxalis corniculata*, *Physalis angulata* L., *Amaranthus spinosus* L., *Cyperus esculentus* L.), as well as liverwort (*Marchantia polymorpha* L.). In our research, the reduction in the aboveground biomass of the herbicide-susceptible biotype after application of the SA meal was approximately 98–100% and, in the case of the herbicide-resistant biotype, it was 73–100% (for concentrations of 1 and 3%, respectively). Dastgheib et al. [6] show that a mixture of terbutryn + terbuthylazine applied at the tillering stage of wheat results in a similar level of efficacy, with more than a 90% reduction in ripgut brome (*Bromus diandrus* Roth) biomass. In our study, the efficacy of herbicide with propoxycarbazone-sodium in aboveground biomass reduction was 35–82% for the resistant and susceptible biotype of rye brome, respectively. Unfortunately, in our research, SA meal also had an inhibiting effect on the initial development of winter wheat—the crop in which *B. secalinus* occurs most frequently [10,13,17,37]. In relation to this, its use as a biological herbicide in the cultivation of winter wheat is impossible. Our finding is supported by [32–34], who also confirm the inhibitory action of *S. alba* seed meal on the growth of various species of crops: maize (plant number and biomass), cucurbits (severely reduced yield) and lettuce (emergence).

This study revealed that meal from lacy phacelia (*Phacelia tanacetifolia*; PT) and from fodder radish (*Raphanus sativus* var. *oleiformis*; RS) at a concentration of 1% limited the development of the aboveground biomass of the herbicide-susceptible biotype of rye brome in the same way as spraying with herbicide. Importantly, in our research, meals from PT1 and RS1 did not result in any significant limitation of germination and development in the mass of aboveground parts of winter wheat compared to the control (C) and the herbicide control (HC). Partly similar results were obtained by Pużyńska et al. [32]. The authors assessed the impact of seed meal from wild radish (*Raphanus raphanistrum* L.) on maize (*Zea mays* L.) and two weed species—barnyard grass (*Echinochloa crus-galli* (L.) P.Beauv.) and redroot pigweed (*Amaranthus retroflexus* L.). As in our experiment, meal from wild radish was not found to have any significant impact on the dry mass of maize shoots. Pużyńska at al. [32] show that meal from wild radish also did not impact on the dry mass of aboveground parts of barnyard grass and redroot pigweed. Many authors draw attention to the allelopathic properties of lacy phacelia and the possibility of its use as a natural product in non-chemical weed management (weeds monocotyledonous: *Sorghum halepense* (L.) Per., *E. crus-galli*, weeds dicotyledonous: *Portulaca oleracea* L., *Chenopodium album* L., *Solanum nigrum* L., *A. retroflexus*, *Convolvulus arvensis* L., *Tribulus terrestris* L., *Sisymbrium officinale* (L.) Scop.) [52–56]. The situation is similar with buckwheat. Most authors focus, however, on the potential for the limitation of the development of weeds by root secretions from this species [57–60]. Our own studies found the emergence of both biotypes of rye grass to be strongly limited after the addition of a higher dose of meal into the soil. The application of this dose of seed meal from buckwheat also resulted in a significant limitation in the length of the aboveground parts of both the weed and the wheat.

In studies by Mioduszewska et al. [61], a limitation on the initial development of wheat after the application of extract from the aboveground parts of buckwheat was also found.

4. Materials and Methods

4.1. Plant Materials

In the experiment, two acceptor species were tested. The first was the crop, common wheat (*Triticum aestivum* L. cv. 'Agil') and the second was the weed, rye brome (*B. secalinus*). Winter wheat seeds were certified and marked as a degree C/1 (certified seed from the first multiplication, obtained after one multiplication of the basic seed). The herbicide-susceptible (S) and -resistant (R) biotypes of rye brome were harvested from winter wheat fields in July 2020. The characteristics of both the S and R biotypes of rye brome are presented in Table 1. The resistant biotype of weed was characterized by a low resistance index ($2 \leq R \leq 4$) to propoxycarbazone-sodium.

Table 1. Characteristics of herbicide-susceptible (S) and -resistant (R) biotypes of rye brome (*Bromus secalinus* L.) used in the pot experiments. ED50 values express the effective dose of propoxycarbazone-sodium (HRAC 2) causing a 50% reduction in plant biomass (ED50).

Biotype	ED50 (g ha^{-1})	Site (Coordinate)
S	16.26	Wrocław (51.132663 N 17.117230 E)
R	48.86	Wielowieś (51.339435 N 16.373906 E)

4.2. Seed Meals and Their Preparation

Qualified seeds of selected crop species (Table 2) were milled the day before the pot experiment was started. All the collected commercial seeds were grounded to meals in a Fritsch Pulverisette 11 laboratory mill (Idar-Oberstein, Germany).

Table 2. Crop species and cultivars used to prepare the seed meals.

Name		Cultivar	Abbreviation
English	Latin		
Common buckwheat	*Fagopyrum esculentum* Moench.	Panda	FE
White mustard	*Sinapis alba* L.	Bardena	SA
Lacy phacelia	*Phacelia tanacetifolia* Benth.	Anabela	PT
Yellow lupin	*Lupinus luteus* L.	Mister	LL
Fodder radish	*Raphanus sativus* L. var. *oleiformis* Pers.	Adagio	RS
Common birdsfoot	*Ornithopus sativus* Brot.	Bydgoska [1]	OS

[1] Variety not included in the national register.

4.3. Herbicide Characteristics

The active ingredient of the herbicide used in the experiment is propoxycarbazone-sodium (70%). According to the HRAC classification, it is classified as belonging to Herbicide MoA Group 2. The active ingredient presents a systemic type of action. Recommended application per leaves. It is a selective herbicide and has the formulation of water-soluble granules (SG).

4.4. Soil Characteristics

A soil that was used in the experiment was formed from light loamy sand underlaid with poorly loamy sand and was classified as IVb quality class (in Poland, equivalent to good rye complex). The topsoil (0–30 cm) was characterized by the following parameters: pH_{KCl} 5.82; P 86.4; K 27.5; Mg 131.0 (mg·kg^{-1} of soil) and C_{org} 0.41%. The soil was taken after harvesting the forecrop of organic forage pea cv. 'Roch'.

4.5. Set-up and Management of Pot Experiments

Two series of pot experiments were carried out during 2020 and 2021. Series I began in November and series II started in March, in a greenhouse at Wrocław University of Environmental and Life Science's Research and Training Station in Swojczyce (Wrocław, Poland). In both, the lighting and thermal conditions were regulated. The first experimental factor was the type of seed meal; the second was the dose of seed meal. Each acceptor (winter wheat and herbicide-susceptible and -resistant biotypes of rye brome) was analyzed individually.

Before starting the experiment, the soil was sieved over 1 cm mesh screens to rid the soil of post-harvest residue. The experiment was set up as a totally randomized design with three pots as replications. Production pots 0.5 L in volume were filled up with a mixture of 500 g of soil and one of the seed meals in an amount of 1 or 3% (*w/w*), separately. The control (C) and herbicide control (HC) pots did not contain any addition of meals. Nine grains each of either winter wheat or either of the biotypes of rye brome (S or R) were sown into soil-filled pots. Fourteen days after sowing, the number of plants per pot was equalized to five if the number of seedlings allowed. The HC treatment was sprayed on the two leaves of unfolded-stage (BBCH 12) rye brome in the spray chamber (APORO Sp. z o.o., Poznań, Poland). The dose of propoxycarbazone-sodium was 56 g ha^{-1}·200 L of water. Experiments were harvested when the plants of winter wheat in the C treatment were at the four leaves unfolded stage (BBCH 14).

4.6. Measurement Range

4.6.1. Winter Wheat

Wheat emergence was counted daily for 14 days after sowing. At the end of each series, the plants were pulled out and counted. The fresh weight of aboveground and belowground parts was determined (roots were washed and dried on a paper towel) using a WTC 2000 scale from RADWAG (Cracow, Poland). The length of the aboveground parts was measured. The next day, the area of the aboveground parts of wheat was measured with using a CI-202 LASER LEAF AREA METER from CID Bio-Science (Camas, WA, USA).

4.6.2. Rye Brome

During the harvest, the plants were pulled out and counted. The fresh weight of above- and belowground parts (roots were washed and dried on a paper towel) was determined using a WTC 2000 scale from RADWAG (Cracow, Poland). On this basis, the efficacy of biomass reduction of the tested treatments in relation to the control treatment (C) was calculated. A minus value of the index indicates an increase in the mass of rye brome with the applied seed meals. The length of aboveground parts of the weeds were also measured.

4.7. Statistical Analysis

Statistical analysis was carried out with using the two-way variance analysis (type of seed meal and dose of seed meal), using Statistica 13.3 software (TIBCO Software Inc., Tulsa, OK, USA). In order to check the normality of the distribution, the Shapiro–Wilk test was performed. The homogeneity of variance was checked using the Levene test. In order to determine and verify the relationships, Tukey's post-hoc test was performed with a significance level of $p \leq 0.05$.

5. Conclusions

The study found that selected seed meals can constitute an alternative to herbicide management strategies for the control of herbicide-susceptible and -resistant (to propoxycarbazone-sodium) biotypes of rye brome in winter wheat. Wheat emergence and initial growth were not affected by the presence of seed meals from common buckwheat (*Fagopyrum esculentum*), lacy phacelia (*Phacelia tanacetifolia*) and fodder radish (*Raphanus sativus* var. *oleiformis*) at 1% concentrations in the soil. The efficacy of these seed meals at the control of rye brome was at the same level as the herbicide or higher. An increase in the concentration of seed meals is not recommended due to the reduction in wheat emergence. Seed meals obtained from lacy phacelia reduced the emergence and initial growth of both biotypes of weeds, but seed meals from common buckwheat and fodder radish limited only the initial weed growth. Furthermore, despite the high efficacy of seed meals from white mustard at reducing emergence of rye brome, they are not recommended for the control of herbicide-susceptible and resistant biotypes of rye brome due to their inhibition of wheat growth. Future experiments should focus on a more comprehensive examination of seed meals in weed management by taking other herbicides, another level of herbicide resistance and other species of crops or weeds.

Author Contributions: Conceptualization: E.P.; methodology: E.P.; formal analysis: E.P. and D.G.-C.; investigation: E.P.; writing—original draft preparation: E.P. and D.G.-C.; writing—review and editing: E.P. and D.G.-C.; supervision: E.P. and D.G.-C.; funding acquisition: E.P. All authors have read and agreed to the published version of the manuscript.

Funding: This research was funded by Wrocław University of Environmental and Life Sciences in Poland, grant entitled "Innovative scientist", grant number N060/0010/20.

Institutional Review Board Statement: Not applicable.

Informed Consent Statement: Not applicable.

Data Availability Statement: Not applicable.

Conflicts of Interest: The authors declare that they have no known competing financial interests or personal relationships that could have appeared to influence the work reported in this paper.

References

1. Spalton, L.M. An analysis of the characters of *Bromus racemosus* L., *B. commutatus* Schrad. and *B.secalinus* L. (Poaceae). *Watsonia* **2002**, *24*, 193–202.
2. Zech-Matterne, V.; Derreumaux, M.; Pradat, B.; Luccioni, P.; Ruas, M.-P.; Toulemonde, F. Should *Bromus secalinus* (rye brome) be considered a crop? Analysis of *Bromus* rich assemblages from protohistoric and historic sites in northern France and textual references. *Veget. Hist. Archaeobot.* **2021**, *30*, 773–787. [CrossRef]
3. Kapeluszny, J.; Haliniarz, M. Selected elements of germination biology of flixweed (*Descurainia sophia* Webb. ex Prantl.) and rye brome (*Bromus secalinus* L.). *Ann. Univ. Mariae Curie-Skłodowska. Sect. E Agric.* **2007**, *62*, 226–233. (In Polish)
4. World Checklist of Selected Plant Families. Available online: http://wcsp.science.kew.org/ (accessed on 13 October 2021).
5. Jun, M.-J.; Liao, G.-I.; Kuo, C.-S. Notes on alien *Bromus* grasses in Taiwan. *Taiwania* **2006**, *51*, 131–138.
6. Dastgheib, F.; Rolston, M.P.; Archie, W.J. Chemical control of Brome grasses (*Bromus* spp.) in cereals. *N. Z. Plant Prot.* **2003**, *56*, 227–232. [CrossRef]
7. Davies, L.R.; Onkokesung, N.; Brazier-Hicks, M.; Edwards, R.; Moss, S. Detection and characterization of resistance to acetolactate synthase inhibiting herbicides in *Anisantha* and *Bromus* species in the United Kingdom. *Pest Manag. Sci.* **2020**, *76*, 2473–2482. [CrossRef]
8. Koch, M.A.; Meyer, N.; Engelhardt, M.; Thiv, M.; Bernhard, K.-G.; Michling, F. Morphological and genetic variation of highly endangered *Bromus* species and the status of these Neolithic weeds in Central Europe. *Plant Syst. Evol.* **2016**, *302*, 515–525. [CrossRef]
9. Georgescu, M.I.; Ionescu, N.; Penescu, A.; Vrabie, F.C.; Săvulescu, E.; Luchian, V.; Popa, V.I. Some observations on the morphology and germination of *Bromus secalinus* L. caryopsis. *Sci. Papers Ser. A Agron.* **2019**, *52*, 513–518.
10. Rzymowska, Z.; Skrzyczyńska, J.; Affek-Straczewska, A. Occurrence and some morphological features of *Bromus secalinus* L. in agrocenoses of the Podlaski Przełom Bugu Mesoregion. *Fragm. Agron.* **2010**, *27*, 102–110. (In Polish)
11. Kącki, Z.; Szczęśniak, E.; Czarniecka, M. *Bromus secalinus* (Poaceae) in Lower Silesia–occurrence and threats. *Acta Bot. Sil. Suppl.* **2011**, *1*, 66–68. (In Polish)
12. Skrajna, T.; Kubicka, H.; Rzymowska, Z. Phenotypic variation in relation to seed storage protein polymorphism in *Bromus secalinus* L. (Gramineae) populations from North-Eastern Poland. *Pol. J. Ecol.* **2012**, *60*, 41–65
13. Stokłosa, A.; Puła, J. Selected morphological traits of cheat (*Bromus secalinus* L.) and cereals grown in mixtures. *Mod. Phytomorphol.* **2012**, *2*, 17–18.

14. Kubicka, H.; Gozdowski, D.; Skrajna, T. Multivariate evaluation of variability of *Bromus* genotypes from north-eastern Polan *Appl. Ecol. Environ. Res.* **2014**, *13*, 809–818. [CrossRef]
15. Adamczewski, K.; Kaczmarek, S.; Kierzek, R.; Urban, M. Germination biology and weed thresholds of rye brome (*Bromus secalinu* L.) in wheat (*Trtiticum aestivum* L.). *Pak. J. Agri. Sci.* **2015**, *52*, 989–995.
16. Stone, A.E.; Peeper, T.F.; Solie, J.B. Cheat (*Bromus secalinus*) control with herbicides applied to mature seeds. *Weed Technol.* **200** *15*, 382–386. [CrossRef]
17. Stone, J.C.; Peeper, T.F.; Stone, A.E. Rotational cropping systems to reduce cheat (*Bromus secalinus*) densities. *Weed Technol.* **200** *20*, 445–452. [CrossRef]
18. Nourozi, M.; Sheidai, M.; Noori, A.; Assadi, M. *Bromus secalinus* L. (Poaceae), a new record for flora of Iran. *Iran. J. Bot.* **2005**, *1* 71–73.
19. Malara, J.; Gancarczyk-Gola, M. Characteristic of species *Bromus secalinus* L. *Pam. Puł.* **2009**, *150*, 337–343. (In Polish)
20. Korniak, T.; Dynowski, P. *Bromus secalinus* (Poaceae)—A vanishing or a widespread weed species of cereal crops in North-Easter Poland? *Fragm. Florist. Geobot. Pol.* **2011**, *18*, 341–348. (In Polish)
21. Mirek, Z.; Zarzycki, K.; Wojewoda, W.; Szeląg, Z. *Red List of Plants and Fungi in Poland*; Inst. Bot. PAN: Kraków, Poland, 200 p. 99. (In Polish)
22. Chmara, R.; Rekowska, E. New locality of *Bromus secalinus* (Poaceae) in central Poland. *Fragm. Florist. Geobot. Polon.* **2020**, *2* 731–733. (In Polish) [CrossRef]
23. Korbas, M.; Strażyński, P.; Horoszkiewicz-Janka, J.; Jajor, E.; Miklaszewska, K.; Danielewicz, J.; Kalinowska, A. *Recommendation for the Protection of Plants for 2020/21. Winter Cereals*; IOR-PIB: Poznań, Poland, 2020; p. 547. (In Polish)
24. Moss, S.; Ulber, L.; den Hoed, I. A herbicide resistance risk matrix. *Crop Prot.* **2019**, *115*, 13–19. [CrossRef]
25. Heap, I. The International Herbicide-Resistant Weed Database. Available online: https://www.weedscience.org (accessed on October 2021).
26. Food and Agriculture Organization of the United Nations. Available online: https://www.fao.org/faostat/en/#data/QCI (accessed on 5 October 2021).
27. Herbicide Resistance Action Committee. Available online: https://hracglobal.com (accessed on 19 October 2021).
28. Das, T.K.; Tuti, M.D.; Sharma, R.; Paul, T.; Mirja, P.R. Weed management research in India: An overview. *Indian J. Agron.* **2012**, *57* 148–156.
29. Shahzad, M.; Hussain, M.; Jabran, K.; Farooq, M.; Farooq, S.; Gašparovič, K.; Barboricova, M.; Aljuaid, B.S.; El-Shehawi, A.M. Zuan, A.T.K. The impact of different crop rotations by weed management strategies' interactions on weed infestation and productivity of wheat (*Triticum aestivum* L.). *Agronomy* **2021**, *11*, 2088. [CrossRef]
30. Głąb, L.; Sowiński, J.; Bough, R.; Dayan, F.E. Allelopathic potential of sorghum (*Sorghum bicolor* (L.) Moench) in weed control: A comprehensive review. *Adv. Agron.* **2017**, *145*, 43–95. [CrossRef]
31. Sturm, D.J.; Kunz, C.; Gerhards, R. Inhibitory effects of cover crop mulch on germination and growth of *Stellaria media* (L.) Vill. *Chenopodium album* L. and *Matricaria chamomilla* L. *Crop Prot.* **2016**, *90*, 125–131. [CrossRef]
32. Pużyńska, K.; Jop, B.; Gala-Czekaj, D.; Synowiec, A.; Bocianowski, J. Effect of allelopathic seed meals on the weed infestation and yielding of maize. *Acta Physiol. Plant.* **2019**, *41*, 193. [CrossRef]
33. Rice, A.R.; Johnson-Maynard, J.L.; Thill, D.C.; Morra, M.J. Vegetable crop emergence and weed control following amendmen with different Brassicaceae seed meals. *Renew. Agric. Food Syst.* **2007**, *22*, 204–212. [CrossRef]
34. Webber, C.L., III; White, P.M., Jr.; Boydston, R.; Shrefer, J. Impact of mustard seed meal applications on direct-seeded cucurbits and weed control. *J. Agric. Sci.* **2017**, *9*, 81–90. [CrossRef]
35. Mehmood, Z.; Ashiq, M.; Noorka, I.R.; Ali, A.; Tabasum, S.; Iqbal, M.S. Chemical control of monocot weeds in wheat (*Triticum aestivum* L.). *Am. J. Plant Sci.* **2014**, *5*, 1272–1276. [CrossRef]
36. Synowiec, A.; Jop, B.; Domaradzki, K.; Podsiadło, C.; Gawęda, D.; Wacławowicz, R.; Wenda-Piesik, A.; Nowakowski, M.M. Bocianowski, J.; Marcinkowska, K.; et al. Environmental factors effects on winter wheat competition with herbicide-resistant or susceptible silky bentgrass (*Apera spica-venti* L.) in Poland. *Agronomy* **2021**, *11*, 871. [CrossRef]
37. Kelley, P.; Peeper, T.F. MON 37500 application timing affects cheat (*Bromus secalinus*) control and winter wheat. *Weed Sci.* **2003**, *51* 231–236. [CrossRef]
38. Vázquez-García, J.G.; Castro, P.; Cruz-Hipólito, H.E.; Millan, T.; Palma-Bautista, C.; De Prado, R. Glyphosate resistance confirmation and field management of red brome (*Bromus rubens* L.) in perennial crops grown in southern Spain. *Agronomy* **2021**, *11*, 535 [CrossRef]
39. Davies, L.R.; Hull, R.; Moss, S.; Neve, P. The first cases of evolving glyphosate resistance in UK poverty brome (*Bromus sterilis*) populations. *Weed Sci.* **2019**, *67*, 41–47. [CrossRef]
40. Adamczewski, K.; Matysiak, K.; Kierzek, R.; Kaczmarek, S. Significant increase of weed resistance to herbicides in Poland. *J. Plant. Prot. Res.* **2019**, *59*, 139–150. [CrossRef]
41. Stankiewicz-Kosyl, M.; Haliniarz, M.; Wrochna, M.; Synowiec, A.; Wenda-Piesik, A.; Tendziagolska, E.; Sobolewska, M.; Domaradzki, K.; Skrzypczak, G.; Łykowski, W.; et al. Herbicide resistance of *Centaurea cyanus* L. in Poland in the context of its management. *Agronomy* **2021**, *11*, 1954. [CrossRef]

2. Stankiewicz-Kosyl, M.; Synowiec, A.; Haliniarz, M.; Wenda-Piesik, A.; Domaradzki, K.; Parylak, D.; Wrochna, M.; Pytlarz, E.; Gala-Czekaj, D.; Marczewska-Kolasa, K.; et al. Herbicide resistance and management options of *Papaver rhoeas* L. and *Centaurea cyanus* L. in Europe: A review. *Agronomy* **2020**, *10*, 874. [CrossRef]
3. Boutin, C.; Strandberg, B.; Carpenter, D.; Mathiassen, S.K.; Thomas, P.J. Herbicide impact on non-target plant reproduction: What are the toxicological and ecological implications? *Environ. Pollut.* **2014**, *185*, 295–306. [CrossRef]
4. Mejri, D.; Gamalero, E.; Tombolini, R.; Massa, N.; Berta, G.; Souissi, T. Biological control of great brome (*Bromus diandrus*) in durum wheat (*Triticum durum*): Specificity, physiological traits and impact on plant growth and root architecture of the fluorescent pseudomonad strain X33d. *BioControl* **2010**, *55*, 561–572. [CrossRef]
5. Kennedy, A.C.; Johnson, B.N.; Stubbs, T.L. Host range of a deleterious rhizobacterium for biological control of downy brome. *Weed Sci.* **2001**, *49*, 792–797. [CrossRef]
6. Reinhart, K.O.; Carlson, C.H.; Feris, K.P.; Germino, M.J.; Jandreau, C.J.; Lazarus, B.E.; Mangold, J.; Pellatz, D.W.; Ramsey, P.; Rinella, M.J.; et al. Weeds suppressive bacteria fail to control *Bromus tectorum* under field conditions. *Rangel. Ecol. Manag.* **2020**, *73*, 760–765. [CrossRef]
7. Synowiec, A.; Kalemba, D. Composition and herbicidal effect of *Heracleum sosnowskyi* essential oil. *Open Life Sci.* **2015**, *10*, 425–432. [CrossRef]
8. Jabran, K.; Mahajan, G.; Sardana, V.; Chauhan, B.S. Allelopathy for weed control in agricultural systems. *Crop. Prot.* **2015**, *72*, 57–65. [CrossRef]
9. Duke, S.O.; Dayan, F.E.; Rimando, A.M.; Schrader, K.K.; Aliotta, G.; Oliva, A. Chemicals from nature for weed management. *Weed Sci.* **2002**, *50*, 138–151. [CrossRef]
0. Gala-Czekaj, D.; Jop, B.; Synowiec, A. The influence of meal from seeds and perianth of Sosnovskyi hogweed (*Heracleum Sosnowskyi* Manden.) on initial growth of maize and two weeds species. *Fragm. Agron.* **2018**, *35*, 29–39. (In Polish) [CrossRef]
1. Boydston, R.A.; Anderson, T. Mustard (*Sinapis alba*) seed meal suppresses weeds in container-grown ornamentals. *Hortic. Sci.* **2008**, *43*, 800–803. [CrossRef]
2. Dhima, K.V.; Vasilakoglou, I.; Gatsis, T.D.; Panou-Philotheou, E.; Eleftherohorinos, I. Effects of aromatic plants incorporated as green manure on weed and maize development. *Field Crops Res.* **2009**, *110*, 235–241. [CrossRef]
3. Saini, R.; Singh, S. Use of natural products for weed management in high-value Crops: An overview. *Preprints* **2018**, 2018100737. [CrossRef]
4. Schappert, A.; Schumacher, M.; Gerhards, R. Weed control ability of single sown cover crops compared to species mixtures. *Agronomy* **2019**, *9*, 294. [CrossRef]
5. Tursun, N.; Işık, D.; Demir, Z.; Jabran, K. Use of living, mowed, and soil-incorporated cover crops for weed control in apricot orchards. *Agronomy* **2018**, *8*, 150. [CrossRef]
6. Dhima, K.; Vasilakoglou, I.; Garane, V.; Ritzoulis, C.; Lianopoulou, V.; Panou-Philotheou, E. Competitiveness and essential oil phytotoxicity of seven annual aromatic plants. *Weed Sci.* **2010**, *58*, 457–465. [CrossRef]
7. Fernández-Aparicio, M.; Masi, M.; Cimmino, A.; Vilariño, S.; Evidente, A. Allelopathic effect of quercetin, a flavonoid from *Fagopyrum esculentum* roots in the radicle growth of phelipanche ramosa: Quercetin natural and semisynthetic analogues were used for a structure-activity relationship investigation. *Plants* **2021**, *10*, 543. [CrossRef] [PubMed]
8. Szwed, M.; Wiczkowski, W.; Szawara-Nowak, D.; Obendorf, R.L.; Horbowicz, M. Allelopathic influence of common buckwheat root residues on selected weed species. *Acta Physiol. Plant.* **2019**, *41*, 92. [CrossRef]
9. Kalinova, J.; Vrchotova, N.; Triska, J. Exudation of allelopathic substances in buckwheat (*Fagopyrum esculentum* Moench). *J. Agric. Food Chem.* **2007**, *55*, 6453–6459. [CrossRef] [PubMed]
0. Szwed, M.; Mitrus, J.; Wiczkowski, W.; Dębski, H.; Horbowicz, M. If phenolic compounds in the soil with buckwheat residues affect the emergence and growth of weed seedlings? *Acta Physiol. Plant.* **2020**, *42*, 154. [CrossRef]
1. Mioduszewska, H.; Klocek, J.; Horbowicz, M.; Wolska, K. Effect of water extracts from tissues of common buckwheat on seed germination and seedling growth of winter wheat and lettuce. *Acta Sci. Pol. Agric.* **2013**, *12*, 45–54.

plants

Article

Volatile and Non-Volatile Allelopathic Characteristics in Thermally Processed Needles of Two Conifers

Chan Saem Gil, Daeun Hong, Shucheng Duan and Seok Hyun Eom *

Department of Smart Farm Science, College of Life Sciences, Kyung Hee University, Yongin 17104, Korea; cskil@khu.ac.kr (C.S.G.); ekdms7275@khu.ac.kr (D.H.); dsc97@khu.ac.kr (S.D.)
* Correspondence: se43@khu.ac.kr

Abstract: With allelopathic composts, potential merits for preventing initial weed infestations have been observed in crop transplantation. However, previous studies have rarely investigated whether high temperatures, generated during composting, decrease allelopathic ability. This study evaluated the thermal allelopathic effect of two coniferous plants (*Pinus densiflora* and *P. koraiensis*) on *Brassica napus* germination and seedling growth using their characterized allelochemical destinations. The 90 °C dry treatment of *P. densiflora* extract exhibited stronger inhibitory effect on germination than its 30 °C dry treatment. In a range from 0.25 to 1 mg mL^{-1}, the germination rate was decreased to 38.1 and 64.3% of control with *P. densiflora* extract dried at 90 and 30 °C, respectively. However, *P. koraiensis* showed potent inhibition of the germination process with no statistical difference in inhibitory effects regardless of the dry temperature. Regarding *B. napus* seedling root growth, the allelopathic effects of aqueous extracts of both conifers were not reduced with the 90 °C treatment, but it was lost in seedling shoot growth. GC-MS/MS confirmed that high temperature treatment drastically decreased volatile contents to 53.2% in *P. densiflora*, resulting in reduced allelopathic abilities. However, a relatively lower decrease to 83.1% in volatiles of *P. koraiensis* accounts for less loss of the root-specific inhibitory effect on *B. napus* seedlings even after 90 °C treatment. Foliar tissues of both conifers with species-specific thermal resistance have potentially valuable functions regarding allelopathic use in horticultural compost processing ingredients, demonstrating their weed control ability during the early cultivation season where crops are transplanted in the facilitated area.

Keywords: allelopathy; bioherbicides; compost processing; coniferous volatiles; *Pinus densiflora*; *Pinus koraiensis*; thermal resistance

Citation: Gil, C.S.; Hong, D.; Duan, S.; Eom, S.H. Volatile and Non-Volatile Allelopathic Characteristics in Thermally Processed Needles of Two Conifers. *Plants* **2022**, *11*, 1003. https://doi.org/10.3390/plants11081003

Academic Editors: Alessia Restuccia and Aurelio Scavo

Received: 15 March 2022
Accepted: 5 April 2022
Published: 7 April 2022

1. Introduction

Pinus (Pinaceae) is the largest genus of extant gymnosperms, widely distributed in the Northern Hemisphere [1]. Among the *Pinus* group, two conifer species, including *Pinus densiflora* Siebold and Zucc., in addition to *Pinus koraiensis* Siebold and Zucc., have been identified. Subsequently, they have been regarded as an essential forestry resource in China, Japan, and Korea. The number of needle-shaped leaves in their fascicle, two needles of *P. densiflora* and five needles of *P. koraiensis*, characterizes these two species [1].

Weed infection is a significant problem interrupting normal crop growth, potentially accounting for approximately 50% of crop yield loss [2]. Besides, weeds not only reduce crop quality but also harbor insect pests and diseases [3]. Although the use of chemical herbicides is increasing to control agricultural weeds, the indiscriminate use of herbicides results in severe ecological and environmental problems, such as soil and water contamination, occurrence of resistant weeds, weed population shifts, and dominance of minor weeds [2,4]. Therefore, current researchers are increasingly focusing on developing eco-friendly herbicides using natural products, and allelochemicals are being verified as an alternative to the use of chemical herbicides worldwide [5,6]. For example, Jabran et al. [7] reported that allelopathic plants can potentially reduce weed pressure, resulting in improved crop yields and reduced synthetic pesticide use. It has also been reported that *P. densiflora* and *P. koraiensis* are allelopathic plants, whose allelopathic effects are derived from non-volatile and volatile compounds including phenolics, flavonoids, and monoterpenes [8]. Therefore, *P. densiflora* can inhibit herbaceous invasion by releasing resin acids into the soil. Afterward, these resin acids are degraded in the soil and transformed into growth inhibitors, such as 15-hydroxy-7-oxodehydroabietate and 7-oxodehydroabietic acid [9]. Studies have further proven the allelopathic activities of *P. densiflora* and *P. koraiensis* needle tissues. As observed, aqueous methanolic extracts of *P. densiflora* needles prevented certain seedlings' initial growth, including cress, lettuce, and alfalfa [10]. In addition, *P. koraiensis* needle water extracts inhibited the germination and growth of *Echinochloa crus-galli*, *Plantago asiatica*, *Achyranthes japonica*, *Ocnothera odorada*, and *Lactuca sativa* [11].

The application of allelopathy in cropping systems can be easily conducted by mixing the allelopathic plant residues with soil, like composts. However, this application remains limited within certain areas due to less information on allelopathic mechanisms and chemical properties. Though the organic composting materials can be encountered to substantial changes by high temperatures which are generated during the compost-producing process [12], only a few studies have investigated whether the allelopathic efficacy of plant tissue is affected by the thermal process. Furthermore, most researchers have mainly focused on compost values as nutritional suppliers to crops [13–15]. Instead, we tried to evaluate the allelopathic efficacy of volatile and non-volatile components in thermally processed needles. Therefore, we conducted this study to identify the allelopathic characteristics of volatile and non-volatile components in two *Pinus* needles against initial seedling growth and to determine whether thermal processes affect the allelopathic efficacy of the needles.

2. Results

2.1. Non-Volatile Assay

The germination rate of *Brassica napus* seeds, treated using needle extracts of *P. densiflora* and *P. koraiensis*, is shown in Figure 1. As shown, the germination rate rapidly decreased to 71 and 48% of the control with 0.25 mg mL^{-1} extract of the *P. densiflora* needle dried at 30 and 90 °C, respectively (Figure 1A). Reduction of the germination rate hardly occurred despite the concentration increase. In contrast, with *P. koraiensis*, the germination rate continually decreased to 57 and 38% of control after using 0.5 mg mL^{-1} extract of the needle dried at 30 and 90 °C, respectively.

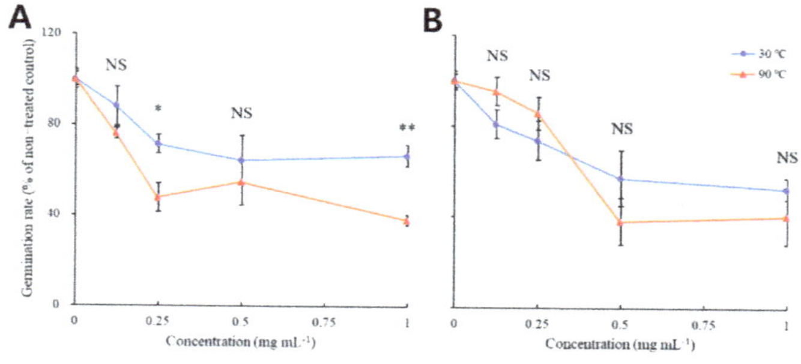

Figure 1. Germination patterns of *Brassica napus* seeds treated with needle extracts of *Pinus densiflora* (**A**) and *P. koraiensis* (**B**) dried at 30 or 90 °C. Asterisks indicate significant difference on Tukey's HSD test followed by * ($p < 0.05$) and ** ($p < 0.01$). NS indicates no significant difference.

Figure 2 shows the growth parameters of *B. napus* seedlings affected by *P. densiflora* and *P. koraiensis* needle extracts dried at 30 and 90 °C. As shown, *P. densiflora* and *P. koraiensis* needle extracts inhibited hypocotyl elongation to 62 and 48% of the control during the non-thermal treatment, using 0.25 mg mL^{-1} of the treatment (Figure 2). Furthermore, hypocotyl length was not further suppressed using treatment quantities above 0.25 mg mL^{-1}. Unlike the inhibition pattern of hypocotyl growth, root growth was dose-dependently inhibited during the treatment with the two coniferous extracts. As observed with 1 mg mL^{-1} of *P. densiflora* and *P. koraiensis* extract treatment, root length was decreased to 40 and 56% of the control during non-thermal treatment, respectively (Figure 2). After the thermal process, the inhibitory effect only disappeared on hypocotyl growth. For example, when 1 mg mL^{-1} of the extract was treated, although hypocotyl elongation was inhibited to 85% in *P. densiflora* and 77% in *P. koraiensis*, root elongation was inhibited to 47 and 42% of the control in *P. densiflora* and *P. koraiensis*, respectively (Figure 2).

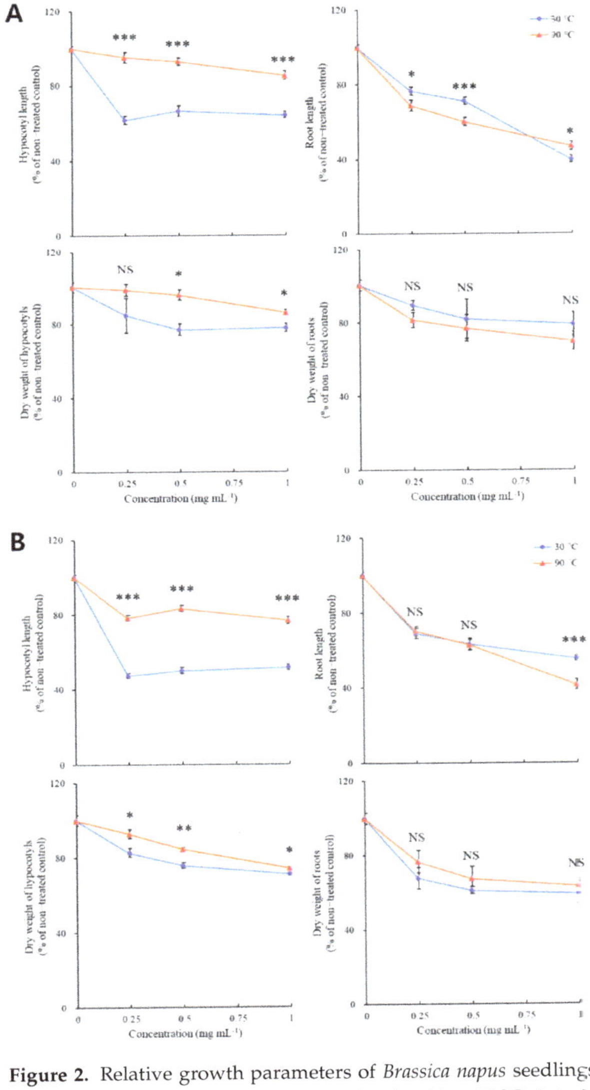

Figure 2. Relative growth parameters of *Brassica napus* seedlings inhibited by needle extracts of *Pinus densiflora* (**A**) and *P. koraiensis* (**B**) dried at 30 or 90 °C. Asterisks indicate significant difference on Tukey's HSD test followed by * ($p < 0.05$), ** ($p < 0.01$), and *** ($p < 0.001$). NS indicates no significant difference.

2.2. *Profiles of Non-Volatiles Using HPLC and LC-MS/MS*

High-performance liquid chromatography (HPLC) detected nine peaks in the *P. densiflora* needle extract (Figure 3A). Among the nine peaks, four components were identified using liquid chromatography coupled with tandem mass spectrometry (LC-MS/MS) (Table 1), including isoquercetrin (peak 2; *m/z* = 463.41), catechin-*O*-glucose-malonate (peak 3; *m/z* = 537.65), chrysoeriol-7-*O*-glucoside (peak 6; *m/z* = 461.49), and kaempferol 3-*O*-rutinoside-7-*O*-rhamnoside (peak 9; *m/z* = 739.57). Thermal treatment reduced most peak areas. Moreover, the 9th peak area was significantly reduced by 26% after thermal treatment at 90 °C. Alternatively, eight peaks were detected as major components in the needle extract from *P. koraiensis* (Figure 3B). Different from *P. densiflora*, after thermal treatment at 90 °C, the area of the 8th peak increased by 16%. Among these eight peaks, seven components were identified (Table 1), including schisandrin (peak 1; *m/z* = 431.91), isosteviol (peak 3; *m/z* = 363.53), lithospermic acid (peak 4; *m/z* = 537.83), quercetin-*O*-xylo-pentoside (peak 5; *m/z* = 579.70), oleuropein (peak 6; *m/z* = 539.89), tiliroside (peak 7; *m/z* = 593.75), and kaempferol 3-*O*-rutinoside-7-*O*-rhamnoside (peak 8; *m/z* = 739.57).

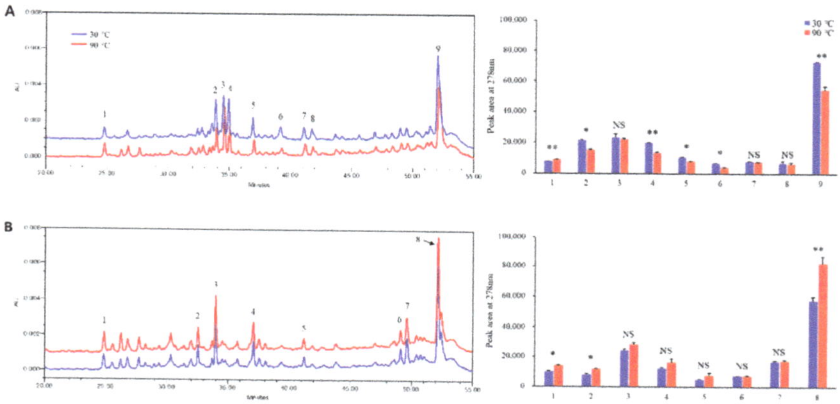

Figure 3. HPLC chromatograms and peak areas detected at 278 nm of separated compounds in needle extracts of *Pinus densiflora* (**A**) and *P. koraiensis* (**B**) dried at 30 or 90 °C. Asterisks indicate significant difference between dry temperatures on Tukey's HSD test followed by * ($p < 0.05$) and ** ($p < 0.01$). NS indicates no significant difference.

2.3. *Volatile Assay*

The allelopathic effects of volatiles from fresh *P. densiflora* and *P. koraiensis* needles on *B. napus* seedlings' growth are shown in Figure 4. We observed that coniferous volatiles inhibited more root growth than hypocotyl growth. Furthermore, although the inhibitory effect on root growth increased with the *P. densiflora* needles' concentration, volatiles did not affect the hypocotyl length of the *P. densiflora* needle, exhibiting 94.2% of control despite the 8 g needle treatment (Figure 4A). Alternatively, with *P. koraiensis*, both root and hypocotyl growth were dose-dependently inhibited. As observed, the dry weight of roots and hypocotyls exhibited 15.6 and 26.3% of control with 8 g fresh needle treatment, respectively (Figure 4B).

Subsequently, the allelopathic capacity of volatiles in the steamed needles of the two conifers was evaluated (Figure 5). After 4 h of steaming, the ability to inhibit hypocotyl growth was similar in both *P. densiflora* and *P. koraiensis* needles. However, the allelopathic capacity to inhibit root growth gradually decreased after one hour of steaming both coniferous needles. After more than one hour of steaming, the inhibitory power of *P. densiflora* disappeared, whereas that of *P. koraiensis* was sustained, inhibiting root length to 97.4% of the control after four hours of steamed needle treatment (Figure 5).

Table 1. Metabolites identified in needle extracts of *Pinus densiflora* and *P. koraiensis* by LC-MS/MS analysis.

Species	No.	RT (min)	[M−H]⁻	MS/MS	Identification
P. densiflora	1	26.40	–	–	NI [1]
	2	34.00	463.41	179.0/271.0/301.1/343.1/417 4/445.2	Isoquercetrin
	3	34.33	537.65	299.3/327.1/328.8/469.1/490.9/515.8	Catechin-*O*-glucose-malonate
	4	34.50	509.54	163.0/179.0/311.1/367.3/385.2/473.3/491.2	NI
	5	35.84	551.88	327.2/329.2/341.1/358.8/359.5/491.1	NI
	6	40.70	461.49	139.1/165.1/193.1/243.1/29.0/298.1/299.0/341.0/342.1/433.2	Chrysoeriol-7-*O*-glucoside
	7	41.40	493.82	315.2/316.4/447.2/448.3	NI
	8	42.70	493.68	259.1/315.2/426.0/447.2/447.9	NI
	9	48.80	739.57	229.0/285.1/286.1/289.1/435.2/453.2/454.2/575.3/593.3/620.3/885.5	Kaempferol 3-*O*-rutinoside-7-*O*-rhamnoside
P. koraiensis	1	26.30	431.91	153.0/223.0/307.3/343.1/385.1/386.2/399.3	Schisandrin
	2	29.60	571.37	316.1/375.3/467.3/525.4/541.3	NI
	3	30.67	363.53	135.0/147.0/165.0/179.1/201.2/221.1/239.0/273.3/315.1/345.1/346.2	Isosteviol
	4	34.30	537.83	163.1/207.4/299.1/327.3/329.2/345.0/477.3/491.1/519.5	Lithospermic acid
	5	43.93	579.70	178.9/255.1/301.1/343.0/433.0/434.2/489.2/561.2	Quercetin-*O*-xylopentoside
	6	46.11	539.89	207.0/331.3/372.9/432.9/492.9	Oleuropein
	7	46.40	593.75	178.9/203.0/285.1/293.1/299.0/300.1/316.2/417.2/447.3/547.1/576.4	Tiliroside
	8	48.80	739.70	229.1/285.0/286.1/289.0/429.1/453.1/454.3/575.1/593.2/594.3/638.9/680.3	Kaempferol 3-*O*-rutinoside-7-*O*-rhamnoside

[1] NI: not identified.

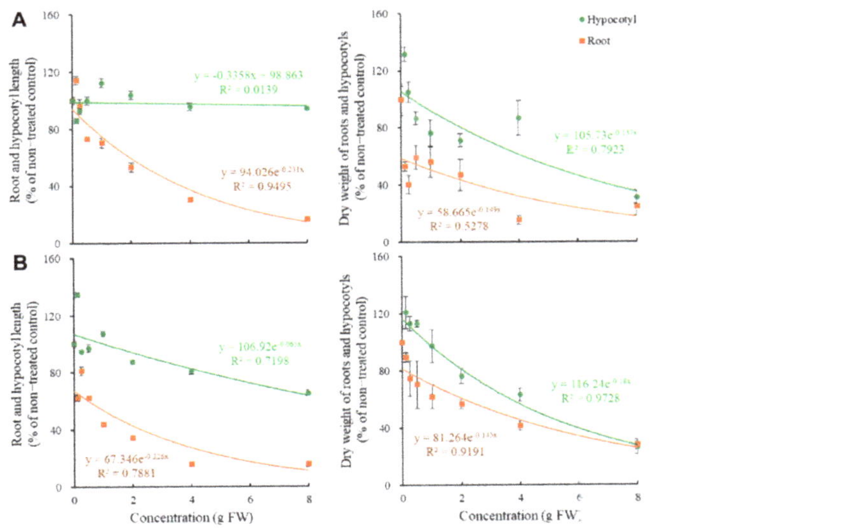

Figure 4. Changes of growth rates of *Brassica napus* seedlings depending on fresh weight needles of *Pinus densiflora* (**A**) and *P. koraiensis* (**B**) in volatile assay.

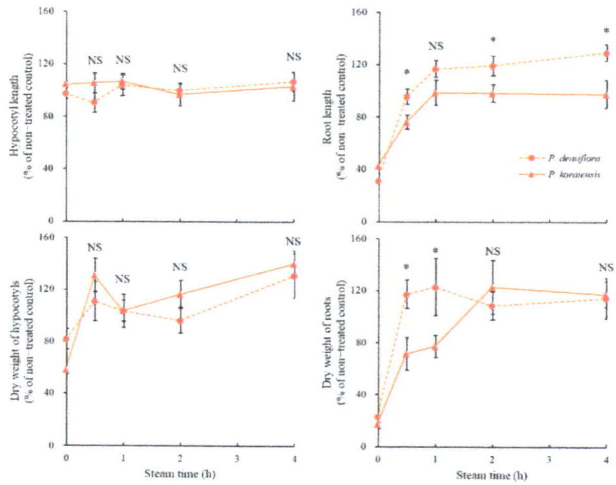

Figure 5. Changes of growth rates of *Brassica napus* seedlings depending on steam time of *Pinu densiflora* and *P. koraiensis* needles. Asterisks indicate significant difference on Tukey's HSD tes followed by * ($p < 0.05$). NS indicates no significant difference.

2.4. Volatile Identification and Quantification using GC-MS/MS

Volatile compounds were identified in *P. densiflora* and *P. koraiensis* needles, afte which changes in the volatile content after 2 h of steaming were analyzed using ga chromatography tandem mass spectrometry (GC-MS/MS) (Table 2 and Figure 6). The results identified 63 and 54 volatiles in the fresh needles of *P. densiflora* and *P. koraiensis* respectively. After steam treatment, 16 volatiles were entirely evaporated in each coniferous needle, and 7 volatiles were the same among the evaporated volatiles. Furthermore although the number of evaporated volatiles was similar, total volatiles' net contents differed between the two steam-processed needles. Compared to the fresh needle of each conifer, results also showed that the total volatile content decreased to 53.2 and 83.1% in steamed *P. densiflora* and *P. koraiensis* needles (Table 2 and Figure 6). Moreover, as shown in Table 2 and Figure 6, α-pinene contained the highest content in both needles, representing 75.9 and 70.7% in fresh *P. densiflora* and *P. koraiensis* needles, respectively. However, the content decreased to 36.0 and 53.9%, respectively, after steaming.

Table 2. Profiles and relative contents of volatile compounds in either fresh or steamed needles o *Pinus densiflora* and *P. koraiensis* by GC−MS/MS analysis.

Compound	RT	*m/z*	*P. densiflora*		*P. koraiensis*	
			Fresh	Steamed	Fresh	Steamed
(3R)-3-phenyl-2,3-dihydroisoindol-1-one	6.52	209.24	0.24 ± 0.02	0.25 ± 0.01	0.14 ± 0.00	0.14 ± 0.00
Ethanol	9.18	46.07	0.32 ± 0.01	0.02 ± 0.00	ND [1]	ND
3,3,5-Trimethylheptane	11.04	142.28	0.11 ± 0.00	0.01 ± 0.00	ND	ND
α-Pinene	11.48	136.23	75.91 ± 0.82	35.95 ± 0.33	70.69 ± 0.10	53.90 ± 0.00
Ethyl vinyl ketone	11.56	84.12	0.25 ± 0.00	0	ND	ND
Hexanal	13.09	100.16	0.46 ± 0.03	0.01 ± 0.00	0.34 ± 0.01	0.01 ± 0.00
pent-1-en-3-ol	14.69	86.13	0.45 ± 0.00	0	0.14 ± 0.00	0.07 ± 0.00
2-Hexenal	16.39	98.14	3.37 ± 0.09	0.01 ± 0.00	2.08 ± 0.06	0
Isoamyl alcohol	17.98	88.15	ND	ND	0.66 ± 0.00	0
cis-2-Penten-1-ol	18.55	86.13	0.20 ± 0.00	0	ND	ND
1-Hexanol	19.41	102.17	0.62 ± 0.05	0	ND	ND

Table 2. *Cont.*

Compound	RT	m/z	*P. densiflora*		*P. koraiensis*	
			Fresh	Steamed	Fresh	Steamed
cis-3-Hexen-1-ol	20.46	100.16	1.34 ± 0.12	0	0.19 ± 0.00	0
2-Hexen-1-OL	21.07	100.16	1.54 ± 0.02	0.01 ± 0.00	ND	ND
1,2,3-Trichloropropane	23.16	147.43	0.48 ± 0.01	1.30 ± 0.03	0.27 ± 0.00	0.36 ± 0.00
Benzyldimethyldecylammonium chloride	26.19	311.90	0.11 ± 0.00	0.10 ± 0.00	ND	ND
6-Fluoro-6-methylheptanol	29.10	148.22	0.36 ± 0.00	0	0.34 ± 0.00	0.30 ± 0.00
3-Nonen-2-one	29.11	140.22	2.72 ± 0.01	2.25 ± 0.02	2.55 ± 0.01	2.25 ± 0.00
Limonene	30.45	136.24	0.05 ± 0.00	0.08 ± 0.00	0.11 ± 0.00	0.21 ± 0.00
α-Amorphene	32.81	204.35	1.52 ± 0.01	1.89 ± 0.01	ND	ND
δ-Cadinene	35.87	204.35	4.47 ± 0.02	6.40 ± 0.06	12.35 ± 0.13	15.29 ± 0.00
9,10-Octalin	36.10	136.23	3.55 ± 0.08	3.52 ± 0.02	8.69 ± 0.10	9.50 ± 0.00
2-(hydroxymethyl) benzoic acid	37.10	152.15	0.42 ± 0.01	0.63 ± 0.02	0.24 ± 0.00	0.29 ± 0.00
Methionol	37.46	106.19	ND	ND	0.19 ± 0.00	0.21 ± 0.00
trans-2-Hexenoic acid	43.90	114.14	0.30 ± 0.04	0	ND	ND
Trace compounds (below than 0.1%)			1.21 ± 0.00	0.77 ± 0.01	1.02 ± 0.00	0.56 ± 0.00
Total			100.0	53.2	100.0	83.1

[1] ND: not detected.

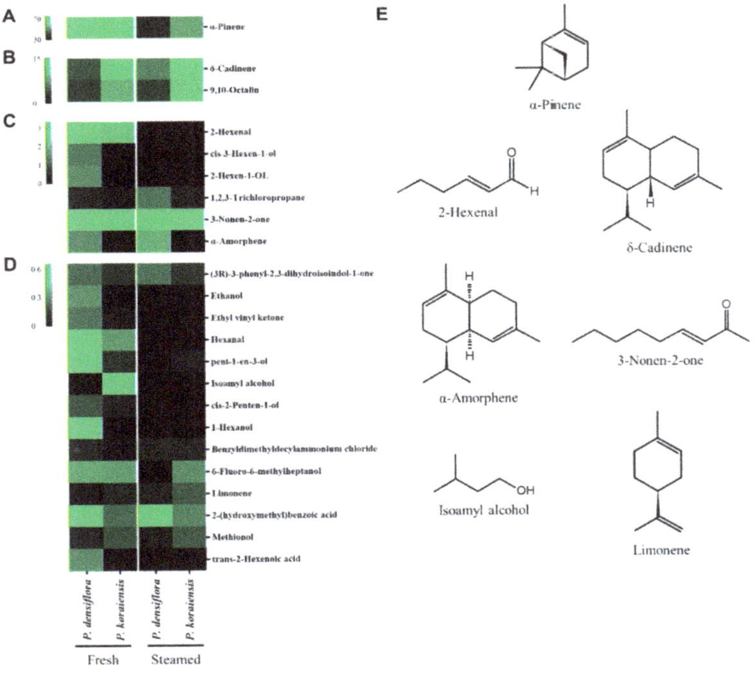

Figure 6. Heatmap describing relative contents of major (**A**); semi−major (**B**); semi−minor (**C**); and minor (**D**); volatiles and allelopathic volatile compounds (**E**) in either fresh or steamed needles of *Pinus densiflora* and *P. koraiensis*.

3. Discussion

This study established that seed germination and seedling growth are inhibited under the allelopathic effects of *P. densiflora* and *P. koraiensis* needles. Notably, their inhibitory effects on seedling growth were higher in root growth than hypocotyl growth. This difference is proposed to be because the roots directly had contact with the extracts, causing them to have a higher sensitivity to allelopathy [16–18]. Our study also showed that thermal treatment did not remove the allelopathic effect of the two conifer needles on root growth. Quantitative HPLC results distinctly showed the destination of each non-volatile allelochemical in coniferous extracts. As observed, most flavonoid contents were significantly decreased in *P. densiflora* needles after thermal treatment, whereas those in *P. koraiensis* needles increased. Several studies have reported a positive correlation between allelopathic efficacy and allelochemical contents [19–22]. Considering the thermal effects, the positive correlation between bioassays and HPLC results was found with the highest dose (1 mg mL^{-1}) in our results. According to the other study, certain allelopathic extracts did not negatively affect seedling growth at low concentration, causing leaf extension and total biomass increase of lettuce seedlings [23]. Furthermore, it was mentioned that dose is a critical factor for exhibiting the inhibitory effects of allelochemicals in rice hull extracts [24]. Therefore, we tentatively propose that enough dose is needed to confirm the effects of allelochemicals on bioassays when extracts are applied.

Our volatile assay identified terpenoids mainly in *P. densiflora* and *P. koraiensis* needles. Results also showed that the primary compound was α-pinene, exhibiting more than 70% of the content in both conifer species. Moreover, the inhibitory effects of the volatiles on fresh foliage revealed a dose-dependent root-specific inhibition. However, unlike the extract treatment in the non-volatile assay, volatiles indirectly contacted the seedling roots. Nevertheless, constituents of essential oil and aromatic volatiles, such as monoterpenes, are easily absorbed in the soil and exhibit allelopathic effects on the rhizosphere of other plants [25–27]. Similarly, it has been reported that volatiles can be released from conifer needles, adsorbed onto a filter paper, and accumulated in high concentrations [26]. Another plausible explanation for root-specific inhibition is a growth characteristic difference between the root and hypocotyl. According to cellular basis, hypocotyl growth relies on elongating each cell already developed in the embryo within the seed. Besides, root growth requires proliferation and elongation of cells [28]. Yet, although volatile monoterpenes did not significantly inhibit cell expansion in both the root and hypocotyl, they inhibited cell proliferation and DNA synthesis in the root apical meristem [26].

The root-specific inhibition of the volatiles was also shown after thermal treatment. As observed, although the inhibitory effect of *P. densiflora* needles was practically eliminated through steam processes for more than one hour, that of *P. koraiensis* needles remained slight despite the increase in steaming time. However, compared to *P. koraiensis*, more volatiles were included in *P. densiflora* needles, and the total content was largely reduced through thermal treatments. After synthesis, monoterpenes are commonly stored in specialized structures, such as resin ducts, oil glands, and secretory cells in conifer needles [29]. The emission of the monoterpenes from these storage structures is related to their volatility and diffusion rate, promoted with temperature increase [30]. Additionally, volatile emission is proposed to be related to the morphological characteristics and chemical composition of storage structures. Moreover, although more resin duct quantities have been discovered in *P. densiflora* than in the *P. koraiensis* needles, they are primarily placed at the external side [1]. Therefore, compared to *P. densiflora*, *P. koraiensis* resin ducts are placed at the inner side surrounded by the hypodermis with thickened cell walls [1]. In contrast, lower nitrogen, lignin, and cellulose contents characterize the chemical composition of the *P. densiflora* needle than the *P. koraiensis* needle [31]. These chemical properties guaranteed a relatively high resistance during thermal treatments and stability in mass reduction when *P. koraiensis* needles were decomposed [31,32].

4. Materials and Methods

4.1. Botanic Materials

Fresh *P. densiflora* and *P. koraiensis* needles were collected during the 2019 summer from a field at the Kyung Hee University Global Campus (N 37°14′36.0″ and E 127°04′52.6″

Yongin, Korea). *B. napus* seeds were purchased from a seed company (Budnara Co., Gwangju, Korea).

4.2. Preparation of Conifer Needle Extracts

Fresh *P. densiflora* and *P. koraiensis* needles were rinsed with distilled water. After removing moisture, the needles were dried in a heat dry machine (KED-066A, C&T Co., Gwangju, Korea) at 30 °C for seven days or 90 °C for three days without light. Then, dried samples were coarsely ground using a commercial grinder and passed through a 100-mesh sieve. Subsequently, the resulting powder (10 g) from each sample was dissolved in 200 mL of 80% aqueous methanol (v/v) and agitated using a shaker (Daewonsci Inc., Bucheon, Korea) at 20 °C for 24 h. Next, the solution was filtered through qualitative filter papers (Whatman No. 2, Maidstone, UK). Finally, the solvent was removed using a vacuum rotary evaporator (Eyela, Tokyo, Japan). To remove residual solvent, the extracts were dried using a freeze dryer (IlShinBioBase Inc., Dongducheon, Korea) and maintained at 15 °C until use for non-volatile assay.

4.3. Non-Volatile Assay

The germination rate and seedling growth of *B. napus* were tested to evaluate the allelopathic efficacy of non-volatiles in each *P. densiflora* and *P. koraiensis* needle extract, dried at 30 or 90 °C. To test the germination rate, we selected 20 *B. napus* seeds, excluding wrinkled and cracked seeds, followed by inoculation in a Petri dish, containing 4 mL distilled water (control) or serial concentrations (0.25, 0.5, and 1 mg mL^{-1}) of each extract, diluted using distilled water. Subsequently, the dishes after seed inoculation were sealed with a parafilm, after which they were maintained in a growth chamber (Daewonsci Inc., Bucheon, Korea), completely randomized under controlled light (16 h fluorescent light per day with 50 µmol m^{-2} s^{-1}), humidity (80%), and temperature (25 °C). Seed germination rate was determined by counting the number of seeds generating 1 mm or longer root daily.

Alternatively, to test seedling growth, 20 *B. napus* seedlings generating 3 mm roots were selected and inoculated in a Petri dish containing 4 mL distilled water and serial concentrations (0.25, 0.5, and 1 mg mL^{-1}) of each extract. On day five after culture, 10 similarly grown seedlings were selected among the 20 seedlings, after which their hypocotyl and root lengths were measured. Then, to measure the dry weight of the two organs, each hypocotyl and root bundle collected from the ten seedlings was entirely dried at 30 °C for 48 h. All treatments were replicated thrice.

4.4. HPLC Analysis

Chemical components in the needle extracts of the two conifers were analyzed using reversed-phase HPLC (Waters 2695 Alliance HPLC; Waters Inc., Milford, MA, USA), coupled with a 250 × 4.6 mm octadecylsilane column (Prontosil 120-5-C18-SH 5.0 µm; Bischoff, Leonberg, Germany). Two solvents were used as mobile phases; water with 0.1% formic acid (solvent A) and MeOH with 0.1% formic acid (solvent B). The gradient flow was as follows: 0–5% of solvent B for 0–10 min, 5–10% of solvent B for 10–20 min, 10–20% of solvent B for 20–30 min, 20–40% of solvent B for 30–50 min, and 40–70% of solvent B for 50–62 min. Furthermore, the flow rate of the mobile phases was 1.0 mL min^{-1}, and the sample injection volume was 5 µL. Then, resulting peaks were monitored at 278 nm using a Waters 996 photodiode array detector.

4.5. LC-MS/MS Analysis

Molecular weights of the flavonoids in the two coniferous extracts were determined using an LC-MS/MS with a Thermo-Finnigan LTQ-Orbitrap instrument (Thermo Fisher Scientific, Waltham, MA, USA) at NICEM in Seoul National University (Seoul, Korea). Then, data acquisition was conducted with Xcalibur™ software (ver. 4.3., Thermo Fisher Scientific Inc., MA, USA). In contrast, MS and MS/MS were operated through an electrospray ionization source in the negative ion mode, recorded in the range of 150 to 2000 *m/z* and

50 to 2000 *m/z*, respectively. Besides, nitrogen gas was used as the sheath gas, whose flo rate was kept at 10 L min^{-1}. The capillary temperature was also maintained at 300 ° nebulizer pressure was set at 45 psi, whereas fragmentation and capillary voltages we 0.2 kV and 4.5 kV, respectively. Additionally, the collision energy was set at 35 Ev, aft which individual compounds were identified by comparing mass data and λ_{max} results previously reported values.

4.6. Volatile Assay

The volatile assay was conducted using a method described in a previous study [33 with some modifications. First, fresh or steam-processed needles were coarsely mashe using a mortar with simultaneous liquid nitrogen pouring. Then, the fresh needles wer respectively, weighed (0.125, 0.25, 0.5, 1, 2, 4, and 8 g) and packaged using one layer c cheesecloth. For the steaming process, fresh needles were placed in boiling water for 0.5, 2, and 4 h, after which 4 g of the steam-processed needles were weighed and packaged wit one layer of cheesecloth. Next, each sample package was positioned above 10 cm from th bottom of a glass bottle (500 mL), after which 20 *B. napus* seedlings, generating 3 mm roo were inoculated on a moistened filter paper (Whatman No.3, Maidstone, UK) placed in th glass bottle. The bottles were tightly sealed with caps and wrapped in a parafilm to avoi any leak of volatiles. Finally, the seedlings were cultured in a growth chamber (Daewons Inc., Bucheon, Korea) under the same conditions as the non-volatile assay. On day five afte culture, seedling growth was evaluated using the same method as the non-volatile assay

4.7. GC-MS/MS Analysis

Fresh or two-hour steamed needles were ground using a mortar with simultaneou liquid nitrogen pouring, after which 1.5 g ground sample and 1 g NaCl were added t the ground samples in a solid-phase microextraction (SPME) amber vial containing 6 u 1,2,3-trichloropropane. Subsequently, an SPME system (Xcalibur™ ver. 4.3, Thermo Fishe Scientific Ind., Waltham, MA, USA) was used to isolate the volatile compounds using a fibe coated with a 65 μm polydimethylsiloxane/divinylbenzene film layer (fused silica 24 Ga Then, analysis was conducted using a GC (Trace1310, Thermo Fisher Scientific Inc., Walthan MA, USA), equipped with a triple quadrupole mass spectrometer (TSQ8000, Thermo Fishe Scientific Inc., Waltham, MA, USA) and a DB-Wax column (60 m × 0.25 mm, 0.50 μm, Agilen Technologies, Santa Clara, CA, USA). Additionally, a helium carrier gas was used at a flow rate of 2 mL min^{-1} and 230 °C inlet temperature. Next, we maintained the ramp temperatur in the GC oven at 40 °C for 5 min, increased the temperature to 120 °C for 8 °C min^{-1}, ther to 160 °C for 2 °C min^{-1}, and 240 °C for 4 °C min^{-1}. The sample was finally held at the final temperature for another 10 min. Afterward, volatile compounds were identified using the NIST/EPA/NIH Mass Spectral Library (ver. 2.0), then the volatiles' quantification was represented as each compound's relative peak area (%).

4.8. Statistical Analysis

Statistical analyses were conducted using the SAS Enterprise Guide (ver. 4.3, SAS Institute Inc., Cary, NC, USA). Then, significant differences among the treatments in these experiments were evaluated using Tukey's studentized range test at $p < 0.05$.

5. Conclusions

The allelopathic capacity of two *Pinus* species showed root-specific inhibition, revealed using dose-dependent needle extracts. In the coniferous needle extracts, flavonoid glycosides were identified as non-volatile allelochemicals. A positive correlation between bioassays and chemical contents was revealed in the application of high doses of the extracts In addition, from the volatile assay results, seedlings were root-specifically suppressed through allelopathic volatiles in the two coniferous needles. Moreover, both coniferous needles contained large terpenoid and α-pinene quantities. However, after the thermal process the decrease in volatile contents was lesser in *P. koraiensis* than in *P. densiflora*. It has been proposed that keeping the allelopathic capacity from thermal treatment is attributed to the chemical and morphological properties of *P. koraiensis* needles. Considering the allelopathic characteristics of the needles on germination and initial root growth, the application of allelopathic compost is considered more beneficial for a cropping system. In addition, it is proposed that coniferous needles have potential as eco-friendly herbicides to control initial weed growth due to their thermal stability and root-specific inhibition.

Author Contributions: Conceptualization, S.H.E.; methodology, C.S.G. and S.H.E.; software, C.S.G.; validation, S.H.E. and C.S.G.; formal analysis, C.S.G.; investigation, C.S.G., D.H. and S.D.; data curation, C.S.G. and D.H.; writing—original draft preparation, C.S.G.; writing—review and editing, S.H.E. All authors have read and agreed to the published version of the manuscript.

Funding: This study was supported by the National Research Foundation of Korea (NRF) grant funded by the Korea government (MSIT) (No. NRF-2019R1A2C1009623).

Data Availability Statement: All data included in the main text.

Conflicts of Interest: The authors declare no conflict of interest.

eferences

1. Ghimire, B.; Lee, C.; Yang, J.; Heo, K. Comparative leaf anatomy of native and cultivated *Pinus* (Pinaceae) in Korea: Implication for the subgeneric classification. *Plant Syst. Evol.* **2015**, *301*, 531–540. [CrossRef]
2. Fahad, S.; Hussain, S.; Chauhan, B.S.; Saud, S.; Wu, C.; Hassan, S.; Tanveer, M.; Jan, A.; Huang, J. Weed growth and crop yield loss in wheat as influenced by row spacing and weed emergence times. *Crop Prot.* **2015**, *71*, 101–108. [CrossRef]
3. Doğan, M.N.; Jabran, K.; Unay, A. Integrated Weed Management in Cotton. In *Recent Advances in Weed Management*; Chauhan, B.S., Mahajan, G., Eds.; Springer: New York, NY, USA, 2014; pp. 197–222. [CrossRef]
4. Guimarães Santos, C.A.; Silva, G.B.L. Daily streamflow forecasting using a wavelet transform and artificial neural network hybrid models. *Hydrol. Sci. J.* **2014**, *59*, 312–324. [CrossRef]
5. Bhadoria, P. Allelopathy: A natural way towards weed management. *Am. J. Exp. Agric.* **2011**, *1*, 7–20. [CrossRef]
6. Cheng, F.; Cheng, Z. Research progress on the use of plant allelopathy in agriculture and the physiological and ecological mechanisms of allelopathy. *Front. Plant Sci.* **2015**, *6*, 1020. [CrossRef] [PubMed]
7. Jabran, K.; Mahajan, G.; Sardana, V.; Chauhan, B.S. Allelopathy for weed control in agricultural systems. *Crop Prot.* **2015**, *72*, 57–65. [CrossRef]
8. Kato-Noguchi, H. Bioactive compounds involved in the formation of the sparse understory vegetation in pine forests. *Curr. Org. Chem.* **2021**, *25*, 1731–1738. [CrossRef]
9. Kato-Noguchi, H.; Kimura, F.; Ohno, O.; Suenaga, K. Involvement of allelopathy in inhibition of understory growth in red pine forests. *J. Plant Physiol.* **2017**, *218*, 66–73. [CrossRef]
10. Kato-Noguchi, H.; Fushimi, Y.; Shigemori, H. An allelopathic substance in red pine needles (*Pinus densiflora*). *J. Plant Physiol.* **2009**, *166*, 442–446. [CrossRef]
11. Singh, R.P.; Ibrahim, M.H.; Esa, N.; Iliyana, M.S. Composting of waste from palm oil mill: A sustainable waste management practice. *Rev. Environ. Sci. Biotechnol.* **2010**, *9*, 331–344. [CrossRef]
12. De Guardia, A.; Mallard, P.; Teglia, C.; Marin, A.; Le Pape, C.; Launay, M.; Benoist, J.C.; Petiot, C. Comparison of five organic wastes regarding their behaviour during composting: Part 1, biodegradability, stabilization kinetics and temperature rise. *Waste Manag.* **2010**, *30*, 402–414. [CrossRef] [PubMed]
13. Hepperly, P.; Lotter, D.; Ulsh, C.Z.; Seidel, R.; Reider, C. Compost, manure and synthetic fertilizer influences crop yields, soil properties, nitrate leaching and crop nutrient content. *Compost Sci. Util.* **2009**, *17*, 117–126. [CrossRef]
14. Afonso, S.; Arrobas, M.; Pereira, E.L.; Rodrigues, M.Â. Recycling nutrient-rich hop leaves by composting with wheat straw and farmyard manure in suitable mixtures. *J. Environ. Manag.* **2021**, *284*, 112105. [CrossRef] [PubMed]
15. Akanbi, W.B.; Togun, A.O. The influence of maize-stover compost and nitrogen fertilizer on growth, yield and nutrient uptake of amaranth. *Sci. Hortic.* **2002**, *93*, 1–8. [CrossRef]
16. Thapa, L.B.; Kaewchumnong, K.; Sinkkonen, A.; Sridith, K. "Soaked in rainwater" effect of *Ageratina adenophora* on seedling growth and development of native tree species in Nepal. *Flora* **2020**, *263*, 151554. [CrossRef]
17. Kumar, M.; Garkoti, S.C. Allelopathy effects of invasive alien *Ageratina adenophora* on native shrub species of chir pine forest in the central Himalaya, India. *J. For. Res.* **2022**, *27*, 53–62. [CrossRef]
18. Bhardwaj, S.; Kapoor, K.S.; Singh, H.P. Studies on allelopathic effects of *Ageratina adenophora* Sprengel (King and Robinson) on some weed plants growing in forest ecosystem. *Int. J. Theor. Appl. Sci.* **2014**, *6*, 1–6.
19. Chung, I.M.; Kim, K.H.; Ahn, J.K.; Chun, S.C.; Kim, C.S.; Kim, J.T.; Kim, S.H. Screening of allelochemicals on barnyardgrass (*Echinochloa crus-galli*) and identification of potentially allelopathic compounds from rice (*Oryza sativa*) variety hull extracts. *Crop Prot.* **2002**, *21*, 913–920. [CrossRef]

20. Al-Huqail, A.A.; Behiry, S.I.; Salem, M.Z.M.; Ali, H.M.; Siddiqui, M.H.; Salem, A.Z.M. Antifungal, antibacterial, and antioxida activities of *Acacia Saligna* (Labill.) H. L. Wendl. Flower extract: HPLC analysis of phenolic and flavonoid compounds. *Molecul* **2019**, *24*, 700. [CrossRef] [PubMed]
21. Hussain, M.I.; El-Sheikh, M.A.; Reigosa, M.J. Allelopathic potential of aqueous extract from *Acacia melanoxylon* R. Br. On *Lactu sativa*. *Plants* **2020**, *9*, 1228. [CrossRef]
22. Alsharekh, A.; El-Sheikh, M.A.; Alatar, A.A.; Abdel-Salam, E.M. Natural control of weed invasions in hyper-arid arable farm Allelopathic potential effect of *Conocarpus erectus* against common weeds and vegetables. *Agronomy* **2022**, *12*, 703. [CrossRef]
23. Wang, K.; Wang, T.; Ren, C.; Dou, P.; Miao, Z.; Liu, X.; Huang, D.; Wang, K. Aqueous extracts of three herbs allelopathical inhibit lettuce germination but promote seedling growth at low concentrations. *Plants* **2022**, *11*, 486. [CrossRef] [PubMed]
24. Berendji, S.; Asghari, J.B.; Matin, A.A. Allelopathic potential of rice (*Oryza sativa*) varieties on seedling growth of barnyardgra (*Echinochloa crus-galli*). *J. Plant Interact.* **2008**, *3*, 175–180. [CrossRef]
25. Müller, C.H. Phytotoxins as plant habitat variables. *Recent Adv. Phytochem.* **1970**, *3*, 106–121.
26. Nishida, N.; Tamotsu, S.; Nagata, N.; Saito, C.; Sakai, A. Allelopathic effects of volatile monoterpenoids produced by *Salv leucophylla*: Inhibition of cell proliferation and DNA synthesis in the root apical meristem of *Brassica campestris* seedlings. *J. Che Ecol.* **2005**, *31*, 1187–1203. [CrossRef] [PubMed]
27. Macías, F.A.; Mejías, F.J.R.; Molinillo, J.M.G. Recent advances in allelopathy for weed control: From knowledge to application *Pest Manag. Sci.* **2019**, *75*, 2413–2436. [CrossRef] [PubMed]
28. Gendreau, E.; Traas, J.; Desnos, T.; Grandjean, O.; Caboche, M.; Hofte, H. Cellular basis of hypocotyl growth in *Arabidopsis thalian Plant Physiol.* **1997**, *114*, 295–305. [CrossRef] [PubMed]
29. Niinemets, Ü.; Loreto, F.; Reichstein, M. Physiological and physicochemical controls on foliar volatile organic compoun emissions. *Trends Plant Sci.* **2004**, *9*, 180–186. [CrossRef]
30. Chen, J.; Bi, H.; Yu, X.; Fu, Y.; Liao, W. Influence of physiological and environmental factors on the diurnal variation in emissior of biogenic volatile compounds from *Pinus tabuliformis*. *J. Environ. Sci.* **2019**, *81*, 102–118. [CrossRef] [PubMed]
31. Chae, H.M.; Choi, S.H.; Lee, S.H.; Cha, S.; Yang, K.C.; Shim, J.K. Effect of litter quality on needle decomposition for four pin species in Korea. *Forests* **2019**, *10*, 371. [CrossRef]
32. Bhusal, N.; Lee, M.; Lee, H.; Adhikari, A.; Han, A.R.; Han, A.; Kim, H.S. Evaluation of morphological, physiological, an biochemical traits for assessing drought resistance in eleven tree species. *Sci. Total Environ.* **2021**, *779*, 146466. [CrossRe [PubMed]
33. Eom, S.H.; Yang, H.S.; Weston, L.A. An evaluation of the allelopathic potential of selected perennial groundcovers: Foliar volatile of catmint (*Nepeta* × *faassenii*) inhibit seedling growth. *J. Chem. Ecol.* **2006**, *32*, 1835–1848. [CrossRef] [PubMed]

plants

MDPI

rticle

\llelopathy in Durum Wheat Landraces as Affected by Genotype and Plant Part

urelio Scavo, Gaetano Pandino *, Alessia Restuccia, Paolo Caruso, Sara Lombardo * nd Giovanni Mauromicale

Department of Agriculture, Food and Environment (Di3A), University of Catania, 95123 Catania, Italy; aurelio.scavo@unict.it (A.S.); a.restuccia@unict.it (A.R.); pcamps@liberc.it (P.C.); g.mauromicale@unict.it (G.M.)

* Correspondence: g.pandino@unict.it (G.P.); saralomb@unict.it (S.L.); Tel.: +39-0954783449 (G.P.); +39-0954783421 (S.L.)

Abstract: Durum wheat is one of the largest cultivated crops across Mediterranean areas. The high demand for sustainable crop productions, especially concerning weed management, is driving the return to local landraces. In the present work, the in vitro allelopathic effects of the extracts of three durum wheat landraces ('Timilia', 'Russello' and 'Perciasacchi') and a modern variety ('Mongibello'), obtained from three different plant parts (ears, stems and roots), were tested on seed germination (G) and mean germination time (MGT) of *Portulaca oleracea* L. and *Stellaria. media* (L.) Vill., two weeds commonly infesting wheat fields. In addition, the total polyphenol (TPC) and total flavonoid (TFC) content of extracts was determined. All extracts reduced G and increased MGT in both weeds compared to the control. The magnitude of phytotoxicity was strongly affected by the influence of genotype, plant part and extract dilution. Overall, the landraces 'Timilia' and 'Russello' showed the highest allelopathic effects, ear extracts were the most active, and the maximum extract dilution induced higher phytotoxicity. Extracts' TPC and TFC corroborated these results. The findings obtained here encourage the use of local landraces as a source of allelochemicals and suggest that they could be left on soil surface or soil-incorporated after harvest for a possible weed control.

Keywords: allelopathy; durum wheat; weed management; seed germination; polyphenols; flavonoids; *Portulaca oleracea*; *Stellaria media*

itation: Scavo, A.; Pandino, G.; estuccia, A.; Caruso, P.; Lombardo, ; Mauromicale, G. Allelopathy in urum Wheat Landraces as Affected / Genotype and Plant Part. *Plants*)22, *11*, 1021. https://doi.org/ .3390/plants11081021

cademic Editor: Tika Adhikari

eceived: 16 March 2022
ccepted: 6 April 2022
ublished: 8 April 2022

1. Introduction

Wheat is the most important grain crop worldwide, native to South-East Asia and widely cultivated since prehistoric times in the temperate zones. Nowadays, the world harvested area is about 215 × 10^6 ha with ~765 × 10^6 Mg of grain [1]. Most of these data refer to the hexaploidy species *Triticum aestivum* L. (bread wheat), while the only tetraploid (2n = 4x = 28) species of economic importance is durum wheat [*Triticum turgidum* subsp. *durum* (Desf.) Husn.]. It is mainly grown in the European Union (EU) on above 2.1 × 10^6 ha with a 7.6 × 10^6 Mg grain production, of which Italy is the main EU producer with 1.2 × 10^6 ha and 3.8 × 10^6 Mg harvested production [2]. Durum wheat is cultivated across the Mediterranean Basin and other semiarid regions, where it is appreciated for its high cooking quality and for the production of pasta, semolina, couscous, flatbread and bulgur [3,4]. In this area, local landraces are specifically adapted to environmental conditions and soil properties, so much so that the pool of Mediterranean landraces contains the largest genetic diversity within the species [5]. These landraces were largely cultivated for centuries, when in the middle of the 20th century they were progressively replaced with more productive and genetically-improved semi-dwarf cultivars. In addition to this, durum wheat yields have been consistently enhanced thanks to the advancements in the agronomic management, in particular herbicide application. Despite some differences between durum and bread wheat in the response to herbicides, both species need considerable amounts of herbicides against monocotyledonous and dicotyledonous weeds [3,6].

However, the irrational use of herbicides caused the development of resistance and shifts i weed populations, the emergence of a substitution weed flora, an important environment pollution and subsequent health hazards [7]. Exploring the application of alternative an sustainable strategies for weed management in wheat agroecosystems has therefore becom mandatory. Within this scenario, allelopathy is a novel tool that is gaining more and mo popularity across the scientific community [8].

Allelopathy, a term firstly coined by the Austrian physiologist Hans Molisch in 193 to indicate the biochemical interactions between all plants, refers to any direct or indirec beneficial or detrimental effect by one plant on another through the release of chemical com pounds into the environment [9]. These chemical compounds, known as allelochemical are secondary metabolites or waste products of primary metabolic pathways produced b living organisms, including plants and microorganisms, belonging to different chemica classes and playing a defensive role for the plant [10]. There are different mechanism by which allelopathy can be exploited to manage weeds: crop rotation with allelopathi species, cover cropping, green manuring, intercropping and use of plant extracts [11]. Th latter, in particular, have been largely adopted alone or in combination with reduced dose of herbicides. Increase of reactive oxygen species (ROS) production, alteration of cell stru ture and membrane permeability, alteration of photosynthesis and respiration, reductio and/or inhibition of seed germination and seedling growth, are extensively documente as common effects of plant extract [12].

Allelopathy in wheat has been deeply studied and demonstrated by a consolidate scientific literature [13,14]. There are many allelochemicals involved in wheat allelopath belonging to three main chemical classes: phenolic acids (e.g., *p*-hydroxybenzoic, feruli syringic, vanillic, *p*-coumaric, etc.), benzoxazinoids (DIMBOA, MBOA, HMBOA, DIMBO glycoside, BOA) and short-chain fatty acids (e.g., propionic, acetic, butyric, etc.) [13,14 The allelopathic traits and the synthesis of these allelochemicals are genetically controlle and characterised by a high polygeneticity. For instance, it is known that the genes codin for benzoxazinoids (DIMBOA) accumulation are located on chromosomes 4A, 4B, 4D an 5B [15]. In addition, Wu et al. [16] identified two major quantitative trait loci on chro mosome 2B related to wheat allelopathy. Allelopathic genetic variability among whea cultivars is very common and the breeding of cultivars with improved allelopathic poten tial is now under investigation [17]. One of the first studies is that of Spruell [18], wh screened 286 bread wheat accessions for their allelopathic effects against *Bromus japonicu* L. and *Chenopodium album* L. Later, Wu et al. [19] firstly evaluated the allelopathic potentia of 92 wheat cultivars against annual ryegrass, and then screened 453 wheat accession from 50 countries, reporting a 10 to 91% range of root growth inhibition [20]. The pro duction and amount of allelochemicals as expression of the allelopathic potential varie in relation to plant parts and plant age. For instance, it was found that the concentratio of benzoxazinoids in wheat seeds was similar to that in foliage and roots [21], wherea Mogensen et al. [22] reported a lower concentration of DIMBOA in roots than in leave Generally, the concentrations of these compounds decline with plant age [22]. The allelo pathic effects of wheat extracts were investigated in laboratory by testing the effects o aqueous and straw extracts on seed germination and seedling growth of selected weeds, a well as under field conditions by evaluating its inclusion in crop sequences and residu incorporation [13,14]. However, most of these studies refer to bread wheat straw and othe plant parts, while durum wheat allelopathy has been very little studied.

Given the well-known effect of genotype and plant part on the allelopathic expressio of plants, the return to local landraces by virtue of their genetic importance and marke demand, and considering also the higher weed suppressive ability of landraces than mod ern cultivars, we hypothesise that durum wheat landraces could be more allelopathic tha modern cultivars, with genotype- and plant part-dependent allelopathic effects. Indeed in our experience, fields of landraces show a lower weed density than modern ones. To test these hypotheses, a systematic study was performed with the aim of screening th allelopathic potential of three durum Sicilian wheat landraces compared to a modern vari ety by testing the allelopathic effects of root, stem and ear extracts on two weed species commonly infesting wheat fields (*Portulaca oleracea* L. and *Stellaria media* (L.) Vill.). Furthermore, the total polyphenol content (TPC) and total flavonoid content (TFC) of durum wheat extracts were determinate to detect possible interactions between extracts phytotoxicity and polyphenols amount.

2. Results

2.1. Allelopathic Effects of Durum Wheat Extracts

From the ANOVA, it emerged that the two target weeds were differently affected by durum wheat extracts (Table 1).

Table 1. *F*-Fisher values of main factors and their interactions resulting from analysis of variance (ANOVA) on final seed germination percentage (G) and mean germination time (MGT).

Source of Variation	Df	*Portulaca oleracea*		*Stellaria media*	
		G	MGT	G	MGT
Main factors					
Wheat genotype (G)	3	0.83 ns	1.45 ns	4.86 **	0.90 ns
Plant part (P)	2	2.74 ns	22.09 ***	28.81 ***	15.28 ***
Extract dilution (D)	2	28.39 ***	58.22 ***	292.44 ***	904.85 ***
Interactions					
(G) × (P)	6	2.21 *	4.50 ***	1.73 ns	1.93 ns
(G) × (D)	6	3.59 **	0.84 ns	3.18 **	1.66 ns
(P) × (D)	4	8.41 ***	11.80 ***	27.01 ***	4.75 **
(G) × (P) × (D)	12	2.83 **	2.28 *	0.79 ns	0.60 ns

Notes: *F*-Fisher values are referred to $\log_{(x+1)}$-transformed data; df: degrees of freedom; ***, ** and * indicate statistical significance at $p \leq 0.001$, $p \leq 0.01$ and $p \leq 0.05$, respectively; ns: not significant.

2.1.1. Allelopathic Effects on *Portulaca oleracea*

The three-way interaction 'wheat genotype × plant part × extract dilution' was significant for both G ($p \leq 0.01$) and MGT ($p \leq 0.05$), with the 'extract dilution' showing the highest contribution to variance (58% and 57%, respectively) (Table 1).

Concerning G (Figure 1), ear extracts caused lower G values than stem and root extracts in 'Timilia' (respectively −12.9% and −7.6%) and 'Russello' (−6.2% and −1.0%), while in 'Perciasacchi' the roots extracts were the most allelopathic (80% vs. 88% of ears and 85% of stems). The lowest seed germination was observed with the 100% ear extract from 'Timilia' (47.5%), followed by the 100% ear extract from 'Mongibello' (66.3%); on the contrary, in 'Russello' the most active extract was that obtained from the ears diluted at 50% (71.3%), while in 'Perciasacchi', it was the 50% root extract (70%). Pooling over wheat genotypes and plant parts, pure extracts (100%) showed a higher inhibitory activity (77.5%) than 50% dilution (80.3%) and the control (92.4%).

Mean germination time showed a similar response to G, with the exception of the significant effect provided by plant part (Table 1). In 'Timilia' and 'Perciasacchi', the ear extracts at 100% caused the highest MGT (4.9 and 3.3 days, respectively), whereas the 100% and 50% stem extracts were, respectively, the most active in 'Russello' (3.1 days) and 'Mongibello' (2.8 days) (Figure 2). Regardless of wheat genotypes and extract dilution, the extracts obtained from ears determined a higher MGT than stems and roots (2.5 vs. 2.2 and 2.1 days, respectively). Moreover, 100% dilution increased more MGT than 50% and 0% (2.7 vs. 2.3 and 1.9 days, respectively).

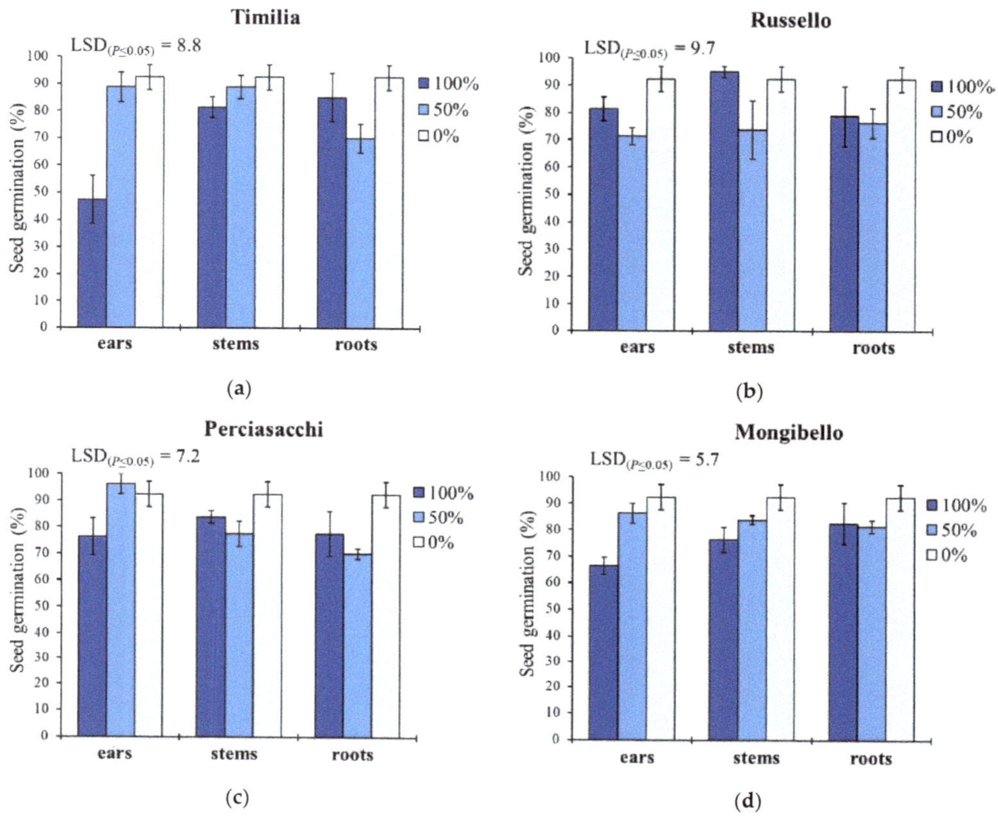

Figure 1. Seed germination percentage of *Portulaca oleracea* as affected by durum wheat genotyp plant part and extract dilution. (**a**) 'Timilia'; (**b**) 'Russello'; (**c**) 'Perciasacchi'; (**d**) 'Mongibello'; Leas Significant Difference (LSD) interaction was calculated with the LSD test at $p < 0.05$. Bars indicate standard error.

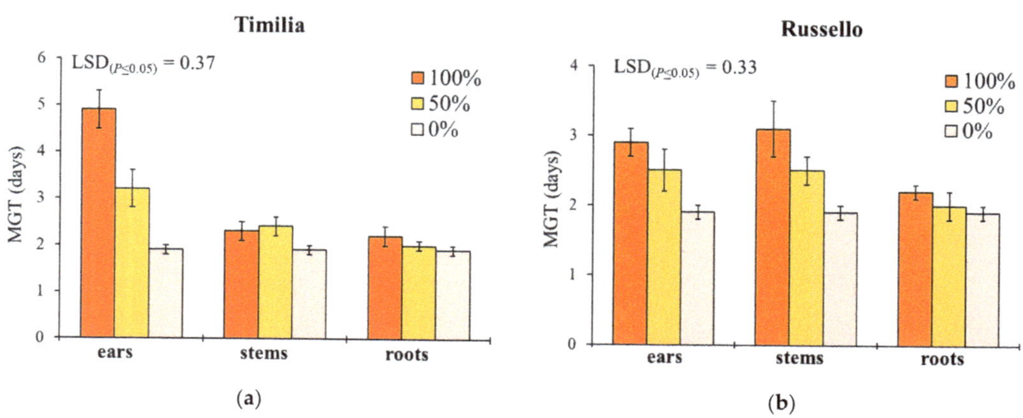

Figure 2. *Cont.*

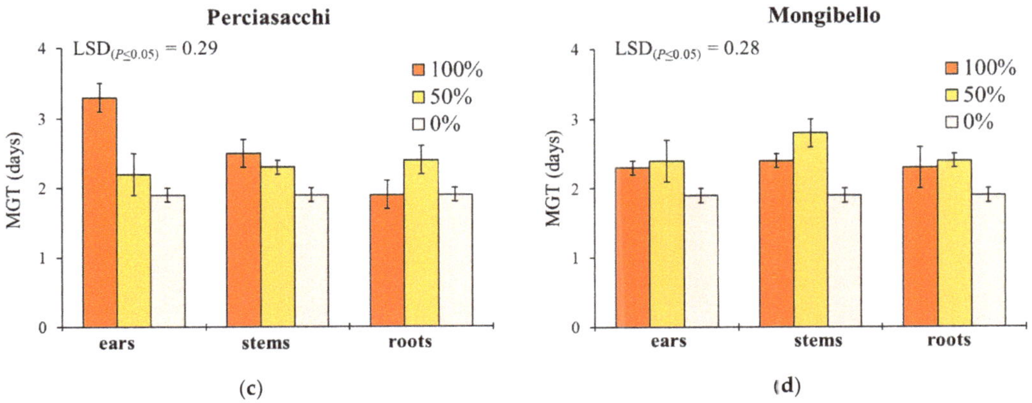

Figure 2. Mean germination time (MGT, days) of *Portulaca oleracea* as affected by durum wheat genotype, plant part and extract dilution. (**a**) 'Timilia'; (**b**) 'Russello'; (**c**) 'Perciasacchi'; (**d**) 'Mongibello'; Least Significant Difference (LSD) interaction was calculated with the LSD test at $p < 0.05$. Bars indicate ± standard error.

2.1.2. Allelopathic Effects on *Stellaria media*

In *S. media*, seed germination was significantly affected by the two-way 'wheat genotype × extract dilution' and 'plant part × extract dilution' interactions (Table 1). Regarding the former (Figure 3a), the trend 0% < 50% < 100% was constant for all the wheat genotypes, with an average reduction of 64% (pure extracts) and 48% (extracts diluted at 50%) of G compared to the control. 'Timilia' extracts at 100% showed the highest allelopathic effect (29.2% vs. 37.5% on the average of the other three genotypes), while 'Perciasacchi' extracts at 50% had the lowest (55.3%). Concerning the 'plant part × extract dilution' interaction, a trend of 0% < 50% < 100% was found for the three plant parts (Figure 3b). Among the maximum dilutions (100%), ear extracts caused the lowest G (19.7%), whereas stem extracts were the most efficient in terms of G reduction among the 50% dilutions (44.4%). Root extracts showed the lowest inhibitory activity at both 100% and 50% dilution.

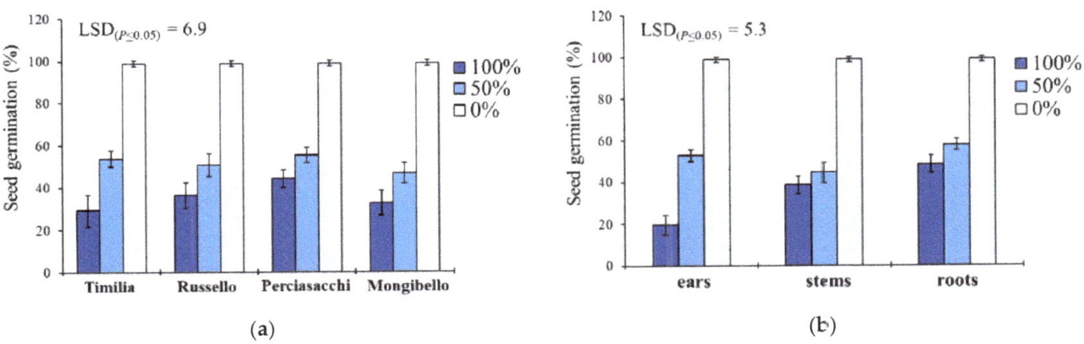

Figure 3. Effect of durum wheat 'genotype × extract dilution' (**a**) and 'plant part × extract dilution' (**b**) interactions on *Stellaria media* seed germination percentage. Least Significant Difference (LSD) interaction was calculated with the LSD test at $p < 0.05$. Bars indicate ± standard error.

'Plant part × extract dilution' was the only significant interaction affecting MGT in *S. media* (Table 1). Pure extracts at 100% better performed than 50% and 0% dilutions in increasing the MGT for both ears, stems and roots (Figure 4). Stem extracts determined the highest MGT at both 100% (8.8 days) and 50% (7.7 days) dilution, followed by ears (8.6 and 7.1 days) and roots (7.8 and 6.6 days).

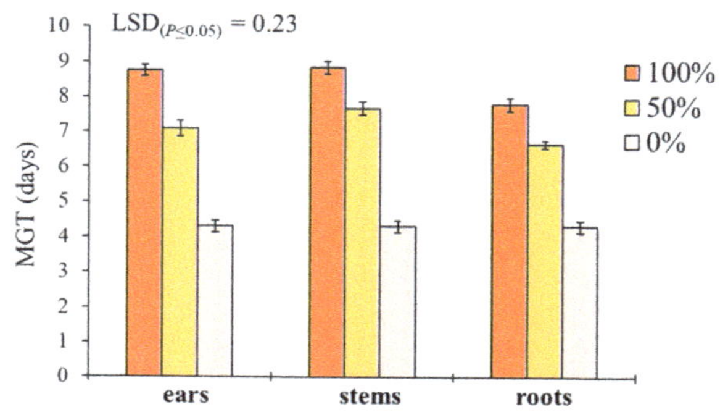

Figure 4. Effect of durum wheat 'plant part × extract dilution' interaction on *Stellaria media* me germination time (MGT, days). Least Significant Difference (LSD) interaction was calculated with tl LSD test at $p < 0.05$. Bars indicate ± standard error.

2.1.3. Synthesis of the Allelopathic Effects

Table 2 shows the allelopathic effect response index (RI) and the synthesis effect (S of main factors for the two target weeds. RI values are not described since they are sti incorporated in SEs. The synthesis effect of wheat genotype was significant for both weed In particular, 'Timilia' showed a significantly higher SE than the other genotypes in *oleracea*. Similarly, significantly higher SEs were obtained by 'Timilia' and 'Mongibell extracts in *S. media*. 'Perciasacchi' extracts showed the lowest SEs in both weeds. Regardin plant part, SE of ears was significantly higher than stems and roots in *P. oleracea*; also, *media* roots showed the significantly lowest SE. Moreover, it can be easily seen that th allelopathic effect was enhanced by increasing the dilution of extracts (0.45 vs. 0.36 SE in *oleracea* and 1.13 vs. 0.87 SE in *S. media*).

Table 2. Allelopathic effect response index (RI) and synthesis effect (SE) of durum wheat extracts o final seed germination (G) and mean germination time (MGT) of *Portulaca oleracea* and *Stellaria medi* RI and SE values are pooled over main factors.

Main Factors		*Portulaca oleracea*			*Stellaria media*		
		RI		SE	RI		SE
		G	MGT		G	MGT	
Wheat genotype	*Timilia*	−0.17 a	0.25 a	0.46 a	−0.58 a	0.45 a	1.03 a
	Russello	−0.14 a	0.21 a	0.40 b	−0.56 a	0.43 a	0.99 a
	Perciasacchi	−0.12 a	0.18 a	0.39 b	−0.50 a	0.44 a	0.92 b
	Mongibello	−0.13 a	0.24 a	0.40 b	−0.60 a	0.43 a	1.02 a
Plant part	*ears*	−0.17 a	0.32 a	0.53 a	−0.63 a	0.46 a	1.08 a
	stems	−0.11 b	0.23 b	0.36 b	−0.57 a	0.47 a	1.04 a
	roots	−0.15 a	0.11 c	0.33 b	−0.45 b	0.40 b	0.86 b
Extract dilution	*100%*	−0.26 a	0.25 a	0.45 a	−0.64 a	0.49 a	1.13 a
	50%	−0.12 b	0.17 b	0.36 b	−0.48 b	0.38 b	0.87 b

Different letters between each column indicate statistical significance at $p < 0.05$ with the LSD test.

2.2. *Total Polyphenol (TPC) and Flavonoid Content (TFC) of Durum Wheat Extracts*

The ANOVA showed that both TPC and TFC values of durum wheat extracts wer significantly affected by both durum wheat genotypes ($p < 0.0001$ TPC, TFC) and plant part ($p < 0.0001$ TPC, TFC), whereas any two-way interaction was significant ($p = 0.1584$ an $p = 0.3459$, respectively). Regardless of plant part, 'Timilia' showed the highest TPC an TFC values (1.04 and 0.69 g kg^{-1} DM, respectively), followed by 'Russello', 'Mongibell and 'Perciasacchi' (Figure 5a). In relation to the plant part, the trend of ears > stems > root was found for both TPC and TFC, with ear extracts showing a +72% of TPC and +286%

of TFC compared to root extracts (Figure 5b). In detail, 'Timilia' showed the highest TPC values in all plant parts (1.33 g kg^{-1} DM in ears, 0.94 g kg^{-1} DM in stems and 0.83 g kg^{-1} DM in roots), followed in decreasing order by 'Russello' (1.22, 0.85 and 0.68 g kg^{-1} DW, respectively, in ears, stems and roots), 'Mongibello' (1.01, 0.73 and 0.57 g kg^{-1} DM) and 'Perciasacchi' (0.84, 0.59 and 0.46 g kg^{-1} DM) (Figure 6). The same trend was found for TFC, with 'Timilia' and 'Perciasacchi' showing, respectively, the highest and the lowest values (Figure 6).

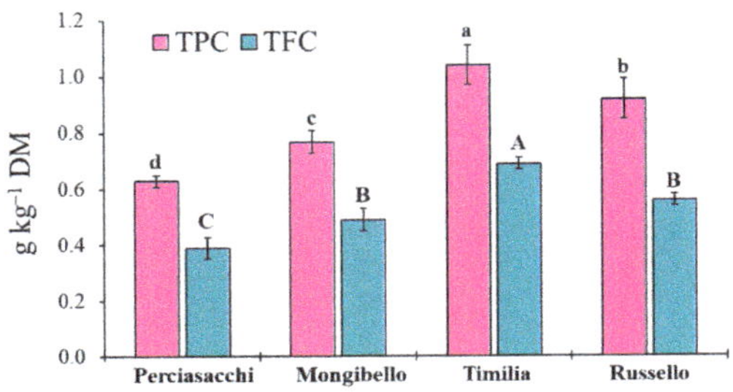

(a)

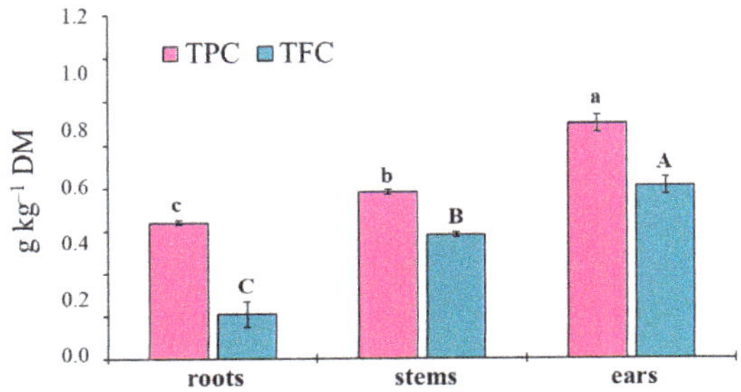

(b)

Figure 5. Total polyphenol (TPC) and total flavonoid (TFC) content in roots, stem and ears of durum wheat extracts in relation to the genotype (**a**) and plant part (**b**). Different letters (a–d) indicate significant differences for the TPC ($p \leq 0.05$). Different letters (A–C) indicate significant differences for the TFC ($p \leq 0.05$). Bars indicate ± standard error. DM: dry matter.

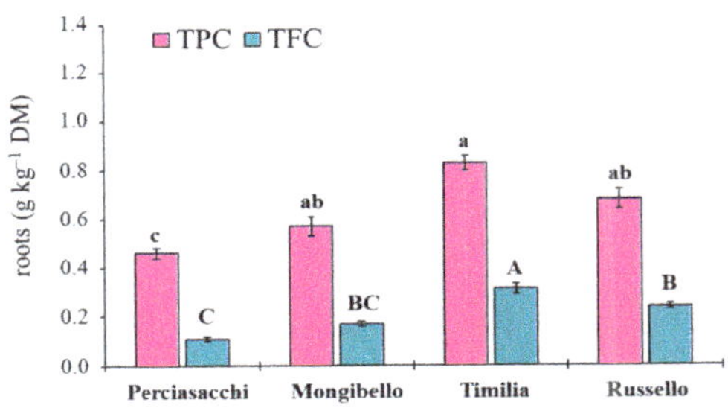

(a)

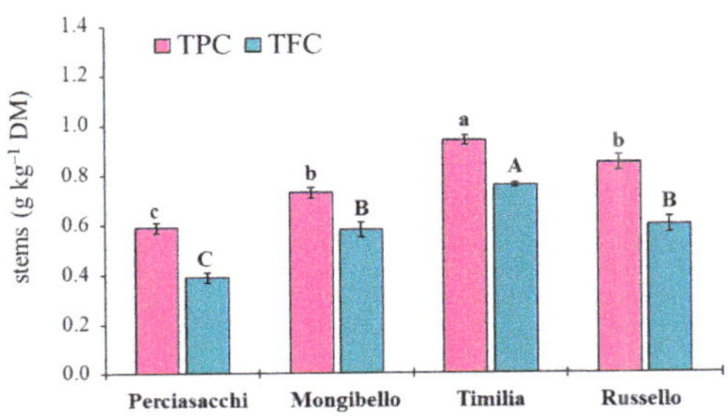

(b)

Figure 6. *Cont.*

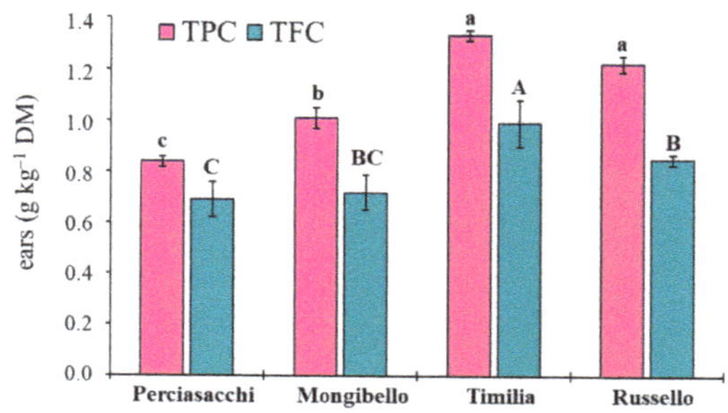

(c)

Figure 6. Total polyphenol (TPC) and total flavonoid (TFC) content in root (**a**), stem (**b**) and ear extracts of durum wheat genotypes. Different letters (a–c) indicate significant differences amon genotypes for the TPC ($p \leq 0.05$). Different letters (A–C) indicate significant differences amon genotypes for the TFC ($p \leq 0.05$). Bars indicate ± standard error. DM: dry matter.

3. Discussion

Most of the research about wheat allelopathy have been conducted on bread wheat an on the allelopathic effects of wheat crop residues, leachates and mulch/cover crop [13,1 Durum wheat allelopathy, on the contrary, is still poorly understood. This study demo strated that durum wheat extracts have a significant allelopathic activity against tw common weeds infesting wheat fields. The allelopathic activity was evaluated in terms c weed G and MGT, which are two well-known secondary expressions derived from primar effects such as the increase of reactive oxygen species (ROS), reduction or inactivation c the physiological activity of phytohormones, alteration of cell membrane permeabilit division and elongation [12]. In particular, plant extracts generally inhibit seed germinatio by disrupting mitochondrial respiration and oxidative pentose phosphate pathways [2 The phytotoxicity level varied in relation to genotype, plant part and extract dilution, a commonly found in many other similar studies. Genotype- and dose-response allelopathi effects are widely reported in the literature. Scavo et al. [24,25], for instance, indicated tha the allelopathic activity of *Cynara cardunculus* L. depends on the botanical variety—wit the wild cardoon being more phytotoxic than cultivated cardoon and globe artichoke—an extract concentration. Allelopathic genetic variability has been thoroughly studied in brea wheat [18–20], while to the best of our knowledge this is the first time in durum wheat. Th allelopathic activity of two Tunisian durum wheat varieties, 'Karim' and 'Om rabii', wa evaluated by Oueslati [26] on seed germination and seedling growth of barley and brea wheat. The same author, investigating the effect of plant part, also found that leaf extract were the most active, whereas root and stems extracts showed no effect in reducing radic length and seed germination.

Here, two local landraces ('Timilia' and 'Russello') showed higher allelopathic effect than a modern variety ('Mongibello') on both target weeds. Moreover, ear extracts provide better results in terms of phytotoxicity than stems and roots. These findings were corrob rated by the phytochemical analysis of aqueous extracts. Indeed, significantly higher TP and TFC values were detected in 'Timilia' and 'Russello' extracts. Moreover, both TP and TFC followed the trend ears > stems > roots, as found for phytotoxic effects. Whea polyphenols mainly include hydroxybenzoic and hydroxycinnamic acid derivatives suc as *p*-hydroxybenzoic, ferulic, syringic, vanillic, caffeic and *p*-coumaric [4,27]. They wer detected in the whole grains and in bran fractions, while no information was available fo other wheat plant parts. In general, these secondary metabolites are produced as a defenc mechanism in adaptation to biotic and abiotic stresses (water-deficit and high intensit of solar radiation in Mediterranean environments). In contrast with Bertholdsson [28 who underlined how bread wheat landraces and old cultivars were less allelopathic tha modern varieties, in this study the landraces 'Timilia' and 'Russello' were more allelopathi than the modern variety 'Mongibello'. This is not strange since agronomic practices such as herbicide application have resulted in the competitive loss of modern cultivars with weeds. Giambalvo et al. [29], in fact, suggested the choice of old and tall landraces such as 'Russello' in weedy and low-N conditions due to their high competitive capacity with weeds. The higher weed-suppressive ability of old landraces compared to modern varieties could be therefore attributable not only to competition, but also to allelopathy, as demonstrated in this study. The decrease of G and increase of MGT mediated by allelochemicals is a strategy adopted by certain crops, such as durum wheat landraces, to win the competition with weeds, thus decreasing the herbicides supply.

4. Materials and Methods

4.1. Location and Agronomic Management of Wheat Field

Durum wheat genotypes were cultivated in the experimental farm of the University of Catania, located in Eastern Sicily (South Italy, 37°25′ N; 15°30′118″ E; 10 m a.s.l.). The soil, Typic and/or Vertic Xerochrepts (Soil Survey Staff, 1999), showed the following physio-chemical properties in the 0–40 cm profile: sand 27%, clay 45%, silt 28% (clay texture), organic matter 1%, total N 1.1 g kg^{-1}, available P_2O_5 10 mg kg^{-1}, exchangeable K_2O 300 mg kg^{-1}, pH 8.1 and cation exchange capacity 169 meq 100 g^{-1}. The climate of the zone is typically semiarid Mediterranean, characterised by ~500 mm of annual precipitations, mild rain winters and hot dry summers.

Durum wheats under study included three Sicilian landraces ('Timilia', 'Russello' and 'Perciasacchi') and a modern variety ('Mongibello'), recently bred by the University of Catania from a 'Trinakria × Valforte' cross. They are autumn-sowing genotypes with late or medium-late maturity, and mean yields in Sicily ranged between 1300–2800 kg ha^{-1}. Plants were arranged in a randomized block design with three replicates, within a 35 × 36.5 m experimental area. Each cv. had a net plot size of 30 m^2, for a total of 12 plots of 10 m^2 with 6 rows spaced 21 cm apart. Sowing was carried out in December 2018 by means of a self-propelled plot seeder (Winterstaiger, Ried, Austria) at the rate of 400 viable seeds m^{-2} to reach a mean target ear density of ~300 ears m^{-2}.

Wheat genotypes were grown by applying a low-input agricultural management. Seedbed preparation was realized with a shallow hoeing (20 cm deep) in early autumn followed by a disk harrow. The fertilization program consisted of 54 kg N ha^{-1} and 108 kg P_2O_5 ha^{-1} before sowing, combined with 26 kg N ha^{-1} (ammonium nitrate, 27% N) at tillering stage. Weeds were controlled by hand and by brush cutter.

4.2. Sampling of Plant Materials and Extracts Preparation

About 30 plants for each variety were randomly sampled in June 2019 at full maturity stage. In the laboratory, the plants were washed with tap water and separated into roots, stems and ears. Extracts were prepared by combining the methodologies proposed by Oueslati [26] and Wu et al. [20], with some modifications. In detail, the three plant parts were finely chopped and dried in an oven at 45 °C up to constant weight. Ten grams of DM of each plant material were soaked with 150 mL of distilled water for 48 h at room temperature (20 °C ± 1) in the dark. Then, the mixtures were filtered through a Whatman no. 2 filter paper and centrifuged at 200 rpm for 15 min at 10 °C to remove debris. Finally, each extract was diluted with distilled water to obtain three final dilutions: 100% (pure extracts), 50% and 0% (distilled water) as control. The prepared extracts were stored in a refrigerator at 3 °C for further uses.

4.3. Seed Collection and Germination Bioassay

The allelopathic effects of the above-mentioned durum wheat extracts were tested on seed germination of *P. oleracea* and *S. media*. The former is a cosmopolitan spring–summer annual weed, therophyte, belonging to the Portulacaceae family; the latter is a Caryophyllaceae member, biennial and hemicryptophyte weed, with an autumn–winter cycle. Both weeds highly infest Mediterranean durum wheat fields, where they exert a severe pressure, respectively, at the end and at the beginning of the crop's biological cyc The seeds of *P. oleracea* were collected around the experimental farm of the University Catania, whereas *S. media* seeds derived from natural populations sited in Calatabiar (Sicily, 37°49′ N, 15°13′ E; 50 m a.s.l.). Mature collected seeds were cleaned from ine materials (debris, pebbles, etc.), selected for size and colour homogeneity with a MS5 Lei stereomicroscope (Leica Microsystems, Wetzlar, Germany), and stored in paper bags room temperature.

Germination bioassays were carried out in 9 cm Petri dishes by moistening a doub Whatman no. 2 layer with 5 mL of root, stem and ear extracts at two different dilutio (100% and 50%). Distilled water was used as a control. There were four replicates of seeds for each extract and dilution. Petri dishes of *P. oleracea* were incubated in continuo darkness at 35 °C and wrapped with an aluminium foil, while *S. media* ones were kept alternating light (dark/light cycle 14/10 h) at 17 °C. These temperatures and photoperio are the optimal conditions for seed germination of each species [24,30]. In both cases, Pet dishes were sealed with parafilm to prevent evaporation of the solution. Seed germinatic was counted daily, considering the seeds as germinated when the radicle protruded ov 2 mm. Germination bioassays ended when no seeds had germinated for 3 consecutive day

4.4. Total Polyphenol Content of Aqueous Extracts

The total polyphenol content was quantified using a modified method proposed b Pandino et al. [31]. About 10 mL of each extract was evaporated under vacuum using rotary evaporator (Buchi rotavapor). The residue was redissolved in 1.5 mL methanol 80 and stirred at room temperature for 1 h, with shaking. The mixture was centrifuged 5000× *g* for 5 min at 25 °C. A diluted aliquot was mixed with Folin—Ciocalteu reagent room temperature for 2 min. Sodium carbonate (5%, *w*/*v*) was added and the mixture wa allowed to rest at 40 °C for 20 min in thermostatic bath. The absorbance was read at 725 nr by a Shimadzu 1601 UV–Visible spectrophotometer (Shimadzu Corp., Tokyo, Japan). Th content was determined on the basis of a standard calibration curve generated with know concentrations of ferulic acid. All data presented are mean values of two independer experiments and expressed as g kg^{-1} of DM.

4.5. Total Flavonoid Content of Aqueous Extracts

Flavonoid content of the extracts was quantified using the aluminium chloride assa method performed by Zendehbad et al. [32]. In brief, 500 µL of redissolved extract wa dissolved in 1.5 mL of ethanol (95%) and 0.1 mL of 10% aluminium chloride. Then, 0.1 m of 1 M sodium acetate were added. The volume was made up to 5 mL with bidistilled wate The absorbance was measured at 725 nm by a spectrophotometer at 415 nm after 30 mi The content was determined on the basis of a standard calibration curve generated wit known concentrations of rutin. All data presented are mean values of two independer experiments and expressed as g kg^{-1} of DM.

4.6. Data Analysis

In order to evaluate the allelopathic effects of wheat extracts on seed germination o the two target weeds, the following parameters and indices were considered:

$$G\ (\%) = \left(\frac{n_i}{N}\right) \times 100 \quad (1$$

$$MGT\ (days) = \left(\frac{\sum_{i=1}^{k} n_i t_i}{\sum_{i=1}^{k} n_i}\right) \quad (2$$

$$RI = \left[\left(\frac{T}{C}\right) - 1\right] \text{if } T < C \quad or \quad RI = \left[1 - \left(\frac{C}{T}\right)\right] \text{if } T \geq C \quad (3$$

$$SE = RI_{|G|} + RI_{|MGT|} \quad (4$$

where: G = final germination percentage; n_i = number of seeds germinated in the ith time; N = total number of seeds used in each Petri dish; MGT = mean germination time; t_i = time from the start of the experiment to the ith observation; RI = allelopathic effect response index; T = treatment value; C = control value; SE = allelopathic synthesis effect. Equation (2) was computed according to Ranal et al. (2009). RI was determined following Williamson and Richardson (1998), with positive values indicating stimulation by treatments and negative values indicating inhibition. SE, which represents the intensity of the allelopathic effect, was calculated as the sum of the corresponding absolute value of RI for germination percentage (|G|) and mean germination time (|MGT|), in accordance with Ma et al. (2020). Laboratory experiments, repeated twice, were arranged in a completely randomized design (CRD) with four replications.

4.7. Statistical Analysis

Data were analysed using analysis of variance (ANOVA) followed by the Fisher's protected LSD test for means multiple comparisons. Statistically significant differences were set at $p \leq 0.05$. Deviations from normality and homoscedasticity were determined before ANOVA, respectively, by graphically inspecting the residuals and with the Bartlett's test. In particular, to meet the basic assumptions for linear models, G and MGT data were $\log_{(x+1)}$-transformed in accordance with Scavo et al. (2020). Mean $\pm$ standard error of untransformed data is presented and discussed. A generalized linear model (GLM) was initially applied considering 'wheat genotype', 'plant part', 'extract dilution' and 'target weeds' as main factors. Considering that the latter factor showed a high significance ($p \leq 0.001$) for all the variables, data were therefore processed according to a generalized linear mixed model (GLMM) with 'target weeds' as a random factor. Two-way ANOVAs were conducted for the second order interactions between main factors, whereas one-way ANOVAs were applied to pooled RI and SE data. Data about TPC and TFC were analysed according to a factorial two-way ANOVA model (4 wheat genotypes $\times$ 3 plant parts). The CoStat® software version 6.003 (CoHort Software, Monterey, CA, USA) was used for statistical analysis.

5. Conclusions

The present research documented the allelopathic effects of selected durum wheat landraces on seed germination of two common weed species infesting wheat (*P. oleracea* and *S. media*). The magnitude of phytotoxicity was related to genotype, plant part and extract dilution. In detail, two landraces ('Timilia' and 'Russello') were more allelopathic than the modern variety 'Mongibello', ears were more active than stems and roots, and seed germination was increasingly inhibited with increasing extract concentration. These findings were supported by extracts' TPC and TFC, since their highest values were found in 'Timilia' and 'Russello' among genotypes, and ear extracts among plant parts. These results suggested that plant residues of local landraces could be left on the soil surface or soil-incorporated after wheat harvest for weed control. Furthermore, durum wheat landraces can be considered potential plants for the possible future production of bioherbicides. However, more research is required in this regard to identify, purify and isolate the allelochemicals involved in durum wheat allelopathy and to evaluate their effects under field conditions on a broader spectrum of weeds and soil seed banks.

Author Contributions: Conceptualization, A.S. and G.M.; methodology, A.S., G.P., A.R., P.C., S.L. and G.M.; investigation, A.S., G.P., A.R., P.C., S.L. and G.M.; data curation, A.S. and G.P.; writing—original draft preparation, A.S.; writing—review and editing, A.S. and G.M.; supervision, G.M. All authors have read and agreed to the published version of the manuscript.

Funding: This research received no external funding.

Institutional Review Board Statement: Not applicable.

Informed Consent Statement: Not applicable.

Data Availability Statement: All data are available via email request to the corresponding autho

Acknowledgments: The authors thank Giovanni Lo Cunsolo for his assistance during seed germi tion trials.

Conflicts of Interest: The authors declare no conflict of interest.

References

1. Faostat. Food and Agriculture Organization of the United Nations. 2021. Available online: http://www.fao.org/faostat/ (accessed on 27 October 2021).
2. Eurostat. European Commission. 2021. Available online: https://ec.europa.eu/eurostat/data/database (accessed on October 2021).
3. Fragasso, M.; Iannucci, A.; Papa, R. Durum wheat and allelopathy: Toward wheat breeding for natural weed management. *Fro Plant Sci.* **2013**, *4*, 375. [CrossRef] [PubMed]
4. Liu, H.; Bruce, D.R.; Sissons, M.; Able, A.J.; Able, J.A. Genotype-dependent changes in the phenolic content of durum und water-deficit stress. *Cereal Chem.* **2018**, *95*, 59–78. [CrossRef]
5. Nazco, R.; Villegas, D.; Ammar, K.; Peña, R.J.; Moragues, M.; Royo, C. Can Mediterranean durum wheat landraces contribute improved grain quality attributes in modern cultivars? *Euphytica* **2012**, *185*, 1–17. [CrossRef]
6. Soltani, N.; Shropshire, C.; Sikkema, P.H. Sensitivity of durum wheat (*Triticum turgidum*) to various postemergence herbicid *Agric. Sci.* **2011**, *2*, 451–456. [CrossRef]
7. Scavo, A.; Mauromicale, G. Integrated weed management in herbaceous field crops. *Agronomy* **2020**, *10*, 466. [CrossRef]
8. Scavo, A.; Abbate, C.; Mauromicale, G. Plant allelochemicals: Agronomic, nutritional and ecological relevance in the soil syste *Plant Soil* **2019**, *442*, 23–48. [CrossRef]
9. Rice, E.L. *Allelopathy*, 2nd ed.; Academic Press: New York, NY, USA, 1984.
10. Whittaker, R.H.; Feeny, P.P. Allelochemics: Chemical interactions among species. *Science* **1971**, *171*, 757–770. Available onlir https://www.jstor.org/stable/1730763 (accessed on 27 October 2021). [CrossRef]
11. Scavo, A.; Mauromicale, G. Crop allelopathy for sustainable weed management in agroecosystems: Knowing the present with view to the future. *Agronomy* **2021**, *11*, 2104. [CrossRef]
12. Radhakrishnan, R.; Alqarawi, A.A.; Abd Allah, E.F. Bioherbicides: Current knowledge on weed control mechanism. *Ecotoxic Environ. Saf.* **2018**, *158*, 131–138. [CrossRef]
13. Aslam, F.; Khaliq, A.; Matloob, A.; Tanveer, A.; Hussain, S.; Zahir, Z.A. Allelopathy in agro-ecosystems: A critical review of whe allelopathy-concepts and implications. *Chemoecology* **2017**, *27*, 1–24. [CrossRef]
14. Ma, Y. Allelopathic studies of common wheat (*Triticum aestivum* L.). *Weed Biol. Manag.* **2005**, *5*, 93–104. [CrossRef]
15. Niemeyer, H.M.; Jerez, J.M. Chromosomal location of genes for hydroxamic acid accumulation in *Triticum aestivum* L. (whea using wheat aneuploids and wheat substitution lines. *Heredity* **1997**, *79*, 10–14. [CrossRef]
16. Wu, H.; Pratley, J.; Ma, W.; Haig, T. Quantitative trait loci and molecular markers associated with wheat allelopathy. *Theor. Ap Genet.* **2003**, *107*, 1477–1481. [CrossRef] [PubMed]
17. Belz, R.G. Allelopathy in crop/weed interactions—An update. *Pest Manag. Sci.* **2007**, *63*, 308–326. [CrossRef]
18. Spruell, J.A. Allelopathic potential of wheat accessions. *Diss. Abstr. Int.* **1984**, *45*, 1102B.
19. Wu, H.; Pratley, J.E.; Lemerle, D.; Haig, T. Evaluation of seedling allelopathy in 453 wheat (*Triticum aestivum*) accessions again annual ryegrass (*Lolium rigidum*) by the equal-compartment-agar method. *Aust. J. Agric. Res.* **2000**, *51*, 937–944. [CrossRef]
20. Wu, H.; Pratley, J.; Lemerle, D.; Haig, T. Laboratory screening for allelopathic potential of wheat (*Triticum aestivum*) accessior against annual ryegrass (*Lolium rigidum*). *Crop Pasture Sci.* **2000**, *51*, 259–266. [CrossRef]
21. Villagrasa, M.; Guillamón, M.; Labandeira, A.; Taberner, A.; Eljarrat, E.; Barceló, D. Benzoxazinoid allelochemicals in whe Distribution among foliage, roots, and seeds. *J. Agric. Food Chem.* **2006**, *54*, 1009–1015. [CrossRef]
22. Mogensen, B.B.; Krongaard, T.; Mathiassen, S.K.; Kudsk, P. Quantification of benzoxazinone derivatives in wheat (*Triticu aestivum*) varieties grown under contrasting conditions in Denmark. *J. Agric. Food Chem.* **2006**, *54*, 1023–1030. [CrossRef]
23. Cao, J.; Dong, Z.; Zhao, H.; Duan, S.; Cao, X.; Liu, H.; Yang, Z. Allelopathic effect of rhubarb extracts on the growth of *Microcyst aeruginosa*. *Water Sci. Technol.* **2020**, *82*, 1092–1101. [CrossRef]
24. Scavo, A.; Restuccia, A.; Pandino, G.; Onofri, A.; Mauromicale, G. Allelopathic effects of *Cynara cardunculus* L. leaf aqueo extracts on seed germination of some Mediterranean weed species. *Ital. J. Agron.* **2018**, *13*, 119–125. [CrossRef]
25. Scavo, A.; Rial, C.; Varela, R.M.; Molinillo, G.M.G.; Mauromicale, G.; Macias, F.A. Influence of genotype and harvest time on th *Cynara cardunculus* L. sesquiterpene lactone profile. *J. Agric. Food Chem.* **2019**, *67*, 6487–6496. [CrossRef] [PubMed]
26. Oueslati, O. Allelopathy in two durum wheat (*Triticum durum* L.) varieties. *Agric. Ecosyst. Environ.* **2003**, *96*, 161–163. [CrossRe
27. Zhang, J.; Ding, Y.; Dong, H.; Hou, H.; Zhang, X. Distribution of phenolic acids and antioxidant activities of different bra fractions from three pigmented wheat varieties. *J. Chem.* **2018**, *2018*, 6459243. [CrossRef]
28. Bertholdsson, N.-O. Varietal variation in allelopathic activity in wheat and barley and possibilities to use this in breedin *Allelopath. J.* **2007**, *19*, 193–202.
29. Giambalvo, D.; Ruisi, P.; Di Miceli, G.; Frenda, A.S.; Amato, G. Nitrogen use efficiency and nitrogen fertilizer recovery of durur wheat genotypes as affected by interspecific competition. *Agron. J.* **2010**, *102*, 707–715. [CrossRef]

Scavo, A.; Pandino, G.; Restuccia, A.; Mauromicale, G. Leaf extracts of cultivated cardoon as potential bioherbicide. *Sci. Hortic.* **2020**, *261*, 109024. [CrossRef]

Pandino, G.; Mattiolo, E.; Lombardo, S.; Lombardo, G.M.; Mauromicale, G. Organic cropping system affects grain chemical composition, rheological and agronomic performance of durum wheat. *Agriculture* **2020**, *10*, 46. [CrossRef]

Zendehbad, S.H.; Mehran, M.J.; Malla, S. Flavonoids and phenolic content in wheat grass plant (*Triticum aestivum*). *Asian J. Pharm. Clin. Res.* **2014**, *7*, 184–187.

MDPI
St. Alban-Anlage 66
4052 Basel
Switzerland
Tel. +41 61 683 77 34
Fax +41 61 302 89 18
www.mdpi.com

Plants Editorial Office
E-mail: plants@mdpi.com
www.mdpi.com/journal/plants

www.ingramcontent.com/pod-product-compliance
Lightning Source LLC
Chambersburg PA
CBHW061815210926
51850CB00006B/713
* 9 7 8 3 0 3 6 5 8 2 9 0 0 *